Exploring
Urban
America

Exploring Urban America

An Introductory Reader

EDITED BY
Roger W. Caves

SAGE Publications
International Educational and Professional Publisher
Thousand Oaks London New Delhi

For information address:

SAGE Publications, Inc.
2455 Teller Road
Thousand Oaks, California 91320

SAGE Publications Ltd.
6 Bonhill Street
London EC2A 4PU
United Kingdom

SAGE Publications India Pvt. Ltd.
M-32 Market
Greater Kailash I
New Delhi 110 048 India

Printed in the United States of America

Library of Congress Cataloging-in-Publication Data

Exploring urban America : an introductory reader / edited by Roger W.
  Caves
    Includes bibliographical references and indexes.
    ISBN 0-8039-5637-1. — ISBN 0-8039-5638-X (pbk.)
    1. Urbanization—United States.   2. Cities and towns—United
States.   3. City planning—United States.   4. Urban policy—United
States.   I. Caves, Roger W.
HT384.U5E97   1994
307.76'0973—dc 20   94-21924

95 96 97 98 99 10 9 8 7 6 5 4 3 2 1

Sage Production Editor: Diana E. Axelsen

To my Mom
Although we will always miss her
she will always be with us in everything we do

Contents

Acknowledgments

This book would not have been completed without the encouragement and assistance of many people. Many of them do not even know they have assisted me in this endeavor. I am most grateful to the myriad people with whom I have talked throughout my career. They have listened to my questions and ramblings and provided me with some good advice. I would like to thank the many members of the Urban Affairs Association for their knowledge and expertise. I benefited a great deal from attending panels at its annual meetings. The knowledge I accumulated from them has been transferred to my students.

The impetus for this book came from a number of sources. As an undergraduate student, I was always taught to ask questions if something was unclear. My graduate professors at Old Dominion University's Urban Studies Program and the University of Delaware's College of Urban Affairs and Public Policy consistently challenged me to seek out additional information. These professors include Wolfgang Pindur, Francis Tannian, Daniel Rich, and Robert Warren. They have taught me well; I hope my students learn as much from me as I did from them.

A number of people have contributed to the completion of this book. First, the project took off after a visit from Harry Briggs, Editorial Director, Sage Publications, Inc., during which he asked whether I had any other project I would like to do. We talked about an urban reader, and he encouraged me to submit a proposal. This book is a direct result of his encouragement. Second, I would like to thank Carrie Mullen, my acquisitions editor at Sage, for her help and sense of humor in this project. Third, I would like to thank the authors who wrote the introductions to the various parts of this book. Their insightful comments add a great deal to the field of urban studies. Finally, I would like to thank a most important group—Carol, Murphy, and little Teddi—for putting up with my different work habits and numerous piles of paper around the house.

Roger W. Caves

Original Sources

The chapters in this book were previously published in the following journals:

Chapter 1: Anthony M. Orum. Apprehending the city: The view from above, below, and behind. *Urban Affairs Quarterly, 26*(4), June 1991.

Chapter 2: Scott Greer. Urbanism and urbanity: Cities in an urban-dominated society. *Urban Affairs Quarterly, 24*(3), March 1989.

Chapter 3: Norton E. Long. The citizenships: Local, state, and national. *Urban Affairs Quarterly, 23*(1), September 1987.

Chapter 4: Cees D. Eysberg. The origins of the American urban system: Historical accident and initial advantage. *Journal of Urban History, 15*(2), February, 1989.

Chapter 5: Richard C. Wade. The enduring ghetto: Urbanization and the color line in American history. *Journal of Urban History, 17*(1), November 1990.

Chapter 6: Mary K. Nenno. Urban policy revisited: Issues resurface with a new urgency. *Journal of Planning Literature, 3*(3), Summer 1988.

Chapter 7: William R. Barnes. Urban politics and urban impacts after Reagan. *Urban Affairs Quarterly, 25*(4), June 1990.

Chapter 8: Robert Warren. National urban policy and the local state: Paradoxes of meaning, action, and consequence. *Urban Affairs Quarterly, 25*(4), June 1990.

Chapter 9: Robert C. Wood. People versus places: The dream will never die. *Economic Development Quarterly, 5*(2), May 1991.

Chapter 10: Susan E. Clarke and Gary L. Gaile. The next wave: Postfederal local economic development strategies. *Economic Development Quarterly, 6*(2), May 1992.

Chapter 11: Norman Krumholz. Equity and local economic development. *Economic Development Quarterly, 5*(4), November 1991.

Chapter 12: Elaine B. Sharp and David R. Elkins. The politics of economic development policy. *Economic Development Quarterly, 5*(2), May 1991.

Chapter 13: Thomas A. Clark and Franklin J. James. Women-owned businesses: Dimensions and policy issues. *Economic Development Quarterly, 6*(1), February 1992.

"UNTIL I GOT MY DEGREE IN URBAN STUDIES, I DIDN'T KNOW WHAT THE HELL WAS GOING ON HERE."

Introduction

Roger W. Caves

We are undoubtedly an urban nation. For the purposes of this reader, *urban* can be equated to a city and the population that surrounds it. Yet we are still left with defining the terms *urban* and *city*. Unfortunately there are no universal definitions (Gruen, 1964). Some scholars claim any definition will vary by discipline. Joseph Russell Passoneau (1963, p. 9) suggested the following:

> To an economist, a city is a large, complex, input-output device. To a sociologist, a city is distinguished from a village by its higher degree of social differentiation and by the wider opportunity it offers by fruitful interaction between diverse individuals. To a political scientist, a group of compact, contiguous, but separately governed suburbs might not be a city, while a sprawling series of communities under a single government might be a city with a distinctive personality.

Others claim that a city cannot be viewed simply as a physical entity. Park (1925, p. 1) adheres to this view:

> The city is, rather, a state of mind, a body of customs, and traditions, and of the organized attitudes and sentiments that inhere in these customs and are transmitted with this tradition. The city is not, in other words, merely a physical mechanism and an artificial construction. It is involved in the vital processes of the people who compose it; it is a product of nature, and particularly of human nature.

Disagreements are certain to continue whenever *urban* and *city* are used. Shall we continue to use the traditional definition of 2,500 inhabitants in a place being defined as an urban population? Definitions of *urban* vary by individual and by country (Palen, 1987). Maybe Glaab and Brown (1967, p. i) are correct when they observe that *urban* possesses a magical quality and that people think it means everything:

> The word *urban* has come in our day to have almost the value of a talisman in academic as well as political life. It lends an air of serious and informed concern to discourse

in which it appears. And yet, other than the fact that *urban* has somehow to do with cities, there is little agreement in ordinary discussion on what it means.

Mumford (1961, p. 3), in the classic *The City in History,* notes the difficulty in a single definition of *city* by suggesting that "no single definition will apply to all of its manifestations and no single description will cover all its transformations, from the embryonic social nucleus to the complex forms of its maturity and the corporeal disintegration of its old age."

Studying the urban arena involves a number of disciplines. There is no denying that urban studies contains a broad range of subject matter. That is, in fact, one reason that it has attracted many of its students. This range has led some scholars to question whether an individual can be an expert on urban affairs. Although Banfield (1974, p. xi) believes the range of subject matter makes it impossible to be an expert on urban affairs, he does note that an individual "can, however, learn enough of several disciplines to make useful applications of some of their major ideas and findings." Moreover, knowledge in a given area may help us make more informed decisions. Murray S. Stedman, Jr., offered the following thoughts on why individuals should study urban affairs:

> One of the most compelling reasons for studying urban affairs is the deeply felt conviction that knowledge has, or at least ought to have, some potential relationship to power. To know something about a social situation is to create the possibility of taking some reasoned action related to it. Knowledge as to how social processes operate brings with it the chance of controlling the end products of such processes. Even in the absence of positive controls over the outcome, knowledge results in an understanding which might otherwise be impossible to achieve. (1975, p. 4)

The purpose of this reader is to introduce students to the field of urban studies. It is intended to offer students insights into the many subfields of urban studies. No attempt has been made to include readings on every urban problem. Such an undertaking would take multiple volumes.

Selection of the Readings

The selection of any readings is inherently a subjective process. After consulting with various individuals teaching or researching in the field of urban studies, the following criteria were used. First, each of the articles had to have been previously published in a Sage Publications journal. Second, each of the articles had to contribute to our understanding of issues in the urban arena and serve to generate discussion in and out of the classroom. I did not want articles that simply rehashed old information that could be found in earlier works. Third, each of the articles had to be readable and not too technical for introductory courses. Finally, each of the selections had to illustrate the multidisciplinary nature of urban studies.

Organization of the Readings

Part 1 "Cities and Urbanism" is introduced by Timothy Barnekov and Daniel Rich of the University of Delaware and contains three selections. In their introduction, Barnekov and Rich discuss the important roles that cities have played over the years, their various functions, and how urbanism has changed over time. Chapter 1 "Apprehending the City: The View From Above, Below, and Behind," by Anthony M. Orum, examines the belief that cities are sovereign territories and how cities have sought out more sovereignty. Chapter 2 "Urbanism and Urbanity: Cities in an Urban-Dominated Society," by Scott Greer, discusses the functions and complexities of cities. Chapter 3 "The Citizenships: Local, State, and National," by Norton E. Long, looks at how cities have changed over the years and the loyalty of their citizens.

Part 2 "Urban History" is introduced by Carl Abbott of Portland State University and contains two selections. Abbott discusses how urban history can help us better understand some of today's issues. He examines the stages of growth, how cities have changed over the years, how we view cities, and how we have designed cities. Chapter 4 "The Origins of the American Urban System: Historical Accident and Initial Advantage," by Cees D. Eysberg, looks at the beginnings of America's urban system and attempts to explain why some areas grew more than other areas. Chapter 5 "The Enduring Ghetto: Urbanization and the Color Line in American History," by Richard C. Wade, examines the consequences of northward migration, especially those faced by African Americans.

Part 3 "Urban Policy," comprised of four selections, is introduced by Robert J. Waste of California State University at Chico. In his introduction Waste acknowledges the fascination that scholars share in developing a national urban policy and the difficulties that have plagued its development. He illustrates those difficulties by tracing how the concept of *urban* has changed for urban policy scholars from the 1970s to the 1990s. Chapter 6 "Urban Policy Revisited: Issues Resurface With a New Urgency," by Mary K. Nenno, traces urban policy issues from 1969 to 1988. William R. Barnes, in Chapter 7 "Urban Policies and Urban Impacts After Reagan," examines national urban policy and aid to America's cities during the Reagan years. Chapter 8 "National Urban Policy and the Local State: Paradoxes of Meaning, Action, and Consequences," by Robert Warren, discusses the need for an intergovernmental partnership in developing a national urban policy. Chapter 9 "People Versus Places: The Dream Will Never Die," by Robert C. Wood, looks at how the forces that influence current urban policy differ from the forces that influenced past urban policy.

Part 4 "Economic Development" is introduced by Norman Krumholz of Cleveland State University and contains four readings. In his introduction Krumholz offers insights into recent economic development activities and discusses questions surrounding who gains and who loses as a result of economic development activities. Chapter 10 "The Next Wave: Postfederal Local Economic Development Strategies," by Susan E. Clarke and Gary L. Gaile, examines the issue of how cities would respond if federal economic development

assistance was reduced. Chapter 11 "Equity and Local Economic Development," by Norman Krumholz, discusses the important distributional question
of who wins and who loses in economic development efforts. Chapter 12 "The
Politics of Economic Development Policy," by Elaine B. Sharp and David R.
Elkins, examines the role of citizen participation in formulating economic
development policies. Chapter 13 "Women-Owned Businesses: Dimensions and
Policy Issues," by Thomas A. Clark and Franklin J. James, discusses questions
surrounding the success of women-owned businesses and the trends taking place
in the United States.

Part 5 "Community Services and Infrastructure" is introduced by Robert
Warren of the University of Delaware and contains three selections. In his introduction Warren examines several questions surrounding who is responsible for
providing infrastructure and the distribution of community services. Warren also
stresses the need to rethink community services and infrastructure by incorporating telecommunications and information technology into an area's infrastructure. Chapter 14 "A Community Services Budget: Public, Private, and
Third-Sector Roles in Urban Services," by Robert Warren, Mark S. Rosentraub,
and Louis F. Weschler, examines the various actors involved in the provision of
services at the local level and proposes a community services budget for which
questions of distributional equity can be asked. Chapter 15 "Infrastructure Policy:
Repetitive Studies, Uneven Response, Next Steps," by Marshall Kaplan, discusses the state of the nation's infrastructure and what can be done to improve
it. Chapter 16 "Needed: A Marshall Plan for Ourselves," by Richard P. Nathan,
examines the nation's public infrastructure needs and calls for the establishment
of a Capital Investment Block Grant that would be used to rebuild the nation's
capital facilities (e.g., roads, bridges, sewers).

Part 6 "Community Development" contains three readings and is introduced by Margaret Wilder of the State University of New York at Albany. In her
introduction Wilder examines the long history, complexity, and objectives of
community development in the United States and also elaborates on some of
the current key issues in the field. Chapter 17 "Government and Neighborhoods:
Programs Promoting Community Development," by Charles Bartsch, notes the
importance of job creation and economic development activities at the local
level. Bartsch also examines how federal and state activities affect localities and
how various techniques can stimulate new economic activity at the local level.
Chapter 18 "Neighborhoods and Public-Private Partnerships in Pittsburgh," by
Louise Jezierski, looks at how neighborhood groups have participated in Pittsburgh's revitalization efforts, how they have questioned Pittsburgh's development policies, and how they have interacted with other involved public and
private actors. Chapter 19 "Neighborhood Associations: Their Issues, Their
Allies, and Their Opponents," by John R. Logan and Gordana Rabrenovic, relates
the need for and roles of neighborhood associations. Using Albany, New York,
as their point of inquiry, Logan and Rabrenovic consider the reasons for the
creation of neighborhood associations, the issues that have gained their attention, the opponents of neighborhood associations, and how they have fared.

Part 7 "Urban Housing" contains two selections and is introduced by Patricia Baron Pollak of Cornell University. Pollak discusses the importance of housing decisions to an individual and to a community. She also discusses the key inter-relationships between housing decisions to other public policy decisions. Chapter 20 "The Limits of Localism: Progressive Housing Policies in Boston, 1984-1989," by Peter Dreier and W. Dennis Keating, provides a discussion of the housing policies of two Boston mayors and looks at which mayor did more to improve the living conditions of Boston's poor and working-class citizens. Chapter 21 "Local Government Support for Nonprofit Housing: A Survey of U.S. Cities," by Edward G. Goetz, discusses the growing importance of nonprofit housing developers in the United States and how localities can assist them in their activities.

Part 8 "Urban Growth" contains two readings and is introduced by Nico Calavita of San Diego State University. In his introduction Calavita discusses why growth is criticized, the conflicts caused by growth, the actors involved in promoting or opposing growth, and the overall success of growth control measures. Chapter 22 "The Growth Machine Versus the Antigrowth Coalition: The Battle for Our Communities," by Ronald K. Vogel and Bert E. Swanson, offers a study of Gainesville, Florida, and looks at how local actors have defined growth and the arguments between progrowth and antigrowth advocates. They examine whether a middle ground consisting of growth management can end the battle between the various factions. Chapter 23 "Growth Politics and Downtown Development: The Economic Imperative in Sunbelt Cities," by Robyne S. Turner, offers a comparative political assessment of alternative downtown development strategies in four cities in Florida.

Part 9 "Urban Education" is introduced by Jeffrey A. Raffel of the University of Delaware and contains four readings. Raffel discusses the various challenges facing our education system and notes the differences of opinion over how to improve urban education, the need for more resources, the calls for changing the governance of schools, and the need to link educational improvement with other areas such as our health and well-being systems. Chapter 24 "Improving Urban Schools in the Age of 'Restructuring,' " by Kenneth A. Sirotnik, examines the importance of linking schools to the community. Chapter 25 "Visions of America in the 1990s and Beyond: Negotiating Cultural Diversity and Educational Change," by Jean J. Schensul and Thomas G. Carroll, analyzes the important issue of cultural and linguistic variation in urban schools. Chapter 26 "The Evolution of a University/Inner-City School Partnership: A Case Study Account," by Andrea G. Zetlin, Kathleen Harris, Elaine MacLeod, and Alice Watkins, discusses how a university and an inner-city school can join forces and improve the quality of education in an inner city. Chapter 27 "Race, Urban Politics, and Educational Policy Making in Washington, DC: A Community's Struggle for Quality Education," by Floyd W. Hayes, III, looks at whether school desegregation and tracking are working in one city.

A concluding chapter provides some closing thoughts on the study of the urban arena and offers some insights into further research in the area. A listing of additional suggested readings is also contained at the end of the book.

References

Banfield, E. C. (1974). *The unheavenly city revisited.* Boston: Little, Brown.

Glaab, C. N., & Brown, A. T. (1967). *A history of urban America.* New York: Macmillan.

Gruen, V. (1964). *The heart of our cities: The urban crisis, diagnosis, and cure.* New York: Simon & Schuster.

Mumford, L. (1961). *The city in history.* New York: Harcourt, Brace & World.

Palen, J. J. (1987). *The urban world* (3rd ed.). New York: McGraw-Hill.

Park, R. E. (1925). The city: Suggestions for the investigation of human behavior in the urban environment. In R. E. Park & E. W. Burgess (Eds.), *The city* (pp. 1-46). Chicago: University of Chicago Press.

Passoneau, J. R. (1963). Emergence of city form. In W. Z. Hirsch (Ed.), *Urban life and form* (pp. 9-29). New York: Holt, Rinehart & Winston.

Stedman, M. S., Jr. (1975). *Urban politics* (2nd ed.). Cambridge, MA: Winthrop.

PART ONE

Cities and Urbanism

Introduction to Cities and Urbanism

Timothy K. Barnekov

Daniel Rich

Cities are more than physical entities bounded in space where large numbers of people settle to live and work. Cities are social institutions; their importance lies in their distinctive contributions to political, economic, and cultural development. As Louis Wirth pointed out in his classic description of urbanism as a way of life, it is from cities that the ideas and practices we call civilization radiate (Wirth, 1938). Historically the key institutional features of civilization took shape and developed through cities. "Cities were proverbially the centers of institutions," wrote Daniel Boorstin, "where records were kept and the past was chronicled, hallowed and enshrined. They were sites of palaces, cathedrals, libraries, archives, and great monuments of all kinds" (1967, p. 121).

Cities were also the centers of the institutions of social change. Industries and entrepreneurs, commercial enterprises and speculators, factories and service centers of all types, artisans and crafters, universities and intellectuals, political parties, interest groups, cultural and religious movements and social innovators, technological inventions and scientists—all of these centered in cities, and the products of their initiative radiated out to the broader society through the transportation, communication, and social exchange networks that were linked together in urban places.

It is not an exaggeration to say that, particularly in the West, the development of civilization may be traced from the ancient period to the modern through a chronicle of the transformation of cities. In this context, Jane Jacobs argues that cities need to be recognized as the source of the wealth of nations (1984). They are the dynamic force that produces and sustains social innovation and creativity "with energy enough to carry over for problems and needs outside themselves" (Jacobs, 1961, p. 448). "The surplus wealth, the productivity, the close-grained juxtaposition of talents" that permit society to support virtually every avenue of advancement, Jacobs proposes, "are themselves products of our organization into cities, and especially into big and dense cities" (1961, p. 448).

Beyond the various functions that cities have performed historically, the ideal of the city as a model of the good human community also has been an important generative force in civilization. As Norton Long states in Chapter 3 "The Citizenships," the ideal of the city haunts the human imagination. In that ideal, citizens "could realize themselves in an active, self-governing life dealing with a full range of affairs of moment. The city was a theater for the exhibition of human excellence. In such a city, the citizen would accept material deprivation in order to possess the rich rewards of an active civic life."

In the modern age, the ideal of the city as the community in which political values and relationships could reach their highest order of expression was challenged, if not fully abandoned. This challenge took place over many centuries. Max Weber claimed that the city as a polity ceased to exist when the citizens were no longer willing to defend the city's walls (Martindale, 1958). But the threat to the political integrity of the city came less from those who successfully scaled the walls of medieval city-states than from those who offered an alternative framework of ideas for recognizing the meaning and value of cities. This alternative framework projected an image of the city as predominantly an economic rather than a political entity. In this view, the city is a center of exchange, a marketplace, a place of commerce, and a place of manufacture. This economic understanding of the city gained popularity just as the process of urbanization was accelerating during the industrial era. The old mythic visions of the city as the natural home of democracy and the center of advanced culture gave way to a new mythic vision of the industrial city as the locus of modern progress. The pace of progress was set by the city's efficiency in technological innovation and application resulting in increased economic growth and greater material wealth. If nothing else, modern experience seemed to confirm the practicality of this viewpoint. Undeniably, cities were the "engines of social change" in the industrial order (Mumford, 1961), so much so that the ideas of industrialism and urbanism were almost inseparable.

In the West the growth of cities as centers of wealth and political power became an essential part of the definition of the modern age. The achievement of these industrial cities, Lewis Mumford proposed, was to "convert power into form, energy into culture" (1961, p. 571). The industrial cities were centers for the application of new technologies requiring the intensive use of energy resources. Odum and Odum (1976) described the transition that was made possible by the efficiency of cities as energy systems:

> Energy flowing as coal and oil made the cities places of concentration of many kinds of work which were organized into new industries. Energy, which had formerly come from solar processes through agricultural systems, more and more began to come from fuels through cities. . . . Western cities became centers of growth promoting activities that accelerated the use of resources for growth. (pp. 153-154)

In the United States, city promotion and speculative investment stimulated economic expansion into new regions. Beyond this, the actual physical building of cities was a fundamental force in the growth of the industrial economy. The industrial nation was an urban nation, and the wealth of modern urban society grew in correspondence with the rise of cities.

In the midst of the growing recognition of urbanism as a facet of modernity, there also arose an intense antipathy to many attributes of city life. Indeed, the teeming masses living under squalid conditions in the industrial cities of Europe and America in the 19th century came to symbolize the social impoverishment and environmental degradation that persisted in the age of industrial wealth. The industrial city may have been a triumph of technology, but in many respects it was also a social and environmental failure. "Considering this new urban area on its lowest physical terms," wrote Mumford, "without reference to its social facilities or its culture, it is plain that never before in recorded history has such vast masses of people lived in such a savagely deteriorated environment, ugly in form, debased in content" (1961, p. 474). The age of urbanism that "boasted its mechanical conquests," said Mumford, "left its social processes to chance, as if the scientific habit of mind had exhausted itself upon machines, and was not capable of coping with human realities" (1961, p. 470).

Beyond the signs of environmental decay and social deprivation, the industrial city was (and still is) for many a place with only transient, instrumental value. The idea of a higher social purpose for the city beyond wealth creation had been forgotten. Critics like Mumford, and a host of municipal and social reformers, lamented the condition and proclaimed the need to construct a new civic vision, a vision of a city that would harness industrial power and technology to support efforts to address social needs. They set about to redesign the industrial city through urban planning, municipal management, sanitary and transportation engineering, social services, and public health systems and to reform the living and working conditions created by the excesses of the industrial order.

Despite efforts at reform, the basic pattern of urbanism during the industrial period persisted, along with its contradictory implications for the development of cities. The result is that cities in America and in many other countries continue to manifest the bifurcated, dualistic characteristics reflective of the cities that shaped the industrial epoch. Modern cities continue to exhibit two contradictory sides.

> Their attractive side includes productive manufacturing, innovative service industries, striking architecture, and experimental programs on the frontiers of social policy. Their unattractive side features slum housing, grinding poverty, widespread crime and attendant social programs that seem unable to cope with people's needs and occasional disorder that threatens the political fabric. (Sharkansky, 1975, p. 71)

While cities increased in economic power and political power throughout the 19th century and early 20th century, they did not increase equivalently in broad-based social acceptance. The dualistic character of the industrial city helped sustain contrasting images and attitudes toward the urban condition. Progressive images focus on the city of "invention" and its proven instrumentality in the creation of new wealth, new culture, and expanded economic freedom. Negative perspectives point to the city as the most apparent symbol of the social disorder and environmental decay brought about by industrial civilization and the callous indifference of its ruling class.

Americans have been particularly ambivalent toward the cities. In some respects, American antipathy to cities reflects an older bias shaped by a predominantly agrarian outlook on the nature of a desirable community. Thomas Jefferson expressed the sentiment in a letter written to James Madison in 1787: "Our governments will remain virtuous for centuries as long as they are chiefly agricultural" (quoted in Stedman, 1975, p. 21). American ambivalence toward cities may also be traced to the fact that American cities developed first and foremost as economic entities, encouraging Americans to think of their cities as places with no history, emerging in an environment "free of vested interests, guilds, skills, and 'No Trespassing' signs" (Boorstin, 1967, p. 121). In the United States, more than in any other society, private institutions played the dominant role in urban change and private decisions largely determined the pattern of American urban development. The cultural tradition that emerged in this context, according to Sam Bass Warner, was a tradition of privatism in which "the first purpose of the citizen is the private search for wealth" (1968). Privatism, above all else, characterized America's "urban inheritance" and meant that "the goal of the city was to be a community of private money makers" (Warner, 1968). Privatism continues to be the dominant cultural tradition affecting urban development in the United States (Barnekov, Boyle, & Rich, 1989). Americans live in one of the world's most urbanized countries, but they regard their urban environment as if it were a wilderness in both time and space (Warner, 1972).

The tradition of privatism sustained a system of American city governance in which public authority remained diffuse. This distribution often meant that the functional politics of American cities—the politics of growth and development of the civic economy—lay outside the domain and beyond the scope of the formal politics of municipal government. Norton Long contends that the emphasis on the city as an open economy contributed to its subordination as a polity (1983). For political leaders, this subordination has often meant that success is measured in terms of achievements in promoting economic development. This subordination also has limited the political commitment that citizens make to the city. As a result, American cities have become, in Scott Greer's words, "communities of limited liability" where there is an absence of investment in the city as a whole, a vision of local identity as transient, and a regard for local political boundaries as impermanent and often trivial (1962, p. 139). Despite the fact that America is predominantly an urban nation, Americans refuse, according to Greer, to accept themselves as an urban people (1963, p. 6).

Ambivalence about urbanism as a way of life and about the importance of cities has been exhibited in the role that American cities play in the governing process and in the treatment of urban issues and city problems by national public policy. Throughout most of American history, the challenges of urbanization were left for each city to address on its own—with meager policy tools and institutional powers. According to Gerald Frug, America chose to have relatively powerless cities and did so through a legal framework that ensured that the authority, and by extension the resources, of municipal government would be substantially constrained (1980). Not only were the cities subordinate to other governments in the federal system, but, in some critical respects, the political authority of cities within that system also was nonexistent. As units of government, cities had no constitutional status whatsoever, and therefore no legal basis for an active relationship with the federal government. Throughout the 19th century and into the 20th, no direct central-local government ties existed in the form of a coherent policy for cities, and, indeed, no such policy ties were thought to be appropriate under the terms of American federalism. Local governments, including city governments, were legally the creatures of the states, and historically they were dependent on state government for their very political existence. Moreover, many state governments were notable—some would say notorious—for their antiurban bias and for providing little support for major urban centers.

It was not until the fourth decade of the 20th century that the federal government acknowledged that urban problems were national problems and that cities played a central role in the prosperity of the American economy. The prestigious National Resources Committee (NRC), appointed by President Franklin Delano Roosevelt in 1937, argued that the neglect of the central role of cities in the national economy, as well as the continual and cumulative disregard of urban policy at the national level, was inhibiting the nation's recovery from the great economic depression of the 1930s. It called for a recognition of the city as "the workshop of our industrial society and the nerve center of our vast and delicate commercial mechanism," and it admonished the federal government to carry out policies that "take into account the place of the urban community in the national economy" (NRC, 1937). The committee warned that "the prosperity and happiness of the teeming millions who dwell [in cities] are closely bound up with that of America, for if the city fails, America fails" (NRC, 1937).

The emergence of explicit federal urban policies represented a belated recognition of the national importance of urban places—an environment in which more than two thirds of the people resided. This in itself was a radical departure from earlier periods of United States history in which the acknowledgement of urban America, much less its support, remained outside the domain of accepted federal responsibilities (Gelfand, 1975).

The recognition of the importance of the cities remained powerful through the 1960s and into the 1970s. Major federal programs were initiated to deal with a range of urban problems. The Housing Act of 1949 authorized the urban renewal program that provided loans and grants to localities for the clearance and development of inner-city slums. President Lyndon Johnson launched the Great Society program in the 1960s and declared "an unconditional war on poverty" (Johnson, 1964, p. 114). Out of this initiative emerged the Jobs Corps, the Community Action Program, Model Cities and other War on Poverty programs, as well as the establishment of a cabinet-level agency to represent urban dwellers: the Department of Housing and Urban Development. While Republican administrations in the early 1970s sought to reduce the federal role in local affairs, federal money continued to be provided to localities through less restrictive "block" grants and a new revenue-sharing program. As a further aid to distressed cities, Congress reoriented the focus of the Economic Development Administration (EDA) in the mid-1970s. From a largely rural development agency, EDA inaugurated the largest antirecessionary public works effort since the Great Depression and gave greater attention to urban areas, particularly those with the most unemployment. Subsequently Congress approved the Urban Development Action Grant Program (UDAG) to stimulate private investment in severely distressed communities. UDAG grants were used to build festival malls, to expand convention centers, and to build public infrastructure that would support business investment and, in general, put a new face on old cities.

But at the very time that the language of public policy was coming to recognize the centrality of cities, the conditions in cities and the problems faced by their residents were undergoing major transformations. New technological innovations encouraged and supported a spatial reconfiguration that diminished both the importance and the capacity for self-control of urban communities. The technology-city nexus that was at the heart of the industrial era in the 19th and early 20th centuries seemed to have broken in the latter half of the 20th century. A cycle of economic decline, already apparent in the 1950s and 1960s, was stimulated by the progressive decentralization of population and firms outward from the older urban cores. The erosion of the central city's economic base was accompanied by a decline of retail activities, a reduction in local public revenues, a deterioration of physical amenities, a decline in the quality of public services, and a decrease in private investment. The question was not whether the transformation was real, but rather how it should be understood and what, if anything, should be done about it to preserve the integrity of cities.

In the late 1970s, many policymakers and urban analysts argued that the transformations underway in the urban landscape were irreversible and that the city was no longer at the center of the process of urbanism. For many, the conclusion was that the city no longer needed to be a focus of national concern. This vision of a new urban reality was declared by yet another presidential commission, this one appointed by President Jimmy Carter. In 1980 the President's Commission for a National Agenda for the Eighties warned that "industrially based urban centers are gradually being unraveled" and that the advantages of agglomeration and central location are being "eroded by technological innovations and new production technologies that have given locational freedom to an ever wider array of industries" (President's Commission, 1980, p. 24). According to the commission, these trends were expressions of the transformation of the United States from an industrial to a postindustrial society, a transformation that defined the path for the economic and technological revitalization of America. The future prosperity of the nation depended on successful accommodation of the forces of postindustrial change. The commission recognized that this change was producing acute economic and social dislocations in hundreds of cities. It acknowledged the distress and despair of those "left behind," but it argued that attempts to restrict or reverse the process "for whatever noble intentions—is to deny the benefits that the future may hold for us as a nation" (p. 66). Rather, the nation must recognize that urban distress resulting from the redistribution of population and economic activity is part of the process of renewal; the nation must "reconcile itself to these redistribution patterns" (pp. 3-4).

The postindustrial transition, the commission suggested, was breaking the economic bond between national and city prosperity. Although urbanization as a process was still important to the development of the American economy, this economy was no longer dependent on the economic growth of the older industrial cities. Indeed, the culture and amenities of city life were not confined to large city centers and were being diffused throughout the country by economic and technological forces. Because of this diffusion, the decline of older industrial centers did not signify a challenge to national prosperity but, rather, was a necessary part of the process of economic renewal. "Cities are not permanent," maintained the commission, "their strength is related to their ability to reflect change rather than to fend it off. . . . In the long run, the fates and fortunes of specific places (should) be allowed to fluctuate" (President's Commission, 1980, p. 65).

The commission's assessment remained influential throughout the 1980s, long after the report itself vanished from the arena of urban policy debate. Although sometimes moderating and occasionally challenging some of the prescriptions of the commission, studies sponsored by prominent institutions such as the Committee on National Urban Policy of the National Research Council (Hanson, 1983) and the Brookings Institution (Peterson, 1985) supported much of the analysis. Thus Paul Peterson proclaimed that the industrial city had become an institutional anachronism and that its demise signaled the emergence of a new urban reality.

> If the great manufacturing centers of Europe and the American Snow Belt developed as by-products of the industrial revolution, their decline is no less ancillary to contemporary technological change. . . . Two to three decades ago urbanists sought to save the industrial city by redeveloping central business districts, or creating model cities that would transform poverty-stricken neighborhoods, or "energizing" citizens to participate in planning their community's future. Few would venture to propose such schemes today. Quite apart from political changes that have occurred in Washington, economic and social changes have moved so far that reversing their direction no longer seems feasible or even desirable. Industrial cities must simply accept a less exalted place in American political and social life than they once enjoyed. Policies must adapt to this new urban reality. (Peterson, 1985, p. 1)

Peterson and other analysts arrived at conclusions that were similar to those of the commission: The best federal policy to assist urban areas is a "non-urban policy"; efforts should concentrate on redressing the social and racial inequities attendant to the emergence of the new urban reality, but federal policies should have "no specifically urban component to them at all" (Peterson, 1985, pp. 25-26). While recognizing that postindustrialism is socially and spatially disruptive, urban analysts warned that it would be folly to try to resist the forces of change or to ignore the beneficial outcomes that ultimately will result from the market-led transition (Bradbury, Downs, & Small, 1981).

A central premise of the new urban reality is that the fortunes of cities are and should be determined by a technologically driven economic transition that is neither the result of public policy nor susceptible to reversal through government interventions.

[The new urban reality] assumes a form of Social Darwinism applied to cities as it has been previously applied, with pernicious consequences, to individuals and social classes. If Cleveland cannot compete with Houston for enterprises and jobs, it follows that it would be folly to try to bail out Cleveland. While some places may prosper at the expense of others, this is not a matter of national concern so long as the consequences are believed to contribute to the efficient spatial allocation of resources in the economy as a whole. (Barnekov, Rich, & Warren, 1981, p. 13)

This premise poses a profound challenge to municipal governance. Because the process of social change is seen increasingly to depend on market and technological forces, the responsibilities of municipalities become largely reactive. In this sense, the value of municipal governance is limited to its ability to facilitate efficiency and productivity in the new postindustrial growth sectors. So viewed, municipalities, at best, can serve as multipurpose service districts or public works authorities that administer decisions of central government and provide the physical infrastructure in locations favored by the spatial choices of private firms (Barnekov et al., 1981). This subordination of local government to markets and technology challenges the very meaning of urban governance. What effective governing authority at the local level can exist if a community's first responsibilities are to market itself as a commodity in a commercial sweepstakes and to refrain from any actions that might modify market decisions or technological demands?

Ultimately the logic of postindustrialism trivializes the city as a political and economic community. What is taken to be distinctive about postindustrial change is that it signifies the declining importance of central places as barometers of and contributors to economic and social progress and, by extension, as political institutions. In a world driven by the idea of postindustrial change, the value of cities becomes ephemeral. Indeed, there is no good reason why any city should exist, except by happenstance of rising or declining economic activity, within a national, and increasingly international, marketplace. The implication is that cities are to be regarded almost entirely as transient economic artifacts whose importance in the larger scheme of economic organization has been and will continue decreasing.

Long summarized the logic of this view and the central question it raises:

In this view aging cities should be treated in much the same way as worn out mines and obsolete factories. They should be phased out as rapidly, and painlessly as possible in the interest of the efficient functioning of the national economy, and resources should not be squandered in a counter-productive effort to keep losers afloat. . . . There can be little objection in principle to phasing out obsolete mines or factories, though the process in practice is fraught with difficulty and suffering since the costs of economic adjustment fall heavily on its victims. An argument can be made and has been made that society, which is the supposed gainer from the readjustment, should share the cost of easing the transition. But one may ask, are cities organizations on a continuum with mines and factories whose fates should be decided by the unhampered play of market forces as they alter the terms of trade and comparative locational advantage? (1983, pp. 21-22)

Under the logic of postindustrialism, this question is itself viewed as irrelevant, a matter that will be decided by the natural forces of change.

In one way or another, the authors of the readings that follow dispute the view that cities have only transient value. In Chapter 2 "Urbanism and Urbanity" Scott Greer discusses the functions of cities as marketplaces of ideas and innovation. He asks, What are the consequences of losing the creative and life-enhancing character of the very center of the city? He concludes that this loss will mean the city no longer will provide "a meeting place for many strands of culture and society" and the result will be greater separation of elites of various sorts and a fragmentation of culture. In Chapter 1 "Apprehending the City" Anthony Orum examines the city as a sovereign territorial unit and, as such, how it exercises control over its residents and affects their daily lives, particularly as the city engages in struggles with rival sovereign powers, including other cities, suburbs, the state, and the federal government. Long criticizes the contemporary views that trivialize the city and argues that society cannot do without cities and their civilizing mission. "The good life of the nation," says Long, "is lived not in the nation but in its cities," and it is only at the scale of the city where it is possible to realize a civilized life. The city, Long concludes, "is the institution that best serves both the homely and the exalted ends of man, and it is the institution that alone can subject the nation-state to the discipline of civilization."

References

Barnekov, T., Boyle, R., & Rich, D. (1989). *Privatism and urban policy in Britain and the United States.* New York: Oxford University Press.

Barnekov, T., Rich, D., & Warren, R. (1981). The new privatism, federalism, and the future of urban governance: National urban policy in the 1980s. *Journal of Urban Affairs, 3*(4), 1-14.

Boorstin, D. J. (1967). *The Americans: The national experience.* New York: Random House.

Bradbury, K. L., Downs, A., & Small, K. A. (1981). *Futures for a declining city: Simulations for the Cleveland area.* Washington, DC: Brookings Institution.

Frug, G. E. (1980). The city as a legal concept. *Harvard Law Review, 93*(6), 1059-1154.

Gelfand, M. I. (1975). *A nation of cities: The federal government and urban America, 1933-1965.* New York: Oxford University Press.

Greer, S. (1962). *The emerging city: Myth and reality.* New York: Macmillan.

Greer, S. (1963). *Metropolitics: A study of political culture.* New York: John Wiley.

Hanson, R. (Ed.). (1983). *Rethinking urban policy: Urban development in an advanced economy.* Washington, DC: National Academy Press.

Jacobs, J. (1961). *The death and life of great American cities.* New York: Random House.

Jacobs, J. (1984). *Cities and the wealth of nations: Principles of economics.* New York: Random House.

Johnson, L. B. (1964). Annual message to the Congress on the state of the union, January 8, 1964. In *Public papers of the presidents: Lyndon B. Johnson, 1963-64.* Washington, DC: Government Printing Office.

Long, N. E. (1983). Can the contemporary city be a significant polity? *Proceedings of the 1983 Annual Meeting of the Urban Affairs Association.* Newark: University of Delaware.

Martindale, D. (Ed.). (1958). *The city by Max Weber.* Glencoe, IL: Free Press.

Mumford, L. (1961). *The city in history.* Orlando, FL: Harcourt, Brace & Jovanovich.

National Resources Committee (NRC). (1937). *Our cities: Their role in the national economy* (Report of the Urbanism Committee to the National Resources Committee). Washington, DC: Government Printing Office.

Odum, H. T., & Odum, E. L. (1976). *Energy basis for man and nature.* New York: McGraw-Hill.

Peterson, P. E. (Ed.). (1985). *The new urban reality.* Washington, DC: Brookings Institution.

President's Commission for a National Agenda for the Eighties. (1980). *A national agenda for the eighties.* Washington, DC: Author.

Sharkansky, I. (1975). *The United States: A study of a developing country.* New York: David McKay.

Stedman, M. S. (1975). *Urban politics* (2nd ed.). Cambridge, MA: Winthrop.

Warner, S. B., Jr. (1968). *The private city: Philadelphia in three periods of its growth.* Philadelphia: University of Pennsylvania Press.

Warner, S. B., Jr. (1972). *The urban wilderness: A history of the American city.* New York: Harper & Row.

Wirth, L. (1938). Urbanism as a way of life. *American Journal of Sociology, XLIV,* 1-24.

— 1 —

Apprehending the City

THE VIEW FROM ABOVE, BELOW,

AND BEHIND

Anthony M. Orum

In this chapter I attempt to put the intellectual footing of the modern American city on firmer theoretical and empirical grounds. I selectively build on the insights of several keen analysts, including Peterson (1981) and Suttles (1968).

Certain elements underlie the theoretical claims I make that should be made explicit at the outset. First, I argue that power animates the drive in cities to dominate other cities, as it does the drive of diverse social groups to dominate the life of the city. By *power* here I mean, quite simply, the effort to impose order on an unruly world, whether that unruliness occurs in the material environment or in the inner workings of the city. To apprehend the city means to understand that cities are, at root, political objects and subjects that are filled at all conceptual levels with power.

Second, I frame my discussion in terms of American cities only, but in such a way that comparisons can be made with cities in other national contexts, as well as in other historical periods. In this, I follow the general methodological dicta of Weber (1949). Part of the problem with some other arguments is that although they are proclaimed to be comparative and historical, they are so abstract that they prevent other researchers from finding suitable points of convergence and comparison (e.g., see Castells, 1983). Finally, my theoretical assertions are intended to delimit a conceptual field of study for the city and to furnish the heuristic energy for empirical studies. However, they are not proposed as a closed conceptual scheme—like that of Peterson (1981), for instance—and thus I do not anticipate they will limit debate on or about the nature of the city today. Indeed, I hope they will serve to reopen it.

AUTHOR'S NOTE: I am grateful to a number of friends and colleagues who have given me the benefit of their wisdom and insight in response to a previous draft of this chapter: William Bridges, Harvey Choldin, Joe Feagin, Richard Johnson, John Johnstone, Stanley Mallach, James Norr, Moshe Semyonov, Neil Smelser, and Bert Useem.

Apprehending the City in Theory

Previous Theories: A Brief Critical Review

To appreciate the theoretical reasons for offering yet another perspective on the city, it is important that some of the prior theories be considered. A number of significant discussions about theories already exist (e.g., see Castells, 1972/1977; Saunders, 1981), but the following discussion will serve to provide the rationale for my propositions.

The pioneering theoretical work on American cities grew out of the fruitful inquiries of sociologists located at the University of Chicago in the 1920s, 1930s, and 1940s. The so-called Chicago school of sociology took the American city as its preeminent object of fascination. Scholars like Burgess (1925/1967) and, later, Wirth (1938) sought to understand the various dimensions of city life. Burgess (1925/1967), in particular, argued that the city should be grasped as consisting of a set of concentric spatial zones in which different segments of the city's social population and industrial base were located. Burgess also helped provide a strong theoretical rationale for the study of the city, arguing that the order and disorder that emerged in cities could best be construed in terms of a biological metaphor. Burgess noted in his seminal article on the spatial organization of the city that one must think of "urban growth as a resultant of organization and disorganization analogous to the anabolic and catabolic processes of metabolism in the body" (1925/1967, p. 53).

This analogy later would help form the framework for the human ecology of the city in which it was thought that the "spatial relationships of human beings are the products of competition and selection, and are continuously in process of change as new factors enter to disturb the competitive relations or to facilitate mobility" (McKenzie, 1925/1967, p. 64). Wirth, a friend, student, and colleague of Burgess, wrote what has been called by Saunders (1981, p. 92), "arguably the most famous article ever to have been published in a sociology journal." Wirth (1938) maintained that the dominant fact of life in the city was its fundamental opposition to life in rural areas. Inspired by the insights of Simmel's (1902-1903/1950) essay on the metropolis, Wirth insisted that the modern urban domain was essentially a place of anonymity and impersonality.

The work of the Chicago school has inspired countless empirical and theoretical analyses. It helped lead to Hawley's (1950) formulation about cities, a work that Berry and Kasarda (1977) regarded as the foundation for the contemporary school of human ecology. Hawley's *Human Ecology* (1950) put into a formal framework the suggestions of Burgess, and, as Burgess did, he relied exclusively on the biological metaphor to capture the salient dimensions of what happened in the life of the city. He distinguished between the biotic and the symbiotic level of the human population in cities and maintained that it was the processes at the biotic level that were most vital to the lives of cities.

In turn, the work by Hawley helped inspire a vast array of the empirical research on cities and populations that has been done during the last four decades (e.g., see Frisbie & Kasarda, 1988). Moreover, with Hawley's formulation and the accumulation of the overwhelming empirical evidence on the nuances of the distribution of people and the forces that seemed to move them, the human ecological paradigm became the prominent paradigm for the study of the city and of the population by sociologists.

But that view has suffered a series of critical attacks in recent years. Castells (1972/1977) maintained in a famous work, published first in French as *La question urbaine*, that Burgess and Hawley had effectively misconstrued the nature of the city (see also Gottdeiner & Feagin, 1988). Castells (1972/1977, pp. 116-121) observed that Burgess had nowhere given a proper conceptual definition to the nature of a city. Although they discussed processes within specific cities such as Chicago, they appeared to have confused the *conceptual* entity of the city with the *concrete* entity of any specific city. The essence of Castells's argument was that the Chicago school, including Hawley, had been guilty of the error of misplaced concreteness. They had speculated about and had researched a concrete phenomenon—cities—but nowhere had they provided a conceptual definition of a city—boundaries, constituent parts, and functions, for example. Later analysts, such as Peterson (1981), would seek to rectify this mistake and help establish the beginnings of a proper conceptual groundwork for the study of the city.

Because of his neo-Marxist orientation to social theory, Castells (1972/1977) also charged that the Chicago school had perceived certain processes to be natural that, in fact, were contrived by men and women. The Chicago school, and later Hawley, had assumed the biological metaphor of the city as a site of alleged natural, virtually Darwinian processes, implying that human beings were no more than prisoners of the fate that biology had consigned them as *Homo sapiens*. (Candidly, this is little different from the substitute argument that originated with Castells [1972/1977], which is that human beings are nothing more than prisoners of the capitalist system of production and distribution. Such vast systems of conceptual determinism leave no room for argument or even for empirical study.) The intellectual issue that remained after Castells's penetrating and influential critique (see Saunders, 1981) was how to put human action and human actors back into the study of the city.

Castells sought to remedy the flaws of human ecology in a number of intellectual projects (Castells, 1983). In *The Urban Question* (1972/1977), for instance, he insisted that cities were, in effect, *collective consumption units,* rather than *production units* in the Marxist sense. Thus those analyzing cities, he maintained, should seek to investigate the real dimensions of housing and other elements of consumption in the structure of city life. Yet however rich and stimulating his new theoretical vantage point proved to be, Castells was unable to demonstrate the empirical value of his claims. His later major work, *The City and the Grassroots* (1983), revealed little, if anything, about how the notion of collective consumption units may profit the empirical study of cities. Indeed, apparently he abandoned this concept altogether in favor of studying social movements in cities.

However fruitful Castells's line of inquiry may prove to be, the criticism that Castells himself made of the school of human ecology applies to his work as well. Castells failed to furnish a clear conceptual definition of the city as a way of embracing, understanding, and furthering his own empirical work. Instead, like many other urban historians (see Gillette & Miller, 1987), he studied a phenomenon *in* cities but failed to associate the phenomenon *with* cities. In fact, such movements apparently could take place anywhere, so little do they concern cities per se.

Peterson (1981), motivated by the effort to demonstrate, contra his political science colleagues such as Dahl (1961), Banfield (1961), and Crenson (1971), that the city was not simply the sum of its individual residents and groups, argued that cities first and foremost must be construed as actors in their own right seeking to compete with other cities in an arena of economic, social, and political resources. Further, he asserted, they try to achieve dominance along each of the hierarchies of this arena. To do so, for example, officials try to make the city into the most economically attractive site possible for prospective human and commercial residents. Yet the extent of such efforts is limited. City governments cannot redistribute income, for instance, because that is a responsibility assigned exclusively to the federal government. However, they can pursue developmental policies that will promote the economic growth and expansion of the city, and, to a lesser degree, they can pursue certain allocational policies, such as improving schools or utility services.

Peterson's scheme is compelling, but unfortunately, as others subsequently noted (e.g., see Swanstrom, 1985), it is also somewhat misguided. Peterson properly observed that cities should be construed as actors in an arena of their own or, in other words, as phenomena sui generis. Although this concept helps solve the problem posed by the weaknesses of the paradigm of human ecology, it does so only partially. Although the problem of human action is addressed, it is solved for corporate entities, the cities. Human action *within* cities—and particularly *against* cities—is, for all intents and purposes, nonexistent. Conflicts between groups are dismissed as somehow misplaced, especially those conflicts of neighborhoods and civil rights activists and the poor, but why they are misguided is not explained. Although such movements represent much of the urban action, they are perceived to be motivated by circumstances over which cities have no control. Yet Peterson fails to show that these struggles are not pertinent to city life—or how they are not, as Swanstrom (1985) recently demonstrated in the case of Cleveland, Ohio. In effect, Peterson's concept of the city has proven to be as limited as human ecologists' and Castells's concepts are.

Although the views of human ecologists, of Castells, and of Peterson represent the most prominent theoretical schemes designed to capture the nature of cities and city life today (but also see Harvey, 1975), they are by no means the only theories about cities in America. Suttles (1968), Hunter (1973), and Fischer (1975), among others, sought to capture the important symbolic and human actions that take place in cities. Fischer (1975) offered a subcultural view of city life, but it did not stimulate much empirical research, partly because he ignored that important dimension of urban struggles, about which the human ecologists, Castells, and Peterson sought to make some sense. Suttles (1968) argued that cities should be viewed as territorial units in which there are boundaries and that groups organize to defend their own turf. Suttles's promising theoretical leads to apprehending the city also failed to generate much research; however, I will demonstrate that Suttles's insights are indeed vital to a proper concept of the city.

Other promising theoretical explorations are now underway, some of which are included in my propositions in the following discussion. Logan and Molotch (1987), for instance, build on Molotch's (1976) important empirical insight into the city as a *growth machine*. Molotch's (1976) theory that cities are like machines that churn out expansion in land and in space has influenced a great deal of urban research during the past decade. Logan and Molotch (1987) have taken the argument a few steps further, noting the importance of entrepreneurs, growth coalitions, and even the contradiction between use values and exchange values in the urban marketplace. But, like the human ecologists and Castells, they have failed to provide a suitable generic concept of the city within which it makes sense to discuss the city as a growth machine and to speak of the role of human action. In other words, their metaphor does as little for them as the biological metaphor does for the human ecologists. If cities indeed are like machines, then how can one properly speak of human action at all? Other work now under way by Feagin (1988), Gottdeiner (1987), and Gottdeiner and Feagin (1988) eventually may help solve some of the errors of this earlier work.

In brief, then, what is missing from previous work is the effort to provide concepts suitable to the obvious historical dimension and variations among cities and to the real struggles that pit human beings against one another. Cities are made; they are not natural. They also are fascinating because of the social and human forces that universally animate them. The issue, then, is how these real, historical matters can be addressed and presented in a form that enables other scholars to study them so that evidence can be accumulated about, as Weber (1949) intimated, the city in history.

Apprehending the City

In my conceptual treatment of the city, I take up the issue where some, such as Peterson (1981), left off. In particular, I seek to convey a more realistic historical assessment of how cities in America work. The sense of history is crucial to any concept of the city because history alone provides insight into the fact that cities—including their political boundaries and their designs for the future—are constructed over time by the energies of men and women and that cities can be deconstructed as well. I also seek to make explicit how human action figures into the life of the city. I believe that to think of life in the city as active and to appreciate the role of human action in the life of the city properly, one must adopt a view that is anchored theoretically by a strong concept of power. I believe it is power that animates the life of the city and those within it. In addition, I offer a set of propositions that are empirically grounded in various works of research. They thus should be empirically falsifiable, satisfying a canon, as Gottdeiner (1987) noted, that seems to elude the efforts of many current analyses of the city.

Proposition 1. A city is a spatially delimited, sovereign territorial system.[1]

Cities are based essentially on two dimensions: land and space. Moreover, these are bounded dimensions. That which establishes the lines of the boundaries represents the limits of the territorial sovereignty of the city. Historically such sovereignty has been created by the cumulative efforts of different men and women; it has not been a natural fact of life, as the human ecologists imply. At a conceptual level, Suttles (1968) is the social scientist who best appreciates the territorial character of the city as a human creation. Suttles argued for a view of the city that construes it in terms of segmented territories, in which are positioned diverse ethnic groups. For such groups, the boundaries *within* the city are absolutely pivotal for life in the city. Here, I take Suttles's concepts a step further and insist that territorial sovereignty is not merely a fact of life *within* the city, but that it characterizes the city qua cities. The key conceptual assertion here is that of sovereignty over the territory of land and space. It is such sovereignty that makes the city into a concept on its own and that made it in history as well (e.g., see Dykstra, 1968).

> *Proposition 2.* Cities, as sovereign territorial systems, engage in political action and struggles against other cities, as well as against rival political sovereignties such as states. Those who dominate the life of the city act as city agents to dominate the lives of other cities and to act against rival sovereignties such as states. There is a continuing quest for territorial dominion by cities, driven by a quest to impose order.

Historians who have studied the creation of cities in America widely underscore the truth of this claim (e.g., see Marcus, 1987). Studies by Dykstra (1968) reveal that the lives of early settlements of people were dominated in large part by the effort to create sovereign territorial boundaries. Writing of Tacoma's failure in the 19th century, Glaab and Brown (1967, p. 124), for instance, observed, "Tacoma's magnificent aspirations were never realized. Again, as elsewhere, the community organization of a rival and the ability of that rival substantially to affect the pattern of railroad connection altered urban ambitions and calculations." One particularly clear illustration of this fact is to be found in the battles that occurred between cities and state legislatures over the phenomenon of home rule (see Beard, 1912; Judd, 1984, pp. 37-41). For example, in Wisconsin the shape of Milwaukee was influenced prominently by the state legislature's rule over the city, a matter that provoked several battles between the city and the state. The state government felt threatened by the growing numbers of people in the city, whereas the city government wished to exercise authority within its own physical realm. In the course of these battles, as in the course of similar battles that took place elsewhere in the United States, the dominant groups in Milwaukee, among them the Association of Commerce and their allies, repeatedly took their case to the state legislature. Sometimes they emerged victorious in the early years; often they did not (Madden, 1972).

Other prominent struggles for territorial dominion are between cities and their surrounding neighbors—sometimes smaller cities, sometimes larger ones. Milwaukee history again provides rich evidence of this type of struggle (Fleischmann, 1986). This history, particularly in the 20th century, has been punctuated often by struggles between Milwaukee and its surrounding suburbs over territorial rights. Histories of St. Louis and Los Angeles, among others, also point to the considerable historical importance of the quest for territorial dominion and how the fate of some cities has hung in the balance over the question of territorial dominion. As Luckingham (1983, p. 315) wrote of Phoenix's phenomenal expansion,

> An aggressive annexation policy helped Phoenix to increase its population sixfold between 1950 and 1980, and enabled the city to increase its physical size in square miles from 17.1 to 324.1 during the same period. . . . As a local official put it, "We wanted to avoid the St. Louis model, where suburbs strangle the city."

Even in the many burgeoning cities of the Sunbelt, one finds such questions of territorial dominion occurring again and again. As Bernard and Rice (1983, p. 22) observed in their review of Sunbelt cities:

> [A] serious challenge . . . occurs when the central city is unable to capture peripheral growth within its corporate limits. . . . The presence of a host of governments . . . diffuses power throughout the metropolitan areas. Sometimes the cost of government is increased, and always public planning is complicated and public responsibility is obscured.

> *Proposition 3.* The fate of cities, as territorial systems, also hinges on their capacity to attain dominion over their external material environment. The capacity of a city, as a sovereign territorial system, to survive depends as much on its effective dominion over its material environment as on its effective success in the battle with rival cities and with states.

The history of American cities is strewn with facts about places that failed to achieve the success envisioned by their founders because they failed to achieve dominion over their material environments. Galveston, Texas, as Feagin (1988) pointed out, rivaled Houston at the turn of the century as the dominant settlement on the Texas Gulf Coast. Then Houston managed to build a canal to the coast and became the principal port there, and its rivalry with Galveston was all but won. Rome and Athens, Georgia, as Wade (1959) once observed, had pretentiously planned to be much like their European counterparts, but efforts to make them into settlements readily accessible to people and to trade failed. Glaab (1961) noted that in the early history of the creation of American cities, Independence, Missouri, had, according to some, the possibilities of becoming a great, booming center, but failed, largely because cities on the Mississippi, such as St. Louis, were better able to take advantage of their proximity to a trade route. Also the major cities of central Texas, including Austin and San Antonio, did not become viable centers until city leaders managed to build a system of dams for electricity and for water along major portions of the Lower Colorado River (Orum, 1987b).

Dominion over the material environment represents a persistent problem for cities, such as the struggles with rival political sovereignties. Cities in California, for example, are endangered constantly by the assaults of various floods and fires. Cities along the banks of Lake Michigan faced comparable dangers with the rising waters in 1986 and 1987, and the energy crisis of 1973 and 1974 posed parallel threats to several American cities. Victory over the material environment, in brief, is absolutely crucial to the vitality of American cities.

Proposition 4. As a sovereign territorial system, a city exercises jurisdiction over the lives of people and institutions that reside within its boundaries.

Cities exercise jurisdiction over the people and the institutions they house. This authority means that the city may enact laws regarding rules of traffic and many other such matters, that, if violated, lead to the prosecution of the offenders. Like any other sovereign territorial system, the city possesses its own array of governmental and police agencies that serve to prosecute violators, as well as a system of prisons to house severe offenders. With the increase in crime in many American cities, these operations of the city, as a sovereign system, have come to play an ever more visible part in everyday life in the city. As a result, they also increasingly have become visible targets for the grievances of discontented urbanites. Occasionally, of course, the forces that serve to impose the jurisdictional sanctions of cities will play extremely salient roles in city, and in national, life as they did in Chicago during the summer of 1968. Histories of the development of American cities also reveal how these inner legal workings of sovereignty have emerged and, in many cases, how difficult it has been for the city to impose order within its bounded domain (Dykstra, 1968; Orum, 1987b). Moreover, as in any other issue of territorial sovereignty, the struggle for the city to impose order never ceases, but is a continuing one between city agents, such as the police, and city dwellers.

Peterson (1981) insisted, in accord with his general view, that cities hold only limited authority in this regard. But to claim that they have limited authority is not to say, as is sometimes inferred, that they have *none.* Indeed, in the course of everyday life in the city, the city agents exercise considerable control over residents and how they pursue their daily lives (see Gurr & King, 1987).

Proposition 5. A city's jurisdiction over its residents is exercised through an administrative machinery and is based on the provision of resources, some of which are provided by taxes on residents. In turn, the city provides important services to meet the needs of residents, including schools and public utilities.

To appreciate properly the role that cities play in the lives of residents in modern America, one must fully acknowledge the inner workings and dimensions of the city. All cities possess administrative machinery, including various kinds of city government, ranging from strong mayor-city council to weak mayor-city manager forms (e.g., see Banfield & Wilson, 1963). In some cities the machinery has grown to unmanageable proportions, such as that in Chicago, often making it difficult to provide direction and governance to the city (Orum, 1988). The operation of this machinery typically is funded by property taxes on residents. And, of course, the movement of wealthy residents and corporations away from the central city in recent times, in many places, particularly those of the Northeast and the Midwest, has created a number of precarious financial situations for the operations of the city (e.g., see Judd, 1984, on New York City and Swanstrom, 1985, on Cleveland). Some analysts, such as Friedland (1983), have argued that these financial crises for cities can properly be construed as struggles produced by the failings of modern capitalism. Yet in presenting such an argument, these analysts, by conceptually reducing the city principally to its internal constituents, fail to appreciate the independent and autonomous workings of the city as a corporate political entity seeking to maintain its own integrity.

The other side of the city's sovereign coin is the services that it provides to meet the needs of its residents. Here, as Peterson (1981), Judd (1984), and others point out, cities battle with one another in order to attract residents. Cities that can furnish better schools or cheaper public utilities hope thereby to attract residents from rival cities. Moreover, in many American cities, particularly those in the areas of industrial decline, the issue of services has become not merely one of the quantity and range of services but also (and primarily) one of the quality of them. This issue has led to a new kind of city boosterism by city agents seeking to provide a distinctive lifestyle above and beyond mere economic considerations.

> *Proposition 6.* Various social groups compete to secure their hegemony over the city and to control it as a sovereign territorial system. Such struggles are patently political, and hegemony can be exercised in many ways, including control over the urban marketplace, but also, and most especially, through control over the meaning and design for the city.

Throughout history, various groups have competed to control American cities as sovereignties (Dykstra, 1968). This fact dominated the early American cities. For example, as Dykstra (1968, p. 365) observed, "The experience of the Kansas cattle towns . . . suggests . . . that social conflict was normal, it was inevitable, and it was a format for community decision-making. . . . [The] external conflict [of cattle towns] with farmers and their internal warfare—business factionalism and moral reform politics—demonstrated anything but a thoroughgoing solidarity" (also see Orum, 1987b, pp. 1-31). Conflict also seems to dominate many American cities today (Coleman, 1957). In the very early periods of the history of American cities, prior to the advent of some system of government and electoral representation, the battles were fought between diverse groups of settlers, such as the reformers and the outlaws. In the 1930s and beyond, as the New Deal policies of the Roosevelt administration began to take hold, such combat took a different form, often between advocates of federal intervention over the city and those who insisted against such interventionist policies (Orum, 1987b). Today such battles sometimes pit ethnic groups against one another, such as in the cities of Miami, where blacks and Hispanics struggle (for some history, see Mohl, 1983), and Chicago, where blacks and whites struggle.

How hegemony develops will vary from place to place; how it is said to operate will vary from analyst to analyst. Many neo-Marxist analysts will point, with considerable accuracy, to the control of the urban marketplace and to the role, in particular, of various factions of the capitalist class (Gottdeiner & Feagin, 1988). Other analysts will point to the role of urban entrepreneurs and market entrepreneurs (e.g., see Logan & Molotch, 1987). Still others will point to the significant role of so-called growth coalitions, alliances of groups of people from both the business and the political inner worlds of the city (Mollenkopf, 1983).

In my view, hegemony, though a complex phenomenon (see Gramsci, 1971), is a matter of control over the design, or system of meaning, of a city. That is, it is control over the way the city is defined in symbolic terms, the cultural architecture for the future of the city, and particularly over how the city is to be viewed by its residents. These designs are autonomous creations and cannot, among other things, be reduced to serving merely as the playthings of any single social group, whether it be the dominant business forces or the dominant political forces. Furthermore, without a proper appreciation of the full dimensions of these designs for the city, one cannot properly appreciate the history of the city—how the inner life, in effect, of the city began and developed and how it may be transformed in the future.

> *Proposition 7.* The dominant meaning, or design, for the city in modern America is that of growth, which entails such virtues as progress, and which in turn entails such other virtues as civilization for the city. In a sufficient and increasing number of empirical studies, however, instances of struggles in American cities demonstrate that growth is by no means the only possible hegemonic design for city life. Such struggles clearly show the historic dimension to the design of growth and undercut the claims of the human ecological paradigm that, in particular, growth is a phenomenon natural to the inner workings of the city.

Studies of the territorial and political development of American cities reveal that they often have been created by so-called visionaries, or people of vision (Chudacoff, 1981). Typically, in the history of these cities, the vision (or, in my terms, the design) for the city has been one of growth. What that has meant, in fact, is the display of a set of symbols that link growth to some heralded virtues for the city, often summed up in terms of progress for the city (Orum, 1987a, 1987b). Johnson, Booth, and Harris (1983, p. viii) put it well: "Regardless of their differences, all the groups that have sought or exercised power in San Antonio share a commitment to progress, a vague but powerful ideological concept that implies a universally beneficial development process. For San Antonians . . . progress has meant population growth, better public services, and economic development." These various designs are displayed in the pages of city newspapers, but they also frequently become the language within which talk about the future is clothed by city agents, such as those associated with the chamber of commerce. Eventually these designs become accepted as natural to city inhabitants.

In the last decade or so, however, these hegemonic designs increasingly have come under attack. Thus, in a wide range of American cities today, struggles take place between environmental groups and the so-called progrowth forces (Abbott, 1987, pp. 214-243; Bernard & Rice, 1983; Johnson et al., 1983). Such struggles are especially evident in cities of the American Sunbelt. Apart from illustrating the universal fact of struggle in the lives of cities, they also reveal that growth is only one of several possible designs for the future of the city. When environmentalists have been successful in their struggle against the city—as in Oregon, Colorado, and Florida, for example—the design for the city has come to emphasize prominently matters of the quality of life and harmony with the material environment, more so than matters of new industry, residents, and the like (see Inglehart, 1977, for his discussion on the changing values between industrial and postindustrial societies).

> *Proposition 8.* Inasmuch as growth is the dominant design for the city in contem-
> porary America, those groups that seek to dominate the life of the city adopt
> it as the banner for their political combat. Indeed, to become dominant as a
> group over the machinery of the city has meant to adopt growth as the watchword
> for struggle. The rich symbolism of growth serves the interests of political actors,
> for it justifies their efforts to increase their power over the number of people
> they control. Likewise, growth serves the interests of the dominant capitalist
> class because the more residents and businesses in a city, the greater the amount
> of profit and production that can occur under the sovereignty of the city.

Molotch (1976) implied that growth is a system of meaning that is hege-monic because it satisfies the interests of many diverse groups seeking to dominate the life of the city (also see Orum, 1987a). Yet, in contrast, this system of meaning must not be construed as something of a plaything, but rather as a system of meaning within which the struggles over the hegemony of the city take place. Thus growth is not so much deployed by those who battle for the dominance of the city, but displayed as the banner under which their struggle is carried out (Gramsci, 1971). "Growth-industry-progress" are the watch-words for those groups that seek to gain power over the machinery of the city. Politicians who wish to be successful in their campaigns often have been con-strained to run under a banner of growth in American cities. Likewise, of course, the symbolism of growth serves the interests of capitalists who wish to dominate the city because the success of its hegemony entails, in a real, material sense, more people who will produce and spend—in short, more profit for those who wish to be dominant in the city.

Accordingly, too, the resistance that those who struggle against the city and its agents face is the resistance to the widespread appeal of this banner to the life of the city (e.g., see Orum, 1987a). Because the symbolism of growth serves so many interests so well, it is naturally profoundly difficult to dislodge as a historic fact of city life.

> *Proposition 9.* Cities use individual actors or groups of actors to carry out their purposes in pursuing dominance over their material environment or dominion over rival political sovereignties, and individual actors as entrepreneurs use cities for their own specific ends as well.

As Peterson (1981) correctly noted, cities cannot be construed exclusively as reducible to their individual constituent parts, whether they be thought of as groups, institutions, or people. They must be viewed properly as units in their own right—as sovereign territorial systems. Yet this view does not demand that one should make the opposite conceptual error of not according the group, institution, or person a reality on its own terms. Thus I have argued here that cities become the object of struggle by groups that attempt to secure hegemony over them.

It is only by according the group and the individual actor their own proper mode of reality that one can apprehend fully how truly powerful people have come to dominate certain American cities and, although using them for their own ends, have produced results for the city that have enhanced the sovereignty of the city. One such obvious example is that of Robert Moses. As Caro's (1975) brilliant biography makes clear, Moses used the machinery and operations of New York to promote his own interest in ever greater power. While working as the city commissioner of parks, he engaged in actions that led to certain material developments, such as the system of bridges, that created an ever-larger territorial dominion for the city of New York. In effect, Moses's own considerable ambitions profoundly affected the territorial ambitions of the city of New York.

Countless other cases in the history of American cities display much of the same degree of power by an individual over the fate of a city. In central Texas, for instance, Lyndon Baines Johnson used the residents' concerns about their ability to survive in the often dry, barren countryside as a means of propelling his own ambitions for national office. Johnson was a master political strategist. As a result of his moves, monies flowed into Texas that eventually led to the conquest of deep environmental concerns, such as the sufficiency of water and electric power, as well as to expanding the territorial dominion of cities such as San Antonio and Austin (Orum, 1987b). In Chicago, the legendary Richard J. Daley operated with much the same flair as Johnson. During two decades, his own considerable political ambitions produced new buildings for the city, new hospitals, and new campuses for the University of Illinois (Banfield, 1961). So too, William Mulholland, turn-of-the-century political mogul in Los Angeles, succeeded in harnessing waters there that eventually would permit that city to expand extensively (Fogelson, 1967).

In all of these instances, the ambitions of truly powerful people redound fortuitously to the fates of the city. It is not enough to label them just *political actors,* as some do. They are, rather, people of sometimes heroic proportions who manipulate the interests of the city to their own advantage.

> *Proposition 10.* Insofar as the dominant social groups in a city succeed in imposing their designs for the city on the subordinate groups of residents, there may be little internal conflict over the control of the city. But when residents, for whatever historical reasons, begin to sense that the dominant meaning for the city encroaches on their own designs, then the city—its agents and their allies—will be the target of politically mobilized efforts designed to challenge the hegemony of those dominant groups and, if possible, to depose them from their positions of power.

Until the mid-1960s the dominant and hegemonic design for most American cities was that of growth. But under the weight of a number of developments, particularly the resistance movements of the 1960s, that design increasingly has been attacked by groups of residents. In large part this assault has happened because the design itself has fostered consequences that have encroached on the territories of residents. The effort to promote new industry and other forms of construction often has led, especially in the larger cities, to the displacement of residents from lands and spaces over which the residents had attained dominion and that they regarded as home. These struggles often have been bitter, and they routinely have pitted the city agents and their proclamations for growth against the residents. As a result, to maintain their territories, residents often have had no recourse but to create neighborhood organizations and to lobby city governments in an attempt to maintain their own dominions.

Like all political struggles and battles, moreover, this struggle between city agents and residents occasionally escalates into events that seem more like war than like democracy in action. The city agents, as the dominant city forces, have been able to construct alliances of real estate agents, city government officials, and leading corporate officers to advance their interests, and alliances of disgruntled neighborhood residents have been constructed from coalitions of neighborhood groups (Johnson et al., 1983; Orum, 1987b). Moreover, a nationwide organization—ACORN—tries to provide intelligence and influence to residents to help them fight their battles. Sometimes residents have been successful in deterring the attempt by city agents to displace them or to encroach on their lands. At other times, however, peculiar coalitions emerge that can bring the city to its knees, as happened in the case of former mayor Dennis Kucinich of Cleveland (Swanstrom, 1985).

> *Proposition 11.* In the United States, the city, as a sovereign territorial system, intervenes between the state—or national government—and the citizen. The state, itself a sovereign territorial system, controls the lives of citizens, in part, by exercising its own jurisdiction over the city, particularly through the provision or denial of resources. In the strongest sense, the city acts as a conduit for the power of the state over citizens. As a result, the city easily can become the

misapprehended target for hostility by residents when the state—not the city—fails to meet the needs of citizens.

In the larger scheme of things, the state, as a sovereign territorial system, does exercise jurisdiction over the city. Peterson (1981), among others, has made a very strong case for this claim (also see Gurr & King, 1987). Such matters, however, are never firmly decided, and battles continue between the national government and cities over their rightful provinces of jurisdiction. Nonetheless, since the New Deal era, the state has managed to hold the upper hand primarily because it gradually has come to control ever greater amounts of resources, particularly in the form of tax revenues. It also has managed to control the fate of many American cities, as the near bankruptcies of New York and Cleveland attest.

Further, because the city intervenes between the national government and the citizen, it becomes the administrative device through which the state will seek to control citizens. As a conduit, the city provides the state a means of its own control over residents; if state agents wish to exercise greater control, they use the city as a means of providing greater sources of welfare or other benefits to citizens (Cloward & Piven, 1971). But as a result of its peculiar position between the state and the citizens, the city—its agents and their allies—can become the misapprehended target for actions that citizens intend to take against the state, not the city.

In historical terms it is often difficult to disentangle these matters. If a group of aggrieved citizens is pounding at the barricades of city hall, it matters little what words they use to voice their displeasure. Moreover, they are pounding at the barricades of city hall precisely because they believe it is the city, not the state, that is responsible for their calamities. Yet, while some struggles genuinely pit the city against its residents, such as that over the hegemony of the designs for growth, other struggles occur when it is the action of the state, not the city, that is in question. In particular, the whole array of issues under the banner of civil rights, poverty, and gender in the 1960s demonstrates this. Prior to the advent of the policies initiated by the federal government, cities—and states— often could resolve disputes over such matters as city busing or the integration of public facilities by their own legislation. If they failed, then the resistance efforts directed against cities were based on realistic assessments by residents. But by the mid-1960s, the decisive action on these matters was in the hands of the agents of the state, not in the hands of the agents of the cities. And it often was the failure of the state agents to act decisively or, in the eyes of residents, appropriately, that prompted various kinds of strikes, riots, and protests across many American cities.

Apprehending the City in Fact: Doing Thick City Histories

The conceptual approach that I have offered here carries with it a particular methodological approach to the study of cities. To examine in complete and full detail the nature of the city as a sovereign territorial unit, with regard to both its external struggles against rival sovereignties and its internal battles, one necessarily must adopt a historical perspective. Only with such a view of the city—in effect, viewing its use of land and space through time—can the city as an entity in and of itself be properly apprehended. Yet no single history suffices. Many, often marvelous, examples of the study of cities in which the authors reveal important insights into the inner workings of cities, such as Warner (1968), Thernstrom (1964), Hershberg (1981), and Conzen (1976), to name but the most prominent, have illuminated hitherto unexplored dimensions of the city's historical landscape. Yet the city as an entity in and of itself has been ignored: how it has developed into a sovereign territory, how it struggles continuously with other sovereign territories, and how battles animate its everyday life, particularly in the late 20th century.

First, then, one must investigate the creation of cities as sovereignties. How have cities been settled, and how profound have the struggles been over their sovereignty? To appreciate these matters properly, the life of the city must be traced back to its beginnings. Next, to follow the course of this life, one must necessarily follow the course of struggles within cities. How, in particular, are questions about the dominion of the city over its material environment raised, and how are they solved? Just a brief review of the histories of some cities in America will reveal how significant such matters as transportation routes, rivers, and a steady supply of water and electricity are to the development of the city and to the lives of its residents (e.g., see Chudacoff, 1981).

Of equal significance, one must examine also how residents settle the inner territories of cities and how hegemonic designs for the entire city come about. To examine these matters properly, one must engage in detailed, primary historical research, attending to newspapers (the prime carriers of such designs), to the personal documents and memoirs of residents, and to documents on the lives of neighborhood and community associations (see the work on Tokyo by Bestor, 1989, for a truly excellent example of this kind of fieldwork). Moreover, because historical documentation on certain key events is often missing, and because the analyst seeks to appreciate the way residents themselves construe the city and its past, oral histories become vital pieces of evidence. Indeed, only through the accumulation of such broad and complete evidence can the full story of the life of the city be told and understood. These guidelines are embraced by an approach that I call *thick history,* following the term used by Geertz (1973) to characterize anthropological ethnographies.

Naturally, this work takes considerable time and even more patience and energy. It is the sort of research that ought to be done by teams of investigators, though team efforts on cities sometimes are doomed for lack of a guiding intellectual approach (see Jackson, McDonald, Zunz, & Hershberg, 1982). Nevertheless, just because this kind of research is so demanding does not mean that social scientists should fail to engage in it. It would be far preferable in terms of time and energy to turn to batteries of compiled statistics and to make claims about city life merely in terms of population size and density, but ease of attack should not substitute for the theoretical relevance and importance of thick histories about cities. Ultimately the propositions I offer can only be further refined and elaborated through such detailed research. Practically, too, city residents will know what it means to live in the city only when they learn the meanings of city life.

Note

1. The chief concern of the critics of this chapter was the claim that cities in America are sovereign territories. It is precisely because they are not sovereign, some have said, that the cities run into trouble. In response, I would argue that cities do exercise sovereignty over their populations even though it may not be a comprehensive sovereignty. Further, the degree of such sovereignty is always contested, and sovereignty ranges over time. Some cities today have won more sovereignty from their state legislatures than they had attained earlier. Some cities today also seek to gain more sovereignty than they have to cope with the difficult problems of poverty and crime that many of their residents face. These variations and permutations on sovereignty do not negate the concept. Rather, they point out precisely why such a concept must be invoked in order to understand the historical battles and struggles that pit the city against other rival sovereignties such as states and the federal government and even suburbs. These are battles fought by agents on behalf of the city against agents of other political entities. Unless the city is understood in this light, an essential dimension of the nature of city life will be missed.

References

Abbott, C. (1987). *The new urban America: Growth and politics in Sunbelt cities.* Chapel Hill: University of North Carolina Press.

Banfield, E. (1961). *Political influence.* Glencoe, IL: Free Press.

Banfield, E., & Wilson, J. Q. (1963). *City politics.* New York: Vintage.

Beard, C. (1912). *American city government.* New York: Century.

Bernard, R. M., & Rice, B. R. (Eds.). (1983). *Sunbelt cities: Politics and growth since World War II.* Austin: University of Texas Press.

Berry, B.J.L., & Kasarda, J. (1977). *Contemporary urban ecology.* New York: Macmillan.

Bestor, T. (1989). *Neighborhood Tokyo.* Stanford, CA: Stanford University Press.

Burgess, E. W. (1967). The growth of the city: An introduction to a research project. In R. E. Park, E. W. Burgess, & R. D. McKenzie (Eds.), *The city* (pp. 47-62). Chicago: University of Chicago Press. (Original work published 1925)

Caro, R. (1975). *The power broker: Robert Moses and the fall of New York.* New York: Random House.

Castells, M. (1977). *The urban question* (A. Sheridan, Trans.). Cambridge: MIT Press. (Originally published as *La question urbaine,* 1972, Paris: Francois Maspero)

Castells, M. (1983). *The city and the grassroots.* Berkeley: University of California Press.

Chudacoff, H. P. (1981). *The evolution of American urban society.* Englewood Cliffs, NJ: Prentice Hall.

Cloward, R., & Piven, F. F. (1971). *Regulating the poor.* New York: Pantheon.

Coleman, J. S. (1957). *Community conflict.* New York: Free Press.

Conzen, K. (1976). *Immigrant Milwaukee, 1836-1860: Accommodation and community in a frontier city.* Cambridge, MA: Harvard University Press.

Crenson, M. (1971). *The un-politics of air pollution: A study of non-decisionmaking in the cities.* Baltimore: Johns Hopkins University Press.

Dahl, R. A. (1961). *Who governs? Democracy and power in an American city.* New Haven, CT: Yale University Press.

Dykstra, R. R. (1968). *The cattle towns: A social history of the Kansas trading centers.* New York: Atheneum.

Feagin, J. R. (1988). *Houston: The free enterprise city.* New Brunswick, NJ: Rutgers University Press.

Fischer, C. (1975). Toward a subcultural theory of urbanism. *American Journal of Sociology, 80,* 1319-1341.

Fleischmann, A. (1986). The politics of annexation: A preliminary assessment of competing paradigms. *Social Science Quarterly, 67,* 128-142.

Fogelson, R. M. (1967). *The fragmented metropolis: Los Angeles, 1850-1930.* Cambridge, MA: Harvard University Press.

Friedland, R. (1983). *Power and crisis in the city*. New York: Schocken.

Frisbie, W. P., & Kasarda, J. (1988). Spatial processes. In N. J. Smelser (Ed.), *Handbook of sociology* (pp. 629-666). Newbury Park, CA: Sage.

Geertz, C. (1973). *The interpretations of culture*. New York: Basic Books.

Gillette, H., & Miller, Z. L. (Eds.). (1987). *American urbanism: A historiographical review*. Westport, CT: Greenwood.

Glaab, C. N. (1961). Visions of metropolis: William Gilpin and theories of city growth in the American West. *Wisconsin Magazine of History, 45*, 21-31.

Glaab, C. N., & Brown, A. T. (1967). *A history of urban America*. New York: Macmillan.

Gottdeiner, M. (1987). *The decline of urban politics*. Newbury Park, CA: Sage.

Gottdeiner, M., & Feagin, J. R. (1988). The paradigm shift in urban sociology. *Urban Affairs Quarterly, 24*, 163-187.

Gramsci, A. (1971). State and civil society. In *Selections from prison notebooks of Antonio Gramsci* (Q. Hoare & G. N. Smith, Trans. and Eds.) (pp. 206-276). New York: International.

Gurr, T. R., & King, D. S. (1987). *The state and the city*. Chicago: University of Chicago Press.

Harvey, D. (1975). *Social justice and the city*. Baltimore: Johns Hopkins University Press.

Hawley, A. H. (1950). *Human ecology: A theory of community structure*. New York: Ronald Press.

Hershberg, T. (Ed.). (1981). *Philadelphia: Work, space, family, and group experience in the nineteenth century*. New York: Oxford University Press.

Hunter, A. (1973). *Symbolic communities*. Chicago: University of Chicago Press.

Inglehart, R. (1977). *The silent revolution: Changing values and political styles among Western publics*. Princeton, NJ: Princeton University Press.

Jackson, K., McDonald, T. J., Zunz, O., & Hershberg, T. A. (1982). Special book review section: Toward an interdisciplinary history of the city. *Journal of Urban History, 8*, 447-484.

Johnson, D. R., Booth, J. A., & Harris, R. J. (Eds.). (1983). *The politics of San Antonio: Community, progress, and power*. Lincoln: University of Nebraska Press.

Judd, D. R. (1984). *The politics of American cities: Private power and public policy*. Boston: Little, Brown.

Logan, J., & Molotch, H. (1987). *Urban fortunes*. Berkeley: University of California Press.

Luckingham, B. (1983). Phoenix: The desert metropolis. In R. M. Bernard & B. R. Rice (Eds.), *Sunbelt cities: Politics and growth since World War II* (pp. 309-327). Austin: University of Texas Press.

Madden, D. R. (1972). *City-state relations in Wisconsin, 1835-1901: The origins of the Milwaukee home rule movement*. Unpublished doctoral dissertation, University of Wisconsin, Madison.

Marcus, A. I. (1987). Back to the present: Historians' treatment of the city as a social system during the reign of the idea of community. In H. Gillette, Jr., & Z. L. Miller (Eds.), *American urbanism: A historiographical review* (pp. 27-47). Westport, CT: Greenwood.

McKenzie, R. D. (1967). The ecological approach to the study of the human community. In R. E. Park, E. W. Burgess, & R. D. McKenzie (Eds.), *The city* (pp. 63-79). Chicago: University of Chicago Press. (Original work published 1925)

Mohl, R. A. (1983). Miami: The ethnic cauldron. In R. M. Bernard & B. R. Rice (Eds.), *Sunbelt cities: Politics and growth since World War II* (pp. 58-99). Austin: University of Texas Press.

Mollenkopf, J. (1983). *The contested city*. Princeton, NJ: Princeton University Press.

Molotch, H. (1976). The city as a growth machine: Toward a political economy of place. *American Journal of Sociology, 82*, 309-332.

Orum, A. M. (1987a). City politics and city growth. In R. G. Braungart (Ed.), *Research in political sociology* (Vol. 3, pp. 223-244). Greenwich, CT: JAI.

Orum, A. M. (1987b). *Power, money, and the people: The making of modern Austin*. Austin: Texas Monthly Press.

Orum, A. M. (1988, September 16). Our governmentless politocracy. *Chicago Tribune*, p. 127.

Peterson, P. E. (1981). *City limits*. Chicago: University of Chicago Press.

Saunders, P. (1981). *Social theory and the urban question*. London: Hutchinson.

Simmel, G. M. (1950). The metropolis and mental life. In K. H. Wolff (Ed. and Trans.). *The sociology of Georg Simmel* (pp. 409-424). Glencoe, IL: Free Press. (Original work published 1902-1903)

Suttles, G. (1968). *The social order of the slum: Ethnicity and territory in the inner city*. Chicago: University of Chicago Press.

Swanstrom, T. (1985). *The crisis of growth politics: Cleveland, Kucinich, and the challenge of urban populism*. Philadelphia: Temple University Press.

Thernstrom, S. (1964). *Poverty and progress: Social mobility in a nineteenth-century city*. Cambridge, MA: Harvard University Press.

Wade, R. C. (1959). *The urban frontier, 1790-1830*. Cambridge, MA: Harvard University Press.

Warner, S. B., Jr. (1968). *The private city: Philadelphia in three periods of its growth*. Philadelphia: University of Pennsylvania Press.

Weber, M. (1949). *On the methodology of the social sciences*. Glencoe, IL: Free Press.

Wirth, L. (1938). Urbanism as a way of life. *American Journal of Sociology, 44,* 1-24.

— 2 —

Urbanism and Urbanity

CITIES IN AN URBAN-DOMINATED SOCIETY

Scott Greer

And in the collective human artifact—the settlement, town or city—men as a group are expressing historically the character and quality of their existence, of the arrangements they have made, on one hand, with the natural world, and on the other, with one another. (Marcus, 1978)

Some believe that in "urban" societies, a special focus on urban sociology or history is obsolete (Martindale, 1958; Thernstrom, 1982). In that belief, the folk insight contained in our continued use of such terms as *metropolis, city, town, village,* and *country* is lost. New York City is not to be mistaken for a small town, no matter how uniform the network television programs. We live in a large-scale society in which organizational networks controlled in urban centers dominate but do not exhaust the possibilities of the institutional structure of the whole. In short, the cities that are the topic of this chapter are in an "urban-dominated society," rather than in a society in which most of life for most people is relatively unaffected by urban centers. In this chapter I focus on the nature of cities in urban-dominated societies.

To understand the urban synthesis, the "city" as concrete experience, we must begin by considering major aspects of cities as constituting distinct parts of the larger society's structure, yet containing their own distinctive local realities. We must understand the functions (or uses) of cities, their reasons for being, their ecological niches, and their holds on existence. We have seen cities appear, and we know of hundreds that have disappeared. But knowing only their survival bases does not explain the nature of the social worlds created from them. Even the most carefully planned, purpose-built cities (and they are few in number) are not fully predictable or even fully knowable, for many reasons. The nature of this mass of humanity evolves and differentiates itself, with purposes and proclivities unknown to the planner, revealed only in the urban ambit where people and social structures, artifacts and ideas, meet in patterns that may be quite foreign to their place of origin. Humans are culture-creating animals, and the city is a veritable hothouse of transplanted cultural complexes and traits. Hybridization and mutation are commonplace and often unnoticed.

Cities in large-scale societies, then, must be studied in situ, as cities-in-environment. They are nodes of organization in large-scale networks of interdependence. Some of their trading partners are as near as the truck gardens that feed part of the population part of the time; some are as distant as half the circumference of the earth. Cities are influenced by their dependency even as they influence their dependents elsewhere.

This approach is forced on us by our expanding knowledge of societies as systems and by our increasing awareness that discrete entities always interact with the environment and are bounded by transactions beyond themselves. Discrete variables tend to become continuous variables, moving from Aristotle's "Yes or No" to "more than-less than." (City-society is comparable to heredity-environment, firms-markets, states and international networks.) The given object of study may be viewed intrinsically, "all other things being equal," but what if the other things are also integral and necessary components of the city itself? "Foreign" markets, governments, and cultural flows impinge on the current metropolis. Awareness of the extrinsic only improves our explanation of the workings of the intrinsic.

The more obvious reason for studying cities as differentiated parts of large-scale organizational networks is simply the course of recent history. With increasing societal scale, the given city becomes more interdependent with other cities on this and other continents, as well as the more ancient hinterland. At the same time, the culture of the city diffuses widely—partly because of this interdependence—and contrariwise, the city population is exposed to a very broad spectrum of cultural influences. It is from these facts that two arguments are made against the continuing importance of studying cities as unique entities: (1) The city no longer controls much of its own destiny, because its polity is subject to decisions that originate far beyond its own boundaries, and (2) because the culture of the urbanites that once set them off sharply from the country folk is now shared, in part, by the remainder of society, urbanites do not differ as significantly in behavior, beliefs, or social choices from the non-urban (Martindale, 1958; Thernstrom, 1982).

The prime function of cities, and perhaps the defining function, is the creation and maintenance of a dynamic order. However the city started, as fortress, *entrepot*, or factory town, it is maintained by its position in a system of order. Two such major systems are found in the United States—market and government. Milwaukee, for example, began as a manufacturing city, a collection of factory towns. Today its manufacturing uses less than one fourth of the labor force, while the remainder is engaged in sales, services, commerce in general (the market), and, surprisingly, almost one tenth in government employment.[1] There are a few exceptions, but by and large, however they began, American cities today are multifunctional.

Any city worthy of its name is a market for products and people, for labor and enterprise, and for money and marriage. It is also a crossroads where strangers meet and new structures are born—from architectures to genetic combinations, from killing disease to cure. Public health services originated first in cities, and they were most needed there; the crowding and intermixture of people were accompanied by the rapid circulation of microparasites. (The easiest things of all to exchange in human society, as T. H. McNeil, 1978, has remarked, are microbes.) The cities of the industrial revolution have been compared to fires, which rapidly burned out quantities of humanity and required a steady supply of new fuel. (Only in the late 19th century, through public health technology, better housing, and diet, did industrial cities become safe enough to hold their own by natural increase.) Modern medicine emerged in the cities where expertise, money, and workers could be organized for the job. New threats to ordinary life also emerged, from epidemic tuberculosis to acquired immune deficiency syndrome (AIDS).

The Intrinsic City

The intrinsic city is, then, a social structure operating in a physical container, a continuous and stable population in a pattern of physical structures and spaces. The social structure is that complex of habits that organizes behavior, allowing the city to produce and reproduce over time with considerable predictability. The two levels of structure interplay, and each is in some degree imposed by extrinsic conditions. The city's terms of trade with its markets and its role in the network of government influence its wealth and its interests; these, in turn, influence the kinds of private and public physical structures and the nature of open spaces the city can afford. (Specific decisions rest on what the city, collectively and segmentally, *will* afford.)

The city as a physical structure, however, is at any time a given fact. The nature of its physical site constrains and channels the nature of settlement, density, and scatteration; the presence of monuments and famous places gives some cohesion to the popular culture of the place. Residential "named places" allow a sort of map of the city by social composition and physical structure, ranging from racy names like The Gold Coast and The Slum to the mundane Downtown or East Side. The structure of the city also affects the health of the population: Some cities are dangerous to human health for reasons as diverse as climate and atmospheric poison to automobile traffic on badly engineered road systems with unenforced rules of the road and inept but adventurous drivers.

In short, the city is not simple. It is a coming together of very complex structures in time and space, and the simple general concept is too simple and too general to be useful.

A major interaction of the two structures in modern cities is the spatial division of labor. Different tasks are sorted out, and the enterprises that perform them are collected in different locations for reasons of convenience and cost, mitigated by historical accidents of all kinds. In the same way, residential sections become differentiated by such attributes of their populations as social

rank, ethnicity, place in the life cycle, and lifestyle. Again distribution is partly
by cost and convenience or by location and housing market. It is also, where
exclusion is the norm, a coercive pattern of segregation by threat or force. Such
segregation in the United States is based most often on race, but any other form
of differentiation may have the same outcome.

One of the consequences of the spatial division of labor is the lack of
integrative community. Residence, market, workplace, and other regular desid-
erata of life may be widely separated, and the larger the city, the greater the
distance to traverse. Thus one major social role (and part of the self) may be
not only separated from others but also invisible to the persons we know in
them. This is the true nature of "urban anonymity." One study indicates that
the majority of the neighbors around the homes of Canadian Members of
Parliament did not recognize them as their MPs (Smith & Zipp, 1983). Equally
striking, many suburbanites in great metropolitan centers do not identify their
neighbor with the notorious gangster who periodically provides front page
news for the metropolitan dailies. The spatial division of labor thus provides
hiding places for one of several roles played by the same person from the others.
Deviants and misfits may be invisible for most of the city, lost from sight in
another role, another place, another community. They are collectively "in the
closet."

The urban place, just because it has a very large population, accumulates
deviants, those far from the statistical norm (and often the moral norm) on any
of a large number of social characteristics. These deviants often are allowed or
forced to concentrate in "their" places. When spatial boundaries coincide with
common threatened values, we find a sense of common fate—in brief, a
protective community. Examples are the homosexual population, artists and
writers, the narcotics industry and culture, homemade religions, and deviant
health care theory and practice. Then, too, that which would be rare, disliked,
and highly visible in a nonurban setting is more commonplace, better organized,
and often less stigmatized in cities. The deviant community, then, makes
proselytizing more likely and attracts new members from outside the urban
area, increasing its numbers. As noted earlier, the coming together of many
cultures and subcultures tends to result in interaction, syncretism, and hybridization. New and previously unknown forms of deviance tend to emerge, from
new vices (e.g., the practice of usury) to new medical approaches (e.g., the germ
theory of disease). And, being carried in the great marketplace, they have a
maximum opportunity to diffuse.

With segregation in space, then, segregation in time becomes more marked.
In peasant villages, "market day" was a necessary and valued institution; in one
day of the week, all business among scattered producers and would-be consumers who "truck and barter" in the village can be accomplished. This day, then,
preempts most market functions from other days of the week.

The results of segregation in time today are many, but one of great
importance is the occupation of most people at a substantial distance from home
in another social order, whether workplace or school. If trouble comes, they
must rely on members of this nonfamilial social order for help, and if it is

inadequate, they may sicken and even die in the midst of a crowd. A man having a heart attack and hanging on to a downtown lamppost to keep from collapsing may need only the medicine available in a nearby pharmacy, but unless an unusual passerby sees him as ill and not drunk, he may be "dead on arrival" at the emergency ward.

So segregated are different social types in residential neighborhoods that when they do come together, they are strangers and may regard each other as almost another species. The fear of violence and the fact of violence, ordinarily prevented through spatial segregation, are vivid in the New York subway and in the Chicago Loop. In this sense our cities also generate interpersonal crime, and our death rate by homicide is the highest among the "advanced" societies. Failing gun control or a police state, and with our weak civic culture, segregation in space and/or time may be a necessity to our cities. In what major American metropolis are you not cautioned against walking in the park at night? Central Park in New York, Golden Gate Park in San Francisco, and the Boston Commons can be dangerous to your health.

Segregation, however, does produce a minimal social order. In the process it also effectively hides those in great hazard from those who might help them. Further, it tends to become self-sealing, as the inhabitants of the neighborhood have for sources of information only those much like themselves. Meanwhile, blacks, Hispanics, and whites of each specific class see only social types like themselves. As residential discrimination declines for the prosperous minority families, the forced community of the ethnic enclave loses its best-educated residents to middle-class, often ethnically integrated, society. The leadership roles are emptied, and valuable sources of information go with them. The public agencies assigned to the area are about all the help that remains.

Cross-cutting segregation, as in the black middle class, historically has provided a bridge, an integrating mechanism. It humanizes strangers and prevents misidentification of them as in Erik Erikson's (1968) phrase "pseudo species." Such bridges appear to be threatened in many cases, leaving greater concentration and segregation by narrowly perceived social types. (The fading working-class identification of Jews is probably an example of a fallen bridge.)

Why do we find such concentration and segregation so specifically marked in the subareas of American cities—concentration by use (residential, work, and transport) and, in residential areas, by social class and ethnic stereotype? It is undoubtedly because of forcible segregation in some cases, as noted earlier, but also often (to an unknown degree, and, therefore, ignored) by preference. Planned integrated neighborhoods can become single ethnic because of black invasion as easily as white. We have worked with too simple and too time-bound concepts in approaching the whole matter.

Or, to take the case of function, why are concentration and segregation so marked in the location of control centers (of commerce, finance, and political dominance) in the largest cities of civil units and in the center of those cities? Why are the centers of the markets and the government usually found there, side by side?

The utilitarian hypotheses from decades of real estate and social ecology theory are well known: The centers are most easily accessible to transportation from the settlement as a whole and most apt to represent the unity of the urban conglomerate. But this kind of argument is extremely susceptible to changes in transportation, and, with the rapid growth of instantaneous communication, is there really a need for all of those clerical workers in those places?

Yet the disappearance of the central city, predicted for decades, is far from coming about. Still, the utilitarian argument has another powerful, though less obvious, application. When "high rollers" among economic heads, among political heads, and between the two want to "cut a deal," they do not want it subject to recorded communication—they want secrecy to maneuver and bargain. They want personal interaction when the "urban power structures" raise their shadowy heads.

As for the armies of technicians, clerks, and other functionaries, they might as well be downtown as anywhere else, from the boss's point of view. For them, too, there are satisfactions in working in the center of world cities, satisfactions that may outweigh transportation costs in getting there and street dangers if they do not "know their way around." At the same time, the greater cost of such location may be considered balanced by urban amenities and the sense of opportunity.

This discussion raises the questions: How far do the cities' inhabitants really want one thing and produce another? And how far do they want what they are getting? Is this a result of a basic conflict over what is wanted, or between what is wanted and what is possible? Or is it because of a weakness in execution that may derive from one or more of such possibilities? Could anyone want the older centers of urban splendor surrounded by dangerous wastelands and largely untraveled at night? It is hard to believe, yet that is what we have produced in a number of once socially thriving "downtowns."

Numbers, Distance, and the Metropolitan World

Claude Brown (1963) once remarked of black-on-black crime that when a man in the Bronx needed a fix, he did not get on a bus and go up to Park Avenue to grab a purse; instead, he turned to the nearest passerby, who was almost certain to be black. Thus even interpersonal crime is highly segregated by race and class in the metropolis. The vast city has room for all kinds of assortment.

When we read of the effects of the earliest metropolitan areas on Victorian novelists and other observers, we are impressed with the concrete, vivid images of the crowds. Charles Dickens wrote about the massive currents of humanity on the sidewalks when work was over, milling in the democracy of foot traffic. Where are these masses now?

We see mostly signs of them in the tens of thousands of automobiles, manifest on major streets and roads out of the center areas, overwhelming at rush hour. They are signs, with little of the concreteness—the human vividness—that the Victorians noted. Yet the denizens of Chicago or New York recognize them at once; they are the masses (including ourselves). In the same way, we see mostly signs of the housed populations. Few, if any, people are visible in these endless blocks of houses.

For many hours a day, the masses are inside, under the roofs of large-scale organizations. They are in markets, factories and other firms, public schools and colleges, shopping centers, jails, and the everlasting housing developments; they emerge briefly at given times of the day in different seasons as they go from structure to vehicle and vice versa. The masses are there, visible or invisible, but they are usually known mostly through their signs.

They are also visible in public assemblies—games, fests, demonstrations and riots, parades, and the like. But even then they are seen by a tiny fragment of the population; the larger audience sees them only through the mass media. For example, they may try to make a riot in order to get it (and them) on television. To this degree the cultures of metropolis and small towns are usually similar; in both, the viewers tend to merge the electronic message with their life, and not the opposite.

Yet always behind the specific scene in the metropolitan area is the awareness of hundreds of thousands of houses and automobiles, signs of people unknown to and unconnected with the individual. But what does one's own neighborhood look like to the ordinary citizen of the metropolis? The citizen sees mass-produced housing; but with domestication and adjustment to the family, the houses and the yards become differentiated. The neighborhood as a commons is a basis for communication and, occasionally, organization; in the process, neighbors develop social relationships. Thus they become individuated to each other, occupying one of a limited number of roles—leader, old citizen, specialist, clown—the ancient *commedia dell'arte* of human communities. As this occurs, so does the creation of a local reality.

So it also becomes possible to "read" a neighborhood by such variation in blocks as housekeeping, socioeconomic status (and the nearer to equality, the finer the gradations made), and perhaps ethnic or religious commitment. Arriving at home is also arriving at its surroundings. The urbanites have returned to something somewhat like the small town from which they or their forebears originally came.

How does one "read" a small town? The main thing that would strike a big-city denizen would be how many people there are, how visible, and how much importance they have in one way or another. They tend to stand out in a landscape not rustling with thousands of unknown people—they are all there is. There is more space around the individual; one is differentiated, and on this basis, one communicates and rapidly also becomes a social individual, as do the new acquaintances seen against their backgrounds. Again what we may call the creation of a local reality is under way.

In the small town one sees the individuals because the stereotype quickly fades to irrelevance in the face of personal and interpersonal detail. The strangers' roles in the occupational, economic, familial, and other orders place them rather quickly. In a way this is a broader scope than the urban, family-centered neighborhood, because in the latter, acquaintance grows slowly and probably does not outreach "neighboring" family to family and, perhaps, local voluntary associations. The process seems to take about 2 years. Yet it is enough to generate social reality and values.

The Urban and the Urbane

Several distinctions are implied in the phrase "the urban and the urbane." *Urban* is simply a demographic measure of clusters of people resident on a site. As we ceased to be a predominantly agrarian society, the process first described by Adna Weber (1899) continued full force, relocating work in urban areas and the bulk of the population in the cities. Those who once would have been in smaller clusters are now in the residential neighborhoods of the metropolis.

Thus much of the metropolis is hardly urbane; it is transposed and reconstituted small-town culture and, to a degree, social structure. The *urbane* refers to the older cultural meaning and the structure that supports it, with connotations of the free port for strangers, the market in all things (especially ideas), the center where fashion, art, and power are aggregated, traded, and transformed.

The metropolitan area may best be seen as the dense, urbane center with its tail towers; the seemingly endless spread of houses surrounding it is now called *suburban*. The two areas lead, to a degree, to separate kinds of life. The activity we associate with the lively and not altogether predictable urban process occurs in the urbane center, but outposts represented by universities, colleges, art or technical institutes, and such, with their surrounding neighborhoods, are pseudopods of the urbane center. They, along with part-time center citizens— the commuters—extend the urbane center outward. As for the square miles of urban small-towners, most of them may be quite unaware at first hand of what goes on "down there." Many urbanites do not see their environment in terms of area maps at all, but in terms of their own itineraries through their part of the metropolis, a form of abstraction as old as the need of humans to remember a route from here to there.

Center and periphery—these are very crude ways of separating the urbane component of the metropolitan areas (and, by extension, metropolitan society). We know that such urbane areas are far from homogeneous (it is their nature), and this means they include a certain number of people who are just looking for a home, as well as survivors of earlier uses of the housing, and the persistence of the urbane character also is precarious; such areas in their urbane character may be more apt to rise rapidly, change, and disperse than the more usual. If this is true, however, the use of smaller units than metropolis or even its center for study is arguable. Named places, census tracts, even blocks should be more accurate as basic units.

If given areas of the metropolis are such crude units for such important variables, why use them? Pragmatically, one reason seems to be the relatively stable nature of residential population over time, as well as certain activities, and the assumption that this applies to the urbane center. Equally important is the belief that urbanity implies a place and a population that provide group reinforcement (negative and positive) and communication, which are the basic protection of new, and therefore deviant, thought and behavior. Some spatial groups nurture innovation, its diffusion and acceptance. All are called, collectively, "cultural creativity." Such a conjunction of place, people, and highly urbane behavior is found in a few places far from the great cities. For example, nuclear physics labs and testing grounds are located in the desert, and not too far away is Taos—a center for the arts of some importance for the past 70 years or so. Both are among the more famous "pseudopod" of urbane populations and institutions, but only pseudopod; material basis and audience remain in the cities.

In this country, the creative and life-enhancing character of the very center of the city seems to be in danger. The tall buildings, heavily guarded against the dangerous streets, and the lack of street life after sundown (except for predatory life, which discourages the rest) are poor accompaniments to what we think of as a relatively relaxed, easygoing, "bohemian" atmosphere. The cheap living quarters and nearness to publishers, theaters, galleries, and the like that created much of the attraction of such areas are often now unsafe for the free-floating intellectuals and artists who are the young beginners at the game. The wise, old heads, often famous, who began the bohemian area are now out of place in the center.

A new structure for the work and play of such social types may be developing in the educational, art, and research institutions and their neighborhoods, in the huge mass of the metropolis. These might take the place of the old, centralized urbane world of the inner city. If so, it will be different from the creative melting pot at the center of cities as we have known them for some 200 years. It may be another phase in the "corporatization" of our social life or "rationalization" by means of bureaucrats and professionals.

One thinks of some probable consequences. First, the elements of the broader culture and society providing and supporting the workers in such areas will be much more highly selected than they were in the past. The urbane area will be not a meeting place for many strands of culture and society but simply one more differentiated and concentrated aspect of urban society. Then, too, the urban masses will be less exposed to an urbane milieu than ever because, in general, it slowly will become less accessible. Finally the separation in space encourages social separation among the elite of various sorts, and this rift may impoverish the whole society. Even hatred is of higher quality if you know your enemy.

Note

1. Although neither a state capital nor a regional administrative center of national government, a major fraction of the labor force (some 40,000 people) is engaged in maintaining governmental control and providing governmental services (Milwaukee Metropolitan Association of Commerce, 1984).

References

Brown, C. (1963). Lecture, Claremont College, Claremont, CA.

Erikson, E. (1968). *Identity, youth, and crisis.* New York: Norton.

Marcus, S. (1978). Reading the illegible. In H. J. Dyos & M. Wolff (Eds.), *The Victorian city: Images and realities* (Vol. 2, pp. 257-276). London: Routledge & Kegan Paul.

Martindale, D. (1958). Prefatory remarks: The theory of the city. In M. Weber, *The city* (D. Martindale & G. Neuwirth, Trans.) (pp. 6-62). Glencoe, IL: Free Press.

McNeil, T. H. (1978). *Plagues and peoples.* Chicago: University of Chicago Press.

Milwaukee Metropolitan Association of Commerce. (1984). *Milwaukee metropolitan fact book.* Milwaukee: Author.

Smith, J., & Zipp, D. (1983). The party official next door. *Journal of Politics, 45,* 958-978.

Thernstrom, S. (1982). The "Dyos phenomenon" and after. In D. Cannidine & D. Reeder (Eds.), *Exploring the urban past: Essays in urban history* (p. 209). Cambridge, UK: Cambridge University Press.

Weber, A. F. (1899). *The growth of cities in the nineteenth century.* New York: Columbia University Press.

3

The Citizenships

LOCAL, STATE, AND NATIONAL

Norton E. Long

The task of urban revitalization is first and foremost concerned with renewing the health of the city as humankind's principal means for living a civilized life. We are all, to a degree, citizens of local, regional, and national governments. Some of us even may aspire to be citizens of the world in the sense of Marcus Aurelius or in a republic of a faith (Sabine, 1961). If we are, we must take care not to deserve Rousseau's jibe against the *philosophers,* whom he accused of indulging in a weak and watery affection, loving the whole world in general in order to avoid having to love anyone in particular (Sabine, 1961). As heirs to Greek political philosophy, our civic culture is imbued with the mythic glory of the ancient city. In the ideal of that city, citizens, full citizens, could realize themselves in an active, self-governing life dealing with a full range of affairs of moment. The city was a theater for the exhibition of human excellence. In such a city the citizen would accept material deprivation in order to possess the rich rewards of an active civic life. Pericles's Funeral Oration, as rendered by Thucydides (1963, Book II, pp. 35-46), has given classic expression to an ideal that haunts the human imagination. After the ideal of the self-governing city was put in limbo by the conquests of Philip and Alexander, according to George Sabine (1961), people had to grow souls in recompense for their lost civic life.

In Sabine's view, people could no longer take seriously the lesser political roles of a subordinate city. This view of the subordinate city as a worthy but essentially trivial affair has characterized both political and economic thought. For a time, the medieval and the Renaissance city revived, to a degree, some of the past glories of the ancient city. But as Max Weber (1962) argued, the modern nation-state battered down the city's walls, and with the loss of its walls, the age of the city was at an end. The Jacobin nationalists of revolutionary France could brook no rival loyalties to their new god, and not only the Jacobins have sought the well-nigh complete subjection of the modern city to the higher

AUTHOR'S NOTE: This chapter is a revision of a paper given at a conference on urban revitalization.

authorities. In Dillon's rule, they are the mere creatures of the state, possessing no inherent powers of their own (Adrian, 1955, p. 142). Feeble attempts at home rule have sought to achieve some semblance of the liberties that cities once fought and bargained for with nobles and kings.

In an address at the University of Indiana, James Wilson (1964) sought to explain social science's neglect of the contemporary city because of the triviality of its concerns. The big issues—war and peace, and the control of the economy — all lay elsewhere. The city, to be sure, is a site of important events (crime, drug abuse, poverty, riots), but as a political entity it is weak, having little real power to affect the lives of its inhabitants for good or ill. And yet the subordinate city does not seem to be condemned inevitably to triviality.

In antiquity, according to Lopez (1967), the cities held the Roman Empire together for centuries. Jones (1940) recounts how the Greek city provided a culture that socialized and inspired civic elites of the most diverse races and religions to join in the devoted management of their cities. These cities lasted for 1,000 years; many founded by Alexander outlasted the successor monarchies of his empire.

From a most important point of view, cities (local communities) are far from trivial. Roland Warren (1963), in *The Community in America,* shows how the vertical institutions of the nation-state, national markets, national transport, national media, national unions, and national professions have eroded the horizontal institutions of the local territorial community and, in doing so, have weakened the normative structure. In his view, only the local institutions can produce and maintain that structure. Without it, the national government is reduced to the status of an occupying army even in the nation's capital. The local community has indeed been weakened by the impact of the outside forces, but that weakening is no source of national strength.

Years ago, John Dewey (1954), in *The Public and Its Problems,* made the rediscovery and re-creation of the local community the precondition of the public finding itself and invigorating our democratic life. As Dewey saw it, the local community could have vigorously pursued common goods shared by all and valued by all just because they were shared by all—the very definition of community and democracy. Whether this state of affairs could be extended to the nation—the mechanically organized great society—was a question Dewey left unanswered. It is still unanswered. The transformation of the Great Society into the Great Community is still more dream than serious agenda.

Aristotle (1958) saw all communities as governed by their ends and teleologically related to one another as the lesser communities and institutions served the purposes of the higher. The hierarchy of the lesser institutions of family and neighborhood led up to the city-state that capped the structure of institutions being characterized by the highest and most embracing purpose of all: the realization of some conception of a good life. Although Aristotle was the tutor of Alexander and thus witnessed the end of the independent city-state, he remained loyal to that institution as the one most suitable for the realization of the full human potential. He never, so far as has come down to us, dealt with what kind of a good life might be attained in the subordinate city and the larger polity.

But if one accepts the validity of Aristotle's political analysis, which finds the conception of a good life embodied in a polity as its energizing principle, the task of examining the conception of the good life in the complex polity of the subordinate city and the modern nation state is inescapable.

Sam Bass Warner, Jr. (1968), in his study of Philadelphia, points out how early on the religious basis of the Puritan and Quaker city of colonial America gave way to the commercial city that has become the norm. The "private city" is made up of individuals, each pursuing his or her own quest for wealth with a minimum of concern for public interests and a common good. The dominant ethos is that of Locke, with its focus on property, individual rights, and a conception of government as a necessary convenience for limited purposes—a man should not be judge in his own cause. As Sabine (1961) remarks, Hobbes and Locke, by a curious and unintended partnership, succeeded in fastening on social philosophy what has become its reigning notion: that private, individual interests are real and substantial, whereas public interests are weak and evanescent. Locke conceived of his philosophy as deriving from Hooker's, whose thinking was still in the classic line of the common good derived from Aristotle by way of St. Thomas. Although Locke thought of himself as a follower of Hooker, his influence mainly led to a radical individualism in which government has little role in pursuing a public interest. In all likelihood, this result may have stemmed from his benign conception, as opposed to that of Hobbes, of the state of nature, and his belief that the pursuit of individual interests mostly would, even if unintendedly, serve the general interest. Sabine (1961) concedes that this idea may have been measurably the case at the time Locke wrote. It certainly seems to have been the case in early Philadelphia.

Very rapidly, however, the increase in scale and the desire to sweat the land turned Penn's plans for his city into a rabbit warren of narrow lots and back-alley dwellings, with inadequate streets and sanitation. Increase in scale, even before manufacturing industry, destroyed the rich community life of the coffee-houses, taverns, and streets, resulting in the weakening of the social organization of the early city. Warner (1968) maintains that well before the Civil War, Philadelphia had ceased to be a community and never became one again. The sheer growth of the city, industrialization—with the attendant business cycle—and ethnic, racial, and religious conflict threatened the peace and civil order of Philadelphia. The "better element," property owners and businesspeople, united to form a metropolitan police force to maintain order. It proved not to be needed; no riots occurred subsequent to its formation. The forces of industrialization, neighborhood segregation, the separation of home and work, the office, and the club movement all served to produce an effective system of social control. Philadelphia became organized again, but not at the scale of the city. A multiplicity of organizations precluded Wirth's (1938) anomie, but from the perspective of the city's overall, common interest, what developed was an organized disorganization, a Tower of Babel of competing and coexisting interest groups without any centripetal structure. The existence of this array of disparate and divergent interests persists in the contemporary city. Douglas Yates (1977), with the sad experience of the Lindsay years in New York, found this "street fighting pluralism" a source of the city's ungovernability. One might equally conclude that the divisions in the city made it governable after a fashion the British colonialists would have understood, but inhibited the city from becoming self-governing.

In Aristotle's political theory, the constitution embodies the ethical purpose of the polity—its conception of the good life. This conception of the good life was exemplified in the lives of the ruling class, and it was because the ruling class best exemplified the ideal that it was legitimate. One would expect that the first citizens of the commercial city would be its successful merchants. They represent its ideal, and they are the most respected and admired members of the polity. Warner (1968) relates the changing scale of the city and its markets to a profound change in its elite and in its elite's values. He finds the outlook and role of Thomas Pym Cope, a merchant prince of the earlier city, a world apart from that of Jay Cooke, the financier of the Civil War. Cope saw his own self-realization in the frame of the city. His role was governmental and general. He interested himself in a wide range of affairs ad diverse as education and the provision of a splendid park and a safe water supply. Cooke's interests were identified only weakly with the city: church membership, charities, a trustee of the university—no governmental role. His concerns far outran the confines of the city, extending to the nation and beyond. Increasingly the holders of top status in the business elite cease to be identified in any serious way with the governing of the city in which, or in the suburbs of which, they reside. Tocqueville (1945, p. 171) remarked that the American manufacturing elite was one of the harshest in history because it came not to govern men but to exploit them.

Robert Dahl (1961), in his study of New Haven, traces the changes in the structure of that city's elite from a time when the hierarchies of politics, economics, and society were in the same hands to the present, when they largely diverge. Warner's (1968) study of Philadelphia shows a similar development. The result of the withdrawal of social and economic notables from overtly dominant political roles was their replacement by lower status "explebes," as Dahl calls them. Warner believes that the low status of the new political breed resulted in a lack of public trust that severely limited their power to undertake important projects and to govern. In time of crisis, the public would demand a return of the notables to positions of power to restore civic order. This demand reveals the public's insistence on regarding the notables as most appropriately representing the polity's ideal of the good life. Indeed, upward mobility in the society to the apex of the social pyramid leads from politics to business to an economically secure position; founding a recognized family is the nearest thing to an aristocracy the democratic society recognizes.

The contemporary city, in many—perhaps most—cases, is far too large to permit the interaction among its members that would encourage the interaction and mutual knowledge and trust that could result in the friendship Aristotle regarded as the hallmark of the true civic community. The town or the small city may permit a kind of personal, individual interaction. The large city is structured by its elites, its system of communications, its churches and voluntary organizations, and its businesses and unions. Only in some of its neighborhoods is the face-to-face interaction a significant reality. What holds the city together is a socialized sense of membership, however weak, and in acceptance of the legitimacy of its governing elite—an elite that comprises others than those holding political office and those in ostensibly political roles.

The city, like the region and the nation, is an artifact of history, and its loyalties are taught in the culture. For some of its inhabitants, the city is a prized possession that they would not readily abandon. For others, their citizenship is merely legal, rather than functional and exhibiting commitment. Like Goths in the Roman empire, these enfranchised aliens use their suffrage as a weapon with which to loot and destroy, rather than as a valued means to preserve and improve. The region and the state, unlike the city, lack the cultural memories of a functional and personally interactive citizenship. They have never had even the illusion of the intimate common action of the town meeting and the voluntary corporation—what Weber (1962) called the "oath-bound community of the burghers." Or, at a deeper level, they lack the memory of their citizenship in what Fustel de Coulanges (1956) saw in the common hearth. The region and the state, as political communities, have citizenship loyalties that stem more from that of the dispersed territorial community undefined by walls. They must look to the tribe and some version of nationalism to give the affect that makes government more than a Lockean convenience. Inhabitants of very few cities since antiquity and the Renaissance have aspired to give a completely independent political life to the city, but as inhabitants of regions, many have aspired to such independence. The American Civil War inhibited such aspirations on the part of the states' inhabitants, though nostalgic memories exist in the legacy of the Confederacy. But regional nationalisms rise ever anew, and some create festering sores in the larger body politic. The nation-state as a community must frequently strive to reconcile the greatest diversity of languages, religions, and ethnicities. The task of reconciling these differences in some overarching common loyalty would have seemed hopeless to the classic Greek political theorist. Yet it was some such task that Alexander undertook when he sought to achieve a union of the hearts between Greeks and Persians (Tarn, 1933).

In classical antiquity, the solution to the problem of creating a common loyalty among the motley members of empires was the institution of the god emperor. The republican institutions of the city were only partially compatible with the requirements of empire. Yet empire was compatible with a Roman citizenship, of which Paul could boast as well as of his valued membership in Tarsus, no mean city. Modern nationalism has given us a substitute for the god emperor as a unifying device that can, at least at times and in places, unify peoples otherwise divided by language, religion, race, regional ties, and history. Without nationalism, and its memories and sense of shared purposes in opposition to enemies without and often within, the state as the regional government would be a mere Lockean convenience incapable of inspiring loyalty sufficient for the most serious tasks of government. In a homogeneous, prosperous country in placid times, the night-watchman state of 19th-century liberalism might have served the calculating, limited citizens that its economist admirers envisioned.

For most of history, no such limited stock company could have survived the hazards of the real world.

The citizenship of a modern nation-state, comprising a wide range of religions, ethnicities, and languages, is driven to seek a common civic culture that can comprehend its diversity without giving fatal offense to the deeply felt elements of its diversity. This drive might well seem to argue for a secular democratic state observing a constitutional separation of church and state and affirming a neutrality as to the lifestyles and values of its constituents, Rajiv Gandhi faced that problem in India in an only somewhat more severe sense than secular democrats elsewhere. When a Muslim woman gains a court judgment awarding alimony, outraged mullahs denounce the invasion of Muslim religious law. The nation's writ must run throughout the land, but in what areas had it better not run lest it tear asunder the fragile fabric of a religiously and ethnically divided people? The vocation of the nation and the creation of a supreme loyalty from the people to that nation threaten religious, linguistic, and particularistic local territorial communities. To many religious fundamentalists, *secular humanist* is a term of opprobrium and a political position they see as a threat. This religious animus goes back at least as far as the papacy's denunciation of Marsilo (see Sabine, 1961, Chap. 15) and, at a later date, the attacks of Bodin and the *politiques* who preferred a religiously neutral state to civil war between faiths (see Sabine, 1961, Chap. 20). "Thou shalt have no other god before me" is an injunction that in the religious view applies to the nation-state. But "Render unto Caesar what is Caesar's" is also in point, as is "My kingdom is not of this world" and the avoidance of the error of Herod.

The lesson of history may well be that a diverse population with differing religions and languages needs a common way of life that can be accepted by all faiths. The Swiss have formed a nation of sturdy strength despite such differences. The United States, from time to time, has been aroused by a sense of common shared vocation that Lincoln eloquently expressed in the face of a fateful division. The civil rights movement tapped a similar vein of public sentiment. From time to time, as Croly (1965) urged, the nation becomes aware of itself as the large public Dewey (1954) envisioned and momentarily transforms the mechanically organized Great Society into a Great Community for the purpose at hand. The broad public, at such times, may have a sense of participation in the production of a shared common good that is valued by all because it is shared by all. But for the most part, the active citizenship of regional and national polities is the province of elites. The broad public participates through mediating institutions and, to the extent the public attends at all, as spectators at a game in which it does not actively play. The role of the broad public is to empower and legitimize the elites who do the governing and to channel their demands through pressure groups and media. The common history, the common language, and the sense of shared common values do transform the members of the nation into a people with a felt sense of shared larger community. The loyalties among local, state (regional), and national polities remain in dynamic tension.

Jefferson saw the hierarchy of loyalties as starting with the primacy of the local territorial community, extending next to the state and then to the nation.

The choice of Robert E. Lee, when confronted with the conflict between his state and his nation, was his state. At the time, that choice might not have seemed strange. That it could occur made civil war possible. At the time of the Civil War, observers wrote of the melancholy fates of federations from those of the Greeks to that of the United States. They may not have been as wrong as history made them seem. The Civil War transformed the United States from the problematic unity of a federation to the secure, if informally recognized, status of a nation. In the relation of states, localities, and nation is a tendency, on the one hand, to seek unity by reducing the lesser governments to the status of administrative areas and, on the other hand, to weaken the central government and, in the extreme, dissolve it. The strength of the parts of the composite unity may contribute to the strength of the whole or, depending on the purposes to which that strength is put, it may, in strengthening some, weaken others.

Perhaps one ought never to have to choose among the demands of local, regional, and national citizenships. But if these roles are vital and the purposes served by the various polities significant, it is scarcely possible that on some issues, for some citizens, the three citizenships should not result in a conflict of loyalties. In the modern world the most critical conflict has been between the city and the higher levels of government. To the extent the city has lost function to higher levels of government, the city as a local territorial community loses its power to produce a normative structure. Just as the city is to a degree dependent on its neighborhoods for its cellular strength, so is the nation dependent on its cities. One might suppose that because the nation-state is the one unit of government for which people in the modern world are supposed to be willing to die, the nation would cap the Aristotelian hierarchy of teleologically related institutions through which the polity's energizing concept of the good life is realized. But in reality, the good life of the nation is lived not in the nation but in its cities. Only at the scale of the city is it possible for the Burkean partnership in all art, all culture—all the things that make a civilized life possible—to be realized (Burke, 1981).

In principle, the city could be made into a humane cooperative serving both the spiritual and material needs of its members. The territorially more inclusive units of government that are now regarded as higher in a legal scale of authority might properly be regarded as instrumentally subordinate to the ends of the city, which alone serves as a theater for the full realization of human excellence. The contemporary city, in America at any rate, is the private city described by Warner (1968) in his study of Philadelphia. Its ethos is dominantly commercial. Its elite is fragmented in terms of the areas of its concerns and frequently suburbanized in its residence. The city's political leaders are of low status and do not command much respect. Neither the elites nor the bulk of the city's inhabitants possess a unifying vision of the city as a species of moral architecture that could structure the pursuit of richly meaningful lives in the emptiness of an existentialist world. The commercial ideal of the economist is one of human atoms combining across the landscape as free-floating factors of production in response to the undirected play of market forces producing an unintended collective beneficence. The end product of Hobbes and Locke is the interest-group liberalism that Theodore Lowi (1969) has rightly condemned. It produces what Yates (1977) has called the ungovernable city of "street fighting pluralism."

The ungovernability of the city does not mean that the existing city is chaos. It is not. It functions as a materialistic ecology uninformed by a unifying purpose save that of material gain and an empty consumerism. In principle, the city is a human creation designed to take humans out of the state of nature and to make a purposive vision of a good life possible. By possessing such a vision, inhabitants of the city can deliberately set about developing a set of roles that, in Plato's sense, will permit them to realize their individual potentials for excellence in the harmony of common action. In doing so, the city members would remove the worst blight of contemporary society—the degrading and degraded roles that afflict the lives of so many of its inhabitants—and burden (or should burden) the conscience of the rest.

Dewey (1954) argued that only in the local community and its revitalization could the public discover itself and, in doing so, realize its shared common purposes, which alone make possible real democracy. Dewey did not say how he saw the modern nation-state being transformed from the present mechanically organized Great Society into the Great Community. Perhaps the answer lies not in the nation-state becoming a single, unified, great community. The community of the nation-state that is desirable may well be the community of communities—something of the sort Althusius envisioned in the Netherlands but with the difference that the nation is both a community of communities and a community of all the individuals of the member communities as well (von Gierke, 1966). In this regard, Aristotle's criticism of Plato's Republic is in point. The nation may become so unified that its constituents cease to have the functional individuality that alone can give the larger structure vibrant life.

The revitalization of the city is first and foremost a spiritual task. The city, more than any other human artifact, has the potential for enabling its members to live full lives in the warmth of shared common endeavor in the pursuit of high purposes. In a world without meaning, it is the mechanism of giving meaningful content to its citizens' lives. It is a time-binding institution that gives Pericles's Funeral Oration contemporary point. It is the institution that best serves both the homely and the exalted ends of humankind, and it is the institution that alone can subject the nation-state to the discipline of civilization.

References

Adrian, C. R. (1955). *Governing urban America.* New York: McGraw-Hill.

Aristotle. (1958). *The politics of Aristotle* (E. Barker, Trans.). New York: Oxford University Press.

Burke, E. (1981). Reflections on the revolution in France. In P. Langford (Ed.), *Works* (Vol. 2). New York: Oxford University Press.

Croly, H. (1965). *The promise of American life* (A. M. Schlesinger, Jr., Ed.). Cambridge, MA: Belknap.

Dahl, R. A. (1961). *Who governs?* New Haven, CT: Yale University Press.

Dewey, J. (1954). *The public and its problems.* Denver, CO: Swallow.

Fustel De Coulanges, N. D. (1956). *The ancient city: A study on the religion, laws, and institutions of Greece and Rome.* Garden City, NY: Doubleday.

Jones, A.H.M. (1940). *The Greek city from Alexander to Justinian.* Oxford, UK: Clarendon.

Lopez, R. (1967). *The birth of Europe*. New York: M. Evans.

Lowi, T. (1969). *The end of liberalism*. New York: Norton.

Sabine, G. M. (1961). *A history of political theory* (3rd ed.). New York: Holt, Rinehart & Winston.

Tarn, W. W. (1933). Alexander the Great and the unity of mankind. *Proceedings of the British Academy, 19.*

Thucydides. (1963). *History of the Peloponnesian War* (B. Jowett, Trans.). New York: Twayne.

Tocqueville, A., de (1945). *Democracy in America* (Vol. 2, P. Bradley, Ed.). New York: Vintage.

von Gierke, O. (1966). *The development of political theory* (B. Freyd, Trans.). New York: Fertig.

Warner, S. B., Jr. (1968). *The private city: Philadelphia in three periods of its growth*. Philadelphia: University of Pennsylvania Press.

Warren, R. L. (1963). *The community in America*. Chicago: Rand McNally.

Weber, M. (1962). *The city* (D. Martindale & G. Neuwerth, Trans.). New York: Collier.

Wilson, J. (1964, November). *Problems in the study of urban politics*. Paper presented at a conference in commemoration of the 50th anniversary of the Department of Government, Indiana University, Bloomington.

Wirth, L. (1938). Urbanism as a way of life. *American Journal of Sociology, 44,* 1-24.

Yates, D. (1977). *The ungovernable city: The politics of urban problems and policy making*. Cambridge: MIT Press.

PART TWO

Urban History

Introduction to Urban History

Carl Abbott

Cities have been the pioneers of growth in the United States. From the beginnings of European exploration and conquest to the Alaskan oil boom, towns and cities have been the staging points for the settlement of successive resource frontiers. From Boston and Santa Fe in the 17th century to Anchorage and Miami in the 20th century, cities have played similar roles in organizing and supporting the production of raw materials for national and world markets. It has been city-based bankers and merchants who have linked individual resource hinterlands into a single national economy.

The United States also has been a pioneer among urbanizing nations. Along with Britain, France, Belgium, and the Netherlands, it was among the first to feel the effects of the urban-industrial revolution. The history of U.S. cities involves the continual invention of new institutions and technologies to cope with massive 19th-century urbanization. We were able to learn, on occasion, from the London Metropolitan Police, French engineers, and German housing programs. In turn, the U.S. experience has been copied in Latin America, Asia, Africa, and even Europe.

A brief review of key topics in American urban history allows us to explore the connections between historical understanding and present issues. It reminds us of the ways Americans have continually adapted their cities to new challenges and "reinvented" local government to cope with changing needs.

Stages of Urban Growth

The making of an urban America has followed the same pattern found in every urbanizing society for the last two centuries. Starting in the late 18th century, one nation after another has experienced an abrupt and accelerating shift from largely rural to largely urban society. After several generations of rapid urbanization, the process tends to level off toward a new equilibrium in which about three quarters of the population live in cities and many of the rest pursue city-related activities in smaller towns.

Urban growth in the United States clearly has followed the three stages of gradual growth, explosive takeoff, and maturity: (a) The era of colonial or premodern cities stretched from the 17th century to the 1810s, (b) the rise of the industrial city dominated a century of rapid urbanization from 1820 to 1920, and (c) the era of the "modern" city runs from 1920 to the present.

The first century of British and Dutch colonization along the Atlantic seaboard depended directly on the founding of new cities, from New Amsterdam (1625) and Boston (1630) to Norfolk (1680), Philadelphia (1682), and Savannah (1733). These colonial towns resembled the provincial market centers in the British Isles. Compact in size and small in population, they linked the farms, fisheries, and forests of the New World to markets in Europe and the Caribbean. With populations that ranged from 15,000 to 30,000 at the time of the American Revolution, the four largest cities were all seaports that dominated the commerce of surrounding hinterlands. New England farmers and fishers looked to Boston; the Hudson River Valley traded by way of New York; Philadelphia took its profits from the rich farms of the Delaware and Susquehanna Valleys; and Charleston centralized the trade of South Carolina and Georgia.

At the first national census in 1790, the 24 recognized cities that counted 2,500 or more people accounted for only 5% of the national population. A generation later, after the War of 1812 and the financial panic of 1819, the 1820 census still counted only 700,000 urban Americans—a scant 7% of the national total.

A century later, the 1920 census found a nation that was 51% urban, giving 1920 as much symbolic meaning for American history as the supposed closing of the frontier in 1890. Between 1820 and 1920, New York expanded from 124,000 people on the lower end of Manhattan Island to a metropolis of 7,910,000 spread across 14 counties. Philadelphia grew from 64,000 to 2,407,000. Los Angeles grew from a few hundred Mexican settlers to 2,319,000 people.

American cities by the end of the 19th century fell into two categories: industrial core, and suppliers and customers. The nation's industrial core stretched from Boston and Baltimore westward to St. Louis and St. Paul, accounting for the overwhelming majority of manufacturing production and wealth. Many of these cities were specialized as huge factories—textile towns, steel towns, shoemaking towns. Their industrial labor force drew from millions of European immigrants and their children, who made up more than two thirds of the population of cities such as Detroit, Chicago, Milwaukee, Pittsburgh, and New York. Cities in the South, the Great Plains, and the Far West were the suppliers and customers. They funneled raw materials to the industrial belt—cotton from Mobile, metals from Denver, cattle from Kansas City. In return, they distributed the manufactured goods of the Northeast.

Cees D. Eysberg, in Chapter 4 "The Origins of the American Urban System," reviews the factors behind the emergence of the urban-industrial core region, posing the question of why New York, rather than New Orleans, became the great gateway to North America. He traces the advantage of Northeastern cities to initial patterns of English colonial settlement in which supposedly less fertile lands along the northern coast were set aside for religious refugees such as

Puritans and Quakers. The resulting middle-class society provided a market for urban services and built early ports such as Boston and Philadelphia. In contrast, the socially stratified plantation society of the Southern colonies had little need for, interest in, or capacity for urbanization. In the 19th century the initial advantage of the Northeast widened as its cities took advantage of revolutions in transportation and industrial production.

The most recent period of urban growth has revolved around the adaptation of U.S. cities to 20th-century technologies of personalized transportation and rapid communication. The 1910s and 1920s brought modern art, modern music, and modern architecture to Berlin, Paris, and New York. The same decades also brought full electric wiring and self-starting automobiles to the middle-class home. George F. Babbitt, the hero of Sinclair Lewis's 1922 best-seller, lived in a thoroughly modern Dutch Colonial house in the bright new subdivision of "Floral Heights" in the up-to-date city of "Zenith." He awakened each morning to "the best of nationally advertised and quantitatively produced alarm clocks, with all modern attachments" (Lewis, 1922). His business was real estate, and his god was Modern Appliances.

The metropolis that Babbitt and millions of real automobile owners began to shape in the 1920s broke the physical bounds of the 19th-century industrial city. In 1910 the Bureau of the Census devised the concept of the "metropolitan district" to capture information about the suburban communities that had begun to ring the central city. By 1990 the federal government recognized more than 300 metropolitan areas with a total population of more than 190,000,000. More than *half* of all Americans live in metropolitan areas with populations of more than 1 million.

What journalists in the 1970s identified as the rise of the Sunbelt is part of a long-term shift of urban growth from the industrial Northeast toward the regional centers of the South and the West—from Detroit, Buffalo, and Chicago to Los Angeles, Dallas, and Atlanta. The causes include the concentration of defense spending and the aerospace industry, the growth of a leisure economy, the expansion of domestic energy production, and the capture of "sunrise" industries such as electronics. As Eysberg also reminds us, this process is creating a better regional balance in a historically unbalanced urban system.

The cities of the post-World War II United States have had the greatest ethnic and racial variety in national history. They have been destinations for a massive "northward movement" as rural Southerners moved north (and west) to cities and jobs. Starting with massive migrations in the 1910s, the African American experience became an urban experience, creating centers of black culture such as Harlem in the 1920s and feeling the bitter effects of ghettoization by the 1930s and 1940s. During the Great Depression and World War II, Appalachian whites joined black workers in Midwestern cities such as Cincinnati and Detroit. Residents of Oklahoma and Arkansas left their depressed cotton farms for new lives in Bakersfield and Los Angeles.

The northward movement has also crossed oceans and borders. After World War II, Puerto Rican immigrants remade the social fabric of New York and adjacent cities. Half a million Cubans had an even more obvious impact on Miami after the Cuban Revolution of 1959. Puerto Ricans and Cubans have been followed to Eastern cities by Haitians, Jamaicans, Colombians, Hondurans, and others from the countries surrounding the Caribbean. Mexicans constitute the largest immigrant group in the cities of Texas, Arizona, Colorado, and California.

By the 1970s and 1980s, Asia matched Latin America as the source of 40% of documented immigrants. Asians have concentrated in the cities of the Pacific Coast and in New York. Los Angeles counts new ethnic neighborhoods for Vietnamese, Chinese, Japanese, Koreans, and Samoans. Honolulu looks for business and tourism to Asia, as well as to the continental United States. A new generation of migrants has revitalized fading Chinatowns in New York, Chicago, Seattle, and Los Angeles.

Richard C. Wade, in Chapter 5 "The Enduring Ghetto," traces some of the long-term consequences of the northward migration, especially for African Americans. He finds the best historical comparison for Northern black neighborhoods not in European immigrant neighborhoods, but in the segregated communities that Southern cities created after the Civil War. The enforcement of segregation, especially in housing and jobs, nullified much of the effect of hard work and education among urban blacks. The result has been the historically rooted and historically explainable crisis of unemployment, crime, and discrimination that causes present-day scholars to write about an urban underclass.

Cities and American Values

These basic facts about the inseparability of urban growth and national growth have clashed repeatedly with the American mythology of national uniqueness and pioneering individualism. One result has been an ambivalent response in which Americans have praised cities with one voice and shunned them with another. Recent public opinion polls show that most Americans would prefer to live in a small town. A few extra questions, however, reveal that they are just as certain that they want easy access to the medical specialists, cultural facilities, and business opportunities that are found only in cities. Over the years the American debate about the value of cities previewed many of the current arguments about the impact of urbanization and modernization in the developing world.

Thomas Jefferson set the tone for American antiurbanism with the strident warning that cities were dangerous to democracy. At the time of Philadelphia's deadly yellow fever epidemic of 1800, Jefferson wrote to a friend:

> When great evils happen, I am in the habit of looking out for what good may arise from them as consolations. . . . The yellow fever will discourage the growth of great cities in our nation, and I view great cities as pestilential to the morals, the health, and the liberties of man. (Jefferson to Benjamin Rush, September 23, 1800, in Lipscomb, 1904, p. 173)

In part, Jefferson feared that American cities inevitably would grow into facsimiles of the London of the 1770s or the Paris of the 1780s as sinks of poverty, scenes of unemployment, and breeders of riotous mobs. He also feared that because city dwellers were dependent on others for their livelihoods, their votes were at the disposal of the rich. Only in a nation of independent farmers could government remain virtuous.

If Jefferson feared first for the health of the republic, many of the antiurban writers who followed feared instead for the morals of the individual. Sensationalizing authors pointed to the moral that "city life crushes, enslaves, and ruins so many thousands of our young men, who are insensibly made the victims of dissipation, of reckless speculation, and of ultimate crime" (*Prairie Farmer,* quoted in Glaab & Brown, 1986, p. 49). Far more serious were religiously based indictments of city life as corrupting of both rich and poor. In *The Dangerous Classes of New York* (1872), Charles Loring Brace warned of the threat posed by the abjectly poor, the homeless, and the unemployed. Jacob Riis used words and photographs to tell the middle class about the poor of New York in *How the Other Half Lives* (1890). A few years later, W. T. Stead used the wonderful title *If Christ Came to Chicago* (1894) to indict the big city as un-Christian because it destroyed the lives of its inhabitants.

The first generation of urban sociologists wrote in the same vein in the 1920s and 1930s. Robert Park worked out a theory of urban life that blamed cities for substituting impersonal connections for close personal ties. As summarized by Wirth in "Urbanism as a Way of Life" (1938), the indictment dressed up Thomas Jefferson in the language of social science. Wirth's city is the scene of superficial relationships, frantic status seeking, impersonal laws, and cultural institutions pandering to the lowest common denominator.

The attack on city living was counterbalanced by sheer excitement about the pace of growth. By the 1830s large numbers of Americans had come to look on cities as tokens of American progress. Booster pamphlets and histories of instant cities drenched their pages in statistics of growth. Boosters counted churches, schools, newspapers, charities, and fraternal organizations as further evidence of economic and social progress. One Chicago editor wrote before the Civil War that "facts and figures . . . if carefully pondered, become more interesting and astonishing than the wildest vision of the most vagrant imagination" (*Review of Commerce for 1853,* p. 1). His words were echoed 70 years later as George Babbitt sang the praises of "the famous Zenith spirit . . . that has made the little old Zip City celebrated in every land and clime, wherever condensed milk and pasteboard cartons are known" (Lewis, 1922, p. 184).

Urban advocates in the 20th century have extended the economic argument to point out the value of concentrating a variety of businesses in one location. The key is external economies—the ability of individual firms to buy and sell to each other and to share the services of bankers, insurance specialists, accountants, advertising agencies, mass media, and other specialists. In *Cities and the Wealth of Nations* (1984), Jane Jacobs argues that cities and hinterlands are natural economic units whose vitality transcends the artificial boundaries of states and nations.

American writers and critics as disparate as Walt Whitman and Ralph Waldo Emerson also have acknowledged cities as sources of creativity. "We can ill spare the commanding social benefits of cities," Emerson wrote in his essay "Culture" in 1844 (Emerson, 1904, p. 155). Nathaniel Hawthorne allowed his protagonist in *The Blithedale Romance* (1852) to take time off from the rigors of a country commune for the intellectual refreshment of Boston. George Tucker, a professor at Jefferson's University of Virginia, wrote in 1843:

> The growth of cities commonly marks the progress of intelligence and the arts, measures the sum of social enjoyment, and always implies increased mental activity. Whatever may be the good or evil tendencies of populous cities, they are the result to which all countries, that are at once fertile, free, and intelligent tend. (p. 143)

Planning Traditions

The nation's ambivalence toward cities has been reflected in America's contrasting efforts to design cities that will both fulfill social ideals and encourage national growth. Since the Puritan settlement of New England in the 17th century, the desire to control social change can be seen particularly in the tradition of "focused" or "closed" communities. A series of planned communities has used urban design and the physical form of the community to promote a better life for a carefully defined and limited set of like-minded residents.

Massachusetts colonial towns were the first focused communities. Designed to preserve the strength and purity of the Puritan experiment in the New World, they have been described as closed communities, freely governed by their "members" but unwelcoming to outsiders. The towns themselves often centered on church, meeting hall, and central square. Salt Lake City and other Mormon settlements in Utah repeated the pattern of religiously focused communities in the mid-19th century.

Company towns offered a different version of planning for closed community. The best examples, such as Lowell, Massachusetts (1822), and Pullman, Illinois (1881), were intended to be livable environments that protected workers from the worst effects of industrialization. At the same time, the beneficiaries of these improved communities were carefully selected and controlled to maintain a stable labor force. Lowell ringed textile mills with dormitories for single women. George Pullman made sure that workers in his rail car factory were sober (by excluding alcoholic drink), serious (by providing a library), and under his control (by renting rather than selling his housing). The community worked well during prosperity and fell apart during depression and labor strife in the 1890s.

A third version of the closed community is the upscale suburb where the upper middle class of professionals and business owners have taken refuge from the cities in which they earn their livings. In the mid-19th century, landscape planners responded to the chaos of booming cities with pastoral suburbs of gently curving streets, large lots, and abundant greenery. Llewellyn Park, New Jersey

(1853), was a refuge for overburdened New Yorkers. Riverside, Illinois (1869), welcomed harried Chicagoans at the end of a 10-mile train ride from Chicago. The automobile in the 1920s brought a new generation of focused suburbs such as Shaker Heights outside Cleveland and the Country Club District of Kansas City. Palos Verdes Estates (1923), draped over a rocky peninsula that juts into the Pacific Ocean south of Los Angeles, came with schools, golf course, swimming club, riding academy, and an art jury to approve the design of each new house.

The irony of focused communities is that economic success usually has brought pressures to open up to a wider range of residents. Boston began in 1630 as a closed Puritan town, for example, but successful seaports are impossible to isolate from the wider world. The Puritan goal of uniformity quickly broke down in a city that drew its livelihood from contact with the rest of the world. In a later period, the growth of Lowell's factories brought Irish immigrants to run the machines and disrupt the social consensus.

It has always made more sense in the United States to plan for open communities that mirror the freedom of movement inherent in the American Constitution and the laws of the land. The form of the open city has followed its function, for it has been planned in the expectation of continual growth. Intended to accommodate all comers, open cities are built piece by piece as space is needed. The consequence has been the characteristic use of the infinitely expandable street grid, with city-building tied inextricably to land speculation and the private market in real estate.

From the start, William Penn defined Philadelphia as an open labor market, choosing a plan of perpendicular streets that could be extended indefinitely as the city grew. The grid had equal appeal to the New Yorkers who mapped out streets for the entire island of Manhattan in 1811. Their north-south avenues and lateral streets supported explosive growth in the 19th century and redevelopment in the 20th. New cities throughout the Mississippi Valley also imitated Philadelphia's layout, building types, and even street names. Travelers in Cincinnati, Pittsburgh, Lexington, Nashville, St. Louis, and other western cities found miniature Philadelphias planned for growth.

The open town was the perfect companion for the railroads that stretched their tracks in straight lines across the prairies and plains between 1840 and 1910. The Illinois Central, the Great Northern, and other lines staked out towns at regular intervals alongside their tracks to collect farm produce for shipment and to serve the farming frontier. Even on the hilly shores of the Pacific, settlers laid gridiron plans over the map with little regard for the landscape—up, over, but seldom around the steep slopes. Seattle washed entire hills into Puget Sound because they got in the way of planning for indefinite growth.

Formal city planning in the 20 century has largely preserved the goal of the open city. The so-called "city beautiful" planners at the beginning of the century were concerned with making the congested industrial city work more efficiently by rearranging its public spaces, streets, and transportation systems. Daniel Burnham prepared his famous plans for San Francisco (1905) and Chicago (1909) for each city's progressive business leaders. The plans were intended to make each city *work* better as an economic machine. The same motivation lay behind the adoption of land-use zoning after 1916 and urban renewal in the 1950s and 1960s.

Livable Cities

Americans had to learn not only to plan their cities but also to live comfortably and safely in the fast-growing communities. In particular, the openness of U.S. cities to continued growth brought the professionalizing of public services. At the start of the 19th century, private companies or amateurs provided everything from drinking water to police protection. In the wooden cities of colonial times, for example, fire fighting was a community responsibility. Householders kept buckets in their houses and responded to calls for help. By the 1820s and 1830s, more cities had added groups of citizens who drilled together as volunteer fire companies, answered alarms, and fought fires as teams. Problems of timely response and the development of expensive steam-powered pumpers, however, required a move to paid fire companies in the 1850s and 1860s. Fire fighters who were city employees could justify expensive training and be held accountable for effective performance.

Fire protection required a pressurized water supply. Residents of colonial cities had taken their water directly from streams and wells or bought it from entrepreneurs who carted barrels through the streets. Boston reached 20 miles into the countryside with an aqueduct in the 1840s. New York surpassed this feat with the Croton Reservoir and an aqueduct that brought fresh water 40 miles from Westchester County to a receiving reservoir in what is now Central Park. Changing theories of disease and the availability of abundant water for municipal cleaning helped cut New York's death toll in the 1866 cholera epidemic by 90% from the 1849 epidemic.

The public responsibilities of 19th-century cities generally fell into the three categories of public health and safety (police, sewers, parks), economic development (street drainage, pavement), and public education. Such expanding responsibilities fueled the municipal progressivism of the early 20th century. Led by local business interests, cities implemented civil service employment systems that based hiring and promotion on supposedly objective measures. The new city manager system placed the daily operations of government under a professional administrator. As the system has spread since the 1910s, city government has become more and more the realm of engineers, budget analysts, and other trained professionals.

Public intervention came later in other areas such as low-income housing. New York pioneered efforts to legislate minimum housing standards with tenement house codes in 1866, 1882, and 1901. Providing the housing, however, remained a private responsibility until the federal government began to finance public housing during the depression of the 1930s. But even massive investments in public housing after federal legislation in 1937 and 1949 failed to reconcile public opinion. Unlike the residents of European cities, American urbanites usually have preferred to believe the worst about public housing and to ignore the viable projects.

A full social agenda for local government waited until the 1960s. Assistance to the poor was the realm of private philanthropy in the 19th century. The crisis of the 1930s legitimated federal assistance for economically distressed individuals, but city government remained oriented to public safety and economic development. By the start of the 1960s, however, criticisms that the urban renewal program of the 1950s had benefited real estate developers at the expense of citizens added to an increasing sense that America's multiracial cities were in a state of crisis. In 1964 President Lyndon Johnson declared a nationwide War on Poverty. On the front lines was the Office of Economic Opportunity, with its Neighborhood Youth Corps for unemployed teenagers, its Head Start and Upward Bound programs to assist public schools, and its Community Action Agencies to mobilize the poor to work for their own community interests. Two years later the Model Cities Program sought to demonstrate that problems of education, child care, health care, housing, and employment could be attacked most effectively by coordinated efforts.

The 1970s and 1980s left U.S. cities with comprehensive social commitments but limited resources. President Richard Nixon simply announced that the urban crisis was over and dismantled programs such as Model Cities before they had a chance to prove themselves. President Ronald Reagan continued the policy of redirecting federal resources to meet the development needs of politically powerful suburbs. Cities and city people absorbed roughly two thirds of the budget cuts in the first Reagan budget. To exaggerate only slightly, cities were left with the expectations of the 1960s and the resources of the 1920s. Or as Mayor Jerome Cavanagh of Detroit complained in 1970, "Some academics now find it stylish to deny that there is an urban crisis at all, let alone one that money can solve. But once—just once—I'd like to try money" (quoted in Clapp, 1984, p. 50).

Learning From Urban History

When we study the history of our cities, it is important to remember what has been tried successfully and what has failed. Politicians and journalists have a tendency to assume that a particular problem is new because *they've* just discovered it. In most cases, of course, there is a long history of efforts to wrestle with the same or similar issues. It saves time if we are willing to learn from past experience with fighting crime, building affordable housing, and a long list of other issues.

A second historical lesson is that variety and diversity in economic activities and people are a key to urban vitality. More than a century ago the British philosopher John Stuart Mill pointed out the value of challenging human beings with unfamiliar attitudes and values and attributed progress to the friction of diversity and pluralism. Time and again the history of American cities carries this hopeful message: The social variety that seems the cause of our greatest problems can also be the source of new ideas.

The third point is that urban problems are an inevitable companion of growth and change. The only problem-free city would be a totally stagnant city. Vital cities, whether colonial Boston or contemporary Los Angeles, are imperfect but improvable. The history of urban America is a repeating cycle in which growth creates problems; citizens and leaders work to devise solutions; solutions allow further growth; and growth triggers new problems. In the process, American cities have become more livable. We have solved or know how to solve many of the physical problems of traffic, pollution, and deteriorated housing. Failures have come from lack of commitment and political will, not from the inherent nature of cities. The longer our comparative perspective, the more credit we will be willing to give to our urban success stories.

References

Brace, C. L. (1872). *The dangerous classes of New York, and twenty years' work among them*. New York: Synkoop & Hallenbeck.

Clapp, J. A. (1984). *The city: A dictionary of quotable thought on cities and urban life*. New Brunswick, NJ: Rutgers University, Center for Urban Policy Research.

Emerson, R. W. (1844). Culture. In *The conduct of life*. Boston: Houghton Mifflin.

Glaab, C. N., & Brown, A. T. (1986). *A history of urban America*. New York: Macmillan.

Hawthorne, N. (1852). *The Blithedale romance*. Boston: Ticknor, Reed & Fields.

Jacobs, J. (1984). *Cities and the wealth of nations: Principles of economics*. New York: Random House.

Lewis, S. (1922). *Babbitt*. New York: Harcourt Brace.

Lipscomb, A. A. (Ed.). (1904). *The writings of Thomas Jefferson*. Washington, DC: Issued under the auspices of the Thomas Jefferson Memorial Association of the United States.

Review of Commerce for 1853. (1853). Chicago: Chicago Democratic Press.

Riis, J. (1890). *How the other half lives*. New York: Scribner.

Stead, W. T. (1894). *If Christ came to Chicago*. Chicago: Laird & Lee.

Tucker, G. (1843). *The progress of the United States in population and wealth*. New York: Hunt's Merchant's Magazine; Boston: Little, Brown.

Wirth, L. (1938). Urbanism as a way of life. *American Journal of Sociology, 44*, 1-24.

Suggestions for Further Reading

Abbott, C. (1993). *The metropolitan frontier: Cities in the modern American West*. Tucson: University of Arizona Press.

Barth, G. (1980). *City people: The rise of modern city culture in nineteenth century America*. New York: Oxford University Press.

Cronon, W. (1991). *Nature's metropolis: Chicago and the great West*. New York: Norton.

Foster, M. (1981). *From streetcar to superhighway: American city planners and urban transportation*. Philadelphia: Temple University Press.

Goldfield, D. (1982). *Cotton fields and skyscrapers: Southern city and region*. Baton Rouge: Louisiana State University Press.

Grossman, J. (1989). *Land of honey: Chicago black Southerners and the great migration*. Chicago: University of Chicago Press.

Hamer, D. (1990). *New towns in the New World: Images and perceptions of the nineteenth-century urban frontier*. New York: Columbia University Press.

Hammack, D. (1982). *Power and society: Greater New York at the turn of the century*. New York: Russell Sage.

Hirsch, A. (1983). *Making the second ghetto: Race and housing in Chicago, 1940-1960*. New York: Cambridge University Press.

Jackson, K. (1985). *The crabgrass frontier: The suburbanization of the United States*. New York: Oxford University Press.

Lotchin, R. (1992). *Fortress California 1910-1961: From warfare to welfare*. New York: Oxford University Press.

Monkkonen, E. (1988). *America becomes urban*. Berkeley: University of California Press.

Rabinowitz, H. (1978). *Race relations in the urban South*. New York: Oxford University Press.

Teaford, J. (1990). *The rough road to Renaissance: Urban revitalization in America, 1940-1985*. Baltimore: Johns Hopkins University Press.

4

The Origins of the American Urban System

HISTORICAL ACCIDENT

AND INITIAL ADVANTAGE

Cees D. Eysberg

The purpose of this chapter is to discuss why the core area of the American urban system developed in the Northeast and not in the South. It is proposed that this development was determined by historical accident and was not a result of environmental constraints. This "coincidence hypothesis" is supported by historical material, and it helps to explain the recent decline of the Snowbelt cities vis-à-vis the rise of the Sunbelt cities.

Before World War II, the national economy of the United States formed one spatial entity dominated by a strong urban-industrial core area that was then commonly described as the Heartland, and now by the term *Snowbelt*. This dominance was expressed both in the consumptive and in the productive sectors of the economy. The dependent national periphery produced the raw materials and the semifinished products that formed the resources for the core. The cities in the peripheral areas served only as specialized regional centers (Yeates & Garner, 1980). The process of concentration of manufacturing activities in the core area continued for a long time as a result of external economies of scale (large markets, a large labor force, well-developed and dense transportation and communication networks). Since 1945, however, the core area has fallen behind in population growth; it first lost people to California and, since the 1970s, to other areas in the periphery.

The traditional relationship of dominance and dependence of core and periphery has evolved into a more balanced relationship. The Sunbelt now demonstrates self-sustaining economic growth (Norton & Rees, 1979). This dramatic shift was caused by the unprecedented rapidity of the development of the American periphery, which in itself was mainly due to its rich natural environment. Most peripheral areas of other countries are much less well endowed, as Vining observed in 1982. Alonso, discussing the same shift in balance between the Snowbelt and the Sunbelt in 1978, observed: "It appears that what is going on is a very long-run equilibration of the national distribution of urban centers, still trying to rectify the original mistake made by the first British settlers when they landed on the upper-righthand corner of our nation's map" (p. 53).

Can we indeed speak in similar terms of a historic geographical mistake with respect to the early development of that impressive urban system that now stretches along the coast from Boston to Baltimore and that comprises the entire region south of the Great Lakes, including such great metropolises as Detroit and Chicago?

In my view, the location and subsequent development of the urban industrial core in the Northeast instead of in the South was a consequence of more or less accidental but crucial human decisions, and these determined the development of two totally different societies in different parts of the country. To state this in a more compact and perhaps more provocative way: The Mississippi could have become the River Rhine of America, and New Orleans could have become the American Gateway, filling a position for the United States that could have been the equivalent of that of Rotterdam in Europe.

It is remarkable that despite the official abjuration of physical determinism as a geographical approach to explanation, most current geography textbooks rely so strongly on environment factors to explain the development of the Heartland in the Northeast. The argument commonly refers to such resources as the falls line in New England, the region's adequate rainfall, and the presence of iron ore deposits in the Lake Superior region and of coal in Northern Appalachia as determining factors.

An article by Carville V. Earle (1977) on the first English towns of North America supports an alternative to this geographical approach. The initial period of development of the American urban system was unique. In Europe, cities had developed gradually as a result of their function as central places providing religious, administrative, and economic functions for the surrounding countryside. In contrast, American cities were founded as new settlements in what was essentially undeveloped country. After Jamestown had been established in 1607, the English established a number of colonial towns on the Atlantic seaboard. The English Crown granted land to "Chartered Companies" or to individual vassals, such as William Penn. Earle describes how the founding of a centrally located coastal town became the first and indispensable colonial act, even before other economic activities were undertaken. The reasons for this not only were found in the attitude of the English to regard towns as nodes of commerce, administration, and defense but also were related to the belief that towns were necessary to save the colonists from slipping into barbarism. Rumors about the lost Roanoke settlers living like "white Indians" fed this idea. However, New Amsterdam demonstrated a more spontaneous development in retaining the trading functions of the original Indian settlement.

Of far greater importance than these notions, however, was the concrete colonial policy based on the perception of the natural environment offered by the Atlantic seaboard. According to the dominant geographical theory of the time, the best colonial lands were those in the Mediterranean latitudes to the south of the Delaware River. On the basis of the assumption that such products as wine, cork, silk, and spices could be procured here, these lands were reserved for the privileged Anglican establishment. The religious refugees—the Puritans and the Quakers—were allowed to settle in the "marginal" zone to the north of the Delaware. This strict division had serious, far-reaching consequences for the first growth incentives for the colonial towns. To the north, the migrants came as families or as groups. A significant proportion of the Protestant immigrants belonged to the wealthy middle class (Rostow, 1975, p. 188).[1] They came in large numbers, especially during periods of economic crisis in England. These first immigrants, when setting up their households in the New World, created an effective demand for various urban services, and thereby stimulated business communities in the towns and provoked general economic growth in early colonial times (Earle, 1977, pp. 44-45).

In contrast, the migration to the south consisted of only a handful of English upper class plantation owners and was numerically dominated by the poor, the "indentured servants." Many poor drifters, convicts, and similar groups characteristic of those tumultuous early capitalistic days preferred the 7 years of slave labor to a life of misery. These poor, predominantly male migrants were shipped directly to the plantations and did not create an effective demand for urban services. In this respect, it is illustrative to consider the contrast of male-female ratios of 1635 for Jamestown and New England, which were given as 6 to 1 and as 1.5 to 1, respectively (Earle, 1977, pp. 44-45).

These first incentives, fueled by a growing effective demand of the immigrants, was of crucial importance to the towns in the North. They turned into an initial advantage and forced a conclusive lead, after which the resulting economies of scale produced self-sustained growth. By then, the growing business community of the towns had become important enough to take initiatives and to develop important trade on which the urban economy could be expanded further.

Pred (1966) provides insight into how "a circular and cumulative process of urban growth" (p. 8) was set in motion in these Northeastern cities. He specified this initial advantage as their local labor supply, their monopoly of many local connections, and financial facilities. Hinterland penetration by transport and route developments caused an expanding pattern of urbanization. By technological and economic changes, initial advantage "evaporated in some localities but lived on in others to trigger repeated cycles of urban growth" (Conzen, 1981, p. 341). (Also, Borchert [1967] characterized three periods in technological development and changed the foundations of urban economy from a commercial-mercantile to an industrial-capitalistic and to a postindustrial one.) But, according to Pred (1966, p. 8), in general there was a great stability in the rank order of large cities. Although Pred takes for granted the initial advantage of the Northeast coastal cities, he did diagnose a remarkably slower urbanization process in the South and a backslide in the urban rank-size of such cities as New Orleans and Charleston as the national urban system developed over time (Pred, 1973, p. 199).

To tackle the problem of the extreme differential in urban development of the North and the South, it is necessary to investigate urbanization in both regions, particularly in the South, in the context of their distinctive socioeconomic and sociocultural development. This approach is in line with David R. Goldfield's (1984) ideas about "new regionalism." This approach brings a possibilistic perspective into historiography versus the traditional self-evidence in history. A very important background for the extreme differences in the urban development of the North and the South was the totally different evolution of the two societies. Sometime after the initial colonization, Southern society began to exhibit some of the traits that are still characteristic of contemporary developing countries in Latin America: These include rigid dividing lines between social classes and a strong dependence on agriculture, leading to a dual society and a dual economy.

The rich English aristocrats founded the plantation economy. Products such as tobacco, rice, indigo, and later cotton were shipped directly from the plantations to England. Most plantations were located on navigable rivers; thus owners bypassed the merchants. They also eliminated the middleman in imports, as they sent directly for industrial products from England. These practices precluded the development of a middle class of merchants, crafters, and industrialists in the South (English, 1977, p. 490). Besides the limited differentiation of intermediate stages between production and final marketing, limited mechanical services were required by cotton production (Goldfield, 1982). Even Charleston, the only colonial town of the South of any importance (which counted an impressive 12,000 inhabitants in 1750), scarcely numbered a middle class of merchants to provide a counterweight to the plantation aristocracy (English, 1977, p. 490).

Nor could a middle class arise from the ranks of the small farmers. It was difficult for small farmers of such cash crops as tobacco and indigo to remain solvent. Prices in the world market have always fluctuated, but there were extra problems. Because of their sparse holdings, these farmers could not introduce crop rotation, and, consequently, the soil became exhausted. They also could not compete with the large plantations, with their rich soil and large numbers of slaves. Only part of the rural population therefore was engaged in marginal subsistence farming.

A portion of these small farmers were the many former "indentured servants" who tended to settle down in the marginal areas of the back country after having worked their 7 years on the plantations. The first black slaves arrived in Jamestown in 1619, and, even though it did not gain dominance until the 18th century, the system of black slavery has had very negative effects on the social development of the South.

Structural deficiencies occurred as well. For example, education existed only for the rich upper class, whose members could afford to employ a tutor. In the economic realm, the plantation retained such a monopoly position as to preclude other economic initiatives. According to Pred (1973, p. 199), "Southern elites had a negative attitude toward manufacturing investment and a tradition of channeling savings into slaveholding" and, referring to Schumpeter, "the existence in a community of vested interests disliking change is considered to be a major impediment to innovation" (Pred, 1966, p. 120). It is important to realize that the urban elite, too, was oriented, economically and mentally, toward the rural agricultural sector. Growth-stimulating multiplier effects in all Southern cities were depressed by the low level of local and regional purchasing power (Pred, 1973, p. 199). And little change occurred under the sharecrop-debt peonage system after the Civil War (Rice, 1984).

By contrast, from the very first, the North reflected the characteristics of a settlement colony, rather than the traits of an exploitation colony. Groups of small farmers and crafters who were well organized socially, religiously, and politically formed village communities all over New England. Here social mobility was possible through rural to urban migration, and this stimulated urban growth.

The regions of New England and the middle colonies should be distinguished. The latter region was characterized by immigrants from a large variety of religious and national backgrounds, but, as in New England, the small family farms dominated.

Throughout the North, elementary education was essentially available to everyone even though the school building was rarely more than a simple log cabin. The result was that the North became a more broadly democratic society, a society that found its basis in the family farm and small industries and in opportunities for social mobility through urbanization.

In the first part of the 19th century, the United States had become one nationally organized society, but within it, three spatially separate economies and societies had come into existence as a result of these divergent paths: (a) The urban North had become a strongly developed capitalist society, (b) the agricultural Midwest had developed into a society of independent farmers, and (c) the South was the region of the slave-based plantation economy dominated by a landed aristocracy (Beard & Beard, 1927-1942, p. 663).

The urban North started to industrialize on the basis of its trade and shipping, while taking advantage of the early tradition of crafts.[2] A period of hectic construction activity ensued, creating the means of communication with the rich agricultural regions to the west of the Alleghenies. The construction of the railways in the 1850s and 1860s facilitated this intrusion; in 1869 the first transcontinental railway was completed. "With the secession of southern states from the Union, federal support for railbuilding in the west came quickly . . . orienting the western territory to Chicago and through Chicago to New York more abruptly and definitively than might otherwise have been the case" (Duncan & Lieberson, 1970, pp. 55-56). In this manner, the growth potential of the West was made subservient to the economy of the North; the region was loosened from Southern control, which had been exercised until then—thanks to the 7,000 km of navigable waterways of the Mississippi, the natural gateway to the Midwest.

The Southern states did not participate so intensively in the transportation revolution. They may have feared the centralization tendencies that could have resulted from a closed transportation network. They may have considered the railways unnecessary for their cotton economy, to which needs the navigable rivers seemed to be more appropriate. Sparse population and the lack of capital may have formed other reasons. But it is obvious that the present Southern growth pole—Atlanta, as a distributing center of Northeastern mercantile interests[3]—

owes its origins to the railway and to the related local business community, an ancestry that was uncommon for the South.

Other sections of the South were part of the Spanish colonial periphery, such as Florida and Texas, or of the French realm, such as New Orleans. These powers provided their colonies with insufficient economic and demographic stimulus from Europe. At the moment that these regions became part of the Union (in 1845, 1819, and 1803, respectively), the skeleton of the American urban system, with its Heartland in the North, was already in existence.

After the South became part of the unitary national spatial-economic system, it was forced into the subservient position of the periphery, with its dominance-dependence relationship with the center. By 1860 the largest cities of the South had not yet begun to function as an interdependent regional subsystem of cities, as shown by their weak economic and informational ties with one another and their "colonial" dependence on New York and other Northeastern centers (Pred, 1980, pp. 113-116, 167). It clearly became the victim of the backwash effects that benefit the core with its attractive economies of scale.[4] In addition to such rational economic laws, the South also suffered from contrived political manipulation by the dominant core after the Civil War. The overtly discriminating "Pittsburgh-Plus" pricing system, for instance, was disastrous for the development of Birmingham as a steel center and thus also hindered the emergence of regional multipliers (Stocking, 1954, p. 73); likewise, the policies of the freight-rate zones discriminated against the entire South (Alexander, Brown, & Dahlberg, 1958). Finally the urban development in the South missed the tremendous stimulus of large-scale immigration, owing to an increasing isolation from Western transportation routes and a negative economic and social image.

The origins of the weak tendencies of urbanization in the South must be sought in British colonial policies and in related social and economic developments. The results, however, are quite cynical; they are exactly the opposite of what the Stuarts had in mind. It seems to be farfetched to look for environmental factors as an explanation for the differences in the paths of development. The South has a relatively mild climate, certainly not the harsh winters of the North. Also in other ways, it is relatively well endowed. It possesses such traditional natural resources as water power (the falls line of the Piedmont), rich forest reserves, and of particular importance to the subsequent phase of economic development, important coal and iron ore deposits (near Birmingham) as a local base for industrialization and urbanization. The presence of so many navigable rivers (not only the Mississippi but also the Arkansas and Red Rivers, and the Alabama, Chattahoochee, and Savannah Rivers) could have provided an initial advantage in the field of transportation if urbanization and commerce had exploited their potential sufficiently. The great resource that waterways can form could have been optimized by hydraulic engineering and thereby could have generated ever-increasing transportation flows in a feedback process. This possibility could have become decisive in the design and the construction of the transportation corridors—whereby in due time, the rivers would have been supplemented by parallel railways and roads. New Orleans could have obtained a more pronounced central location in the transportation network that includes the Mississippi, the Ohio, and the Missouri Rivers. This position could have become equivalent to that of Chicago within the railway network of the North.

In the past few decades, the tremendous development of trucking and the more diffuse highway system, bolstered by the air transport system, has countered the dominant east-west bias of the railway era. These recent transport developments seem to have been important conditions in mobilizing the South and in making a partial correction of a "historical mistake."

The domination of the northeastern part of the United States in the national urban system was not an a priori given; instead, it was caused by an arbitrary colonial policy and by the resulting specific differences in social and economic development between the North and the South. Of course, this conclusion has important repercussions on our interpretation of recent developments in the eastern part of the Sunbelt.

Notes

1. At that time, this was also true for Protestant refugees elsewhere—that is, the French Huguenots—whose departure strengthened the creative capacity of Holland, Britain, and Germany and weakened France in a critical historical period of industrial development. Rostow (1975) shows how significant specific minorities can be for the dynamic efficiency of a society.

2. How important a local business community can be as an autonomous growth factor in urbanization and regional economic development is revealed in the development of Boston and its hinterland. Lampard (1955) has argued that in the early 19th century "the impetus toward the integrated factory came from the embattled commercial metropolis of Boston" (p. 118). While mercantile capital in New York, Philadelphia, and Baltimore was being diverted into building routes to the interior and organizing agriculture staple crops in the interior, Bostonians were financing the early industrial development in New England because they were isolated from the main axis of westward development.

3. According to Goldfield (1982, pp. 118-121), Atlanta's reaction to Northeastern "colonialism" was accommodation and cooperation as a distribution center for the South.

4. Especially the emergence of large multiplant corporations and the growth of stock markets, at the end of the 19th century, increased the mobility of capital to the core where investments were more profitable (Rust, 1975, p. 24).

References

Alexander, J. W., Brown, S. E., & Dahlberg, R. E. (1958, January). Freight rates: Selected aspects of uniform and nodal regions. *Economics Geography, 34,* 1-18.

Alonso, W. (1978). The current halt in the metropolitan phenomenon. In C. L. Leven (Ed.), *The mature metropolis.* Lexington, MA: Lexington Books.

Beard, C. A., & Beard, M. R. (1927-1942). *The rise of American civilization: Part 1. The agricultural era.* New York: Mcmillan.

Borchert, J. R. (1967, July). American metropolitan evolution. *Geographical Review, 57,* 301-332.

Conzen, M. (1981). The American urban system in the 19th century. In D. T. Herbert & R. J. Johnston (Eds.), *Geography and the urban environment: Vol. 4. Progress in research and applications.* New York: John Wiley.

Duncan, B., & Lieberson, S. (1970). *Metropolis and region in transition.* Beverly Hills, CA: Sage.

Earle, C. V. (1977). The first English towns of North America. *Geographical Review, 67,* 34-50.

English, P. W. (1977). *World regional geography: A question of place.* New York: John Wiley.

Goldfield, D. R. (1982). *Cotton fields and skyscrapers: Southern city and region, 1607-1980.* Baton Rouge: Louisiana State University Press.

Goldfield, D. R. (1984, February). The new regionalism. *Journal of Urban History, 10,* 171-186.

Lampard, E. E. (1955). The history of cities in the economically advanced areas. *Economic Development and Cultural Change, 3,* 81-136.

Norton, R. D., & Rees, J. (1979). The product cycle and the spatial decentralization of American manufacturing. *Regional Studies, 13,* 141-151.

Pred, A. R. (1966). *The spatial dynamics of U.S. urban-industrial growth, 1800-1914: Interpretive and theoretical essays.* Cambridge, MA: MIT Press.

Pred, A. R. (1973). *Urban growth and circulation of information: The United States system of cities, 1790-1840.* Cambridge, MA: Harvard University Press.

Pred, A. R. (1980). *Urban growth and city systems in the United States, 1840-1860.* Cambridge, MA: Harvard University Press.

Rice, B. R. (1984, November). How different is the Southern city? *Journal of Urban History, 11,* 115-121.

Rostow, W. W. (1975). *How it all began: Origins of the modern economy.* New York: McGraw Hill.

Rust, E. (1975). *No growth: Impacts on metropolitan areas.* Lexington, MA: Lexington Books.

Stocking, G. W. (1954). *Basing point pricing and regional development.* Chapel Hill, NC: University of North Carolina Press.

Vining, D. R., Jr. (1982). Migration between the core and the periphery. *Scientific American, 247,* 37-45.

Yeates, M., & Garner, B. (1980). *The North American city.* San Francisco: Harper & Row.

5

The Enduring Ghetto

URBANIZATION AND THE

COLOR LINE IN AMERICAN HISTORY

Richard C. Wade

I return in this chapter to where I began. More than three decades ago I wrote my first article for a learned journal. It described the violent destruction of a black ghetto in Cincinnati in 1829 and appeared in the *Journal of Negro History*. Contemporaries, of course, did not use the word *ghetto*, but called the area "Little Africa." It had grown rapidly after 1800, and in a quarter of a century comprised 10% of the Queen City's population. Huddled on low land along Columbia Street and Western Row, more than 2,000 blacks lived in wooden shacks and shanties interspersed between frame tenements 10-12 feet high. It was the town's most congested spot, though a few good houses and a church and a school attested to the creation of fledgling community institutions. Whites had complained of its continuous growth, a purported high incidence of "crime, noise and lewd behavior," and the number of runaway slaves who had sought safety and anonymity in the free black neighborhood. Whites sought to break up the colony by enforcing the Ohio Black Laws, which would have forced more than half of the blacks out of the city.

The first ultimatum was given about July 1, 1829; authorities later extended the deadline until September. In the interim, sporadic raids took place and reached a climax on the weekend of August 22 when, as a newspaper put it, "some two or three hundred of the lowest *canaille*" descended on the blacks, bent on terror and pillage. By Saturday evening the black residents, despairing of official protection, armed to defend themselves. Fighting broke out about midnight. One of the raiders was killed, and several people were injured. The police arrested 10 blacks and 7 whites. Later the mayor released all of the blacks and fined the whites $700. But the damage had been done. The combined

AUTHOR'S NOTE: An earlier version of this chapter was presented as the presidential address to the Urban History Association at its meeting in San Francisco, 1989.

pressure of the city and the attacks of the mob destroyed Little Africa. Within a year more than 1,000 blacks had left, many to Canada.

My second article a few years later, published in the *Journal of Southern History,* concerned the Denmark Vesey rebellion in Charleston, South Carolina, in 1822. Urban slavery's primary housing purpose was to prevent the kind of black concentrations that developed in Cincinnati. Yet the census of 1820 showed that the black population had approached that of the whites and included a growing number of freedmen. City authorities worried about the continuous contacts between the slaves and the freed. In the summer they heard rumors that an armed rebellion was about to take place. In a preemptive strike, constables arrested almost every black in sight of the local armory and rounded up scores of others. Ultimately, after prolonged but secret trials, 35 were executed. Slave owners whose blacks were spared sold their young males out of the city as expeditiously as possible. By the decade's end the proportion of blacks had been substantially reduced.

These two episodes seemed important to me because I had come to believe then, as I do now, that the relationship of blacks and whites in our cities was both a central theme of American history and also the central test of American democracy. That seems almost self-evident today. And in an optimistic country that had just defeated racism in a world war, when the armed forces broke the old color line, and when the Supreme Court declared racial separation of school children to be unconstitutional, an integrated America seemed the natural and inevitable result of the promise made in Philadelphia 200 years ago. Moreover, the frontier of this new freedom would be in the cities because whites and blacks there were thrown together, rubbed elbows, worked side by side, and mingled in the streets and stores. These numberless contacts and the constant proximity made obsolete the easy spatial segregation of the countryside. Urbanization would be the authentic melting pot.

Moreover, the surge of blacks from the South to northern cities made this happy prospect attainable in my lifetime. Trains, buses, and automobiles were the modern underground railroad; they would be like the steamships that brought the "homeless, tempest-toss'd" immigrants to our shores. And, like them, these poor but ambitious folk would be incorporated into our metropolitan life and, in time, would find their share of the democratic dream.

The hardest fact of my life is that this great expectation has not been fulfilled. Instead of moving into the broad sunshine of equal rights, we have wandered into the dark streets and dangerous alleys of ever-growing ghettos, deepening pockets of hopelessness and despair, and dangerous prospects of impending disorder. Instead of incorporation, we find separation; instead of mutual respect, we find mutual fear; instead of hope for a better future, we find expectations of increased contention. The dream seems, not just deferred, but denied.

Yet this failure is the citizen's lament; explaining how it happened is the historian's duty. The greatest miscalculation of the postwar generation stemmed from the widespread assumption that the reception of blacks in northern cities would be like that of the immigrants who came before them. That process was well known and understood. The newcomers would congregate in the city's center. Soon they would begin to discover their numbers, find jobs, get some education, and begin the movement upward in status and outward in residence.

This system had served urban America well; indeed, its foremost academic exponents had themselves been its beneficiaries. It had permitted millions of people with different backgrounds, nationalities, languages, and religions to weather the shock of transplantation and ultimately to be incorporated into America's metropolitan mainstream. Recent literature has invested this past with a certain charming nostalgia, however, usually written 30 years and 30 miles from the old neighborhood. Historians know that the process was more often pitiless and painful and usually involved wretched housing, irregular jobs, inadequate schools, high crime rates, rampant epidemics, broken families, and endemic disorder. Yet this cruel rite of passage had worked in the long run; indeed, it was the genius of the American system.

The old ghetto, no matter how difficult the life, was tolerable because it was thought to be temporary. Even in the worst areas, everyone knew some who had made it out, if only to a hot-water flat or a triplex. An uncle or a fellow parishioner or a schoolmate had moved to a better neighborhood where his or her children would have better schools and improved prospects. How this system worked was never very clear, but policymakers in every city found it simpler to step aside, apply only modest public policy, and let private forces work their will. This immigrant analogy was the operating assumption in northern cities when the black migration arrived at their gates in the 20th century.

In retrospect, there was a more appropriate analogy—southern cities in the decades following the Civil War. There the end of slavery forced a new set of race relations onto Dixie. Emancipation had put freed blacks on their own. Mostly rural, with no land, no property, and few skills, they drifted to the cities. By 1880 they composed a considerable and permanent minority in each. From the beginning, even during the presence of federal troops and before the appearance of Jim Crow laws, southern cities developed highly segregated societies. Given the racial attitudes of local white leaders, it is not hard to see why. Municipal governments had somehow to accommodate the black influx within their established boundaries. Both races had to share the same space; postwar economies had to be adjusted to a new labor force; and public facilities had to be allocated. And all this in the context of growing congestion, where continuous contacts between white and black meant that neither was often out of sight or out of mind of the other.

Municipal leaders in Dixie erected an elaborate scaffolding of segregation to provide a public etiquette for the behavior of the races. It was official policy to have two courts, two jails, two schools, two hospitals, two almshouses. Moreover, ordinances required separation in public facilities and even private conveniences. By the 1890 census, this racial policy resulted in clearly defined black neighborhoods with their own churches, schools, shops, recreational spots, and social groups. Most of the housing was substandard, streets remained unpaved, and these areas were the last to get water, sewers, and transportation. Few had fire or police boxes. Yet, inevitably, they expanded as newcomers arrived, and blacks, scattered across other parts of the city, gravitated to the established black communities. A nascent middle class, made up largely of ministers, teachers, shop owners, and an occasional lawyer and doctor, struggled to sustain what the community had and to handle the relations with white society. By the turn of the century, the black ghetto as we now know it was everywhere a part of the urban South.

This development foreshadowed the subsequent black reception in northern cities more than the immigrant experience. To be sure, there had always been blacks in the urban North and West, but their numbers were relatively small, and there they were only one, albeit the most conspicuous, among many minorities. Public policy seldom addressed their concerns, and the color line, though always present, was never as thick as in Dixie's cities. Yet mini-ghettos almost always popped up; some were evanescent, but increasingly more were permanent—the first harbingers of things to come in the 20th century.

The great black migration to the North began in the century's first decade and grew rapidly until, in the post-World War II period, its numbers approached earlier immigrant figures. At first it appeared that the general expectations were correct; the blacks did indeed concentrate at the center where they inherited the worst housing, poor schools, irregular jobs, high mortality rates, and ubiquitous crime and disorder. But, unlike the old ghetto, the new ghetto did not disperse, but rather simply oozed outward into contiguous areas. As the numbers grew, the new ghettos converted abandoned warehousing districts to residential use, moved through working- and middle-class districts, and later even spilled beyond municipal boundaries. But generally the suburban route of their white predecessors was filled with mine fields that led to mini-ghettos instead of integrated communities. By 1960 the black ghetto was an American institution found in every region of the country.

The new black ghetto differed from the old immigrant ghetto, however, in two important regards. The first was its impact on the emerging black middle class. This group was, after all, among the most successful groups that urban America had ever produced. In 1950 the census listed 7.8% as middle class; by 1960 that number had grown to 17.5%; by 1970 it had reached almost 30%; and 10 years later it approached 40%. Few other groups had developed a middle class that rapidly. As the ghetto expanded, successful blacks moved into good housing and better neighborhoods on the outer edge of the black frontier. Yet they still were denied the single most important symbol of American success—the right to live wherever you want, commensurate with your resources, accomplishments, and ambitions. And these middle-class blacks understood the single reason why. They, after all, had done all that society had asked of them—stayed in school, avoided trouble, gotten an education, and earned a good income. Yet they still were denied the full residential choices enjoyed by whites of every background. The reason was clear—it was the color of their skin; they were the indelible immigrants. As a result, successful blacks were forced to turn back to their brethren in the inner city. What they found there was a new generation, bred in the ghetto and knowing little beyond its squalor and despondency. Block upon block of wretched housing, abandoned buildings, littered vacant lots, and dilapidated schools sat like fortresses behind high fences. Boarded-up shops, deserted by failing and frightened owners, signaled the quarantine of the area by commercial interests. Churches alone retained some dignity in the neighborhoods' tatterdemalion state.

Cold statistics contained the human price of the ghetto: high unemployment, with nearly half of the young adults without jobs; more than half living in substandard housing; nearly three quarters of the families with only one parent; a 40% drop-out rate in schools; infant mortality figures that look like those in the Third World; half of the population and three quarters of the children living below the poverty line; and crime rates triple the city average. Whatever help came from the outside arrived at 9 in the morning and disappeared at 5 in the afternoon. The night belonged to the bold and the reckless. No nostalgia about the old neighborhood, the redemptive qualities of poverty, or the promise of governmental action could conceal the meanness of these streets, the deprivation of these people, and the poverty of their expectations. In the 1960s, with the young in the vanguard, these areas exploded in a feckless assault on some of the symbols that they thought represented their confinement. Like middle-class blacks in nicer neighborhoods, they, too, were finding the ghetto to be intolerable because it appeared permanent.

A shocked nation groped for an explanation of this mutiny. The President's Commission on Violence was followed by the famous *Kerner Report*. The first identified the ghettos as the continuing flash points of disorder; the second was more precise, as it warned of the emergence of two societies—one white, one black—separate and unequal. There followed a rash of "landmark" legislative acts, scores of "significant successes" in the courts, and a Niagara of editorial comment urging action to reduce tensions. But, as Mr. Dooley once observed, Americans have a penchant for "the short distance crusade." By the 1980s the public's ardor had waned, civil rights and poverty became election liabilities, and a new white generation grew up with only a hazy idea of what had occasioned rebellion in the first place. The reform of the 1960s had been a tide without a turning.

Neglect, however, only deepened the problem. The census of 1980 saw the index of residential segregation higher than ever before. The suburban black figures grew, but so did the mini-ghettos that received them. Even with a robust national economy, unemployment of black males was double that of whites; and among young adults the numbers were twice again higher. Ghetto schools continued to fail. In 1990 New York State identified the 39 most troubled schools in the city: 37 were in Harlem, the South Bronx, Jamaica, and Brooklyn. A similar report found the city's worst hospitals in the same areas.

But beyond the statistics, the lives of those who lived in the ghetto withered. Crime, always endemic in poor areas, exceeded even expected levels. Outsiders saw the danger only when it occasionally invaded surrounding communities. Meanwhile, residents watched fearfully as streets fell to the predators; schools became unsafe; and housing projects, once the safest redoubts, succumbed to local gangs. Drugs magnified and aggravated every dimension of disintegration. Neither the police from outside nor the churches and civic institutions inside could contain the increasing menace. Life in the ghetto was nasty, brutish, and often short.

These developments bore out the grim prophecies of those who had analyzed the explosions of the 1960s. But it generally was believed that the relative quietude of the following two decades marked at least acquiescence within the ghetto and a consequent lessening of the danger to society in general. Yet it slowly became clear that the pathologies of the inner cities had consequences for those who did not live there. The antisocial behavior seeped out to other neighborhoods and downtown; crime became ubiquitous, and the costs of welfare and police and fire protection overwhelmed city budgets. There simply was no way to ignore the ghetto. Like Banquo's ghost, it hovered over the prosperous America of the 1980s.

As conditions worsened, a new group of scholars, journalists, and think-tankers, largely neoconservatives, suddenly discovered what everyone had warned about in the 1960s—that not only does the ghetto constitute a threat to its own people but it also disturbs the general tranquility of our cities. These authors usually do not use the word *ghetto,* which smacks of race, but rather they embrace the trendy term *underclass,* which seems color-blind. Yet in identifying this underclass in 800 computerized census tracts, they have found that nearly every one of the infected areas is black or Hispanic. Uncomfortable with the explosive connotation of race, they invented their own definition of underclass neighborhoods: those where half drop out of school, where over half of the families are headed by a single parent, where half are on welfare, and where half of the adult males have no regular connection with the labor market. If their definition is somewhat contrived, their solutions are of the most touching modesty—more public money spent on children and early education, aid to families trying to get out of the area, and vouchers to girls who do not get pregnant while in school. As desirable as these proposals are, they scarcely will dent the deeper problem.

Yet we should be grateful to these authors, even if they have arrived by slow freight, for identifying this question as the central issue of our time and for carrying it to a broader audience. Still the dilemma is neither elusive nor enigmatic. The people who live in the ghetto, even when successful, believe their confinement to be permanent and their inclusion in white society to be a fading dream. The first generation of blacks had hope and expectations; the second generation held on to fragile possibilities; the third generation knows neither that hope nor those possibilities. For it has never known anything else but the chaos of broken homes, the debilitation of idleness, the curse of random violence, and the disintegration of neighborhoods. Even the strong weaken and the vulnerable capitulate. Yet no generation of Americans has been willing to settle peacefully for a slice of the slum. I doubt this one will.

Here, I suppose, the historian must stop. But the responsibilities of citizenship do not. History does not dictate, but it does suggest. Obviously, in the short run, the ghetto needs jobs more than mere compassion, for there is nothing like useful and gainful employment to establish a sense of self-worth, to organize and discipline life, and to build a family's future. It also needs more of what it does not get: better schools, more day care, improved services, greater police protection, and neighborhood clinics.

In the longer run, however, we should be thinking of some kind of universal national service that will take our young people out of their familiar environment, whether it be the inner city or the crabgrass white ghettos of the suburbs, and mix them up in a common purpose. After all, society that confers so much on its middle-class white children has a right to ask them to give something back when they reach 18. Moreover, it is a public imperative to get young adults out of difficult neighborhoods so that they can see what is available to those who are willing to strive and to learn skills that will give them entry into the world of work.

Beyond jobs and national service, blacks and whites will have to create a new civil rights movement. More than a quarter century ago, an earlier generation put together a genuinely interracial coalition that successfully dismantled the historic segregation of the South and significantly reduced discrimination in the North. With the national legislation of the 1960s, the two parts of the movement found new goals. The blacks headed for "empowerment," and the whites gravitated to environmental issues. The former is essentially urban, the latter suburban. The blacks emphasized unity over ideology, while environmentalists created an ideology to reduce divisiveness. If we are to reduce tensions, we will require the painful and incremental rebuilding of the bridges that made the civil rights movement so successful.

But none of this will work if we do not begin to break up the ghettos, because it is the concentration of the underclass in inner-city areas that presents the problem and the danger. Public policy for 40 years has tried to stop discrimination in housing, and governments at all levels have enacted some form of fair housing laws designed to spring successful blacks and Hispanics from the ghetto. Success has been modest, however, because government has always acted defensively, awaiting complaints from the aggrieved, thus assuring resisting communities that there is no enthusiasm or urgency about dispersing the ghetto. Unless public commitment to open housing goes beyond litigation to educational and aggressive programs, the cancer will spread, endangering not just adjacent areas but also the whole metropolitan society.

Still the ghetto endures, mocking our most fundamental ideals. Its persistence is to our generation what Thomas Jefferson called "a fire bell in the night". Its intractability has led to a sense of fatalism on both sides. Kenneth B. Clark, the author of *Dark Ghetto*, whose brief so influenced the Supreme Court in the Brown case, has said sadly and simply, "I have given up". Others are willing to accept the permanence of the ghetto and to seek measures to make the "societies" tolerable neighbors. Indeed, a *New Yorker* cartoon caught the dark side of their thinking. It depicts a Manhattan couple on a high-rise terrace. The husband says, "The problem of the ghettos? The ghettos, my dear, are a solution, not a problem." Others think frankly (though seldom write) of a "garrison city," and ponder strategies to limit the damage.

But it is the beginning of wisdom to understand that this question is not just an American problem. In Europe the same issue is increasingly dominating public policy. The "guest workers" in Germany, the Turks in Belgium, and the returning colonials in France, Great Britain, and the Netherlands raise the problem of large numbers of new minorities living in Old World cities. Our dilemma is important because we are the first, as well as the oldest, nation to confront the most explosive issue of our time. There are 5 billion people on the planet, three quarters of whom are nonwhite. If any society is to find racial accommodation on a basis of justice and decency, it is most likely to occur here. And that is perhaps as it should be, because it was here that that promise of democracy was made to the world more than 200 years ago in Philadelphia.

PART THREE

Urban Policy

Introduction to Urban Policy

Robert J. Waste

We are a nation of cities. According to the 1990 census, most Americans—77.5%—live in cities, and 50% of the entire U.S. population lives in one of the 39 largest metropolitan areas with populations of 1 million or greater. In the 1980s, 90% of U.S. population growth took place in metropolitan areas (U.S. Bureau of the Census, 1991). *Urban policy* is the study of programs designed to aid these cities, the politics associated with federal efforts to aid cities, and the politics within cities themselves. Most of the readings in Part 3 focus on efforts by the federal government in the 1960s to the 1990s to fashion a clear and effective urban policy, an effort that has been characterized by a series of successes and failures, programs begun and programs terminated, periods of great expectations of progress to be made on the urban front only to be followed all-too-often by dismal results and lowered expectations. Making sense out of this on-again off-again federal effort to aid cities is no small task.

After a careful reading of the contributions in Part 3, readers should have a much clearer picture of why it is so difficult for the federal government to make policy affecting cities and why, despite the incredible difficulty associated with such efforts, the effort to design and enact an effective national urban policy remains for many scholars and politicians a crucial national goal. Indeed, for many urbanists such as former Undersecretary of Housing and Urban Development Robert C. Wood, "The Dream [of a successful national urban policy] Will Never Die." Whether the dream of a national urban policy should live or die and—if it lives—in what shape and with what program elements and with what dollar amounts dedicated to attacking the problems of America's cities—is the core debate in the current field of urban public policy. The readings in Part 3 serve as an excellent introduction to that debate.

Other Dreams, Other Debates

Although resolving whether or not the U.S. government should have one unified national set of programs and policies directed at urban areas is the core debate in the urban policy field, it is far from the only debate in the field. Other key lines of inquiry include the following questions: (a) How should we conceptualize American cities? What is the best way to define and understand the modern American city? (b) How shall we describe the politics of American metropolitan areas? (c) What are the key problems facing American cities? (d) What works and what does not in terms of urban policies and urban programs? Although a full discussion of each of these four areas would extend well beyond the space available in this brief introduction, a quick orientation to the first of these issues—how best to conceptualize modern American cities—is a useful background topic to address prior to considering the specific arguments raised in the four readings in Part 3.

Eleven Views of Cities

Modern American cities have changed a great deal from World War II to the present. Urbanists have attempted to keep up with these rapid changes with a series of descriptions to fit the changing times. These descriptions have included (1) "Sunbelt" or "Frostbelt" cities (Abbott, 1981; Bernard & Rice, 1983; Phillips, 1969), depending on the location of the sprawling metroplexes of the postwar period in either Southern or Northern regions of the United States; and (2) "ungovernable cities" (Wirt, 1974; Yates, 1977), cities that seemed to many urbanists to be characterized by strong problems and weak governments. This pessimistic view was countered rather quickly by (3) a more sanguine view dubbed the "governable city" (Ferman, 1985). Pessimism returned with the description of (4) "urban reservations" (Long, 1977) to denote inner-city ghetto neighborhoods such as South-Central Los Angeles or Harlem in New York City populated by the "truly disadvantaged" (Wilson, 1987, 1989), an "urban underclass" (Jencks & Peterson, 1991) so poor and cut-off from the more affluent suburbs and higher income sections of the city as to constitute "separate societies" (Goldsmith & Blakely, 1992; Massey & Denton, 1993; Massey & Eggers, 1993).

One influential political economy perspective (5) presented the "limited city" view in which "in the pursuit of a city's economic interests . . . [the modern American city is prevented from] making allowance for the care of the needy and unfortunate members of the society. Indeed the competition among local communities all but precludes a concern for redistribution" (Peterson, 1981, pp. 37-38). Many urbanists (Henig, 1992; Swanstrom, 1993; Waste, 1993) now reject the mechanistic "limited city" view in favor of viewing cities as (6) local political "regimes" (Elkin, 1987; Stone, 1989; Stone, Jones, & Longoria, 1992), which—contrary to the "limited city" view—may put in place social welfare policies and programs. Regime theorists refer to such cities as "opportunity-expansion regimes" (Stone, 1993). A popular alternative perspective views cities as (7) "growth machines" (Molotch, 1993), cities driven by a coalition of investors and land owners seeking to intensify the value of land holdings by promoting policies of rapid development.

A pluralist alternative to the regime or growth machine view recently (Schumaker, 1991; Waste, 1989, 1993) rejected the view of cities as either "limited cities" or mechanical captives of economic forces. It also rejected the view of cities as controlled by narrow coalitions such as those depicted in either the "regime" or "growth machine" view. In this more recent (8) "stepladder city" perspective (Waste, 1993), city policymakers have the ability to enact and frequently do enact (Anton & Flaum, 1992; Brian & Pohlmann, 1992) six types of policies and programs. In this view, metro policymakers are not "limited"; they may choose to enact any of the six types of policy, but the process is not without risk. As in climbing a stepladder with a heavy load and balanced on unsure footing, each step up the city policy stepladder tends to involve more difficulty and more risk for city policymakers. Thus local conditions—in short, the local policy *ecology* (Waste, 1989)—and "local politics" matter very much (Stone, 1993; Stone & Sanders, 1987a, 1987b; Swanstrom, 1993; Waste, 1993). Some policymakers and some metro areas will elect to go up the policy stepladder; others will not. Much of the explanation for whether they choose to engage in risky but rewarding social welfare policy making—for example, enacting programs such as the mini-Job Corps of New York City—has as much to do with the ecology of city policy making (Waste, 1989) as with "limits," local "regimes," or "growth machines."

Other popular descriptions such as (9) "the dependent city" (Kantor with David, 1988; Morgan & Hirlinger, 1993) attempted to convey the increasing reliance of cities on state and federal financial assistance or (10) the increasing Balkanization of American cities into high- and low-employment and -income areas or high- and low-density residential sectors. These labels included (10) "dual cities" and "quartered cities" (Beaureguard, 1992; Marcuse, 1988; Orum, 1991), "edge cities" (Garreau, 1991), and suburban bedroom communities representing the secession of the successful (Reich, 1992).

Each of these descriptions of the modern American city has merits and demerits, advocates and detractors among urban scholars. The most recent label to describe modern U.S. cities is regional in emphasis. In this view, cities are seen as systems operating within a framework that is both national and international in scope. Thus labels such as "cities" or even "urban" should give way to (11) "metro" and "metropolitan areas" in which the physical, social, and political fate of the inner and suburban city are inextricably tied together (Goldsmith & Blakely, 1992; Peirce, 1993a, 1993b). In this most recent view, cities need to be seen as organic, interdependent wholes in which ghetto poverty policies and suburban or exurban land-use, transportation, employment, and taxation policies are all part of one immensely large metro policy. In this view, "if downtowns like New York, Detroit, or Chicago die, the suburbs will be imperiled by their demise" (Blakely & Ames, 1992, p. 40).

As syndicated urban affairs columnist Neal Peirce has written:

> In a suburban nation, there's no payoff in talking about cities or using the word "urban." But in reality, many of the answers to poverty, homelessness, crime and credit

shortages will have to be forged in the "real" cities of the 1990's—the metropolitan regions, or citistates, where the vast majority of Americans live. (Peirce, 1993a, p. B7).

Secretary of Housing and Urban Development Henry Cisneros has argued in favor of much the same view: "Interwoven destinies tie the fate of the inner cities to their entire metropolitan areas. . . . Upon the shoulder of these citistates rests the future of our nation's competitiveness in the global economic marketplace" (Peirce, 1993a, p. B7).

If this view—a view to which I subscribe—is accurate, then the age-old search of urbanists for a viable national metro policy that is both politically possible and of real assistance to America's metro areas is far more crucial in the 1990s than ever before. What progress has the national government made toward enacting such a policy? What key issues face a coherent national metro policy? What progress has been made from the Reagan presidency to the present, and what still remains to be done to meet the multiple challenges facing U.S. metro areas? Fortunately these very questions are precisely the focus of all four of the contributors in Part 3. We turn next to a brief introduction to each reading and a brief bibliographical section outlining further readings for those readers interested in expanding beyond the issues considered in the four readings that follow.

Mary K. Nenno and "Urban Policy Revisited"

Chapter 6 is important for several reasons. First, as the widely respected Associate Director for Policy Development of the Washington, DC-based National Association of Housing and Redevelopment Officials (NAHRO), Mary K. Nenno is an experienced inside-the-Beltway observer of Washington attempts to construct national urban policy. Second, Nenno summarizes urban policy issues and programs from 1969 to 1988 and, in so doing, illustrates (a) the critical role played by Senator Daniel Patrick Moynihan (D-NY) in formulating national urban policy; (b) the process by which urban policy concerns have become institutionalized since 1977 as a result of the 1977 Housing and Community Development Act, which requires that the president submit an Urban Policy Report to Congress; and (c) the ways some issues have remained constant from the 1970s to the present in America's cities and the critical role of new issues such as increased immigration, the restructuring of American industry, the increasing concentration of poor people in high-poverty neighborhoods in cities, and the "new entrepreneurship" now present in many American cities.

Nenno's argument that urban issues likely will emerge on center stage after 1988 has already proven to be true. What remains to be seen is whether the actors and intergovernmental agencies identified by Nenno in the closing section of her chapter and the outlines of national urban policy she sketches out will also emerge in the 1990s as an accurate view of national urban policy actually enacted by the federal government. Whether the federal government embraces such a proposal or not, the challenges facing urban policymakers are perhaps nowhere better summarized in one place than Nenno's "Chart 4: A Resurfacing of Urban Issues," which is a summary of the challenges facing urban policymakers.

William R. Barnes and "Urban Policies and Urban Impacts After Reagan"

Despite the fact that every president since Richard Nixon has had both a domestic and a national security council to advise him, the domestic council has never had an obligation to consider urban issues. Although, as Nenno observed, since 1977 every president has had to submit an Urban Policy Report to Congress, some presidents had opted to place the problems of cities on the policy back-burner in favor of other domestic concerns or in favor of pursuing large macroeconomic approaches such as stimulating foreign trade and increasing the competitiveness of American employers. In Chapter 7, William R. Barnes argues that urban policy and aid to the cities became a policy "noncategory" in the Reagan years due to "New Federalism," deficit spending, and non-city spending priorities but that urban policy took a slight turn for the better during the Bush presidency. Bush launched an effort to support child care in urban areas, appointed Jack Kemp, a highly visible Secretary of Housing and Urban Development (HUD), and supported the Kemp effort to enact Urban Enterprise Zone (EZ) legislation. Barnes concludes that Bush programs tended to increase metro suburbanization, to aid high-income over low-income residents, and to encourage macroeconomic policies operating by both an "invisible hand" and "an invisible foot." Barnes offers a political explanation for the lack of emphasis on urban problems by the Bush administration that is tied to a lack of support for Republican presidential candidates in large cities such as New York, Chicago, and San Francisco. Although the Barnes hypothesis is a strong analysis of antiurban sympathies by the Republicans at the national level at the time that Barnes wrote, more recent municipal elections in New York City, Los Angeles, and San Francisco in which conservative law-and-order candidates won mayoralties may point to an interesting update to Barnes's argument. Republicans may have more incentive to strike a clear urban policy note than ever before—not only because of conservative suburban voters but also because of voters within metro city limits, who are beginning to display a distinctly conservative mood in several of America's largest metro areas.

Barnes provides a lively discussion of what he labels "the search for the real *urban*." Illustrating the difficulty of distinguishing the purely urban from domestic policy at-large, he constructs and deconstructs two widely held explanations of "urban policy." In the final analysis he sides with Harold Wolman, who has argued that in the early 1990s "the effects of implicit urban policy on urban areas far outweigh those of explicit urban policy" (Wolman, 1986, p. 317).

Robert Warren and "National Urban Policy and the Local State"

In Chapter 8, Robert Warren argues that the checkered history of national urban policy, a policy that, he argues, can be traced back "at least to the 1930s," has tended to be characterized by an "unsatisfactory top-down" quality of the primacy of national policy and national policymakers. Such an approach tends toward centralized solutions and policy making at the expense and to the detriment of local policy-making institutions. Warren would have the state and local policymakers play a larger role in constructing urban policy in what he labels a "dual local-national strategy." Warren tempers his suggested strategy with the experience of a long-term and shrewd urban observer when he notes: "Addressing urban policy seriously is like simultaneously trying to solve a half-dozen puzzles with the pieces mixed and some missing. A number of critical matters have not been considered adequately or considered at all."

Warren emphasizes the role of state and local governments in attacking the urban problem—indeed, even in defining at least the *local* urban "crisis," or "problem," much less the preferred local "solutions." This is sage advice; as Osborne and Gaebler have argued in *Reinventing Government* (1993), much of the productive dialogue about both what government should do and how it should be done is taking place at the subnational levels, in state and local government. To ignore this dialogue, to ignore this intergovernmental partnership is to construct a national urban policy that, as Warren notes, has a far too "unsatisfactory top-down" character.

Robert C. Wood and "People Versus Places"

Part 3 begins with a reading on the role of one of the giants of national urban policy making—Senator Daniel Moynihan—and concludes with a reading from another of the giants of the field, former Secretary of Housing and Urban Development in the Johnson Administration, Robert C. Wood. Like Moynihan, Wood was both a highly respected academic and a high-level national urban policymaker, so his comments must be given extra weight in terms of the experience base from which he writes. Wood argues that even though a coherent national urban policy still may not be said to exist in the early 1990s, the pressure for such a policy—driven in the 1960s by poverty, urban decay, and urban violence—is in the 1990s tending to be pushed and pulled by very different pressures. These include suburban sprawl, land use, and land management, although one might well add after Wood's writing, the emergence of crime as the chief national, metro, and local issue in the country.

Wood distinguishes between the 1960s-era policies aimed at providing transfer payments to individuals and the emerging 1990s policies aimed at regulating the place in which people may or may not require transfer payments and government programs. In contrast to Warren's call for increasing the role of the local metro governmental partners, Wood sees local government as parochial and as a force to be reckoned with by the state and federal government. Wood argues spiritedly not to abandon the vision of top-down national urban policy. "The conclusion is, for the 1990s, to try government again." Otherwise, if local metro governments are left to their own devices—or, in Osborne and Gaebler's terms, to their own "reinventions"—local government will end up in a parochial battle of the acronyms, pitting NIMBYs against TOADS against LULUs.

Conclusion

It remains for the reader to determine which vision of the local metro partnership with the federal makers of national urban policy should carry the day. Nenno and Barnes describe accurately the formulation—at times, the lack of formulation—of national urban policy from the 1970s to the present. Warren and Wood engage us in the heart of the contemporary debate about what shape the national urban policy should take and why. The reader is advised to look at both the heart and the periphery of the national urban policy debate. Wood and Warren both make a persuasive case for different views of what the heart of national urban policy should be. The reader should be reminded also, however, of Wolman's argument that the largest urban effects may well come from policies and programs not generally considered urban at all. If Wolman is correct—and I believe that he is—then the careful student of urban policy is well advised to study not only the "heart" of the national urban policy debate but also the "periphery" as well. The readings in Part 3 are a sound introduction to that literature.

References

Abbott, C. (1981). *The new urban America: Growth and politics on Sunbelt cities.* Chapel Hill: University of North Carolina Press.

Anton, T. J., & Flaum, A. R. (1992, Winter). Theory into practice: The rise of new anti-poverty strategies in American cities. *Urban News, 5,* 1, 4-7.

Beauregard, R. A. (1993). Descendants of ascendant cities and other urban dualities. *Journal of Urban Affairs, 15,* 217-229.

Bernard, R. M., & Rice, B. R. (Eds.). (1983). *Sunbelt cities: Politics and growth since World War II.* Austin: University of Texas Press.

Blakely, E. J., & Ames, D. L. (1992). Changing places: American urban planning policy for the 1990s. *Journal of Urban Affairs, 14,* 423-446.

Brian, A., & Pohlmann, M. (1992, Winter). Free the children: Fighting poverty in Memphis. *Urban News, 5,* 1, 8-9.

Elkin, S. L. (1987). *City and regime in the American Republic.* Chicago: University of Chicago Press.

Ferman, B. (1985). *Governing the ungovernable city.* Philadelphia: Temple University Press.

Garreau, J. (1992). *Edge city: Life on the new frontier.* Garden City, NY: Doubleday.

Goldsmith, W. W., & Blakely, E. J. (1992). *Separate societies: Poverty and inequality in U.S. cities.* Philadelphia; Temple University Press.

Henig, J. R. (1992). Defining city limits. *Urban Politics Quarterly, 27,* 375-395.

Jencks, C., & Peterson, P. E. (1991). *The urban underclass.* Washington, DC: Brookings Institution.

Kantor, P., with David, S. (1988). *The dependent city.* Glenview, IL: Scott, Foresman.

Long, N. E. (1971). The city as reservation. *The Public Interest, 25,* 22-32.

Marcuse, P. (1989). Dual city: A muddy metaphor for a quartered city. *International Journal of Urban and Regional Research, 13,* 697-708.

Massey, D. S. (1990). American apartheid: Segregation and the making of the underclass. *American Journal of Sociology, 95,* 329-357.

Massey, D. S., & Denton, N. A. (1993). *American apartheid: Segregation and the making of the underclass.* Cambridge, MA: Harvard University Press.

Massey, D. S., & Eggers, M. L. (1993). The spatial concentration of affluence and poverty during the 1970s. *Urban Affairs Quarterly, 29,* 229-315.

Molotch, H. L. (1976, September). The city as a growth machine. *American Journal of Sociology, 82,* 309-330.

Molotch, H. L. (1993). The political economy of growth machines. *Journal of Urban Affairs, 15,* 29-53.

Morgan, D. R., & Hirlinger, M. W. (1993). The dependent city and intergovernmental aid: The impact of recent changes. *Urban Affairs Quarterly, 29,* 256-275.

Orum, A. M. (1991). Apprehending the city. *Urban Affairs Quarterly, 26,* 589-609.

Osborne, D., & Gaebler, T. (1993). *Reinventing government: How the entrepreneurial spirit is transforming the public sector.* New York: Penguin.

Peirce, N. R. (1993a, December 21). A blueprint to put cities back together. *Sacramento Bee,* p. B7.

Peirce, N. R. (1993b). An urban agenda for the president. *Journal of Urban Affairs, 15,* 457-467.

Peterson, P. E. (1981). *City limits.* Chicago: University of Chicago Press.

Phillips, K. (1969). *The emerging Republican majority.* New Rochelle, NY: Arlington House.

Reich, R. (1991, January 20). The secession of the successful. *The New York Times Magazine,* pp. 16-17, 42-45.

Sanders, H. T., & Stone, C. N. (1987a). Competing paradigms: A rejoinder to Peterson. *Urban Affairs Quarterly, 22,* 548-551.

Sanders, H. T., & Stone, C. N. (1987b). Developmental politics reconsidered. *Urban Affairs Quarterly, 22,* 521-539.

Schumaker, P. (1991). *Critical pluralism, democratic performance, and community power.* Lawrence: University of Kansas Press.

Stone, C. N. (1989). *Regime politics: Governing Atlanta, 1946-1988.* Lawrence: University of Kansas Press.

Stone, C. N. (1993). Urban regimes and the capacity to govern: A political economy approach. *Journal of Urban Affairs, 15,* 1-28.

Stone, C. N., Jones, B. D., & Longoria, T. (1992). Human capital investment as a research focus. *Urban News, 5,* 10-11.

Swanstrom, T. (1993). Beyond economism: Urban political economy and the postmodern challenge. *Journal of Urban Affairs, 15,* 55-78.

U.S. Bureau of the Census. (1991, September). Metropolitan areas and cities. *1990 Census Profile, 3,* 1.

Waste, R. J. (1989). *The ecology of city policymaking.* New York: Oxford University Press.

Waste, R. J. (1993). City limits, pluralism, and urban political economy. *Journal of Urban Affairs, 15,* 445-455.

Wirt, F. (1974). *Power in the city: Decisionmaking in San Francisco.* Berkeley: University of California Press.

Wolman, H. (1986). The Reagan urban policy and its impacts. *Urban Affairs Quarterly, 21,* 311-335.

Yates, D. (1977). *The ungovernable city.* Cambridge: MIT Press.

Suggestions for Further Reading

Beauregard, R. A. (1993). *Voices of decline: The postwar fate of U.S. cities.* Cambridge, MA: Blackwell.

Caves, R. (1992). *Land-use planning: The ballot box revolution.* Newbury Park, CA: Sage.

Cummings, S., & Koebel, C. T. (Eds.). (1992). Toward an urban policy agenda for the 1990s [Special issue]. *Journal of Urban Affairs, 14.*

Galster, G. W., & Hill, E. W. (Eds.). (1992). *The metropolis in black and white: Place, power, and polarization.* New Brunswick, NJ: Rutgers University, Center for Urban Policy Research.

Kaplan, M., & James, F. J. (Eds.). (1990). *The future of national urban policy.* Durham, NC: Duke University Press.

Logan, J. J., & Swanstrom, T. (Eds.). (1990). *Beyond the city limits.* Philadelphia: Temple University Press.

Peirce, N. R. (1993). *Citistates: How urban America can prosper in a competitive world.* Arlington, VA: Seven Locks.

Wong, K. W. (Ed.). (1992). *Research in urban policy: Vol. 4. Politics of policy innovation in Chicago.* Greenwich, CT: JAI.

— 6 —

Urban Policy Revisited

ISSUES RESURFACE WITH A NEW URGENCY

Mary K. Nenno

When Daniel Patrick Moynihan (U.S. senator from New York) became the director of President Richard Nixon's newly established President's Council for Urban Affairs in 1969, he began a process of looking at the urbanization of the United States in a new national perspective. His charge was to develop "coherent, consistent patterns of activities that the national government should encourage or discourage in responding to the urbanization of American life" (*Journal of Housing,* 1969, p. 118). This first attempt to rationalize the national government's role and responsibility did not produce a continuing national policy, nor did the more deliberate attempts of the Carter administration in 1978 and 1980. Under the administration of President Ronald Reagan, urban policy was sublimated to overall national economic and fiscal strategies.

Today, 25 years later, the issues raised by Moynihan are surfacing again with a new urgency, and urban policy is again demanding a place on the national agenda. Ironically a key element in the Nixon urban policy—total reform of the welfare system—is now pending passage in the United States Congress. The second and third elements of this early policy—reform of the intergovernmental system, and reforms to expand the range of opportunities for all Americans—have been rapidly germinating and evolving.

Our understanding of the urban policy of the future has already assumed new dimensions—an understanding that it applies to rural and undeveloped areas as well as urban areas, a recognition of the importance of public-private partnerships, and emerging definitions of government responsibilities at federal, state, and local levels.

What Moynihan saw in 1969 and 1970 was a capsule of issues that increasingly demanded attention in 1988. He summarized them in 10 points, many remarkably consistent with the urban policy elements set forth in the 1970 and 1977 Housing and Community Development Acts (*Journal of Housing,* 1970, pp. 68-69; see Charts 1 and 2):

Chart 1   Past Approaches to National Urban Policy

Comprehensive provisions of national legislation or executive action.

1969-70   Daniel Patrick Moynihan (now U.S. senator, New York), Director of the
            Council of Urban Affairs (later Counselor to the President) in the Nixon
            White House, was ordered to develop "coherent, consistent patterns" of
            activities that the national government should encourage or discourage in
            responding to the urbanization of American life. President Nixon proposed
            a "national growth policy" in his 1970 State of the Union message. The
            president also called for "a total reform" of the welfare system.

1970   *National Urban Growth and New Communities: Title VII of the Housing
            and Urban Development Act of 1970* (PL 91-609; 84 Stat. 1791): Defines
            the purposes of a national urban policy. Charges the president with the
            submission of a report on urban growth every 2 years. Directs the Secretary
            of Housing and Urban Development to encourage the formulation of plans
            and programs under the Section 701 Comprehensive Planning Program for
            guiding urban growth. Amends the Urban Renewal Program (Title I of the
            Housing Act of 1949) to make possible the use of functionally obsolete
            and uneconomic land.

1977   *National Urban Policy and New Community Development: Title VI of the
            Housing and Community Development Act of 1977* (PL 95-128;
            91 Stat. 1149): Changed the emphasis of the 1970 act (above) from
            urban growth policy to *national urban policy;* findings and purposes were
            modified or expanded to include energy conservation, population distri-
            bution, and tax base considerations. The biennial report on urban policy,
            required in February every even-numbered year beginning in 1978, was
            expanded to include assessment of rapid growth areas (see Chart 2,
            Purposes of a National Urban Policy).

1. Poverty and social isolation of minority groups in central cities
2. Enormous imbalances between urban areas in the programs that affect them
3. Inadequate structure and capacity in local government
4. Fiscal instability in urban government
5. Lack of equality in public services among jurisdictions within metropolitan areas
6. Migration of people displaced by technology and from central cities to suburban areas
7. Inadequate structure and capacity of states in their indispensable roles in
   managing urban affairs
8. Ineffective incentive systems to encourage state and local governments and
   private interests to help implement the goals of federal policy
9. Adequate information, as well as extensive and sustained research on urban
   problems
10. An insufficient sense of the finite resources of the national environment and
    the fundamental importance of aesthetics in successful urban growth

Chart 2   Purposes of a National Urban Policy

The 1977 Housing and Community Development Act (PL 95-1286; 91 Stat. 1149)

The Congress declares that the National Urban Policy should—

1. favor patterns of urbanization and economic development and stabilization that offer a range of alternative locations and encourage the wise and balanced use of physical and human resources in metropolitan and urban regions, as well as in smaller urban places that have a potential for accelerated growth;

2. foster the continued economic strength of all parts of the United States, including central cities, suburbs, smaller communities, local neighborhoods, and rural areas;

3. encourage patterns of development and redevelopment that minimize disparities among states, regions, and cities;

4. treat comprehensively the problems of poverty and employment (including the erosion of tax bases, and the need for better community services and job opportunities) that are associated with disorderly urbanization and rural decline;

5. develop means to encourage good housing for all Americans without regard to race or creed;

6. refine the role of the federal government in revitalizing existing communities and encouraging planned, large-scale urban and new community development;

7. strengthen the capacity of general governmental institutions to contribute to balanced urban growth and stabilization; and

8. facilitate increased coordination in the administration of federal programs to encourage desirable patterns of urban development and redevelopment, encourage the prudent use of energy and other natural resources, and protect the physical environment.

Although individual efforts have been made and some progress can be claimed on each of the above issues, what is still lacking is an accepted national consensus and a coherent national policy and strategy to deal with these issues. What has been germinating and evolving over the past 25 years is a growing understanding of the impact of urbanization on the economic, physical, and social structure of American life. Today urbanization affects industry, cities, and individuals in direct, immediate ways.

Past Approaches to Urban Policy

The 1982 president's report on urban policy (U.S. Department of Housing and Urban Development, 1982; see Chart 3) cited three stages of federal involvement in urban affairs. The first stage, lasting until about 1935, was described as a passive one, limited to ad hoc activities. The second stage, beginning in the mid-1930s, was seen as one of "cooperative federalism" involving a selective federal role in managing the economy and providing federal grants-in-aid to stage and local governments. The third, emerging around 1960, was described as a "marble-cake federalism," a confusing swirl of federal, state, and local programs. Along the way the federal government tried several approaches to urban problems: (a) *functional* approaches to problems such as housing, crime, or education; (b) *structured* approaches intended to improve the division of responsibility; (c) *targeted* approaches, directed at the problems of a particular sector; and (d) *revenue sharing* or *block grant* approaches dealing with local resource needs while minimizing federal intervention in local affairs. The 1982 report concluded that most of these efforts were unsuccessful—"some spectacularly so"—as were efforts at reform, which inadvertently increased the size, scope, and intrusiveness of the federal grant system. The alleged solution, set forth in both the 1982 and 1986 reports by the Reagan administration, was for the national government to pursue national economic strategies comprising tax cuts, reduced federal government spending, regulatory relief, and monetary restraint. These actions would "create the conditions" under which individuals, businesses, and communities could productively pursue their own interests, thus generating urban well-being. At the same time, the administration's "urban policy" would realign governmental responsibilities, placing more authority and responsibility, along with selected revenue sources, at the local level (U.S. Department of Housing and Urban Development, 1982, pp. 45-46).

Some students of urban affairs, in 1982, read past history differently. Goldsmith and Jacobs (1982), of the Cornell University Department of City and Regional Planning, cited two periods of history in which there was "partially coherent" centralized urban policy. The first period was in the economic crisis of the 1930s, when the housing sector collapsed and the federal government intervened by using housing construction to stimulate the economy and—with the help of the later postwar boom (VA loans and the interstate highway program)—house and suburbanize vast numbers of people. The second period was during the social crisis of the 1960s, when the federal government channeled a good deal of money into cities, providing considerable relief. At the same time, the authors note that other actions by the federal government in these same periods directly conflicted with urban policy objectives, and numerous others were left totally unconsidered. Goldsmith and Jacobs find two dimensions to past urban policy—policies that reinforced the private economy, and policies that mopped up problems left over by the private economy.

The Carter Administration Initiatives

Buoyed by the passage of the national urban policy provisions of the 1977 Housing and Community Development Act (PL 95-128; 91 Stat. 1149), the Carter administration mounted an intensive effort to develop and implement a national urban policy and strategy (see Charts 1, 2, and 3). Its first national urban policy report of 1978 established a comprehensive partnership of federal, state, and local action built around five major goals: (a) implementation of urban policy, (b) economic health of cities, (c) city facilities and infrastructure, (d) preservation of urban neighborhood environments, and (e) status of urban residents (Nenno, 1978). Federal activity was spread across a dozen federal departments and agencies and involved $6.5 billion in direct federal outlays over three fiscal

Chart 3   The President's Urban Policy Reports 1978-1988

1978   *(President Carter)* Delineated a five-part policy: to strengthen the implementation of urban policy, to restore the economic health of cities, to
strengthen total city facilities and infrastructure, to improve the neighborhood environment of cities, and to improve the status of urban residents.
President Carter: "It is in our national interest not only to save our cities
and urban communities, but also to strengthen them and make them more
attractive in which to live and work. . . . The federal government alone
does not have the resources or knowledge to solve all of the problems of
our urban areas . . . it must do its share . . . but its more critical responsibility is to act as a catalyst to encourage investment and contribution from
the states, local governments, private sector and individuals . . . a new partnership between the federal government and these groups should be developed." (August 1978, HUD-CPD-328, 147 pages)

1980   *(President Carter)* Building on the "principles" of the 1978 policies, the
1980 report documented 9 urban policies leading to 19 specific legislative
initiatives, executive orders, and administrative actions; 16 of the legislative
initiatives were approved by the Congress and involved 9 federal departments and agencies. The 1980 report confirmed the Carter administration
commitment, working in partnership, "to strengthen urban economies,
expand job opportunities and job mobility, promote fiscal stability, expand
opportunity for those disadvantaged by discrimination and low income,
and encourage energy-efficient and environmentally-sound urban development patterns." (August 1980, HUD-583-1-CPD, 247 pages)

1982   *(President Reagan)* The approach of the 1982 urban policy report, the first
of the Reagan administration, was based on two principles: (a) that the
foundation for urban policy is an Economic Recovery Program including
tax cuts, reduction in the rate of federal government spending, regulatory
relief, and monetary restraint in order to restore economic vitality to
American industry and create jobs for workers; and (b) that an evolving
program of "New Federalism" would create a major realignment of responsibilities between federal, state, and local governments, with more authority
and responsibility at the local level. The position of the Reagan administration was that individuals, businesses, and communities will realize
greater and longer lasting benefits if the federal government "creates the
conditions" under which all can productively pursue their own interests.
In particular, the administration recommended "enterprise zones," providing tax incentives to private businesses to invest in depressed urban areas.
(August 1982, HUD-S-702-2, 74 pages)

1984   *(President Reagan)* The 1984 report reconfirmed the approach of the 1982
report and stated: "There is increasing evidence that this strategy is sound
and is working." The report cited evidence that inflation was reduced, the
gross national product was increased, employment was rising, and that
states and localities were moving from a budget deficit to a budget surplus.
Furthermore, federal assistance to the needy was more effectively targeted,
new "block grant" approaches were providing more state and local flexibility, infrastructure needs were regarded as manageable, and serious crime
was down. (August 1984, HUD-9-09-DRR, 75 pages)

(continued)

Chart 3   Continued

1986   (*President Reagan*) The 1986 report reiterates that an era of renewed
            prosperity and stability in the nation's cities was being achieved. In addition
            to the elements of administration policy cited in 1982 and 1984, a fourth
            element of administration policy was added in 1986—"encouraging self-
            sufficiency by those able to work." Housing assistance should move away
            from construction programs to a program of housing vouchers and rent
            supplements, enabling families to move into the private market toward
            self-sufficiency. The intent was to develop a new strategy for meeting the
            legitimate subsistence needs for the poor, while reducing dependence on
            government support. (December 1986, HUD-1068-PDR, 54 pages)

1988   (*President Reagan*) The 1988 report continues the previous Reagan admin
            istration theme that national economic growth, strengthening the authority
            of state and local governments, stimulating public-private partnerships,
            and encouraging family self-sufficiency are the ingredients of an urban
            strategy. Federal assistance should be shifted from "place-specific" urban
            aid to state governments or to direct assistance to needy people and
            families. (October 1988, HUD-, 230 photocopied pages)

years, plus an estimated loss of $4.9 billion in federal revenue due to supporting
changes in the tax law. A significant component was "incentive" funding for
private investment, state and local governments, and neighborhood groups. The
largest component of direct outlays (68%) was to restore the economic health
of cities. But a major emphasis of the Carter strategy in both the 1978 and 1980
reports was not federal dollar outlays, but using federal decisions as mechanisms
to implement urban policy—including location of federal facilities, federal
procurement, simplification of federal assistance mechanisms, and improved
data, research, and analysis. Significantly, Carter's 1978 urban policy program
established a federal Inter-Agency Coordination Council and required all federal departments and agencies to submit as part of their legislative programs an
analysis of how each initiative affected communities. The 1980 urban policy
report, the last of the Carter administration, presented a comprehensive analysis of urban conditions and cited 19 legislative initiatives and executive actions
designed to implement the 1978 urban policy objectives. The Carter initiatives
(in the words of Robert C. Embry, the Assistant Secretary for Community
Planning and Development of HUD, who had a major role in the development
of the Carter urban initiatives) were:

> An analysis and strategy for the federal government—a statement of what the federal
> government thinks the problem is, and what it is going to do about it. . . . The central
> thrust was not to spend more money, to establish new programs, to usurp local
> functions. It was to make federal decision makers aware of the impact of their
> actions, and where possible, to have their decisions strengthen, rather than weaken,
> distressed places and distressed persons. (Nenno, 1982, pp. 168-172)

A significant result of the Carter urban policy initiatives was to encourage the number of states undertaking urban strategies on a statewide basis, mostly resulting from state legislative actions taken in the late 1960s and early 1970s. In his 1980 publication prepared for HUD, Charles R. Warren documented these strategies in ten states: California, Connecticut, Florida, Massachusetts, Michigan, New Jersey, North Carolina, Ohio, Oregon, and Pennsylvania (Warren, 1980). Three of these states (Massachusetts, North Carolina, and Oregon) developed "vertical" strategies with direct participation and involvement of local government officials; five states (California, Connecticut, Michigan, New Jersey, and Pennsylvania) adopted "horizontal" strategies characterized by local government participation of a more advisory and reactive character. Florida and Ohio developed combinations of these strategies. All of these state strategies were generally broadly based, covering a wide range of issues. The four major themes were economic development, growth management, urban revitalization, and fiscal reform. Each of these states also addressed the better coordination of the state's functional programs.

With the advent of the Reagan administration in January 1981, a basically new approach to urban issues was adopted, as described above. The programs and administrative vehicles of the Carter urban initiatives were dropped. Some of the basic themes of urban policy, however, dating back to the Moynihan years and reinforced under the Carter administration—such as changes in the intergovernmental system—reemerged in the Reagan years, but with a different focus.

A Resurfacing of Urban Issues

As the Reagan era drew to a close, persistent signs were that the issues surrounding the urbanization of the United States were resurfacing with a new urgency. Despite the assurances of the 1984 and 1986 urban policy reports of the Reagan administration, it was clear that macroeconomic policy and disparate individual actions had not resolved urban problems. Evidence was mounting that some urban conditions had deteriorated to a crisis stage. Five issues, identified in Chart 4, were of paramount concern (a similar but shorter statement of issues appears in Committee on National Urban Policy, 1982).

Changing Population and Household Structure

Recent demographic scholars have tracked the changing composition of the American population, including the increasing immigration from developing countries and the dramatic shifts in the American native population's household structure. It is estimated that between mid-1979 and the year 2000, the total annual immigration into the United States will rise from 1.1 million to 1.8 million persons, of which 98% will be from developing countries (Committee on National Urban Policy, 1982). These new immigrants will enter labor markets across the country, attend schools, and look for health and human services.

Chart 4   A Resurfacing of Urban Issues

The following five issues represent key elements to be addressed by national urban policy in the 1990s and are discussed more fully in the text of this chapter.

Changing Population and Household Structure
New levels of immigration into the United States, plus rapid changes in the household composition of the native population, have deep implications for urban policy. These impacts are particularly evident when broken down by income level and employment.

The Restructuring of American Industry
Shifts in the economy based on knowledge-based industry and telematics are changing the way strategies for physical, economic, and social revitalization of urban communities are addressed. This has both national and international dimensions.

Urban Poverty, Education, Welfare, and Housing
The most troublesome trend in urban change is the growing concentration of poor people in cities. Federal income maintenance programs have not affected the roots of poverty, nor has national macroeconomic policy. There is a new "permanence of poverty" that needs to be recognized with new policies and experiments, and a new national commitment to the impoverished.

Technology and Physical Development Processes
The modern American city was shaped by the inventions and technology of the last two decades of the 19th century and by the post-World War II inventions. The technology of the future is now on the horizon—innovations in structural materials, wall surfaces, transit, heating and cooling, communications, and waste disposal. We must accelerate research and applications of this new technology while maintaining our current infrastructure and reforming our building construction techniques.

Public Responsibilities, the Intergovernmental System, and the New Partnerships
Efforts to implement a complete transformation of the federal system in the United States over the past 25 years have not succeeded, but important incremental changes have occurred. The maturing of state government in urban affairs and the new "entrepreneurship" of city governments are important dimensions for the future. Reform of the total process will not come quickly, but escalating urban needs will stimulate action. Local governments, for the first time, could become equal partners in the federal system. The federal government retains a major role, providing a national strategy and financial support to respond to urban change.

In this same period, the indigenous American population will continue the rapid change in household composition that characterized the 1970s. By the end of the century, the U.S. population likely will exceed 106 million households, an increase of 21 million households over the 85 million households in 1983 (Turner, 1987, pp. 13-14). Single individuals will be the fastest growing household type and will account for almost one third of all households in the year 2000. The share of "young elderly" households (age 65 to 74) will decline, while the proportion of "older elderly" (age 75 or over) will expand to about half of all elderly persons. The number of married couples—with or without children—will increase slowly, with an actual decline for married couples under 35 years of age.

At the same time, families in single-headed households will grow by almost half, with the majority headed by women between the ages of 30 and 55 (Turner, 1987). These trends in household composition have deep implications for economic and social policy. Their impact is particularly evident when types of households are analyzed by income level and employment (see the following two sections).

The Restructuring of American Industry

Royce Hanson, a longtime student of urban policy, sums up the transformation of the United States industrial system based on the impact of three pervasive forces:

1. The economic shift—from an economy based on making things to one based on knowledge—is changing what is done and where it is being done.
2. Telematics—the marriage of telecommunications with computers—is transforming the way things are done and the concept of place itself.
3. Demographic changes—especially the aging of the population, women in the labor force and as heads of households, the growth of the urban underclass, and immigration and internal migration patterns—are reshaping who does the work.

Hanson concludes that these forces and their impacts also are changing how we think about strategies for physical, economic, and social revitalization of urban communities (Hanson, 1986). Agricultural, extractive, and manufacturing jobs have declined drastically as "knowledge work" has boomed. The movement toward services has accelerated the shift from blue-collar to white-collar jobs, and this trend will be even more pronounced by the year 2000. Although cities historically have been closely related to their geographic regions, the future of any given city is now determined more by what it has done and is capable of doing than by where it is located. A number of cities, such as New York, Los Angeles, Chicago, and San Francisco, are transcending their regions and operate instead as "world cities" in a global economy.

This shift to a service economy for many workers also affects the fiscal stability of cities with economies based on manufacturing, mining, or agriculture, where revenues are declining and needs are rising. Most cities are now in a period of fiscal experimentation as the old assumptions on which public finance was based have eroded or disappeared. Many cities have resorted to new systems of pricing services, contracting out for them, or even abandoning certain functions. It is not happenstance that the top item on the agenda for both mayors and governors is economic development.

The impact of telematics on industry is becoming profound. The principal effect to date is the dispersal of back-office operations to the suburbs, to smaller, remote cities, and even to localities in foreign countries. But these changes could well lead to a redeployment of economic activities within metropolitan areas and among urban areas throughout the world. The most knowledge-intensive activities are being concentrated in a relatively few, large, diverse "command-and-control" cities, linked to the world by international airports, telematics, and distribution facilities. Branch offices, back-office activities, production plants, and large-scale consumer services are being located in many subordinate centers (Hanson, 1983, pp. 5-6). National macroeconomic policy, for the most part, is concerned with aggregate levels of employment and efficiency and has been relatively indifferent to how it affects particular places. Increasingly the need to link the national economy with both sectoral and local economies is recognized. As Ralph Widner, a leading urban analyst, has pointed out:

> Economic policy has been dominated by concerns for the aggregate performance of the economy. . . . But, as is the case in many of the other advanced industrial countries, economic policy must also be concerned with two other aspects of national economic performance: the performance of major sectors within the aggregate economy; and the geographic consequence of national policy and technological change. . . . It is time to recognize that sound economic policy depends upon the balanced recognition of all three aspects of economic performance: aggregate, sectoral, and geographic. (Nenno, 1982, p. 169)

Furthermore, as the Carter administration urban policy initiatives of 1978 and 1980 anticipated, it is clear that actions of the federal government itself in the areas of taxation, construction of capital facilities, location of federal employment, and purchasing of supplies and services directly affect regional economies.

Urban policy traditionally has not been closely related to national economic policy because it is place oriented. But the nation's urban areas supply most of the physical infrastructure and social institutions on which the economy is built. Their economic and fiscal health is not separate from that of the nation (Hanson, 1983).

Urban Poverty, Education, Welfare, and Housing

Looking at urban change and poverty in 1988, the Committee on National Urban Policy of the National Research Council found the most troublesome trend to be the growing concentration of poor people in central cities (McGeary & Lynn, 1988, pp. 11-12). Central cities, in general, have developed a very high poverty rate: 19% in 1985, compared with the national poverty rate of 14%. But even more serious, in poverty areas within central cities (census tracts in which 20% or more of the population is below the poverty level), the poverty rate increased to 37.5% in 1985, up from 34.9% in 1975. Both the number and the percentage of poor people living in census tracts with the highest concentration (40% or more) of the poor are increasing. This growing poverty is occurring even in command-and-control cities with improving economies, such as New York and Boston, indicating that economic growth alone is not sufficient to improve the situation. The Committee concluded:

> Urban poverty appears to be a major by-product of the changing economic functions of large cities. . . . Always a serious problem, it appears to be getting worse, and it

is being compounded by long-term changes in the structure of the national economy and in metropolitan and regional demographic patterns. (McGeary & Lynn, 1988, p. 6)

A further significant finding of the committee was that "federal income maintenance programs have reduced poverty, but they do not seem to have affected the roots of poverty" (p. 23). Urban poverty continues to be a persistent national problem because it is concentrated, isolated, and entrenched; it is related to the cultural environment and to the failure of public institutions, particularly the educational system, to address the needs of those who are caught in this pattern. The impoverished are becoming isolated from mainstream society. As Hanson (1986, p. 17) observes: "Functional illiteracy and the absence of work habits render a large block of inner city youth unfit for employment in even the lowest level jobs in the legitimate economy."

The gap between the educational requirements of available jobs and unemployed persons in central cities is growing. The number of jobs available to persons with less than a high school education has dropped dramatically in most large cities since 1970. This mismatch between jobs and people results not only in personal tragedies for the individuals involved but also in serious consequences for the cities themselves, because their future development in a knowledge-based economy depends not on proximity to resources or transportation, but on the quality of their labor forces.

American cities are now at a disadvantage with cities in a number of other economically advanced nations, among them Japan, because their labor forces are not adequately trained in mathematics, the sciences, computers, and languages—including English. Social ills related to poverty, such as crime, poor health, and homelessness, also have a degrading effect on the quality of community as well as personal life.

If income maintenance programs do not get at the roots of poverty, the public welfare system certainly should play a pivotal role in working with the impoverished. The Reagan administration's strategy of trying to break dependency by reducing welfare benefits did not produce this result, nor did the previous administrations' strategy of benefit increases between 1955 and 1975. This experience, among other factors, triggered the basic reform of federal welfare embodied in 1988 national welfare reform legislation (H.R. 1720; S. 1511) (Kosterlitz, 1986; Stevens, 1988). The new welfare approaches would provide greater support for education, job training, day care, and other services for families that undertake a structured effort to lift themselves to self-sufficiency. This is a significant philosophical change, but it still falls short of establishing national standards for welfare benefits to balance the widely varying benefit levels among the various states, or linking welfare assistance to other major systems providing additional support assistance, such as housing, to low-income families (Nenno, 1987b; Newman & Schnare, 1988; Peterson & Romm, 1988).

The housing dimensions of poverty have been dramatized over the last 10 years by the increasing number of homeless persons, particularly in large central cities. Surveys undertaken by the United States Conference of Mayors have documented the continuing growth of homelessness, hunger, and poverty in large central cities—in 1987 alone, the demand for emergency shelter increased in 26 large cities by an average of 21% (Reyes & Waxman, 1987). Access to decent and affordable housing, however, is a problem that goes well beyond homelessness. The Joint Center for Housing Studies of Harvard University summed up the situation in a 1988 report: "America is increasingly becoming a nation of housing haves and have nots. . . . Continuing high housing costs limit the ability of low and moderate income households to improve their standards of living as many households struggle to secure even minimally adequate housing" (Apgar & Brown, 1988, p. 1). Despite 5 years of strong construction activity and increasing vacancies, the report concluded, the supply of low-cost rental housing continues to shrink as units are lost to abandonment or are upgraded for higher income occupants. Real rents (measured in constant 1986 dollars) have moved up sharply since 1981 and now stand at their highest level in more than two decades. At the same time, the incomes of renter families have been dropping, with an increasing number forced to pay large proportions of their incomes for rent. In addition, the number of inadequate dwelling units, though down somewhat in percentage terms, remains high in absolute terms. In 1983, 4.5 million owners and 5 million renters continued to occupy structurally inadequate housing—2.3 million of these units were in central cities. At that time, 26% of renter households with real incomes below $5,000 occupied structurally inadequate housing; in absolute terms, this means that the number of low-income households living in such conditions increased from 1 million in 1974 to 1.3 million in 1983.

Designing effective antipoverty policies, The Ford Foundation (1988) observed, is complicated by the changing and interrelated causes of poverty and by the difficulty of knowing at any point in time the kind of poverty a family is facing. But the "new permanence of poverty" cannot be ignored: "Over the next decade, new policies and experimental programs will be needed to restore economic and social mobility to the persistently poor. What we need now is recognition of the challenges faced by the new poverty and a national commitment to building a stronger future for all Americans" (The Ford Foundation, 1988, p. 2).

Technology and Physical Development Processes

The development patterns of cities are shaped by the technology of the times. The modern American city was born during the last 20 years of the 19th century, stimulated by a remarkable set of inventions that made the physical reality of the modern city possible. These included steel-structured buildings, elevators, electricity, central heating, flush toilets and sewer systems, telephones, automobiles, and subways. The next 40 years were spent redesigning cities to incorporate these inventions (Hammer, 1984, p. 58). World War II also produced new technologies—new products, new processes, new materials—that were applied to the exploding postwar demand for housing, schools, factories, hotels, office buildings, shopping centers, and an unprecedented range of new consumer products. This postwar technology boom also created and built nuclear power stations, linear accelerators, deep-dish astronomical observatories, and space-launching facilities.

A number of professionals involved in the design and development of the built environment foresee a third era of technology inventions and innovations comparable to those that created the modern city a century ago. Even if these new innovations do not have a comparable impact, the professionals argue, they could create new dynamics, forces of change in working and living patterns that may solve some current urban problems.

John Eberhard of the National Academy of Sciences foresees a future in which "urban construction will be made up of integrated assemblies that contain no separate structural system, no pipes or tubes for water or waste, no roads or subway tunnels for movement, no wires for communication, no personally operated vehicles, and no dependence on fossil fuels (Eberhard, 1988). Today's research, he believes, already is approaching innovations in structural materials, wall surfaces, transit, heating and cooling, communications, and waste disposal that will create such future possibilities. But he warns that these things will not happen easily because the United States is still using "hundred year old inventions to build our cities, and has institutionalized these earlier inventions in building codes, architectural specifications, departments of city governments and schools of engineering." The challenge, Eberhard believes, is a dramatic, immediate program of construction research, along with an equally forceful effort to introduce new findings into the worlds of education and practice.

As the new technology evolves over the next 15 years, more immediate trends and forces are shaping urban communities that need to be recognized and addressed. Significant among these is the repair and maintenance of the nation's basic infrastructure—roads, bridges, subway systems, and utilities. Urban infrastructure supports economic development and is not being repaired or replaced fast enough to prevent its continued deterioration. The problems are particularly acute in older, fiscally stressed cities in the Northeast and Midwest and in some fast-growing cities in the South and West. The concern is that the nation's infrastructure may be inadequate to sustain national economic growth (McGeary & Lynn, 1988).

On a broader scale, the Council on Development Choices in the 1980s (created by HUD in 1980 and administered through the Urban Land Institute) identified 14 trends and forces shaping communities in the 1980s that reflect many of the urban issues already discussed in this chapter (Council on Development Choices in the 1980s, 1981). These include energy cost and supply, increasing competition for land, and protecting the natural environment. It is already clear that the interplay of these forces is raising tensions, such as the conflict between forces for new development and those striving to preserve the existing environment. The council identified the problem in realizing success as that of "creating a pervasive climate for reform of development policy" to affect the thousands of men and women who influence the choices and opportunities for other Americans through the choices they make as members of planning and zoning boards and as private practitioners in development finance, architecture, and construction. The Council (1981, p. 103) concluded:

> The era that began after World War II is drawing to a close. . . . America may allow itself to become preoccupied simply with endless patchwork changes to mitigate outmoded policies. Alternatively, we can deliberately set a new course tempered by an appreciation of the full cost to society of wasting capital and other resources and invigorated by new preferences to suit the changing composition of our population.

Moving to a new technological era for cities of the future while reforming current development policy and practice is an integral part of a revitalized national urban policy.

Public Responsibilities, the Intergovernmental System, and the New Partnerships

Among the 10 principles of national urban policy set forth by Moynihan in 1969 were four that relate directly to the role and capacity of state and local governments, including their relationship to the private sector. This issue had been a pervasive one for almost two decades, in the national urban policy proposals of the Nixon, Carter, and Reagan administrations. It is clear that the capacity of the public sector at all levels of government is a bedrock requirement for the effective implementation of urban policy.

Traditionally urban assistance in the forms of housing assistance, community development, and redevelopment funding came from the federal government directly to cities. During the 1960s two specially targeted programs supplemented early categorical assistance—the Model Cities Program, designed to focus resources on distressed areas and distressed persons in urban areas, and the War Against Poverty under the Economic Opportunity Act. Although these "special purpose" programs produced incremental improvements, it was clear that physical development strategies alone could not address the multifaceted problems of urban areas and that local coordination of multipurpose federal assistance programs such as Model Cities was a monumental task (Real Estate Research Corporation, 1974).

The federal program reforms of the 1970s moved in the direction of federal "block grants" keyed to natural purposes but with broad discretion for local governments in program administration. Chief among these reforms was the Community Development Block Grant Program enacted in 1974. The concept of revenue sharing to cities was also given a trial for several years. In 1981 Congress folded 57 categorical grants-in-aid programs, mostly in health and social services, into 9 block grants. The Reagan administration, in 1983, proposed additional block grants called "megablocks," consolidating many categorical programs, general revenue sharing, and eight of the nine blocks enacted in 1981. No political consensus could be reached on further block grant development or decentralization, despite continuing independent studies such as the

Robb-Evans proposal, made by governor Charles Robb of Virginia and Senator Daniel Evans of Washington, in the study called "To Form a More Perfect Union" released in December 1985 (Evans & Robb, 1985).

One result of the changes to date, however, has been to stimulate the role of the states, both in traditional areas of state functional responsibility such as health and social services and in new areas such as community development. State governments have been given more flexibility and discretion in these areas (Nenno, 1987a). Charles R. Warren (1985, pp. 25-28) foresees a further "sorting out" of the responsibilities and resources of federal, state, and local governments over the next decade, with a fundamental emphasis on the changing relationships between state and local governments. Changing the traditional "antiurban" posture of many state governments will be a key component in any change.

Beginning in the mid-1970s and expanding rapidly in the 1980s, there was a dramatic upsurge in the activity of both state and local governments in urban activities, including economic development, community development, and housing. Assuming these functions is part of a maturing process for these levels of government extending back 45 years—part of a general renaissance in their capacity (Nenno & Brophy, 1982, p. 190). The president of the National Conference of State Legislatures said in 1984 on the occasion of the 25th anniversary of the Advisory Commission on Intergovernmental Relations, "Longer sessions, better educated legislators, more capable staff and computerized management have prepared the states to take their rightful place in the federal system (Advisory Commission on Intergovernmental Relations, 1985a, p. 16). Alice Rivlin (former Director of the Congressional Budget Office) added, "The conversation about federalism has not yet caught up with the rising level of competence and professionalism at state and local levels" (Advisory Commission on Intergovernmental Relations, 1985b, p. 77).

Beginning in 1981, the deep cutbacks by the Reagan administration in domestic assistance programs accelerated an already developing trend for states and localities to initiate new efforts and to adopt novel strategies for dealing with urgent urban needs. In the areas of housing and community development alone is a rising tide of state and local public-private partnerships, housing trust funds, linkage strategies, negotiated development, tax increment activities, and other innovations (Nenno & Colyer, 1988). In particular, the growing number of public-private partnership arrangements in the housing and development field, including the growth of cities and states as "entrepreneurs," has spawned an entirely new way of defining public functions. Kirlin and Marshall (1988, pp. 348-349) identify a bifurcation in local government functions:

It is not much of an overstatement to argue that a bifurcation in policy processes is occurring, with two distinct sectors emerging. One sector is focused on traditional service delivery through tax-supported public employees (e.g., public education). The other sector is focused on economic growth/infrastructure provision/land use and increasingly achieves its purposes without the expenditure of tax funds and with less extensive use of public employees. The two sectors are not only different in goals, resource base, and dependence on public bureaucracies, they also differ in participants, beneficiaries, and political styles. The two sectors are competitors for resources and space on the public policy agenda. . . . They also complement each other and interact in several ways.

The emergence of these "entrepreneurial" functions in government, particularly at the local level, introduces a basically new element into the federalism debate. In particular, local government may use this development to claim equal standing with the federal government and states in the federal system. The resolution of all of the elements into a coherent intergovernmental system has a long way to go, but such a consensus remains integral to addressing urban needs in the 1990s. Also, students of planning and public administration are concerned about a blurring of the distinctions between public and private functions.

Urban Policy for the 1990s

When President Bill Clinton took office in January 1989, he was faced with a revised demand for action on national urban issues: a demand for a new domestic policy—an agenda that had been substantially set aside by the Reagan administration in its two terms of office. The last Reagan administration report on national urban policy, issued in 1988, did not break much ground beyond the reports already issued in 1982, 1984, and 1986. Although acknowledging the continuing existence of such urban problems as drug trafficking, low academic achievement in public schools, high unemployment and skill deficiencies among youth and the disadvantaged, declining condition of infrastructure, environmental pollution, and deterioration of neighborhoods, the 1988 Reagan report rejected a direct federal role. As a basic principle, it stated:

> The hope for ameliorating remaining urban ills lies with local communities, families, and state and local governments, where the power, creativity, authority, resources and knowledge of local needs are concentrated. . . . The federal role should be primarily one of providing basic family-oriented financed assistance that enhances the capacity and authority of families and aids states and localities in responding flexibly to family needs. This includes programs such as housing vouchers and assistance to states for job training. Federal assistance should be shifted from "place-specific" urban aid to direct assistance to needy people and families or to state governments. (U.S. Department of Housing and Urban Development, 1988)

The seeds of this national urban policy are contained in the experience of the 25 years since the issuance of the Nixon/Moynihan agenda, with some newly developing dimensions.

A National Urban and Development Policy

It is becoming increasingly clear that what has been called an "urban" policy or an "urbanizing" policy does not relate to central cities or dense urban areas alone; the changing forces in motion affect rural areas and the diminishing number of undeveloped areas as well. Changing economic structures, demography, and development patterns are nationwide in scope. This range may necessitate a new name for the policy, as well as a national development perspective

A National Research and Demonstration Program

We do not know all we need to know about the impacts and the interrelationship of the changes that are taking place. If comprehensive, workable policies are to be developed, they must be supported by a significant, focused national information and research effort. This research effort should be accompanied by a meaningful range of experiments and demonstrations to accelerate the application of new knowledge. Our educational system, at all levels, must begin to incorporate the new knowledge and experience and its applications for future leaders.

Public-Private Partnership Roles and Resources

Urban policies involve a complex web of public and private actions intersecting and affecting each other. Each, operating completely independently of the others, can succeed in achieving only limited objectives. The experience of the last 15 years has generated promising public-private coordination, particularly in economic and housing development and urban reinvestment. Much of this activity has taken place at the local level, with an increasing number of statewide partnerships. This same concept should be explored and extended on a national basis. Only the combination of public and private resources can hope to address urban needs; public funding alone will be far too limited.

New Dimensions of Public Responsibility

Each level of government in our federal system has responsibilities in advancing a new urban development policy for the nation.

The Federal Government

Currently the federal government's fiscal capacity has serious resource constraints because of the ever-increasing national debt. However, the federal government has prime leadership responsibility for establishing overall national development guidelines and for providing the incentive funding to support levels of government and the private sector in performing their responsibilities in harmony with national goals and strategies. The beginnings of such guidelines can be seen in the Carter urban policy initiatives of 1978 and 1980. But an Interagency Coordinating Council will not suffice, nor will loose coordination of federal assistance programs. Federal government resources must be joined and targeted in new ways, particularly in joining federal income maintenance assistance to employment and housing initiatives. The federal government has the responsibility for developing strategies and resources to ensure a balanced development across the country. It also has the responsibility for establishing a comprehensive research and demonstration effort in cooperation with state and local governments, universities, and private corporations. Federal fiscal and tax policies should be geared to support regional and local economies, as well as national macroeconomic goals.

State Governments

State governments, building on earlier initiatives, should be increasingly active in developing statewide urban and development strategies similar to those advocated at the federal level. State functions can be coordinated and resources can be targeted to local governments and to private sectors under the partnership concept. States can also contribute important research on particular state needs. Educating and training public personnel to carry out these new responsibilities should be a primary function of state governments.

Local Governments

The implementation and administration of comprehensive national and state urban development strategies falls at the local level, and particularly on local government. Over the last 15 years, local governments have substantially increased their capabilities in these areas, especially in partnerships with the private sector. Two prime requirements for local government in the 1990s will be to establish local planning strategies—both short and long range—to implement new initiatives and to train additional public personnel to carry out the increasingly more complicated functions.

In the past, explicit and comprehensive national urban policy has not been achieved, primarily because it is inherently complex, cutting across the individual endeavors of public entities, private businesses, and individuals in a broad spectrum of our national life. To harmonize and harness these forces is an awesome task. It is not likely that we will see complete success anytime soon. But the experience of the past 25 years—and the urgency of current needs— could point to a new national consensus, lending hope that new efforts will bring it closer to reality.

References

Advisory Commission on Intergovernmental Relations. (1985a). *The condition of American federalism.* Washington, DC: Author.

Advisory Commission on Intergovernmental Relations. (1985b). *Emerging issues in American federalism.* Washington, DC: Government Printing Office.

Apgar, W. C., Jr., & Brown, H. J. (1988). *The state of the nation's housing, 1988.* Cambridge, MA: Harvard University, Joint Center for Housing Studies.

Committee on National Urban Policy, National Research Council. (1982). *Critical issues for national urban policy: A reconnaissance and agenda for further study* (First annual report of the Committee on National Urban Policy). Washington, DC: National Academy Press.

Council on Development Choices in the 1980s. (1981). *The affordable community: Adapting today's communities to tomorrow's needs.* Washington, DC: Urban Land Institute.

Eberhard, J. P. (1988, June 26). Building the city of tomorrow. *Washington Post.*

Evans, D. J., & Robb, C. S. (1985). *To form a more perfect union: The report of the Committee on Federalism and National Purposes, the National Conference on Social Welfare.* Washington, DC: National Conference on Social Welfare.

The Ford Foundation. (1988). The new permanence of poverty. *The Ford Foundation Letter, 19*(2), 1-16.

Goldsmith, W. W., & Jacobs, H. M. (1982). The improbability of urban policy: The case of the United States. *Journal of the American Planning Association, 48*(1), 53-66.

Hammer, P. G. (1984). Technology and urban revitalization. In *Technological trends and the building industry.* Washington, DC: National Research Council, Advisory Board on the Built Environment.

Hanson, R. (Ed.). (1983). *Rethinking urban policy: Urban development in an advanced economy* (Report prepared for the Committee on National Urban Policy, National Research Council). Washington, DC: National Academy Press.

Hanson, R. (1986). *Urbanization and development in the United States: The policy issues.* Washington, DC: National Association of Housing and Redevelopment Officials.

Journal of Housing. (1969). Washington news (prepared by Mary K. Nenno). *Journal of Housing, 26*(3), 117-119.

Journal of Housing. (1970). Washington news (prepared by Mary K. Nenno). *Journal of Housing, 27*(2), 67-70.

Kirlin, J. J., & Marshall, D. R. (1988). Urban governance: The new politics of entrepreneurship. In M. G. H. McGeary & L. E. Lynn, Jr. (Eds.), *Urban change and poverty* (pp. 348-370). Washington, DC: National Academy Press.

Kosterlitz, J. (1986, December 6). Reexamining welfare. *The National Journal,* pp. 2926-2931.

McGeary, M.G.H., & Lynn, L. E., Jr. (Eds.). (1988). *Urban change and poverty* (Report prepared for the Committee on National Urban Policy, National Research Council). Washington, DC: National Academy Press.

Nenno, M. K. (1978). President Carter's 1978 urban policy initiatives: A NAHRO description and analysis. *Journal of Housing, 35*(5), 217-225.

Nenno, M. K. (1982). The president's urban policy report: Findings and observations. *Journal of Housing, 39*(6), 168-172.

Nenno, M. K. (1987a). States respond to changing housing needs. *Journal of State Government, 60*(3), 122-127.

Nenno, M. K. (1987b, March 31). Welfare reform proposals neglect critical housing component. *NAHRO Monitor.*

Nenno, M. K., & Brophy, P. C. (1982). *Housing and local government.* Washington, DC: International City Management Association.

Nenno, M. K., & Colyer, G. S. (1988). *New money and new methods: A catalog of state and local initiatives in housing and community development.* Washington, DC: National Association of Housing and Redevelopment Officials.

Newman, S. J., & Schnare, A. B. (1988, January 5). Housing and welfare: A logical link. *The New York Times.*

Peterson, P. E., & Romm, M. C. (1988). The case for a national welfare standard. *The Brookings Review, 6*(1), 24-32.

Real Estate Research Corporation. (1974). *Future national policies concerning urban redevelopment.* Washington, DC: HUD.

Reyes, L. M., & Waxman, L. D. (1987). *The continuing growth of hunger, homelessness, and poverty in America's cities: A 26-city survey.* Washington, DC: U.S. Conference of Mayors.

Stevens, W. K. (1988, June 22). The welfare consensus. *The New York Times.*

Turner, M. A. (1987). *Housing needs to the year 2000.* Washington, DC: National Association of Housing and Redevelopment Officials.

U.S. Department of Housing and Urban Development. (1978-1988). *The president's national urban policy report.* Washington, DC: HUD. (The report has been published biennially since 1978)

Warren, C. R. (1980). *The states and urban strategies: Executive summary.* Washington, DC: HUD.

Warren, C. R. (Ed.). (1985). *Urban policy in a changing federal system* (Report prepared for the Committee on National Urban Policy, National Research Council). Washington, DC: National Academy Press.

7

Urban Policies and
Urban Impacts After Reagan

William R. Barnes

The years since 1978 have not been an encouraging time for advocates of federal urban policy. From the announcement and subsequent dissipation of the Carter urban policy to the public disclosures of scandals at the Department of Housing and Urban Development, the federal government has not been an important urban policymaker. Similarly the urban impacts of federal actions have not been an important criterion in policy development or program evaluation. The federal government nonetheless has importantly affected the shape and condition of urban America throughout these years.

Regarding urban policy and urban impacts in the Bush administration, continuity with the Reagan era seemed more likely than major change because key determinants were aligned: the interests of electoral constituencies, urban policy ideas, ideology, and the actual effects of relevant federal actions. The continuity and alignment are barriers to urban policy development and to achievement of the more important goal: a domestic policy with an urban consciousness.

Continuity in Policy Ideas

In its first urban policy report, the Reagan administration declared that it sought to "reduce the influence of the federal government in domestic affairs so that other more effective centers of decision-making can flourish" (U.S. Department of Housing and Urban Development, 1982, p. 57). That same report declared that "the foundation for the administration's urban policy is the Economic Recovery Program," the title given to Reagan's overall economic policy effort (p. 1).

The Bush administration began with basically the same commitments. A compilation of Bush campaign statements reveals little by way of goals regarding

AUTHOR'S NOTE: The views expressed here are my own and are not necessarily shared by NLC or its members. I appreciate the helpful comments, on an earlier draft, offered by Michael Pagano and Éva Dömötör and the research assistance provided by Susan Newton and Vidella Hayes.

urban America directly and few program promises that reflected urban concerns. In response to a candidate questionnaire during the campaign, Bush (National League of Cities Institute, 1987, p. 50) declared that "the state of American cities reflects the state of the economy." Furthermore, he said:

> President Reagan's "New Federalism," his effort to shift much of the responsibility for local needs from Washington to our states and cities, has started a process with potentially far reaching results. I'm confident that we are laying a good foundation, not only to return to the proper roles of state and local governments under the constitution, but also to deal effectively with local problems.

If one takes these ideas seriously—and one should—then the combination of the New Federalism and an economic focus on the national rising tide renders urban policy a noncategory. When urban policy is a noncategory, a concern with urban impacts is unlikely to flourish.

The second major legacy that Bush seemed likely to carry forward from the Reagan years was the politics of deficit budgets and deficit policy making. Bush's campaign pledge of "no new taxes" and his apparent desire, at least in 1989, to fulfill that commitment ensured that the federal government would continue to function deep in the red. Reductions in projected defense spending did not alter this situation significantly in the short run.

To date, the major demonstrable effect of these immense federal deficits has not been economic. The major visible result has been to hinder or derail policy making.

Urban programs were particularly hard hit in the budget cutting of the 1980s, and this did not change under Bush. Indeed, in the Reagan years, budget cuts were the primary means by which the New Federalism was implemented. Aside from a few programs, there was never a substantial devolution of functions from the federal government to states or local governments. Instead the federal government simply stopped or decreased doing certain things. The possibility that state and/or local governments would pick up the functions or fill the funding gaps was sufficient for the rhetorical and budgetary purposes of the New Federalism effort. The original Bush budget, offered in February 1989, would have been even more devastating to urban programs than the budget that Reagan left when he departed office.

Change

The Bush administration, however, was not simply, nor only, a continuation of the Reagan administration. There was change as well as continuity. Among these changes were (a) Bush's sense of seriousness about government rather than hostility to it, (b) a few initiatives that emerged from the campaign that clearly were relevant to the urban agenda, and (c) the appointment of a highly visible Secretary of the Department of Housing and Urban Development (HUD).

Bush arrived in the White House with a reputation as a government man and specifically as a man of the federal government. This was a distinct contrast with Reagan, who ran his election campaigns essentially against the federal government, even in 1984 when he already held the White House. It also distinguished Bush from Carter, who similarly ran against inefficiencies and unethical behavior inside the Beltway. This reputation seemed based not on a tendency to public sector activism, but rather on an attitude toward the institutions of government. In 1989 that attitude appeared to be composed of respect for the institutions of government and an emphasis on prudence in policy making. Thus columnist Broder (1989) speculated that the Bush era would be a period of consolidation and that Bush would, in this respect, be like Eisenhower. These attitudes could translate into a problem-solving posture that would be conducive to program development.

In appointing Jack Kemp, representative from Buffalo, New York, to be Secretary of the Department of Housing and Urban Development, Bush not only co-opted a former rival for the Republican nomination for president but also automatically gave HUD and HUD programs more access to public attention than had been true for 8 years under Secretary Samuel Pierce. Kemp immediately declared that a "war on poverty" would be an important focus of his tenure (Shafroth, 1988, p. 1). His avowed priorities were home ownership and affordable housing, economic development through enterprise zones, ending homelessness, resident management and homesteading, fair housing, and drug-free public housing. Kemp had been demonstrably delayed in pursuing his intentions and ambitions by the revelation of various scandals at HUD under the previous administration ("Kemp, Cranston Unveil," 1989, p. 390). Budget and policy constraints, plus the need for Kemp to attend to and resolve the program issues that underlay these scandals, left the extent to which Kemp would be able to carry out new urban policy development in question.

It was speculated that Kemp probably would get (at last) an enterprise zones (EZ) program, a proposal he had advocated for a decade. This program would be useful, depending on the specific incentives that were provided and the targeting that was used. A Bush urban policy with an EZ program as the centerpiece would have more effect on symbolic politics than on urban conditions.

During 1989 the Bush administration also made several initiatives that clearly were relevant to urban policy concerns, including the following:

1. Bush announced in September 1989 a national program related to illegal drugs. Much of the money involved simply rearranged existing programs, and much else was based on assumptions about additional state and local expenditures.

2. He offered a major proposal with regard to clean air. Conflicts within the administration resulted in important shifts in the stringency of the program that Bush originally had announced.

3. Bush followed through on his campaign promise to support child care. Conflict between Bush's interest in a tax program versus the Democratic Congress's interest in an expenditure program reflected some difference in targeting.

4. Toward a "clean bill" for the savings and loan bail-out legislation, the White House opposed efforts to add requirements that savings and loans provide stable funds for low-income housing. Those requirements were written into the laws, and it remains to be seen how effectively the requirements will be enforced.

5. Seeking to be the "education president," Bush convened an education summit with governors. It was unlikely that this summit would result in additional federal monies for education or in any substantial change in the balance of intergovernmental responsibilities.

Some of these initiatives were not the traditional stuff of urban policy. But, for example, *drugs* was rated most frequently as an important community problem in a survey of city officials in December 1988 (Peterson, 1989). A survey of city hall officials showed that child care tops their concerns about the needs of children in cities; low-income housing is first in their concerns about the needs of families (Born, 1989, p. vi). The view of these kinds of programs as nonurban may reflect a lag in the perception of analysts behind the evolving urban policy agenda.

More than a decade of inaction at the federal level may indeed have caused the politics of these issues to crystallize at state and local levels. Those local officials now must act as spokespersons for the needs of their citizens, rather than as directors of institutions who are mainly concerned with programs for which monies flow through their corporation. This is surely an optimistic assessment, but the beginnings of some such transformation occurred during the 1980s. If that transformation continues, local support for a new urban policy agenda could evolve.

Urban Impacts: Constituency and Ideology

The balance of continuities and changes from the Reagan years into the Bush administration provided good reason to expect little interest on the part of the Bush administration in urban policy per se and, consequently, little interest in the issue of the urban impacts of nonurban policies. In addition, there was no reason to expect any substantial change in the general nature of urban effects under the Bush administration, as compared with those during the Reagan years.

Two researchers who searched for "Reagan's real urban policy" and for the effects of "nonurban policies as urban policies" described the direction and nature that probably was expected of urban effects of the activities of the Bush administration. Glickman (1984, p. 476) found that

Reagan's *real* urban policies have remarkably consistent effects. . . . First, changes in the tax laws directly aid firms and taxpayers in growing areas and the suburbs. . . . Second, changes in the composition of the budget reduced social programs that most affect declining areas.

Therefore, the real urban policy rewarded key Reagan constituencies in important ways. Voters in growth areas (such as the South, Southwest and the suburbs) that backed the President have benefitted from the Reagan economic program in two ways: first, because of Reaganomics class bias and second, because their regions have been aided in the interregional battle for federal funds and new jobs. Although rewarding the wealthy was an unabashed goal of the administration, the urban effects were basically incidental and unplanned.

Mills (1987) concluded that the "clearest effect" of the range of programs and policies he analyzed was "that most have favored high income residents." He concluded further that "it seems almost certain that the net effect of government programs has been to induce excessive suburbanization" (p. 568).

One can read these two assessments either in terms of the implicit analyses about political constituencies being served or in terms of program assessments. Either way, one has little reason to expect substantial positive impacts, for example, around redistributive concerns or around the central-city revitalization concerns of President Carter's urban policy. Policies, programs, and indirect effects seem well shaped to constituencies; outcomes will benefit middle to upper income groups, the suburbs, and the Sunbelt.

The Reagan-Bush orientation also can be understood as an effort to ensure or establish the smooth workings of economic and political marketplaces in metropolitan areas. This view pretty much takes them at their word. On the one hand, they preferred to leave decisions to the private market, which they saw as autonomous and smoothly operating only if it could be protected from public tinkering and intervention. They believed in the beneficial functioning of the "invisible hand." In addition, they wanted to remove federal intervention from a similarly autonomous mechanism at the local level in which people "vote with their feet" to determine the mode of governance and the mix of services provided by state and local governments. They believed in the beneficial functioning of "the invisible foot."[1] But of course the functioning of the invisible hand and the invisible foot precisely benefited the constituencies of the Reagan-Bush coalition. Thus there was a happy confluence of political interests with the allegedly natural functioning of these economic and political systems.

When these systems failed to function in substantial ways or they began to function to the benefit of other interests to the detriment of the coalition's interests, these ideological beliefs were shelved and the Bush administration responded in various ways, including an urban policy. That policy spoke to the concerns of these interests, not necessarily at all to the concerns of central cities or the poor or older cities that were the basis of Carter's urban policy coalition.

There was little political basis for expecting this alignment to change. The situation was summed up in the *National Journal* as "more problems, less clout" for big cities (Kirschten, 1989, p. 2026). A similar report in *Governing* led with a headline declaring that "a shrinking urban bloc in Congress plays defense" (Ehrenhalt, 1989, p. 21). These analyses identified urban policy with the interests of older central cities. They overlooked the changing nature of urban America, which includes suburbs that contain more population and more jobs than their central cities and new kinds of urban communities in Austin, Phoenix, and San Jose, as well as the older urban centers such as Detroit, Cleveland, and Newark. The new "technocities" and "technoburbs" (Fishman, 1987, p. 184)

Table 7.1 1988 Presidential Voting Patterns by Selected Cities

City[a]	% 1988 Total Vote[b]		% Black and Hispanic[c]	
	Republican	Democrat	Black	Hispanic
New York, NY	32.8	66.2	25.23	19.88
Chicago, IL	29.8	69.2	39.83	14.05
Philadelphia, PA	32.5	66.6	37.84	3.77
Baltimore, MD	25.4	73.5	54.80	.97
San Francisco, CA	26.1	72.8	12.73	12.28
Washington, DC	14.3	82.6	70.32	2.77
Boston, MA	33.2	65.2	22.42	6.41
New Orleans, LA	35.2	63.6	41.12	1.34
Denver, CO	37.1	60.7	12.03	18.76
St. Louis, MO	27.0	72.5	45.55	1.22
United States	53.4	45.6	12.2[d]	6.45

NOTE: a. Large cities for which vote totals are reported by municipal jurisdiction.
b. Data furnished by Alice McGillivary, Election Research Center.
c. Data furnished by U.S. Bureau of the Census.
d. Unpublished data from U.S. Bureau of the Census.

turned their backs on central cities locally and at the state level just as the federal government turned away nationally.

This alignment of constituencies, ideology, and policy effects also reflected the pattern of Democratic and Republican voting in the presidential election. In 1988 the total presidential vote split 53.4% for the Republican ticket and 45.6% for the Democratic. The central areas that contain the 32 largest cities, in contrast, reversed that split: 45.5% Republican and 53.5% Democratic. (These data probably understate the Democratic preference in the central cities because the votes are mostly reported by county and thus do include some suburbs.) Table 7.1 shows George Bush's lack of electoral obligation to some of the nation's largest central cities and their disproportionately minority populations. Vote counts do not necessarily mirror what Ferguson and Rogers (1981, p. 6) call the "hidden" election involving conflict and coalitions among "pivotal interest groups." The patterns of such groups, as well as the political constellations of the Congress, may be different and even countervailing but are beyond my scope and capacity here.[2]

The loosening grip of older central cities on our intellectual and political frameworks may also be seen as both cause and effect of the emphasis on federal programs that help people rather than places. This distinction was crystallized in the urban policy report of the President's Commission for a National Agenda for the Eighties (1980). Dearborn (1989, p. A21) argued that as the federal government turns its back on large cities but problems accumulate and demand attention, "urban problems ultimately are likely to land on the state's doorsteps." He predicted that "additional assistance may come in the form of new and enlarged state programs directed at people with problems not at city governments" and that this will "mean an expansion of state governments and diminishment of city governments."[3]

Urban historians[4] are reinterpreting the development of metropolitan America with a focus on the development of suburbs and the suburban ideal. In this view, it is precisely the extraction of the suburb and its goals from the complex web of the urban system that has created the class- and race-segregated metropolitan areas of 20th-century America. As the black middle class moves out from central cities, race is meshed with class and with images of physical decay in the central area to form the antithesis of the suburban ideal. A similar phenomenon is occurring with regard to Hispanics in some cities, especially in the Southwest.

Those suburbs and the new cities created in the image of the suburban ideal were the main constituency of the Reagan-Bush coalition. Their lack of interest in urban policy was essential to their definition. Here is President Bush on August 8, 1989, in remarks to the National Urban League: "In many respects—let's face it—urban America offers a bleak picture—an inner city in crisis. And there is too much crime, too much crack. Too many dropouts, too much despair, too little economic, too little advancement—and the bottom line, too little hope" (The White House, 1989, p. 1). Here is urban America as poor, black, isolated, without power. In this formulation, urban policy is about what George Bush and his constituency were not.

Next Steps

Although virtually all federal actions may be relevant, not all federal actions are equally significant in terms of their potential impact on urban areas, residents, and governments. The Bush administration faced the following important issues —in addition to the budget, tax legislation, and the items already mentioned— that illuminated and defined its intentions in this regard:

1. Federal housing programs were in disarray. Congressional timetables called for major legislation in 1990. Meanwhile, low-income housing needs piled up, and poverty and near-poverty were all too quickly transformed into homelessness.
2. In 1991 authorization for the basic federal surface transportation programs expired. The administration was called upon to offer proposals for reauthorization.
3. Law required that the president submit his first national urban policy report in 1990.
4. At this writing, the Congress and/or the Environmental Protection Agency seems likely to suggest new federal roles on solid waste.
5. Even with the repeal of the Section 89 regulations about health care benefits and Medicare catastrophic insurance, some follow-up reformulation of the health care costs issue seemed likely.

The Search for the Real Urban

The search for the real urban policy of Reagan or Bush revealed the ultimate fruitlessness of arbitrary distinctions between urban and nonurban federal actions. A recitation of the categories used by Glickman (1984) and Mills (1987) illustrate this point. Mills noted that his selection of programs to analyze is "inevitably somewhat arbitrary." His selection included the following: government procurement, the U.S. system of local government, national transportation policies, housing programs, urban development programs, antipoverty programs, capital taxation, and the 1986 tax reform act (p. 562).

Glickman suggested that the "real urban policy" of any administration consists of four parts. The first three are what he called "explicit" or "nominal" urban policy: place-targeted urban programs in functional areas, the federal system of intergovernmental grants, and people-oriented social programs, particularly income transfers that serve urbanites. "The fourth element consists of the urban effects of economic and other non-urban policies." The fourth element "dominates the others," and "any comprehensive analyses of the real urban policy must centrally consider economic programs as well as nominal urban policies" (pp. 471-472). Wolman, in a separate effort to assess the impacts of the Reagan urban policy, stated: "It is likely that in aggregate the effects of implicit urban policy on urban areas far outweigh those of explicit urban policy" (1986, p. 317).

Thus the interest here is in the entire array of domestic policies and indeed other policies such as trade. "Urban policy" may be a useful basket for carrying a jumble of program initiatives. More importantly, however, the goal is probably more aptly stated as a domestic policy with an urban consciousness.

The politics of disengagement and polarization described earlier work against such an approach and go far beyond the capacities of policy analysis to overcome. Nonetheless, policy analysis has a chore in the eventual bridging of the disengagement. That chore is to create frameworks for understanding the connections between ghetto and suburb and between old and new cities. In short, the task—which needs to be undertaken politically, as well as analytically—is to reconstitute the whole urban community that encompasses these dangerously and increasingly distant parts.

This chore shifts the focus from urban policy to urban reality. Markusen (1980, p. 103) noted critically that the urban impact analysis "heralds a historical shift of presumed culpability for urban problems from the private sector toward the public sector." Her comment foreshadowed Ronald Reagan saying that government is part of the problem rather than part of the solution. There is, of course, enough "culpability" to go around: hand and foot, visible and invisible. We need usable analytic and political frameworks that encompass both federalism and capitalism, both public and private sector factors at work in creating and sustaining urban problems.

The short-run outlook is that the nation's communities will continue to be affected by federal actions that are and are not labeled urban and that most frequently have not been examined in terms of their urban impacts. Some of these effects will be substantial and significant; some will be negative. The current pattern of urban impacts will likely persist. Although some new federal programs may be undertaken and the urban policy agenda may broaden at the local level, it is hard to find the political or policy source for a full-fledged federal urban policy in the early 1990s.

Notes

1. This provocative and insightful phrase was used by Burton (1970, p. 146) in testimony to a Congressional committee. The context for the remark was a critique of the Tiebout model for understanding public finance.

2. Alice McGillivary, from the Election Research Center, provided these data upon my request. I very much appreciate her assistance.

3. Dearborn (1989) predicts that "urban fiscal crises are only a recession away from happening."

4. See especially the studies by Fishman (1987) and Stilgoe (1988). See also Jackson (1985) and Keating (1988).

References

Born, C. E. (1989). *Our future and our only hope: A survey of city halls regarding children and families.* Washington, DC: National League of Cities.

Broder, D. S. (1989, September 13). Eisenhower lives. *Washington Post,* p. A25.

Burton, R. N. (1970). *The metropolitan state: A prescription for the urban crisis and the preservation of polycentrism in metropolitan society.* U.S. Congress Subcommittee on Urban Affairs of the Joint Economic Committee, 91st Congress, 2nd session, October 13-15.

Dearborn, P. M. (1989, May 10). Hard times ahead for cities. *Washington Post,* p. A21.

Ehrenhalt, A. (1989, July). As interest in its agenda wanes, a shrinking urban bloc in Congress plays defense. *Governing, 10,* 21-25.

Ferguson, T., & Rogers, J. (Eds.). (1981). *The hidden election: Politics and economies in the 1980 presidential campaign.* New York: Random House.

Fishman, R. (1987). *Bourgeois Utopias: The rise and fall of suburbia.* New York: Basic Books.

Glickman, N. J. (1984, Autumn). Economic policy and the cities. *Journal of the American Planning Association, 50,* 471-569.

Jackson, K. T. (1985). *Crabgrass frontier: The suburbanization of the United States.* New York: Oxford University Press.

Keating, A. D. (1988). *Building Chicago: Suburban developers and the creation of a divided metropolis.* Columbus: Ohio State University Press.

Kemp, Cranston unveil wide-ranging proposals for reform, new programs. (1989, October). *Housing and Development Reporter, 17,* 390.

Kirschten, D. (1989, August). More problems, less clout. *National Journal, 21,* 2026-2030.

Markusen, A. R. (1980). Urban impact analysis: A critical forecast. In N. J. Glickman (Ed.), *The urban impacts of federal policies* (pp. 103-118). Baltimore: Johns Hopkins University Press.

Mills, E. S. (1987). Non-urban policies as urban policies. *Urban Studies, 24,* 561-569.

National League of Cities Institute. (1987). *Election '88: Presidential candidate questionnaire responses.* Washington, DC: Author.

Peterson, D. (1989, January 16). Municipal officials worry, but are optimistic on future. *Nation's Cities Weekly, 12,* 7-10.

President's Commission for a National Agenda for the Eighties. (1980). *Urban America in the eighties: Perspectives and prospects.* Washington, DC: Government Printing Office.

Shafroth, F. (1988, December 19). Kemp tapped to lead Bush housing initiative. *Nation's Cities Weekly, 11,* 1, 11.

Stilgoe, J. R. (1988). *Borderland: Origins of the American suburb: 1820-1939.* New Haven, CT: Yale University Press.

U.S. Department of Housing and Urban Development. (1982). *The president's national urban policy report.* Washington, DC: Government Printing Office.

The White House. (1989, August 8). *Remarks by the President to the National Urban League Conference.* Washington, DC: Office of the Press Secretary.

Wolman, H. (1986, March). The Reagan urban policy and its impacts. *Urban Affairs Quarterly, 21,* 311-335.

8

National Urban Policy and the Local State

PARADOXES OF

MEANING, ACTION, AND CONSEQUENCES

Robert Warren

Few projects have been pursued with as little success as has the formulation of a national urban policy for the United States. Efforts to establish one can be traced back at least to the 1930s, and they continue today. During this period the terms *urban* and *urban problems* have been redefined continually with wide variations and contradictions in their logic and programmatic content. The result has been the mystification of both the ends and the means of a national urban policy.

The Bush administration continued a strategy adopted during the Reagan presidency of conducting a de facto urban policy while a formal one was eschewed. Initiatives under Reagan that had urban implications, according to Hicks (1983, p. 360), were derived from a "general commitment to a program of economic recovery and revitalization," rather than urban policy. Similarly Wolman (1986, p. 313) found "a relatively minor and residual federal role toward urban areas" during the 1980s. The actual urban impacts of these de facto policies, however, have been significant and intended—the subordination of spatially organized communities and the welfare of their residents to market processes to further the goal of national economic efficiency (Clark, 1983, p. 156).

A number of proposals have been made during the last decade, largely by academics, calling for modifications in or alternatives to existing policies. Scholars concerned with ameliorating or opposing national urban-related policies differ considerably among themselves in programmatic terms and, more basically, about the extent that the public sector can or should intervene in market processes in the interests of localities and their residents. Even with their differences, academics and the Reagan and Bush administrations all had worked from a common understanding that a postindustrial restructuring of the world economy was occurring that had significant impacts on local communities and their governments (Castells & Henderson, 1987; Fainstein & Fainstein, 1987; Friedland, 1983; Hanson, 1983; Hicks, 1983; Savas, 1983; Tabb, 1982; Yago, 1983).

Innovations in telecommunications and information technology and the capital deepening of production processes have made it possible for firms to organize on a multinational scale and to pursue profit-maximizing strategies in world markets. Two critical dimensions of this process are (a) the ability of firms to decentralize production and administrative service functions spatially on a global scale and (b) a new international division of labor. In combination, these factors have facilitated an increasing centralization and mobility of capital and undermined the ability of labor to bargain effectively over wage issues and conditions of employment (Harvey, 1987, p. 261). The increasing spatial mobility of capital and reductions in a number of urban-related governmental programs have made cities increasingly dependent on their local economies for public revenues. Although different language is used by the administration and critics, one result can be characterized as the commodification of cities.

National policies have facilitated the creation of a place-market in which communities, like commodities, are "bought, consumed, and discarded when no longer exploitable" (Hill, 1983, p. 18). American communities compete with one another and with foreign sites to maintain or gain firms and the jobs and tax revenues they are presumed to provide. Cities attempt to create a favorable business environment through low taxes and tax concessions, allocative rather than redistributive programs, and limited public regulation.

The more serious the local revenue problem is, the greater the pressure to invest public resources to keep businesses and attract new enterprises (Harvey, 1987, p. 264). The resulting "bitter struggles" among communities to attract firms are fueled by local leaders and labor who believe that failure to do so will mean the loss of socioeconomic viability for their communities (Cox & Mair, 1988; Hudson & Sadler, 1986, p. 188). An unintended consequence for workers, who support this competition in defense of communities and jobs, has been the reinforcement of capital hegemony by legitimizing both wage labor and the free withdrawal or relocation of employment by firms (Hudson & Sadler, 1986, p. 189).

Whether and to what degree a change should be made in the current national support of this restructuring of the space economy has been a focus of debate for longer than a decade. Some scholars suggest policies to reduce the vulnerability of cities to capital mobility but do so only within a logic that reflects an uncritical acceptance of the neoclassical and technological determinism assumptions that underlay Reagan's New Federalism. Others have proposed federal programs to reestablish an earlier liberal-democratic national commitment to cities that deal more with symptoms than causes. Even when critics have focused on the inherent destructiveness of pursuing urban policies that subordinate use to exchange values and social to economic space, some of their solutions are partial or problematic. As a result, the mystification of meaning and paradoxes of federal action that have characterized national urban policy debates and programs, particularly since World War II, have increased, rather than been reduced.

If the meanings of competing urban policies and the values that they are designed to further are to be made explicit and realistically assessed, the framework of discourse must be reformulated. An authentic debate about urban policies cannot occur without a clarification of the meaning of *urban*, of the value to be placed on the integrity of spatially organized communities, and of the relationship between capital and local states and the degree to which both are assumed to be subordinate to or autonomous from the national state. As indicated by these questions, I argue that national urban policy must be built from the "realities of place and community" situated within both the apparatus of the national state and the evolving world political economy characterized by "flexible accumulation" (Harvey, 1987, p. 281).

In the discussion that follows, I consider factors that have limited the success of resolving contradictions of means and ends in the formulation and application of urban policy, review the current de facto national urban policy and proposed alternatives, and explore the dimensions of an urban policy with social and economic justice as goals that would enhance the legitimacy and viability of urban places in relation to both capital and national state.

A Policy in Search of Form and Meaning

Since the 1960s, frequent calls have been made for a national urban policy. Not until the Carter presidency, however, was an explicit, though abortive, effort made to formulate one (Glickman, 1983, p. 83). This lack of a formal policy has not prevented waves of national programs intended to solve urban problems and many others that have had significant impacts on urban places and residents. Yet, after almost half a century, the central problems that have been addressed by established national programs—poverty, racial segregation, political marginality, adequate housing, environmental degradation, and uneven intra- and intermetropolitan development—are still endemic in the United States.

Problem Solving by Abandonment

The failure to resolve these problems is not because of the absence of an explicit urban policy or programs. It is more a reflection of the contradictions, ambiguities, and partial responses to basic questions of the relationship among capital, labor, and the state and social equity that have characterized a wide range of urban-related federal programs. These anomalies have been perpetuated by the failure in mainstream discourse to move away from the assumption that the local state's function is to carry out strategies of "crisis avoidance in order to sustain the continued reproduction of capitalist relations" (Clark & Dear, 1984, p. 136).

During the last 45 years, wide shifts in the content of federal programs have occurred as the definition of the urban problem being addressed has changed. Compare, for example, national responses to urban renewal and infra-structure replacement and expansion needs in the 1950s, racism and economic marginality in central cities in the 1960s, municipal fiscal stress in the 1970s, uneven urban and regional economic development in the 1980s, and, currently, drugs.

These transformations in meaning and content have not occurred because the prior problems were solved. Rather, one "urban crisis" has been displaced on the national agenda by another, ostensibly because a new regime with different ideological commitments or spatial electoral base gains office, civic insurrection explodes and subsides, or as public opinion is manipulated to accept a "war on something" as the appropriate national priority in dealing with the problems of urban places and residents.

The work of many scholars and the media facilitates these solutionless shifts and supports their legitimization. For example, referring to the publication record of a major urban research institution, Wade (1989, p. 37) commented that the tendency is to chase "today's headlines" and produce "shelves full of marginally useful information and inconclusive judgments." Embedding issues such as structural poverty and racism as components of urban policy allows inadequate solutions to be adopted and then abandoned with political impunity as the next socially constructed urban crisis appears (Gittell, 1985, p. 13; Smith, 1988; Weiher, 1989).

Urban, Place, and Local State

The term *urban* has had no consistent meaning. Actual or proposed national programs with urban targets, from Model Cities to enterprise zones, have been, at various times, directed toward selected neighborhoods, selected cities, cities in general, metropolitan areas, and subsets of their population. These swings between emphasis on place and people and their delimiting eligibility criteria have frequently meant that some areas and populations that should have benefitted in terms of policy goals were left out and that others that should not have been were included, particularly through capitalization (Edel, 1980). In addition, policymakers have assumed, without an adequate rationale, that place- and people-oriented urban policies are mutually exclusive (Glickman, 1983).

Federal programs directed to urban places commonly have delivered grant and revenue-sharing dollars and imposed conditions and regulations on a variety of local public entities—counties, special districts and authorities, and municipalities. However, most major programs and much of the debate that has been generated about federal policies have defined *urban* in spatial and organ-izational terms as communities, cities, or central cities, with little attention to operational or theoretical meanings.

Quite recently in the United States two important clarifying terms—*place* and *local state*—have gained some currency in discussions. Until the infusion of these concepts, the necessary interrelated issues of the integrity of spatial communities as socioeconomic and political entities within larger economic and governance systems had been absent, peripheral, or partially treated in the main-stream discussions of urban policy.

Clark (1983, p. 60) observed that the issue of what authority local states do or should have vis-à-vis state and national governments is "rarely considered directly," and the meaning of local autonomy has remained "incredibly opaque." One reason for this neglect has been a contradiction between an unquestioned acceptance by both liberal and conservative theorists that the "American ideal of decentralized democracy" existed and a lack of interest in analysis of the actual initiative and immunity available to localities (Clark & Dear, 1984, p. 131).

The first concept, *place,* requires dealing with the implication that stable local communities are necessary as the spatial manifestation of social arrange-ments that provide meaning and identity to residents through their history and interaction with the macro order (Agnew, 1987, p. 230).

Use of the second concept, *local state,* moves the discussion beyond an ambiguous vision of free-standing municipalities and cities connected to state and national governments through benign models of intergovernmental relation-ships. Although definitional differences with the term *local state* exist (Broadbent, 1977; Cockburn, 1977; Saunders, 1979; Johnston, 1982; Kirby, 1982), use of the concept requires recognition that the public sector, at whatever scale, cannot be understood apart from capital and that locally organized public authority cannot be understood outside its role as part of the apparatus of the national state (Clark & Dear, 1984, p. 132).

In this analysis I adopt the position that the national state exercises relative autonomy from capital and that, in turn, local states are "semiautonomous concentrations of authority" in relationship to both capital and the national state, authority that can be used in the pursuit of a variety of interests (Gurr & King, 1987, p. 50). The degree of autonomy varies at any given time with how legal authority and politico-economic resources are distributed among place, national state, and capital.

Federal Effects Without an Urban Policy

Consistencies have occurred in the consequences of federal policies despite the changing goals, content, and spatial focus of urban-related programs. These consistencies, in large part, have been the result of the urban effects of national programs that are not explicitly labeled *urban policy.* A tolerance for urban effects without a national urban policy has made it easier to ignore the contradic-tions between programmatic goals and achievements, particularly when class and race are issues. A range of writers, spanning several decades, have found both that no adequately articulated national urban policy exists and that federal programs have consistently influenced the spatial organization of people and activities and whether the basic needs of urban populations are being met. As Hill (1983, p. 6) argued, the "recognition of the latent as well as manifest urban effects of government policy is a necessary element of urban theory."

In assessing the prevailing myth that the massive post-World War II move-ment of population to the suburbs was the result of free market choices reflecting the locational preferences of individuals, Jackson (1985) found that spatial biases in a number of federal programs encouraged "scattered development in the open countryside" and denied federal funds to individuals and local governments that would have aided the revitalization of central cities: "Suburbanization was not an historical inevitability created by geography, technology, and culture, but rather the product of government policies. In effect, the social costs of low-density living have been paid by the general taxpayer rather than only by suburban residents" (p. 293).

Nearly 20 years earlier, Dyckman (1966, p. 36) pointed out that "market choices" of economically able households that move to the suburbs are strongly influenced by federal agencies that are able to alter market terms relating to the availability of credit, interest rates, and capital gains taxes even though they are not guided by a national urban policy. Similarly, in 1970, Moynihan (1970, p. 6) concluded that although no urban policy existed, there was "hardly a department or agency of the national government whose program does not have important consequences for the life of cities and the people in them."

More recently, Markusen (1980, p. 104) characterized the adoption of the Urban Impact Analysis process by the Carter administration as an acknowledgement that the urban consequences of present, as well as past, national policies may be "severe, counterproductive, unintended and/or inscrutable, even to its architects." Wolman (1986, p. 317), in reviewing the Reagan record, also found that "in aggregate the effects of implicit urban policy on urban areas far outweigh those of explicit urban policy." After looking at the effects of these implicit policies, Logan (1983, p. 88) asserted that they have reinforced market processes that lead to disparities in the life chances of different individuals and communities by spatially segregating races and classes and clustering jobs in some areas but not others.

The Deconstruction of Place

One positive aspect of the Reagan administration's willingness to trade off the social and governing integrity of communities to achieve national economic goals was that it forced those wishing to contest such policies to consider the current urban crisis regarding structural and class questions, as well as to confront the issues of how place is to be valued and what role place-centered local states should play in governance.

The deconstruction of place as a stable node of socioeconomic and political functions through national policies during the 1980s was grounded in neoclassical theory and technological determinism. One of the first systematic statements of this position came in 1980 from Carter's President's Commission for a National Agenda for the Eighties (1980). The commission found cities to be obsolete as mechanisms by which the postindustrial society can be maintained and advanced. It warned that attempts to restore declining industrial cities to "the influential positions that they have held throughout the industrial era" would not only fail but also have negative effects on efforts to revitalize the national economy. More generally, the commission (p. 4) stated:

> The nation can no longer assume that cities will perform the full range of their traditional functions for the larger society. They are no longer the most desirable settings for living, working, or producing. They should be allowed to transform into more specialized service and consumption centers within larger urban economic systems.

These assumptions were reflected fully by the Reagan administration. From the start, its policies were premised on the belief that national economic development goals could be met only if "urban areas and the economic functions they perform" were transformed to meet the requirements of "powerful and deep rooted structural changes in the national and international economy" (Hanson, 1983, p. 1). Hicks (1983, p. 362) observed that within the policy framework of the Reagan regime, cities were "less usefully understood as places than as environments conducive to the flourishing of certain [economic] activities, and even less conducive to the nurturance of others."

Savas (1983, p. 450) carried the matter to its logical conclusion: Decline is an "age-old process to which all cities are subject." Consequently it is the fate of cities, however great at one time, to become "a pile of archaeological artifacts" when they lose their economic role. In the meantime, national policy should provide enough fiscal assistance for those "left behind" so that the resistance of declining communities to their economic transformation can be minimized (Hanson, 1983, p. 171).

Proponents of current federal policy assume, with little justification beyond appeals to market efficiency, that communities and their residents cannot legitimately oppose the effects of programs of the national state. Thus Hicks (1983, p. 361) was able to state that market-forced shifts in people, jobs, capital, and income out of older cities and even entire regions should not be opposed because they raise the wealth of the entire nation.

The function of the national and local states from this perspective is to facilitate, rather than constrain, the mobility of labor and capital in the interests of the efficient organization of the space economy. Consequently a critical element of Reagan's policy was to prevent local states from contributing to a spatial mismatch between jobs and people by using their authority and resources to "anchor" unemployed or inefficiently employed workers (Kasarda, 1983, p. 23). One intent of cuts in national welfare and urban programs has been to increase the negative incentives for workers to move to areas with labor needs by making it difficult for cities to provide services or jobs that would impede the relocation of workers who have become redundant in the local economy (Clark, 1983; Friedland, 1983; Yago, 1983).

In cities, then, economies, populations, and public expenditure patterns must be adapted to accommodate the spatial decisions of firms. The resulting social costs at the local level are justified by the aggregate economic growth on a macroscale. The devaluing of localities as mechanisms for collective decision making obviously follows from this logic. Hicks's (1983, p. 161) vision of a successful national economic policy included "new larger scale urban entities that have enhanced economic, if ever less political integrity."

At the core of the Reagan policy was the assumption that the United States is composed of clusters of labor, maintained and reproduced through the production of social consumption goods by local states. The location and duration of these *places* are determined by decisions of capital. The movement of capital, in turn, is dictated by the profit-maximizing opportunities and competitive market pressures on firms that are created by the dominant technology at any given time. In a system in which the public goal is to facilitate private investment decisions regardless of their effects on cities, the community disappears as an entity in national policy (Barnekov, Rich, & Warren, 1981; Logan, 1983, p. 75).

Policy Analysis and Solutions From the Mainstream

The absence of an explicit national urban policy for 12 years obviously has not prevented a variety of scholars from commenting about the effects of federal programs and, in many cases, to offer alternatives. During the 1980s, influential writers with neoclassical, public choice and liberal-democratic perspectives adopted all or much of the Reagan framework and, even when they sought to change the impacts of capital mobility, proposed policies that would reinforce the hegemony of capital and the national state at the community level.

Rethinking Urban Policy, the work of the Committee on National Urban Policy of the National Research Council and edited by Hanson (1983), provided an academic rationale for the policies of the Reagan administration. The committee assumed that urban policy "should understand and reinforce the market" because the public sector has only marginal influence over market forces (Hanson, 1983, p. 172). A priority was given to encouraging the flow of capital and labor to growth sectors and promoting the new urban infrastructure systems necessary to serve a restructured space economy.

The committee recognized that significantly uneven urban development can reduce economic efficiency and generate political opposition to change (Hanson, 1983, p. 4). Consequently the adjustment problems of "the people and places left behind" should be addressed, but in ways that do not "resist or offset" national macroeconomic policy. To achieve this balance, the "transition" of declining places should be facilitated by using public jobs to offset structural unemployment and federal and state fiscal transfers to maintain "a minimum level of public services and facilities" with local tax rates near the national average. Finally all welfare programs would be shifted to the federal level.

Beyond the "natural" dominance of the market, the influence of local private sector interests in public decision making, particularly in lagging communities, would be reinforced through federal and state actions. When local leadership is weak, state government would intervene to induce business leaders to join public officials in local development efforts (Hanson, 1983, p. 10). Private influence would be institutionalized in municipal infrastructure and development projects through the requirement of "substantial evidence of public-private cooperation" as a condition of localities receiving funds from a proposed national urban economic development bank, which would centralize federal capital leverage programs (Hanson, 1983, p. 176).

In *City Limits* Peterson (1981) considered the influence of capital mobility at the local level from a perspective that is closer to Tiebout's (1956) public-choice model than a world-system framework. Cities compete to maximize their economic position. Their tax rates must be high enough to provide public goods and services at a level that will retain mobile citizens and firms and attract new capital but low enough to remove incentive for moving to other communities.

In this place-market environment, rational elected officials must adopt *developmental* policies that "enhance the economic position of the city" or *allocative* policies that provide necessary public goods and services without affecting the attractiveness of the community to the average taxpayer. Powerful disincentives exist for a third option—*redistributive* policies—that negatively affects the local economy by driving out better-off residents and attracting the poor from less generous communities.

According to Peterson, policy should be designed to insulate communities from intercity competition that distorts local politics. A national revenue-sharing program would be used to provide fiscal stability by equalizing per capita resources among local governments and requiring that they meet federally mandated minimum standards of urban services. Welfare programs would be transferred to the federal government. City access to capital markets would be insured by federal guarantees of all local indebtedness.

Peterson (1981) believes that these policies would revitalize local politics so that redistribution could become a legitimate issue and "a new range of local policies" would become possible (p. 222). Yet they also would create substantial federal control over municipalities. In exchange for revenue sharing and the elimination of categorical grants, localities would be required to meet federally determined minimum standards in their production of local public goods and services and be fiscally penalized if they improved their tax base by a proportionate withdrawal of federal funds.

The discipline of the private market over local access to capital for infrastructure and development projects would be replaced with federal control. This control would create a relationship (as would the Committee on National Urban Policy's suggested national urban economic development bank) between the national government and local states with leverage similar to the extraordinary control the World Bank and International Monetary Fund exercise over Third World countries in the interests of capital.

Further, no federal constraints are placed on capital mobility. Although communities with declining economies will be able to maintain minimum levels of service, they and their residents still will absorb the social costs of the spatial restructuring that is occurring. This idea is particularly relevant because Peterson (1981) offered a caveat about the actual benefits of reintroducing redistribution issues into city politics. He commented that even if all of his policy proposals were adopted, they would not produce an "egalitarian utopia" wherein "the needs of low-income groups are carefully met by liberally minded public officials" (p. 222).

Kantor, in his book *The Dependent City* (1988), attempted to ameliorate the social costs for communities "losing" in the competition for capital. He found that the urban crisis is caused by "severe urban economic rivalry to induce capital investment," which has imposed a highly dependent position on cities within the market order. Citizen control over community development decisions is limited and biased in favor of "many wasteful public policies that enhance the privileged position of business and other revenue providers to the neglect of others" (p. 398). Kantor's solution is a "national policy for the liberation of cities" based on a revitalization of the historical American liberal-democratic system. His policy goal is to "more equitably regulate the social costs of urban development than the market does and ensure that 'loser' communities are compensated and, if possible, recycled in the process of economic change" (p. 406). As in the two cases above, federal fiscal transfers would be used to reduce city reliance on local revenue sources, and welfare programs would be shifted to the national level.

The effect of his proposals, however, appears to legitimize a market-driven spatial restructuring of the nation by using public authority to internalize the social costs of capital mobility, rather than by having public intervention in the market. Kantor believes that a competitive position for the country in world markets requires "considerable freedom of capital location" even if unequal development is the result. To ensure continued capital mobility, the costs of compensating declining places would be absorbed not only by the "businesses that precipitate them" but also by the "winner communities" and even the general public.

An Alternative Urban Policy

The emphasis on capital mobility and the dependency of local economies on private sector decisions, which characterized urban policy dialogue in most of the 1980s, virtually disappeared in the first wave of post-Reagan assessments of the urban condition. Four of the initial articles published in 1989 reflected a fragmentation in mainstream analysis. Brintall (1989) reported that the urban crisis was over and that the need for a national urban policy was largely dissipated. Shalala and Vitullo-Martin (1989) agreed that the urban crisis was over for the most part but then called for a range of national urban programs associated with pre-Reagan Democratic administrations. Two other scholars, Waste (1989) and Weiher (1989), found the urban crisis not ended but exacerbated because of increasing inner-city poverty. However, they provided no plausible policy solutions.

These more recent discussions move us no closer to demystifying the issue of national urban policy and allowing basic questions of ends and means to be confronted than did Hanson, Peterson, and Kantor. In the aggregate, mainstream analysis gives creditability to Hudson and Sadler's (1986, p. 183) argument that urban policy cannot be addressed effectively without radically redefining the problem and the solution. After reviewing the experience of a number of industrialized countries, including the United States, Glickman (1983, p. 318) came to the similar conclusion that "urban policy has been wrong-headed—it has attacked the symptoms of the urban malaise rather than root causes."

Ends and Means

The Reagan administration rationalized the imposition of extraordinary social costs on places and people by asserting that such spatial restructuring was necessary to achieve greater efficiency in the national economy. Clark (1983, p. 165) characterized this position as one of "impossible arrogance" because it implies that an empirical case can be made for the ability of the "free" market to achieve efficient solutions without providing the evidence for and denying the validity of alternative visions of community and society.

Molotch and Logan (1985, p. 163) argued that state constraints on capital mobility on behalf of use values imply little about the effect on the macro-economy because the locational outcomes and urban agglomerations that result from the decisions of mobile capital are no more "inherently connected to efficiency" than to "social justice." Glickman (1983, p. 308) noted that the cost-benefit analysis used to support the efficiency claimed for profit-maximizing spatial decisions of capital fails to account for the "social costs borne by both people and places." Similar points are made by Friedland (1983, pp. 47-48), Tabb (1982), and Yago (1983, p. 118). In responding to the assumption of neo-classical ideology that the blame for urban problems rests with undue governmental intervention, Markusen (1980, p. 115) stated that there is better evidence of a causal relationship among social malfunctions and chaos of the market, highly concentrated economic power, and the indifference of capital toward externalities created in the production process. Similarly Clark (1983, pp. 161-165) argued that economic efficiency cannot be proven superior to spatial equity by appealing to economic models. To do so confuses ends and means.

Selecting a value or criterion to guide public policy is a political decision (Tabb, 1982, p. 125). The means to implement that value must be consistent with it. Social justice is an alternative to capital accumulation adopted in *this* discussion as the goal of urban policy. If urban policymakers are to pursue this value effectively, it must also be an overall goal of the national state. Clark and Dear (1984, p. 191) concluded that the "achievement of social justice is actually the *only* true rationale for the state." To posit this end, it must be assumed that there is a sphere of relative autonomy in state-capital relationships. The implications of this assumption for urban policy are clear. As Tabb (1982, p. 6) put it, the "social control of investment in accordance with human needs is the prerequisite of a satisfactory urban policy" and must replace state programs that sanction private profit maximization "without regard for the social cost or the social needs." In more programmatic terms, Logan (1983, p. 88) argued that national urban policy should reverse, rather than promote, the spatial segregation of races and social classes, the concentration of jobs in certain areas, and the inequalities in life chances among communities.

Operationalizing Place and Local State

The status and the function of the local state present a complex set of theoretical and operational problems. The integrity of community, in and of itself, and the relationships of places and their residents to capital, the national state, and the goal of social justice must be sorted out if an urban policy is to have coherence.

There has been increasing recognition of a need to situate place within the larger economic and political systems of which it is a part so that an understanding can be reached of the "reciprocal relations between the process through which localities and places acquire socially endowed meanings, with how these meanings influence what people do and with how what they do influences how they interpret the space around them in terms of meaningful places" (Hudson & Sadler, 1986, p. 175).

Agnew (1987, p. 230) proposed that place must be understood in terms of the following dimensions: (a) as *locale,* the objective element of local social arrangements; (b) as a subjective *sense of place;* and (c) as *location,* which refers to the impact of the economic and political macro-order. Thus, for Agnew, it is in places that the power of people to act as human agents is both realized and channeled within limits laid down by locally dominant practices within which are incorporated extralocal, structurally determined influences or powers that can direct and facilitate, as well as constrain and obscure, agency.

The need for the relative autonomy of place from both the national state and capital increasingly is treated as a normative requirement and necessary means to achieve social justice. Harvey (1987, p. 280) commented that political ideology on the left has shifted toward a " 'feasible' decentralized socialism . . . drawing much more inspiration from social democracy and anarchism than traditional Marxism." The view that the organization for radically changing economic relationships in the United States must extend beyond the workplace to the politics of the community has growing support (Katznelson, 1981, p. 194). Smith and Judd (1984, p. 190) proposed an alternative set of ideas and practices anchored in *placeness* that would provide an ideology for "the defense of community against the transformation of meaningful places by rapid growth and capital accumulation."

Building an urban policy that empowers communities to act with relative autonomy within the state apparatus requires some degree of initiative and immunity at the community level (Gurr & King, 1987). Agnew (1987, p. 228) stated that only when the power of the state is defined as "contingent rather than complete" can the relative autonomy of place be accepted.

Transnational Policy Flows

The issue of national-local state relationships has been confounded in American discussions by suggestions that European models, especially those of Great Britain, of urban and regional development and fiscal transfers be adopted (Hicks, 1983, p. 362; Kantor, 1988, p. 406; Peterson, 1981, p. 220; Robertson & Judd, 1989, p. 316). Recent assessments of these policies, however, indicate that, particularly in Britain, localities and their residents are even more exposed to external private and public exploitation than are those in the United States.

In his review of urban policies in a number of industrialized countries, Glickman (1983, p. 309) noted that nearly all have adopted policies to reduce uneven spatial development that are place oriented and that include aid to distressed regions, reductions in interregional income inequalities, greater efficiency in the spatial organization of the national economy, and reducing population concentrations in major metropolitan areas. In terms of results, Glickman found little to suggest that similar policies would solve problems of uneven development in the United States.

Efforts to target aid to lagging places generally have foundered as the result of political pressure for increasing the number of communities aided, which diluted already-low funding commitments. The power of the state has been proven to be limited in efforts to direct private investment, particularly in the high-growth service sector, to declining areas. Finally there have been few attempts or successes in integrating place-oriented urban policies with larger and better funded social policies such as income transfers, employment, and health and welfare, or with monetary and fiscal policy.

Looking at Great Britain specifically, Keating (1989, p. 533) found an urban system in which "extraordinary living standards are enjoyed by some classes and regions, while others remain in recession." Similar conclusions are reached by Hepworth and Robins (1988). These conditions are not simply the result of technical failures in national urban policy. They are strongly reinforced by reductions in the autonomy of local states (Barnekov, Boyle, & Rich, 1989, p. 219). In addition to imposing spending caps on cities to force cutbacks in welfare programs, the Thatcher government radically changed the basis of financing local government with a poll tax that, like user charges, was based on benefit rather than ability to pay and that favored the better off (Bramley, LeGrand, & Low, 1989).

When local governments resisted Thatcher's policies to reduce social expenditures, "the offending political units [were] demolished" (Kirby, 1985, p. 215). As Gurr and King (1987, p. 188) reported, a number of local Labour governments that opposed national policies of retrenchment were intentionally eliminated by the Thatcher government in 1986 when Parliament abolished the metropolitan councils and the Greater London Council.

A Dual Local-National Strategy

Urban policy must be designed to reverse the "unsatisfactory top-down" urban policies that have been produced by both central and local governments (Robson, 1987, p. 213). Harvey (1987, p. 279) suggested that decentralization and deconcentration, combined with concern for the qualities of place and space, can "create a political climate in which the politics of community, place, and region can unfold in new ways." Yago (1983, p. 127) similarly called for new locally based institutions in both the public and private sectors, which would give rise to a new urban agenda. Hoggett (1988, p. 229) suggested that initial innovations are necessary but not sufficient without a "theory of 'permanent revolution' " applied to the structure of the state.

Brindley, Rydin, and Stoker (1989, p. 185), drawing on experience in Britain, argued that "conventional representative democracy at the local level and local authority-led participation programs may not always be the most appropriate means of involvement and certainly do not exhaust the democratic options." A range of organizational forms should be explored beyond traditional formal structures. Castells (1983, pp. 320, 322), for example, proposed using urban social movements to organize an alternative governing structure based on "a network of cultural communities defined by time and space, and politically self-managed toward the maximization of use values for their residents; this use value is always decided and re-examined by the residents themselves."

These new forms and institutions must not only produce internal democratic participation in the local use of public authority and production but also rationalize the relationship of place to both the national state and the capital. Spatial decentralization and democratization of power must occur in the firm, as well as in the state. In discussing an alternative urban policy, Tabb (1982, p. 127) proposed a threefold strategy. The first strategy is that "counterinstitutions" are necessary to respond to local needs with control devolved to the neighborhood level. The second strategy would be to refine, develop, and adopt much wider concepts of economic democracy. The third strategy is the renewal of a liberal-labor political coalition revitalized by neighborhood, minority, women's, and consumer groups.

Glickman (1983, p. 322) provided a more detailed policy based on a "new social contract" designed to produce a "broad-based democratic restructuring of the economy" at both the local and national levels. Four basic components include (a) the establishment of economic democracy, (b) a national commitment to full employment, (c) community rejuvenation, and (d) equity in the public and private sectors. He placed the highest priority on economic democracy, which means the "democratic control of capital" through the restructuring of the work site and relationships among workers, managers, and government. Harvey (1987, p. 280) expanded the possibilities by suggesting community-worker coalitions to effect plant buyouts and, assuming a supportive state apparatus, establishment of cooperatives in service provision, housing, and production.

Drawing on his critique of European national urban policies and past American experience, Glickman concluded that place- and people-oriented programs must be integrated as part of a single policy. This point also was made by Edel (1980) and Wolman (1986, p. 330). Further, the people-oriented elements must be directed toward *eliminating* poverty, rather than ameliorating it. The lack of such a commitment in past national policy has not only continued but also exacerbated class, race, and gender income differences within and among cities. Programs that Glickman called for to address poverty and structural unemployment include a national full-employment program with guaranteed jobs, democratic reindustrialization strategies, and long-term control of wages and prices. The tax system, pay for work, and access to the housing market would all be restructured at the national level in the interest of equity, particularly in relationship to minorities and women.

Macromanagement of the economy, oriented to place, would include providing selective tax cuts to foster research and development, encouraging investment by firms in depressed areas through credit subsidies, and targeting countercyclical and long-term grant programs to economically stressed communities. Finally Glickman argued that a capital infrastructure program is necessary for general need and, along with quick-start standby programs, as a countercyclical measure.

Gaps in the Dialogue
and Prospects for Action

Addressing urban policy seriously is like trying simultaneously to solve a half-dozen jigsaw puzzles with the pieces mixed and some missing. A number of critical matters have not been considered adequately or considered at all.

The role of state governments has been improvidently ignored in the space available. The degree of decentralization that I suggest in this chapter is an issue to be debated with others who believe "fractionated" decision systems impede social justice within metropolitan areas, as well as within the national state. To support the integrity of place through macropolicies is not to assume that the spatial system is at an end state. Rather, means must be devised to allow spatial dynamics to be played out through collective decisions that recognize use and place value and the social costs of profit maximization.

If economic democracy and other forms of collective action outside the traditional structure of the state are to be serious policy goals, the implications of worker and government participation in the control or ownership of firms and volunteer action at the community level will require a reformulation of our understanding of the boundaries of governance.

A quite different loose end is the question of the position of the national state in the world system. For example, I have not discussed the policy implications of world cities in relationship to the state and national urban system in which they are situated (Friedmann & Wolff, 1982). Two related and more basic issues are the extent to which and how a national government can constrain capital in the world system in its own interest or that of communities.

Taylor (1985, p. 29) suggested that the state is a filter between the world market and the local community. According to him, the precise effect of global processes can be "reduced or enhanced by the politics of the nation-state in which it is located." Others, however, offer less sanguine views and suggest that there can be considerable variability in the control that national states can exercise over transnational capital.

For this issue to be relevant, the state must be willing to act in the interests of localities. How a national urban policy with this commitment is to be achieved is far from clear. In the United States, neither major political party would be receptive to the alternative urban policy proposed in this chapter. Brindley et al. (1989, p. 185), reflecting on the British experience, believe that a powerful and coherent formulation of a "market-critical" approach to urban policy is needed to coalesce communities in quite different circumstances. This proposal requires resolving dilemmas in the relationship of place- and class-based strategies (Hudson & Sadler, 1986).

The likelihood that such a formulation will emerge from the national political process in the United States is small. In the foreseeable future, the form of the project may be a continuing collective effort to demystify the discourse, construct a more fully elaborated local-national policy, and work for its application from the bottom up in receptive communities and through social movements (Fisher & Kling, 1989).

References

Agnew, J. (1987). *Place and politics*. Boston: Allen & Unwin.

Barnekov, T. K., Boyle, R., & Rich, D. (1989). *Privatism and urban policy in Britain and the United States*. New York: Oxford University Press.

Barnekov, T. K., Rich, D., & Warren, R. (1981, Fall). The new privatism, federalism, and the future of urban governance. *Journal of Urban Affairs, 3*, 1-14.

Bramley, G., LeGrand, J., & Low, W. (1989, July). How far is the poll tax a "commodity charge"? The implications of service usage evidence. *Policy and Politics, 17*, 187-205.

Brindley, T., Rydin, Y., & Stoker, G. (1989). *Remaking planning*. London: Unwin Hyman.

Brintall, M. (1989). Future directions for federal urban policy. *Journal of Urban Affairs, 11*(1), 1-19.

Broadbent, A. (1977). *Planning and profit in the urban economy*. New York: Methuen.

Castells, M. (1983). *The city and the grass roots*. Berkeley: University of California Press.

Castells, M., & Henderson, J. (1987). Techno-economic restructuring, socio-political processes, and spatial transformation: A global perspective. In J. Henderson & M. Castells (Eds.), *Global restructuring and territorial development* (pp. 1-17). Newbury Park, CA: Sage.

Clark, G. L. (1983). *Interregional migration, national policy, and social justice*. Totowa, NJ: Rowman & Allanheld.

Clark, G. L., & Dear, M. (1984). *State apparatus*. Boston: Allen & Unwin.

Cockburn, C. (1977). *The local state: Management of cities and people*. London: Pluto.

Cox, K. R., & Mair, A. (1988). Locality and community in the politics of local economic development. *Annals of the Association of American Geographers, 78*(2), 307-325.

Dyckman, J. W. (1966). The public and private rationale for a national urban policy. In S. B. Warner (Ed.), *Planning for a nation of cities* (pp. 23-42). Cambridge: MIT Press.

Edel, M. (1980). "People" versus "places" in urban impact analysis. In N. J. Glickman (Ed.), *The urban impacts of federal policies* (pp. 175-191). Baltimore: Johns Hopkins University Press.

Fainstein, N. I., & Fainstein, S. S. (1987, Spring). Economic restructuring and the politics of land-use planning in New York City. *Journal of the American Planning Association, 53*, 237-248.

Fisher, R., & Kling, J. (1989, December). Community mobilization: Prospects for the future. *Urban Affairs Quarterly, 25,* 200-211.

Friedland, R. (1983, September). The politics of profit and geography. *Urban Affairs Quarterly, 19,* 41-54.

Friedmann, J., & Wolff, G. (1982, September). World city formation. *International Journal of Urban and Regional Research, 6,* 309-344.

Gittell, M. (1985, September). The American city: A national priority or an expendable population? *Urban Affairs Quarterly, 21,* 13-19.

Glickman, N. J. (1983). National urban policy in an age of economic austerity. In D. A. Hicks & N. J. Glickman (Eds.), *Transition to the 21st century* (pp. 301-343). Greenwich, CT: JAI.

Gurr, T. R., & King, D. S. (1987). *The state and the city.* Chicago: University of Chicago Press.

Hanson, R. (Ed.). (1983). *Rethinking urban policy.* Washington, DC: National Academy Press.

Harvey, D. (1987, December). Flexible accumulation through urbanisation: Reflections on "post-modernism" in the American city. *Antipode, 19,* 260-286.

Hepworth, M., & Robins, K. (1988, July). Whose information society? A view from the periphery. *Media, Culture and Society, 10,* 323-343.

Hicks, D. A. (1983). Urban and economic adjustment to the post-industrial era. In D. A. Hicks & N. J. Glickman (Eds.), *Transition to the 21st century* (pp. 345-370). Greenwich, CT: JAI.

Hill, R. C. (1983, September). Market, state, and community: National urban policy in the 1980s. *Urban Affairs Quarterly, 19,* 5-20.

Hoggett, P. (1988). A farewell to mass production? Decentralization as an emergent private and public sector paradigm. In P. Hoggett & R. Hambleton (Eds.), *Decentralization and democracy.* Bristol, UK: University of Bristol, School for Advanced Urban Studies.

Hudson, R., & Sadler, D. (1986). Contesting work closures in Western Europe's old industrial regions: Defending place or betraying class? In A. J. Scott & M. Storper (Eds.), *Production, work, territory* (pp. 172-193). Boston: Allen & Unwin.

Jackson, K. T. (1985). *Crabgrass frontier: The suburbanization of the United States.* New York: Oxford University Press.

Johnston, R. J. (1982). *Geography and the state.* New York: Macmillan.

Kantor, P. (1988). *The dependent city: The changing political economy of urban America.* Glenview, IL: Scott, Foresman.

Kasarda, J. D. (1983, September). Entry-level jobs, mobility, and urban minority unemployment. *Urban Affairs Quarterly, 19,* 21-40.

Katznelson, I. (1981). *City trenches.* New York: Pantheon.

Keating, M. (1989, June). The disintegration of urban policy: Glasgow and the New Britain. *Urban Affairs Quarterly, 24,* 513-536.

Kirby, A. (1982). The external relations of the local state in Britain: Some empirical examples. In K. R. Cox & R. J. Johnston (Eds.), *Conflict, politics, and the urban scene* (pp. 88-106). New York: Longman.

Kirby, A. (1985, December). Nine fallacies of local economic change. *Urban Affairs Quarterly, 21,* 207-220.

Logan, J. R. (1983, September). The disappearance of communities from national urban policy. *Urban Affairs Quarterly, 19,* 75-90.

Markusen, A. R. (1980). Urban impact analysis: A critical forecast. In N. J. Glickman (Ed.), *The urban impacts of federal policies* (pp. 103-118). Baltimore: Johns Hopkins University Press.

Molotch, H., & Logan, J. R. (1985, December). Urban dependencies: New forms of use and exchange. *Urban Affairs Quarterly, 21,* 143-169.

Moynihan, D. P. (1970). Toward a national urban policy. In D. P. Moynihan (Ed.), *Toward a national urban policy* (pp. 3-25). New York: Basic Books.

Peterson, P. (1981). *City limits.* Chicago: University of Chicago Press.

President's Commission for a National Agenda for the Eighties. (1980). *A national agenda for the eighties.* Washington, DC: Government Printing Office.

Robertson, D. B., & Judd, D. R. (1989). *The development of American public policy.* Glenview, IL: Scott, Foresman.

Robson, B. (1987). The policy framework. In B. Robson (Ed.), *Managing the city* (pp. 211-215). Totowa, NJ: Barnes & Noble.

Saunders, P. (1979). *Urban politics: A sociological interpretation.* London: Hutchinson.

Savas, S. S. (1983, June). A positive urban policy for the future. *Urban Affairs Quarterly, 18,* 447-453.

Shalala, D. E., & Vitullo-Martin, J. (1989, Winter). Rethinking the urban crisis: Proposals for a national urban agenda. *Journal of the American Planning Association, 55,* 3-13.

Smith, C. J. (1988). *Public problems.* New York: Guilford.

Smith, M. P., & Judd, D. R. (1984). American cities: The production of ideology. In M. P. Smith (Ed.), *Cities in transformation* (pp. 173-196). Beverly Hills, CA: Sage.

Tabb, W. K. (1982). *The long default*. New York: Monthly Review Press.

Taylor, P. J. (1985). *Political geography*. New York: Longman.

Tiebout, C. (1956, October). A pure theory of local expenditure. *Journal of Political Economy, 64*, 416-435.

Wade, R. C. (1989, Winter). The Reagan revolution: Much ado about nothing much. *Urban Resources, 5*, 35-37.

Waste, R. J. (1989, Winter). Federal urban policy in the 1990s: Déjà vu and disaster. *Urban Resources, 5*, 21-24, 33.

Weiher, G. R. (1989). Rumors of the demise of the urban crisis are greatly exaggerated. *Journal of Urban Affairs, 11*(3), 225-242.

Wolman, H. (1986, March). The Reagan urban policy and its impact. *Urban Affairs Quarterly, 21*, 311-335.

Yago, G. (1983, September). Urban policy and national political economy. *Urban Affairs Quarterly, 19*, 113-132.

9

People Versus Places

THE DREAM WILL NEVER DIE

Robert C. Wood

For scholar and politician alike, fashioning a national urban policy has always been a slightly disreputable affair. At least ideologically, city builders—developers, bankers, political bosses—typically have been suspicious characters, and city building has never seemed a virtuous undertaking. Americans have preferred as folk heroes the frontiersman, the sturdy yeoman, and the cowboy. Furthermore, the learned disciplines generally have given urban studies little attention. Largely subsumed under the subfield of public finance in economics, and relegated to the morality plays of civic reform in political science, students of cities were second-rate citizens in the academic ranks.

To be sure, the 1950s and 1960s saw a flurry of interest as Walter Isard (1956) initiated his ambitious subfield in regional sciences and Raymond Vernon (1963) and Edgar Hoover carried out their comprehensive inquiries in the New York metropolitan region economy. Briefly, interdisciplinary urban centers waxed in the late 1960s and then waned rapidly in the 1970s. Still, as a major source of disciplinary paradigms, urban studies never seems to have caught on.

Up With People

The 1980s have been especially cruel on the field. Conceptually, the decade began with the explicit verdict of President Carter's President's Commission for a National Agenda for the Eighties (1980) that urban issues were best forgotten. From the findings of its Panel on Policies and Prospects for Metropolitan and Non-Metropolitan America, the commission concluded: "Contrary to conventional wisdom, cities are not permanent. . . . An often-noted 'urban renaissance' within cities, while enriching and laudable, seems not to be taking place on anything like the scale suggested in popular commentary . . . we forget that cities, like all living things, change" (pp. 65-67). Accordingly, advocating such a major shift from urban policy to social policy—people instead of places—the commission signed off on the panel's flat-out declaration that "there are no 'national urban problems,' only an endless variety of local ones. Consequently, a centrally administered national urban policy that legitimizes activities inconsistent with the revitalization of the larger national economy may be ill advised" (p. 99).

Thus, with perfect labor mobility enshrined as macroeconomic's chosen instrument, the commission's learned treatise confirmed President Ford's 5-year-old judgment about New York and other American cities: They could or should "drop dead."

That conclusion, that there was not then and never had been an urban problem, brought one of the few rejoinders from Carter's lame duck administration. Marshall Kaplan, a Housing and Urban Development (HUD) deputy assistant secretary, authorized a vigorous protest asserting that cities were nationally important and continued to deserve federal help.

By 1990, however, even Kaplan had come around. Coediting with Franklin James *The Future of National Urban Policy* (1990), he joined the conclusion that "the era of national largess concerning city problems appears to the writers in this book to be over. Urban initiatives no longer appeal" (p. 351). Further, the consensus is a powerful one. "The marriage between the academic, the politician, and the citizen has never been as solid. Everybody doubts; no one believes or disbelieves strongly. Jane Jacobs provides understandable legitimacy to the comprehensive policy doubters when she decries the lack of wisdom and theory about economic cause-and-effect relationships" (p. 352).

Speaking for himself, Kaplan finds that "the Federal government does not have a theory of city development and city decay. . . . None of the models put forth to explain urban problems and/or city trends stand up well under sustained scrutiny." Hence his counsel is "to figure out how non-urban policies that are likely to win favor in the Congress—welfare reform, infrastructure assistance, education help—could best benefit cities . . . guarantee cities a fair shake or a favored position" (pp. 181-184).

Kaplan is not alone in this "people" preference. In the same book, economist Robert Reischauer, Director of the Congressional Budget Office, reaches the same conclusion, perhaps in even more emphatic terms. Pointing out the arrogance and naivete of the policymakers of the 1960s ("social scientists and political activists who were drawn to Washington by Presidents Kennedy and Johnson") who put cities as places close to the top of the national agenda, he argued that "the 1960 to 1978 era was an aberration. . . . There never existed a neatly defined or coherent national urban policy" (Kaplan & James, 1990, p. 228). Generously, Reischauer excused the idea-brokers for their failure to sustain a coherent policy. Failure "was inevitable given the inherent complexity and diversity of the Federal system and the economic turmoil that characterized the last fifteen years." But he warned against "renewed efforts to establish a national urban policy." That would, he believed, "be futile" (Kaplan & James, 1990, p. 234).

So Ronald Reagan did the right thing. His studied indifference to urban affairs, his new New Federalism enhancing the states' authority and discretion, his savaging of HUD, his tolerance of scandal and sleaze, his reluctant acceptance of the minimal people-oriented social programs for the "truly needy" on which the Congress insisted, turned out to be truly in the public interest. The 1981 budget and tax acts and the 1983 banking deregulation act—which created, in Anthony Downes's felicitous phrase, "money-driven markets instead of demand-driven ones" (Downes, 1985, p. 14)—ensured that the market would reign supreme. So arose the entrepreneurial city in the old central business district —peaking in 1986 when 1 billion square feet of new commercial office space came on line in 22 major American cities. So appeared the new "urban villages: office, industrial, retail, housing, and entertainment focal points amidst a low density city escape" (Lockwood, 1986, p. 11). All was done by virtue of the "nonurban" national policies and the sharp deliberate withdrawal of the public sector, which Carter's commission had recommended and Kaplan and Reischauer applaud.

Second Thoughts on the Vanishing Act

Yet neither American cities nor American social policy seem well off. The average vacancy rate in those shining new towers of the 22 top entrepreneurial cities averaged over 20% in September, 1990, with New England's Hartford, Connecticut, on top with 28.1% and Texas's Dallas runner-up at 26.9% ("Coldwell Banker," 1990, p. 29). The taxpayers' bill for the savings and loans debacle reached a new estimate of a half-trillion dollars. One metropolitan area after another was pockmarked by abandoned shopping centers and vacant upscale condominium developments that only months ago were designed as an antidote for dreary suburban lives.

Something seems awry in the "people in the market place plus a safety net for those who can't hack it" social policy for our times. Despite the contemptuous dismissal of urban-oriented research ("we do not need scholarly treatises to remind us of the lamentable by-products of urban deconcentration for the older industrial cities"), the study of urban life in *places* has continued (President's Commission, 1980, p. 68). Nor have the policy issues of urban education, the urban underclass, the urban homeless, or the urban neighborhood vanished. In November 1990, big-city mayors met again in New York City with an agenda for Washington. As the metropolitan spread city retrenches, the suburban defensive policies of the not-in-my-backyard (NIMBY) stand were revealed as major place barriers to the provision of affordable housing. The 101st Congress concluded with a major housing program designed to pull the industry out of its most severe depression since the Great Depression and to develop some 8 million new and rehabilitated units by the year 2000.

The Case for Place

Why did the 1980s dismiss the cities as phenomena not worth studying and as places not requiring national policy? Put more simply, why did we dismiss *land* as irrelevant to our pursuit of quality community life? Second, and more important, why do cities seem to be coming back on the public agenda?

A short answer to the first question is that modern economics, as the unquestioned lead discipline in modern social science, is not very comfortable with land either as a research subject or as a policy topic. This unease has historical origins. The principal theorists, J. H. von Thunen and David Ricardo, did their work in the early 19th century, a half century before the discipline, either in Germany or England, was considered to have arrived. Indeed, Keynes was to conclude in 1933 that "the complete domination of Ricardo's approach for a period of a hundred years has been a disaster to the progress of economics" (Sills, 1968, Vol. 13, p. 511).

Moreover, the chief policy advocate for land use and land taxation was suspected to be no genuine economist at all. Henry George was a school dropout, an itinerant journalist, advocating his single-tax reform on site value in *Progress and Poverty* at a time when "the academic profession of economics was inchoate in the Eastern United States and had no existence whatever in California"— where George lived (Sills, 1968, Vol. 6, p. 153). Perhaps the fact that his major work has been the best-selling treatise on economics by an American (if textbooks such as Samuelson's are not included) has also not endeared him to the guild.

Fascinated by the challenge to manage national economies and international trade, to sustain capitalism intellectually as well as militarily in the face of the communist threat, the discipline has barely tolerated "spatial economics." Indeed, in one of few texts devoted to urban economics, author James Heilbrun (1981) somewhat ruefully observed that "traditional economic theory omits any reference to the dimensions of space by treating all economic activity as if it took place at a single point. It refers to consumers and producers, firms and industries, but not to distance or contiguity, separation or neighborhood" (p. 2).

The same lack of respect is typically accorded the property tax in the annals of public finance. Conventionally condemned as inequitable, regressive, and prone to inept administration (although empirical research in recent years drastically modifies these criticisms), the property tax is tolerated on historical rather than theoretical grounds. It has, since colonial days, just been "there" in America.

It is not just the comparative indifference of economics, until recently subsuming land under a single production function, that deters land analysis. There has also been a general intellectual sloppiness about the concept of *place*. The other major social sciences, torn by internal conceptual disputes, still divided as to their philosophical roots, have been unclear as to the propriety of urban studies as well. The professional schools of planning and architecture are essentially aesthetic and subjective in their orientation. So, not subjected to rigorous study, land analysis typically proceeds by simplistic classification schemes and wide variations in meaning—ranging from "property" and "capital" to "nature" and "space." And the urban place has all too often been treated as a stage across which "issues" parade. Poverty, racism, hopelessness, crime, and addiction are prefixed by the adjective *urban* and are too often treated as problems or crises peculiar to the city.

We can do better than ignoring land and place in a scholarly sense or relegating it almost contemptuously to a subordinate part of a discipline or treating it with journalist hype. We have the basis for both study and policy if we make clear that a large number of people inhabiting limited space under conditions of high density and high velocity of interaction create *people's* behavioral attributes and the attributes of the *space* that are measurably distinctive. Examples of this are the incidence of mental illness, the choice options in shopping, environmental pollution, the range of clusters of occupation, education, the degree of racial heterogeneity, and the frequency of specified crimes. All of these have been shown to be qualitatively and quantitatively distinctive in urban settings. Consequences as well as attributes are distinctive as well—gridlock, accordion expressway accidents, fire—and all carry the potential of a multiplier effect.

Once the case for different kinds and intensities of urban needs, human and environmental, is established, then the inadequacy of "people policy" becomes clear. Effective policy does not allocate resources or authorize transfer payments or establish delivery systems by space-indifferent indexes. Effective policy targets them and calibrates according to differential needs, some dispersed, but others clustered. No city, for instance, has a dominant constituency as the farmer represented in rural life. Accordingly, diverse urban peoples and their needs are not capable of resolution without attention to institutions and physical environment. Urban schools cannot be replaced by vouchers, nor court orders mandating segregation be obeyed by mandating free choice. Public housing, project or scattered site, requires attention to architectural form and neighborhood conditions.

Whatever the abstractions required to activate economic models, people live in places. They are always somewhere—in schools, in church, in neighborhoods, in the office, in the factory, at home. People without places, far from being calculators of pleasure and pain, rationally responding to invisible market hands, purely passionless, ceaselessly pursuing and achieving their self-interest, are in fact lost. They are homeless, jobless, alone. Losers. They require special public programs precisely because they are without place.

Policy Priorities for Place

If one rejects the counsel of the 1980s—that all one needs is transfer payments to individuals (because bureaucracies are hopelessly inept but the market behaves with textbook precision)—then the question becomes, What place policies do we need? Which government should do what?

For openers, one acknowledges our present dire straits. Although the housing needs of the 1990s total some 8 million units, new and rehabilitated, our major metropolitan markets are clearly depressed. Especially in the central cities, as commercial vacancies continue, the linkage policy San Francisco pioneered will be suspended for some time. Moreover, with all levels of government registering deficits, this is no occasion for fashioning new deep-dish subsidies. Perforce, policymakers need to turn instead to the second major instrument of public authority—regulation. In fact, in housing we are long overdue in adopting a regulatory strategy. Until we come to grips with land costs so far as affordable housing is concerned, all of our interest subsidies, mortgage guarantees, and new construction technologies will not yield prices even the middle-class American can afford.

The first regulatory target is obvious: The states need to override local defensive zoning practices that add up to NIMBY. Benjamin Chinitz' (1990) observation that local growth management efforts are additive to national and international environmental efforts may be persuasive, but simultaneously he acknowledges that they push housing prices up (p. 7). They also push people out or prevent them from coming in, to an extent now violating state constitutional rights, as New Jersey's succession of Mount Laurel decisions clearly signals. If access to affordable housing is a public goal, equal in weight to environmental and amenity objectives, then the growth-management policies coming into vogue in recent years can only be carried out at the state level, and it is the state that needs to determine how "fair shares" of low- and moderate-income housing can be allocated throughout the metropolitan region. As Connecticut's Condominium Development Acquisition Pilot Program demonstrates, a distressed suburban real estate market, overloaded with new condominium construction, provides genuine opportunities for "instant" scattered-site public housing. The key factor is that the state take the initiative in growth management—that is, in place policy.

For central cities, the acronyms are TOADS and LULUs, as the Rutgers team of Greenberg, Popper, and West have informed us (1990). According to their inquiries in 14 cities, the spread of TOADS (Temporarily Obsolete Abandoned Derelict Sites) is assuming epidemic proportions. They join LULUs (Local Unwanted Land Uses) in posing serious threats to the quality and safety of central cities and inner suburbs, the gray areas that Raymond Vernon discovered 35 years ago. Derelict buildings, some productive, others unwanted, vacant parcels of land, and TOADS infect neighborhood after neighborhood, with rates of abandonment reaching 20% of all city structures.

When I considers the Rutgers colleagues' findings, tentative as they may be, the inclination to read Henry George again is strong. However condescendingly his broad-brush propositions have been treated, no matter that his disciples, the Georgists, occasionally take on cult coloration, site-value taxation seems to have considerable appeal in the face of TOADS and LULUs. Carefully drafted and professionally administered, the separation of land and development values in assessment seems precisely designed to overcome the artificial speculative costs and inefficient noncommercial uses of central city land. Thoroughly respectable economists, such as Dick Netzer, endorse its application generally, and Pennsylvania cities, such as Pittsburgh, actually have used the policy for some years. One does not have to concur with Rand McNally's designation of Pittsburgh as America's "most livable city" or accept Georgist principles of "synergistic surplus" to conclude that a close examination of the Pittsburgh experience is in order as we go about coping with the process of urban decay and abandonment.

In short, all three acronyms—NIMBY, TOADS, LULUs—seem to capture the urban issues of the 1990s. They speak to matters of place as well as people. They imply public action, not the aggregation of private choices. They involve government authority in the form of regulation, rather than subsidy and incentive. They suggest that our lighthearted abandonment of national urban policy in the 1980s did not turn out so well. Then the conclusion is, for the 1990s, to try government again.

References

Chinitz, B. (1990, Winter). Growth management: Good for the town, bad for the nation? *American Planning Association Journal*, p. 7.

Coldwell banker commercial real estate. (1990, October 2). *Boston Globe*, p. 29.

Downes, A. (1985, August). Tax reform: What about real estate? *Urban Land*, p. 14.

Greenberg, M. R., Popper, F. J., & West, B. (1990, March). The TOADS: A new American urban epidemic. *Urban Affairs Quarterly, 25*, 3.

Heilbrun, J. (1981). *Urban economics and public policy* (2nd ed.). New York: St. Martin's.

Isard, C. W. (1956). *Location and space-economy*. New York: John Wiley.

Kaplan, M., & James, F. (Eds.). (1990). *The future of national urban policy*. Durham, NC: Duke University Press.

Lockwood, C. (1986, November 26). The arrival of the urban village. *Princeton Alumni Weekly*, p. 11.

President's Commission for a National Agenda for the Eighties. (1980). *Report of the Panel on Policies and Prospects for Metropolitan and Non-Metropolitan America*. Washington, DC: Government Printing Office.

Sills, D. S. (Ed.). (1968). *International encyclopedia of the social sciences*. New York: Macmillan.

Vernon, R. (1963). *Metropolis 1985*. Garden City, NY: Doubleday.

PART FOUR

Economic Development

Introduction to Economic Development

Norman Krumholz

During the 1970s and 1980s, economic development in many U.S. cities consisted largely of downtown real estate development. If a city had a new festival market shopping mall, new office towers, and a new hotel—preferably with an atrium and an outside elevator—it was considered to be revitalized. If circumstances warranted and the local public-private partnership was sufficiently powerful and well organized, cities might also add a publicly financed new stadium (preferably domed), an aquarium, and a cleaned-up waterfront. These were the essential elements in city "renaissance," and they emerged unplanned and piecemeal, a project at a time (Frieden & Sagalyn, 1989).

The redevelopment of urban downtowns during the 1980s was engineered largely by an old team: the "growth coalition" of politicians and business leaders who led the fight for urban renewal in the 1960s. This time, the coalition was wearing the cloak of "public-private partnerships" (Porter & Sweet, 1984). The switch was based on a change in methods used by the cities to fund urban and economic redevelopment.

The old urban renewal and slum clearance programs of the 1960s had been based largely on federal funds. Cities were funded to prepare plans, acquire and clear land in blighted areas, expunge any cloud on ownership titles, and sell the land to any developer who would agree to build in accordance with the city's plan. The rules somewhat limited local discretion, but the federal subsidies were attractive. The federal government absorbed two thirds of the cost of land acquisition and clearance, and frequently the only cost to the city was a road, sewer, or other site improvement.

In the 1970s the urban renewal program, having displaced in its projects large members of mostly poor and minority people, became politically controversial and was replaced by the Community Development Block Grant (CDBG) and the Urban Development Action Grant (UDAG) programs. Both programs provide much more local discretion over how the federal funds actually would be spent. As federal aid to the cities fell during the Reagan administration, and as local discretion increased, cities began to promote economic development to try to make up for lost federal revenues. In the process, economic development became one of the big urban news stories of the 1980s, and a group of successful public entrepreneurs called "messiah mayors" began to emerge (Teaford, 1990).

The logic of local economic development efforts was based on the assumption that growth and new development automatically would provide jobs for the city's unemployed residents and new net tax returns to the public treasury. With this improved tax base, the city would be able to make improvements in the basic infrastructure and to improve city services as well. The result would be an improved local business climate that would encourage still more private investment. Extensive public subsidies clearly were justified to support these desirable objectives.

Generally the method of operation in city after city was similar. Cities contributed part of their CDBG and UDAG funds to big downtown projects, abated property taxes, floated industrial revenue bonds, negotiated tax increment financing deals, built streets, sewers, and other expensive capital improvements, and provided whatever public subsidies were necessary to encourage private investment in downtown development. Often these arrangements resulted in land deals, financing schemes, and tax breaks so complicated that only the handful of attorneys working on the contracts could understand them, thus raising serious problems of public accountability.

Further problems of accountability and oversight often arose from the implementing structure used in each project. To speed the development process, cities supported the formation of new, quasi-public, nonprofit development corporations. As a result, a sports authority would be responsible for building the new stadium or arena, and a waterfront development corporation, run like a private corporation but empowered to receive and expend both public subsidies and private investments, would build waterfront facilities (Eisinger, 1988). Not surprisingly, it became less and less clear over time just what was public and what was private. Few seemed to care; the process was "working," at least in terms of new buildings.

Harborplace in Baltimore, Quincy Market in Boston, Renaissance Center in Detroit, Peachtree Plaza in Atlanta, Pike Place Market in Seattle, Bunker Hill in Los Angeles, Horton Plaza in San Diego, Union Station in St. Louis, Gateway in Cleveland, and dozens of similar projects transformed the skylines of U.S. cities. In many cities the building of these "big-bang" projects was accompanied by some degree of gentrification as upper income, younger professionals with jobs in the city displaced the poor residents of older, but still fashionable, urban neighborhoods. This ousting happened on a broad scale in Boston, San Francisco, Chicago, Brooklyn, and Washington, DC, and on a smaller scale in other cities. The local media were very supportive, acting as cheerleaders and finding colorful positive copy in pictures and stories of new construction, redevelopment, and historic preservation.

Amid all of the new construction and the obvious regeneration of the business district, nagging questions of equity, purpose, and effect remained. First, the spillover effect in the neighborhoods surrounding downtown seemed negligible. Second, local economic development did not seem to be satisfying its stated purpose, which was to generate new jobs for unemployed city residents and tax increases with which to address other city problems, including decaying infrastructure and inadequate public services. But "success" in downtown development by the public-private partnerships did not seem to translate into lower poverty and unemployment rates for city residents (Squires, 1989). The number of Americans living below the federal poverty line grew to 31.7 million, or 12.9% of the population in 1990, from 27.4 million, or 12.4% of the population in 1980. In 1980 about 16.5% of the population in central cities lived in poverty; a decade later, the figure was 18.7%. Despite the tangible successes of downtown-focused public-private partnerships, suburban job growth during the 1980s ran far ahead of job growth in central cities.

Other questions surfaced around such issues as who gains and who loses as a result of downtown development; how many job do these projects actually produce; are they permanent or temporary; who gets the jobs—city residents or suburbanites; should public funds be engaged in real estate development or job generation; how is the public interest served when the city trades uncollected property taxes, which provide social and educational services for lower income people, for physical development that provides benefits for higher income people; what of investment in preserving manufacturing jobs versus investing in service jobs; should we not emphasize new opportunities for minority and female businesses, rather than the market and economic efficiency; and should these public-private corporation not be accountable to the public because they are spending public money?

Cleveland in 1993 represented a compelling compendium of these issues. The city had been hailed in the local and national media as a "comeback" or "renaissance" city rising phoenix-like from the depths of the 1978 default (Magnet, 1989). With great regularity, the city's civic and political leaders congratulated themselves for the new hotels and office buildings in downtown, the new stadium for the baseball Indians, the arena for the basketball Cavaliers, the Rock 'n' Roll Hall of Fame, and the restoration of three old theaters in Playhouse Square—all of which were financed extensively by public funds.

At the same time, by many indices, Cleveland was in deep trouble ("Local Yearbook," 1993). Cleveland's population dropped 11.9% from 1980 to 1990; it is down to 505,616, about the size the city was in 1905. Cleveland, once the nation's 6th largest city, now ranks 25th in size. In terms of personal income, Cleveland ranks next to the rock-bottom, 99th out of 100. Only the residents of Hialeah, Florida, have lower incomes. The city is on the high end, however (3rd highest in the nation), in percentage of households receiving public assistance. Here it ranks only behind Detroit and Newark. Cleveland ranked 7th highest in percentage of unemployment. In 1990, Cleveland's black unemployment rate was greater than 20%, the highest of any major city in the nation, while the nonemployment rate (unemployed, discouraged and not looking for work, or in jail) for black men aged 25 to 54 hovered near 50%. Only three cities in the United States—Detroit, New Orleans, and Miami—have higher levels of poverty than Cleveland; over 40% of all Cleveland families live under the poverty line, and the number is rising rapidly. Finally Cleveland stood 96th out of 100 in percentage of persons over 25 who are high school graduates.

The readings in Part 4 raise some of these vexing questions. They are reports of studies that deal with a mix of issues regarding both shaping policy and implementing it.

In Chapter 10, "The Next Wave," Susan E. Clarke and Gary L. Gaile report on their national study of local economic development activities in American cities with populations over 100,000 in 1975. Their findings reflect data from 1978 to 1989 from the 101 cities responding. The reading focuses on the question of how cities with economic development programs would respond to significant cutbacks in federal aid for economic development; would they be inclined toward more or less risk taking?

As federal resources waned during the Reagan administration, local governments faced with tighter budgets and lack of staff did not cut back on their economic development efforts, but sought new ways to generate the revenues necessary to continue or expand them. They turned to other revenue sources. These included larger shares of local general funds, more CDBG funds shifted to economic development, tax money from other development projects diverted via tax increment financing, and other devices that let cities channel future revenues into the development process outside usual budgetary procedures. They also engaged extensively in partnerships with private developers, and they used new quasi-public corporations for project implementation, rather than depended on traditional line agencies. Because more private sector orientation was included, it is not surprising that the new local strategies reflected a shift toward more market-based or entrepreneurial approaches. Among important features of the shift were a greater tolerance for risk and the use of market criteria (e.g., return on investment), rather than political criteria, in setting priorities for the investment of public funds.

Few examples were reported in which these new initiatives were threatened by other local political actors. Perhaps cloaking these entrepreneurial, risk-taking activities in the language of jobs, growth, and wealth dissuades political opposition from risk taking.

Not surprisingly, the mayors and administrators of economic development programs appreciated the career-enhancing opportunities offered by these project successes. Finally the study suggests that cities using more risk-accepting, entrepreneurial tools seem to have higher average job and firm growth than other, more risk-adverse cities.

In Chapter 11, "Equity and Local Economic Development," Norman Krumholz raises the distributional questions of who gets and who pays as a result of economic development policies. Krumholz notes that planning for economic development is not as it seems. Instead of setting goals, weighing strategies, and undertaking technical studies for economic development as outlined in approved textbooks, public sector practitioners seem to concentrate largely on marketing, public relations, and sales. In the process, they frequently become willing extensions of their private-sector partners, using identical criteria and even identical language. Nor do they ordinarily seem to be deeply interested in ensuring that jobs in economic development flow to the city's unemployed, or that projects throw off net new taxes for the public fisc, or that new development in downtown is balanced or linked with development in poor and working-class neighborhoods.

This reading concludes by offering the *Chicago Development Plan, 1984* as a model for local economic development planning that would incorporate issues of social justice and result in a more equitable distribution of costs and benefits from the economic development process. The Chicago Plan selected five goals for economic development: (a) increased job opportunities for Chicagoans, (b) balanced growth between downtown and neighborhoods, (c) neighborhood assistance, (d) enhanced citizen participation, and (e) pursuit of a regional, state, and national agenda.

In Chapter 12, "The Politics of Economic Development Policy," Elaine B. Sharp and David R. Elkins deal with the question of citizen involvement in economic development. They report on a national sample survey of city chief executive officers done in 1984.

The study suggests that cities differ widely with respect to citizen involvement in economic development. Further, the wide variety of mechanisms for citizen involvement includes the use of (a) appointed advisory committees, (b) open meetings for public hearings, (c) citizen surveys, and (d) elected neighborhood committees. Their analysis points out that in communities suffering from high property tax stress, higher levels of citizen involvement may dampen the probability that the city will choose to use such tools as tax abatements. Abatements for the well-to-do can be contrasted unfavorably with the tax burden on other taxpayers. Conversely, such conditions increase the likelihood that economic development tools such as loan guarantees that minimize apparent tax costs will be used. The authors conclude that a city's propensity to use various economic development tools can be forecast from the political pressures of high property taxes and the extent of citizen participation.

Thomas A. Clark and Franklin J. James study trends in the scale of business ownership by women in Chapter 13, "Women-Owned Businesses." Using 1987 figures from the U.S. Census of Business, they verify that during the last 20 years, the number of women-owned businesses grew rapidly. The number of women-owned businesses may have grown by a factor of six since 1970, and their gross receipts have grown by a factor of 50. However, the researchers find that most women-owned businesses are "small and undercapitalized" and that many are not full-time operations. "Full 89% of women's businesses are started or acquired for less than $10,000." Many women business owners have responsibilities for the care of others, and business schedules often are shaped around these activities. These factors constrain the probability of success and, as a result, the payoffs of business ownership of women have been relatively insignificant so far. Taking into account data concerning experience and other qualifications, the authors suggest that potential earnings for most women may be higher in management positions on payrolls, rather than in self-employment.

The study concludes on a controversial note suggesting that there is little economic rationale for new public efforts to stimulate women-owned businesses but that the federal government should continue to provide advocacy and research on women-owned business as part of the effort to promote equal opportunity within business in general.

According to the American Economic Development Council (AEDC), economic development is "the process of creating wealth through the mobilization of human, capital, physical, and natural resources" (AEDC, 1984, p. 18). No mention is made of how this wealth is distributed with respect to the interest of the population at large. The overall objective is one that all can support, but distributional issues should receive more attention. Indeed it seems probable that the downtown-focused economic development strategies described in Part 4 have contributed to, rather than reduced, tensions in our inner cities. Most of the new office towers, convention centers, stadiums, and hotels have been funded, in whole or in part, by public tax funds, yet they have not produced the promised taxes necessary to improve public services or jobs within the skill levels of inner-city residents. In the interests of equity and perhaps the long-term maintenance of social peace as well, planners, politicians, and academics should be working on acceptable alternatives to the current model of economic development.

References

American Economic Development Council (AEDC). (1984). *Economic development today.* Chicago: Author.
Eisinger, P. K. (1988). *The rise of the entrepreneurial state: State and local economic development policy in the United States.* Madison: University of Wisconsin Press.
Frieden, B. J., & Sagalyn, L. B. (1989). *Downtown, Inc.: How America builds cities.* Cambridge: MIT Press.
Local yearbook 1993. (1993, July). *Governing.*
Magnet, M. (1989, March 27). How business bosses saved a sick city. *Fortune,* pp. 106-110.
Porter, P. R., & Sweet, D. C. (1984). *Rebuilding America's cities: The road to recovery.* New Brunswick, NJ: Rutgers University, Center for Urban Policy Research.
Squires, G. D. (Ed.). (1989). *Unequal partnerships: The political economy of urban redevelopment in postwar America.* New Brunswick, NJ: Rutgers University Press.
Teaford, J. C. (1990). *The rough road to Renaissance: Urban revitalization in America: 1940-1985.* Baltimore: Johns Hopkins University Press.

Suggestions for Further Reading

City of Chicago Department of Economic Development. (1984). *Chicago works together: 1984 development plan.* Chicago: Author.
Clavel, P., & Wiewel, W. (1991). *Neighborhood and economic development policy in Chicago: 1983-1987.* New Brunswick, NJ: Rutgers University Press.
Fainstein, N. I., & Fainstein, S. S. (1987). Economic restructuring and the politics of land-use planning in New York City. *Journal of the American Planning Association, 53*(2), 237-248.
Judd, D., & Ready, R. L. (1986). Entrepreneurial cities and the new policies of economic development. In G. E. Peterson & C. W. Lewis (Eds.), *Reagan and the cities* (pp. 209-248). Washington, DC: Urban Institute Press.
Krumholz, N., & Forester, J. (1990). *Making equity planning work.* Philadelphia: Temple University Press.
Luria, D., & Russell, J. (1981). *Rational re-industrialization: An economic development agenda for Detroit.* Detroit: Widgetripper Press.
Markusen, A., Learner, J., Patton, W., Ross, J., & Schneider, J. (1985). *Steel and southeast Chicago: Reasons and opportunities for industrial renewal.* Evanston, IL: Northwestern University, Center for Urban Affairs and Policy Research.

Marris, P. (1987). *Meaning and action: Community planning and conceptions of change.* London: Routledge & Kegan Paul.

Mier, R. (1993). *Social justice and local development policy.* Newbury Park, CA: Sage.

Mier, R., Moe, K. J., & Sherr, I. (1986). Strategic planning and the pursuit of reform, economic development, and equity. *Journal of the American Planning Association, 52*(3), 299-309.

10

The Next Wave

POSTFEDERAL LOCAL

ECONOMIC DEVELOPMENT STRATEGIES

Susan E. Clarke

Gary L. Gaile

Recent changes in economic conditions and national policies mark out a new local terrain. In the past, American communities were able to rely on some measure of aid from state and national government. Although this external aid was substantially less than in other industrialized countries, evidence is strong that it allowed local governments to offset the effects of structural economic changes (see analysis in Ladd & Yinger, 1989). As such aid wanes, local governments are seeking new ways to generate sufficient revenues. In this chapter we assess the policy choices American local officials are making in response to resource changes, particularly the withdrawal of federal funds for local economic development activities in the 1980s. We draw on findings from a national survey of local economic development officials in 1989 to describe the postfederal policy context, local policy responses to changing resources, and the possible economic effects of current strategies.

The New Centrality of Locality

A volatile local resource base is a fundamental, enigmatic feature of this new setting. There is little reason to hope that local resource dilemmas will improve in the near future. Independent of who controls the White House, Congress, or your state capital, the effects of global competition, recessions,

AUTHORS' NOTE: This research was supported, in part, by the Center for Public Policy Research, University of Colorado, and the Economic Development Administration, U.S. Department of Commerce, Grant 99-07-13709. The statements, findings, conclusions, and interpretations are those of the authors and do not necessarily reflect the views of the Economic Development Administration.

increasing poverty and crime, changes in military priorities, and environmental concerns are apt to dominate political agendas in the next decade. The potential for a community-oriented policy response is weaker than in the past, thanks to ideological interpretations of economic restructuring processes that trivialize the role of cities in the national economy and undercut the rationale for national urban policies (these implications are drawn out in Barnekov, Boyle, & Rich, 1989). Neither the perceived crises nor the political constituency concerns that drove urban policy more than two decades ago (Brintnall, 1989) are as salient now as international crisis and competition issues. In the absence of alternative interpretations of economic trends, national and state policymakers will be reluctant to seemingly jeopardize economic productivity goals by directing resources to communities. Thus there is every sign that a chilly climate for cities will persist in the near future (see Walters, 1991).[1]

Ironically these broader trends may enhance local politics in the coming years. If contemporary economic change processes are uneven and imperfect, rather than single, uniform, monolithic, global processes, there may be greater potential for local political discretion and choice (Logan & Swanstrom, 1990, p. 30). And, as Mayer (1989, p. 1) points out, many of the conditions necessary for the attraction and expansion of new production processes and complexes cannot be organized centrally by either the national, state, or multinational corporation. In these settings, local governments play critical roles in coordinating the economic and political resources necessary for national economic growth and development. This claim for the new centrality of locality is not necessarily a more optimistic view or one promising more local autonomy. But it does emphasize that place-specific differences can be competitive advantages (Logan & Molotch, 1988); that is, cities matter. By default, if not design, current global trends and national policies promote a new economic localism and the potential resurgence of local politics (Clarke & Kirby, 1990).

Why anticipate that this new localism will engender significantly different economic development activities? The direction of local policy change is problematic, given the withdrawal of federal resources previously supporting local efforts to reduce investment uncertainty and risk. At a minimum, there is a possible reversion to caretaker roles and more risk-aversive, less active local agendas. And even if global economic trends do create a need for greater local policy coordination, local officials now must resort to their own revenues to provide such incentives and to entertain greater risk in joint projects.[2]

The Contemporary Policy Context

In a national study of local economic development efforts, we analyzed patterns of local economic development activities from 1978 to 1989. We examined whether cuts in federal resources affected the level of local economic development efforts and whether local efforts relying on own-source revenues, rather than federal resources, appeared to be relatively risk aversive or risk taking.[3] Here we report on the new local policy context—local strategies undertaken in

Table 10.1 Revenue Sources Most Important for Economic Development Activities (percentage of sources mentioned)

	Before 1984	Since 1984
Community Development Block Grant	20	18
General fund	16	25
Urban Development Action Grant	10	—
Economic Development Administration	9	7
Tax increment financing	8	10
Program income	—	7
Private/corporate contributions	—	8
Number of sources mentioned	174	226

the absence of federal resources. The study includes all American cities with populations above 100,000 in 1975; these were the cities primarily targeted for federal economic development aid, and these are the cities now coping with its withdrawal. Data on current and historical local economic development activities were collected for the 178 cities above the population threshold. Officials in all cities above the 1975 population threshold were contacted to collect information on current economic development strategies; the findings here reflect responses from 101 cities.[4]

Resource Change and Local Policy Efforts

Federal program funds prompted the acceleration of local economic development activity in the late 1970s and 1980s. As Table 10.1 shows, by the mid-1980s, federal program funds were the most important revenue sources for local economic development. All together, 49% of the revenue sources for local economic development important before 1984 involved federal program funds. Community Development Block Grant (CDBG) funds received most mentions (20%) although Urban Development Action Grant (UDAG; 10%) and Economic Development Administration (EDA; 10%) funds were also important. But even during this period of federal attention, local general funds constituted 16% of the sources mentioned.

In the late 1980s, the volatility of federal resources, as well as absolute cuts in the level of resources, forced many local officials to turn to nonfederal resources to support local economic development activities. How did this change in resource availability affect local economic development efforts? Of the 237 problems mentioned by local officials as constraints on local economic development efforts, resource issues stand out: Lack of funds, tight budgets, and lack of staff were mentioned most frequently (24%) as a major problem.[5] But city officials also report a tremendous increase in public and private cooperation. For 75% of the cities, this is a dramatic increase in public/private activity compared with 5 years ago.

With the loss of federal program funds, local officials turn to three other revenue sources to support economic development activities: (a) relying increasingly on local general funds, (b) shifting CDBG funds to economic development purposes, and (c) using revenues from successful redevelopment projects (Table 10.1). The first two suggest reallocation of funds across functional areas; in the absence of federal funds previously used to subsidize local economic development efforts, cities turn to general funds or to the declining pot of CDBG funds. Cities increasingly devote CDBG funds to economic development: The median use of 10% in fiscal year 1988 is double the median reported by these cities for 5 years ago. Greater reliance on tax increment financing and program income from previous redevelopment projects signal attempts to make economic development a self-financing local enterprise.

Overall, local economic development efforts are expanding in the postfederal era. The local focus clearly has shifted to the state level, although there is no perception that state programs have compensated for the federal absence or offset federal program cuts.[6] The findings suggest a more politicized, more fragmented local setting in the wake of federal withdrawal; increased competition over general fund revenues, CDBG funds, and tax increments makes consensus more difficult, a situation only compounded by basic problems of aging infrastructure and land shortages.

Shifts in Local Economic Development Policy Orientations

The new local policy strategies reflect a shift away from conventional economic development orientations toward market-based, or entrepreneurial, approaches.[7] Two features distinguish market-based economic development strategies from more conventional approaches: (a) the focus on facilitating value-creating processes by private investors and (b) the investment and risk-taking role adopted by local officials (Vaughn & Pollard, 1986). Variously labeled "generative development," "enterprise development," and "entrepreneurial" strategies, these new approaches center on public policies that encourage wealth creation, rather than subsidize locational decisions or employment strategies.[8] As such, they demand a new range of local roles and responsibilities.

Because this new generation of complex, entrepreneurial, local policy initiatives operates at the grassroots, local strategies are less visible and less easily characterized than nationally designed local economic development programs. These initiatives are not embodied in specific programs or agencies but, rather, in particular policy tools or instruments sharing certain common features that distinguish them from previous approaches (see Clarke & Gaile, 1989). Greater tolerance for *risk* is one of these features; others are differences in *purpose* (stimulating new enterprise, rather than stabilizing or protecting) (Dubnick & Bardes, 1983); in *focus* (using government authority to shape market structure and opportunity, rather than influencing the functions of individual businesses) (Sternberg, 1987); in *criteria* (using market criteria such as maximizing rates of return, rather than political criteria in setting priorities for allocation and investment of public funds); in *finance* (leveraging public and private resources, rather than relying on one or the other); in *public roles* (relying on joint public-private ventures for implementation of economic development projects, rather than bureaucratic approaches); in *administrative ease* (administering through quasi-public agencies, rather than line agencies); in *decision processes* (involving negotiated decisions on a case-by-case basis, rather than juridical, standardized decision processes); and in *linkages* (establishing contractual, contingent relations with those affected, rather than linkages based on rights or entitlement).

This new generation of policy approaches is rooted in a legacy of diverse federal programs encouraging local discretion, investment orientations, and greater risk taking. The early National Development Council traveling seminars on packaging "deals," the UDAG leverage ratios and kicker features, and the introduction of CDBG float loans and Section 108 guarantees were particularly important in stimulating new entrepreneurial practices. Over time, these entrepreneurial strategies began to diffuse across cities and over programs. The federal urban legacy, however, was ambivalent: It promoted certain redevelopment goals, encouraged large commercial and industrial projects, initiated new local development institutions, and endorsed innovative financing techniques. When cities were cut loose from federal ties, they retained some of the entrepreneurial aspects of past federal programs but geared them to local ends. Local entrepreneurial approaches remain crucially dependent, however, on state-enabling legislation and fiscal discretion authorities. Many of the more entrepreneurial tools require positive state action, sometimes including changes in state constitutional limitations on local fiscal powers.[9] Ironically, although the extensive local use of these tools is often attributed to the withdrawal of federal oversight, federal tax code changes, and the deregulation climate for private investment, their ultimate range may be limited by state governments.

In 1989, local officials reported notable policy shifts prompted by cuts in federal economic development programs. These shifts included changes in local objectives, smaller projects, more diverse projects, and smaller public shares of development costs but little change in sectoral, spatial, or minority targeting efforts in most cities. Nearly 40% of the cities responding claimed that federal cuts resulted in more diverse economic development objectives, in many instances including more housing and social programs responsive to local priorities. About one third (29%) of the cities saw these cuts as negative, leading to restricted, narrower local objectives. Most cities now are involved in smaller projects (55%), and many reported that the nature of local projects is now more diverse (38%), often because the choices are market driven, but more often because they are now more open to local political and organizational pressures. Public shares of development costs have decreased in over 46% of the cities responding. Targeting efforts appear to be unchanged in a majority of cities, although the percentage of cities claiming to do more minority (17%) and low-income (20%) targeting is similar to the percentage claiming to do less (15% and 21%, respectively).

Local officials believe that their current policy orientations differ in important ways when compared to the past: Current policies are oriented more toward risk taking, job growth, downtown development, job creation, local concerns, market feasibility, and aiding local firms, when compared with policies of 5 years ago. They also are more likely to see the city's role as a public developer engaged in contractual relations, rather than responding to entitlement obligations. This brief profile intimates a more entrepreneurial orientation, emphasizing job creation and indigenous growth and accommodating more risk taking by local officials.

Local Policy Choices
When Using Nonfederal Resources

These impressions of shifting local policy orientations are borne out by comparing the clusters of tools adopted recently with those reported in use before 1980. Local officials reported whether and when they first had used 47 strategies employing nonfederal resources. At least 50% of the cities reported using 25 core strategies; of those 25, 7 were first used after 1980 by at least half of the cities. As listed in Table 10.2, planning, management, and marketing strategies are prominent and pervasive before 1980. Land acquisition, building demolition, and the use of infrastructure to support development are also classic local development tools. Direct federal influence is evident in the importance of tax-exempt bonds, historical tax credits, and program income. Federal programs also spread or diffused more widely the use of revolving loan funds, local development corporations, enterprise zones, strategic planning, below market rate loans, and community development corporations. Thus the current menu of local economic development strategies is shaped by past federal initiatives and residual local powers.

Patterns of Policy Change

But some of the most heavily used local strategies are relatively recent innovations. In Table 10.3, strategies used before 1980 are sorted out from those first used in the postfederal era. The first wave of policies—those reported as first used before 1980 by at least 30% of the cities responding—rely heavily on cities' ability to regulate and facilitate development by land-use controls, public services, and provision of infrastructure (Table 10.3). Only limited, traditional financial tools entailing debt or cheap loans are noted; there is scant evidence of the revenue-generating financial tools or higher risk city roles that characterize recent entrepreneurial approaches.

In the second wave—tools adopted since 1980—cities are characterized by a stronger investment and entrepreneurial approach; the use of revolving loan funds, below market rate loans, and program income indicates generation of revenue streams independent of federal programs and tax revenues. Further, both enterprise zones and tax increment financing districts let cities reorganize their local fiscal structure to channel future revenues into allocation procedures outside the usual budgetary processes. In this sense, revenues come into the city

Table 10.2 Most Frequently Used Economic Development Strategies
(percentage of cities that ever used strategy)

Strategy	Percentage
Comprehensive planning	93
Capital improvement budgeting	91
Marketing and promotion	86
Infrastructure as in-kind development contribution	83
Land acquisition and demolition	80
Revenue bonds	79
Strategic bonding	74
Revolving loan fund	73
Streamlining permits	73
Selling land	69
Industrial parks	68
Below market rate loans	67
General obligation bonds	65
Local development corporations	63
Annexation	62
Historical tax credits	60
More metropolitan and regional cooperation	58
Tax increment financing	56
Industrial development authorities	55
Enterprise zones	55
Use program income for economic development	55
Special assessment districts	54
Community development corporations	52
Land leases	52
Trade missions abroad	50

NOTE: Italics indicate majority of cities that first used strategy after 1980.

on the basis of public investment decisions, rather than of tax policies or federal programs, and future resources may be allocated with little public notice or accountability. Finally, attention to markets and business start-ups—two hallmarks of entrepreneurial approaches—also characterize this new postfederal era.

Looking ahead, entrepreneurial orientations predominate (Fosler, 1991, also uses the wave analogy to describe innovative local development strategies). Although a wide range of programs are reported as "under study," those mentioned most often share a number of traits. The importance of business incubators as a recent and future strategy underscores the salience of recent approaches encouraging new business start-ups, rather than conventional "smokestack chasing" strategies. There is also a clear interest in spatial reorganization of the local tax and resource base. The interest in greater metropolitan and regional cooperation on development issues suggests that economic regionalization issues may be more manageable than political consolidation efforts. Reports of

Table 10.3 Patterns of Use of Local Economic Development Strategies
(percentage of cities that used strategy)

	Before 1980	After 1980	Under Study
Comprehensive planning	84		
Capital improvement budgeting	77		
Revenue bonds	69		
Land acquisition and demolition	67		
Infrastructure as in-kind development contribution	64		
Selling land	59		
General obligation bonds	59		
Industrial parks	57		
Annexation	53		
Marketing and promotion	47	39	
Strategic planning	46		10
Industrial development authorities	40		
Local development corporations	38		5
Special assessment districts	37		11
Land leases	37		5
Land banks	34		10
Below market rate loans	33	34	
Enterprise funds for public services	33		
Historical tax credits	31		
Streamlining permits		59	7
Enterprise zones		52	
Revolving loan funds		45	
Business incubators		40	21
Trade missions abroad		39	
More metropolitan and regional cooperation		36	14
Use program income for economic development		33	
Tax increment financing		32	12
Foreign trade zones			10
Export and promotion			10
Equity participation			10
Taxable bonds			9
Tax abatements—targeted at new business			7
Equity pools: public-private consortiums			7
Venture capital funds			5
Linked deposits			5
Sale-leasebacks			5

the extensive use of regional economic development associations further attest to the viability of a regional approach despite assumptions of interjurisdictional competition.

This viability is complemented by an interest in carving the local tax base into foreign trade zones, special assessment districts, tax increment financing districts, and enterprise zones. In each instance, these spatial arrangements tend to reduce the revenue base allocated by normal budgetary processes; tax revenues from these arrangements often are dedicated to debt service and further a real redevelopment, rather than enhance the city tax base. Finally it appears that local officials are adopting a more business oriented approach toward use of local assets; potential land-use control strategies emphasize the flexible management of land and its exchange value, rather than permanent transfers or sales. Similarly there is a growing inclination to view public capital in terms of its investment potential, with the greater public risk taking that that implies.

Changes in Institutional Arrangements

In most cities, mayors take the lead in promoting economic development. Many cities also indicate that deputy mayors have special responsibilities in this area. Economic development line agencies reporting to the mayor and to the mayor/city manager or city council have prime responsibility for policy formulation in most cities; over 50% place prime responsibility in economic development line agencies that are separate from community development or planning agencies. Thus the design of economic development policy rests with elected officials or those appointed by them. In contrast, economic development line agencies often share implementation responsibilities with special authorities or quasi-public organizations. Cities often turn to organizations such as citywide development corporations or to special authorities such as redevelopment agencies and port authorities to carry out economic development projects. Many of these organizations have special financing authorities or resources not available to line agencies; further, they bring expertise and resources from the private sector that are especially germane to economic development projects.

In the absence of federal programs, additional place-specific means of coordinating private and public sector commitment will be critical. In this sample, 63% of the cities reported using local development corporations. Cleveland credits its recent turnaround, for example, to institutional coordination of the business community, neighborhood groups, and local government agencies. The institutional arrangements include Cleveland Tomorrow and its offshoot, the Neighborhood Program Inc., a citywide development corporation, the Cleveland Foundation, the city's Neighborhood Partnerships Program, and a network of burgeoning neighborhood local development corporations.

Some hope for disadvantaged groups in American cities rests in community development corporations (CDCs) and local development corporations (LDCs). Some cities, such as Miami, view these organizations as an alternative delivery system for carrying out economic development activities in the postfederal era. They may become pivotal as cities reorganize their tax base into discrete areas;

LDCs and CDCs in those areas may claim some voice in the development promoted in those areas, as well as in the allocation of revenues generated by these new arrangements. Yet this importance presumes an autonomous capacity and technical knowledge base not yet apparent in most such groups. At this stage, most LDCs and CDCs are capitalized by public funds and staffed by city agencies and private organizations; many have displaced older voluntary organizations or share power with these groups in an uneasy alliance.

Analyzing the Effects of Local Choices

Attempting to assess the effects of local economic development strategies is a quagmire of good intentions and bad measures. An interest in finding out whether policies have the desired effect is laudable, but there is little consensus on appropriate measures of success or impact (see Hatry, Fall, Singer, & Liner, 1990). For many city officials, project completion is unique enough to be a salient and sufficient success measure (Frieden & Sagalyn's [1989] interviews with local officials underscore this perspective). Few have the means or interest in monitoring project-specific revenue and job impacts, so there is little systematic data on project effects, especially those without federal funds or monitoring requirements.

Indeed, most local officials adopt a portfolio mentality in which they assume that some projects, such as a convention center, are "loss leaders" but that the overall local policy effort contributes to local economic growth (Frieden & Sagalyn, 1989, p. 270). Nevertheless analysts persist in measuring the success of local policy efforts in terms of changes in local per capita income and local employment per resident. Not surprisingly, there is little evidence that local economic development policies affect either of these measures. As any local official will point out, the former is a measure of wealth beyond the control of local efforts and often an artifact of jurisdictional boundaries. There is little evidence that changes in residential wealth, as reflected in income measures, are associated with changes in a city's fiscal well-being (Ladd & Yinger, 1989).

Jobs are another matter. Jobs for city residents are important goals for city officials for many reasons. Most local economic development strategies emphasize job creation and retention as their primary justification. Trying to shape a city's employment structure seems more tangible than influencing the distribution of wealth; it involves negotiations with developers and merchants in which the city can bring something to the table. Cities can negotiate jobs for city residents and set-asides for local workers and contractors.[10] These concessions will not necessarily alter the overall employment structure, but they do display city initiative and generate local revenues. Unfortunately employment growth may not be associated with income growth because of commuting patterns, wage differences, and overlying jurisdictional boundaries. Thus cities with healthy job growth also may have impoverished populations (Ladd & Yinger, 1989, p. 17). Further, increases in employment may be mixed blessings because they entail higher service costs that may not be borne by commuting workers (Ladd

Table 10.4 New Economic Development Tool Usage Index and City Growth Features

City Growth Features	Correlation Coefficient
Job growth rate, 1987	.20
Fast-growing companies as a percentage of new firms, 1987	.19
City tax per capita, 1985	−.18
Property tax per capita, 1985	−.18
City government employees per 10,000 population, 1985	−.18
City government general expenditures per capita, 1985	−.16

NOTE: All coefficients significant at .05 level.
SOURCES: New Economic Development Tool Usage Index—An additive index measuring whether a city had used the following tools for the first time in 1980 or later: revolving loan fund, venture capital, net cash flow participation, interest subsidies, equity participation, equity pools established by private-public consortiums, or use of program income for economic development purposes. Usage measures are based on national survey of local economic development officials.
Job growth rate—Change in private employment between January 1983 and March 1987. Bureau of Census and Bureau of Labor Statistics, as reported in *Inc.*, March 1988, p. 76.
Fast-growing companies—The percentage of all companies founded between January 1983 and July 1987 by its percentage of employment growth during the same period. Companies with an index of 20 or higher were classified as high growth. As reported in *Inc.*, March 1988.
City fiscal data—County and city data book.

& Yinger, 1989, p. 290). But for political and fiscal reasons, job-related measures are the primary concern of local officials and is the success indicator we use in the following preliminary assessments.

Because of the newness of these postfederal strategies, it is too early to determine conclusively whether they have brought about net changes in local jobs and revenue. Given this necessary disclaimer, the last two tables compare local growth features of cities adopting specific tools and those that have not. For Table 10.4, local policy activity is measured with an additive index of specific entrepreneurial tools: whether a city first used revolving loan funds, venture capital, net cash flow participation, interest subsidies, equity participation, equity pools established by private-public consortiums, or used program income for economic development purposes in 1980 or later.[11] From data in Table 10.4, we suggest that, controlling for region, cities using entrepreneurial strategies have higher job growth rates and higher proportions of fast-growing new firms than cities without these tools. Further, they are operating with significantly lower taxes, lower expenditures, and lower levels of city government employment than nonadopting cities.

In addition, there is some modest support for associating use of specific policy tools with these city growth features. Statistical analyses comparing group means (*t* tests) of cities that had ever used a particular tool with those that had never used a particular tool (Table 10.5) show several instances of tools whose use appears to effectively distinguish among communities with fast economic growth and those with slow growth. Not all of the tools associated with distinctive

Table 10.5 Economic Development Tool Usage and City Growth Features;
Means of Standardized Residuals From Trend Surface Analysis

Policy Tools	City Growth Features	Ever Used Tool	Never Used Tool
Marketing and promotion	Increasing job growth rate	.22	−.58
Cash flow participation		.69	−.09
Selling land		.24	−.31
Equity pools unded by public-private consortium	Fast growing companies as a percentage of new firms	.21	−.07
Cash flow participation		.51	−.17
Marketing and promotion		.13	−.53
Community development corporations		.58	−.20
Enterprise zones	Increasing relative business growth rate, 1980-1986	−.15	.01
Interest subsidies		1.23	−.07
Land acquisition and demolition		1.15	−.06
Land leases		1.07	−.05
Loan guarantees		.83	−.03
Infrastructure as in-kind contribution	Increasing business birthrate	−.06	.68

NOTE: Difference of means per t test significant at .05 or greater.
SOURCES: Job growth rate—Change in private employment between January 1983 and March 1987. Bureau of Census and Bureau of Labor Statistics, as reported in *Inc.*, March 1988, p. 76.
Fast growing companies—Young companies enjoying high growth rates. The percentage of all companies founded between January 1979 and July 1983; for each company, an index was calculated by multiplying its absolute growth in employment between January 1983 and July 1987 by its percentage of employment growth during the same period. Companies with an index of 20 or higher were classified as high growth. As reported in *Inc.*, March 1988.
Relative business growth rate—Change in business growth rate, 1980-1986. As reported in *Inc.*, March 1988, p. 76.

growth differences fit a strict entrepreneurial definition; interestingly they also represent a cross-section of capital, land, and institutional strategies. Although neither Table 10.4 nor Table 10.5 should be interpreted as attributing these growth features to specify policy choices, the findings do make a case for further exploration of these new local strategies.

Conclusions

Most American cities in the postfederal period are making increased use of relatively entrepreneurial strategies that entail risk of own source revenues and substantial opportunity costs. There are concomitant shifts in policy orientation: the ascendance of market feasibility over social criteria, the resurgence of downtown as the locus for redevelopment, a trend toward reliance on nonprofit organizations rather than line agencies for implementation, and the redefinition of city responsibilities as a public developer. We find some evidence that use of these entrepreneurial strategies distinguishes communities with job growth, new firm formation, and fast-growing firms from those less fortunate. This is not necessarily a causal relationship, but it does indicate a clear difference in the public investment climate in communities adopting market-oriented strategies and those with more conventional orientations.

Although this analysis indicates that entrepreneurial approaches are becoming more characteristic of city economic development strategies, several factors guard against anticipating a wholesale conversion to this orientation. Most significantly, an entrepreneurial policy orientation capable of comprehensively dealing with the economic transformation issues impacting local governments is beyond the organizational and institutional capacities of American local government. Entrepreneurial policies present cross-cutting distributional and redistributional policy issues that are not easily dealt with through extant policy-making processes. The benefits and costs of these approaches are less visible and possibly more long-term than most economic development issues; this feature, in addition to the obvious fiscal and political advantages, may partially explain why these policies are so often the province of joint public-private or off-budget institutions. In the European context, this need for reducing certainty and increasing consensus on complex economic policies often results in the establishment of corporatist procedures and institutions, but in the American setting the result is the institutionalization of off-budget arrangements. Few current American political institutions are geared to the consensual, cooperative decision processes demanded by these approaches.

Even if the institutional and organizational capacity could be established, there is rarely a compelling consensus on entrepreneurial goals or effective strategies for achieving them. The entrepreneurial emphasis on wealth generation quickly breaks down into distributional issues. Curiously enough, the interest group politics characteristic of distributional politics are less evident in the emergence and spread of this new approach. There is no coherent voice from the business community nor any clear dissent from other groups. Older manufacturing interests clearly benefit more from traditional subsidy orientations; newer growth sectors and smaller, newer businesses are more likely to gain from the market facilitation slant of entrepreneurial strategies. But in few instances have the actions of other interests thwarted entrepreneurial initiatives. To some extent, cloaking entrepreneurial approaches in the symbolic language of "growth," "wealth," and "jobs" dissuades informed discussion or active opposition to these risk-taking activities.

Despite this lack of organized opposition to entrepreneurial approaches, interviews with local officials revealed a great deal of conflict over these policies. This local conflict appears to have two dimensions: (a) the appropriate city role in economic development and (b) the balance of executive and legislative voice in the formation and implementation of economic development policy in the postfederal era. Federal economic development resources obscured ideological issues over the appropriate degree of local activism on economic development issues; because outside external resources were at risk, it was possible to move ahead without resolving disagreements over local roles. With the withdrawal of federal resources, local economic development issues once again center on whether cities should return to "caretaker" roles and a more risk-aversive stance or pursue more activist, risk-taking roles.

Presumably this traditional debate became a moot point in the face of federal revenue losses and increasing service demands, but it is a surprisingly salient issue in many cities. Local officials explain it in terms of divergent political constituencies: Mayors and administrators appreciate the "politics of announcement" and the career-enhancing, credit-taking opportunities offered by economic development activities. Council members are concerned about immediate constituency needs and are less willing to risk their own source revenues on ventures with real opportunity costs and, at best, long-term payoffs. These conflicts in institutional incentives, coupled with a lack of consensus on goals and tools, may contribute to a stalemate on further initiatives and to increased local political conflicts over future economic development strategies.

Notes

1. Like Brintnall, Walters (1991) stresses the shifting political constituencies and coalitions that work against future efforts at formulating *urban* policy. He takes particular note of the salience of urban versus suburban hostilities for both national and state policymakers.

2. An insightful series of case studies on how local officials in different cities respond to these dilemmas is available in Frieden and Sagalyn, 1989.

3. Findings from the larger study are reported in Clarke and Gaile, 1990. The study includes comparisons of city economic development strategies when using federal economic development funds with tools used in the absence of federal resources. The analysis indicates that needy cities were more likely to use risk-taking entrepreneurial tools in federal programs, whereas better-off cities are more likely to use such tools when relying on nonfederal resources.

4. Field visits to 15 cities with distinctive federal program participation profiles used unstructured open-ended interviews with local officials to identify current approaches and those under study. From these discussions, we developed a mail survey instrument to systematically collect information from the 178 cities meeting the 1975 population criterion. In each city a government official responsible for economic development was identified and asked to provide this information. The inquiries focused on the use of 47 economic development tools that do not require federal resources, the timing of policy adoption, and the assessment of policy effectiveness. Of the 178 cities in this population, usable responses were gathered from 101 cities, for a response rate of 57%. A statistical comparison of this subpopulation of respondents with the overall population indicated that the respondent subpopulation was not significantly different. For more detail see Clarke and Gaile, 1990.

5. State laws restricting local fiscal powers and fiscal authority also are perceived as serious barriers; they constitute over 9% of the problems noted. Fragmented public voice is also a concern; 9% of the problems cited involved the lack of local political consensus and leadership, and the absence of focus for economic development efforts. Cities also are limited by land availability problems; these issues constituted over 7% of all problems mentioned. Aging infrastructure and the lack of amenities or perceived image problems limit many cities. Finally federal restrictions on program eligibility, fund uses, environmental impacts, and other guidelines, as well as the loss of federal funds, are seen as constraints (6%) but rank behind lack of resources, state laws, and land availability issues.

6. State aid is problematic. With the exception of state enterprise zones, traditional state programs addressing infrastructure needs (11%) and providing financing options (10%) are seen as most useful of the 169 programs nominated. State enterprise zone programs were mentioned more often (17%) than any other

state program. Although other innovative programs such as business incubators and high-tech programs were noted (5%), they are not perceived to be as useful as conventional programs such as infrastructure assistance, revolving loan funds, linked deposit schemes, job training programs, and marketing and promotion efforts, including international trade promotion. The diffuse endorsement of state programs, and the 9% specifically mentioning that state programs are not useful, indicates that state governments have yet to fill the gap left by the absence of federal funds.

7. For a cogent description of these orientations, see Eisinger, 1988. For an empirical analysis of state economic development policy shifts, see Clarke and Gaile, 1989.

8. These distinctions are traced by several observers, including the Committee for Economic Development (1986); Bowman (1987a, 1987b); Eisinger (1988); and Bingham, Hill, & White (1990).

9. For example, the strategies reported as least used by over 75% of the cities responding include those that would interfere with private investor decision making, compete with private investment capital, or require significant state actions loosening local fiscal constraints. This includes mechanisms allowing cities to pool public and private capital or to influence the use of private investment funds, some of which face state or local legal restrictions. Mechanisms for pooling capital receive high ratings as future strategies currently "under study," though it is currently something cities may not do (in many cities, pension funds are controlled by state personnel systems), or lack the capacity or private sector cooperation to do (zero coupon bonds, linked deposits, venture capital funds, public-private consortium equity pools).

10. For a national study analyzing the conditions under which such local linkage policies are adopted, see Goetz, 1990.

11. We explicitly incorporated and controlled for the expected spatial variation in local entrepreneurial activity and its outcomes through use of trend surface analysis techniques. Analysts and local officials are well aware that a given policy effort in a region characterized by distress will not have the same impact on outcomes as a similar level of effort in a region characterized by economic prosperity; that is, the same strategy applied in Buffalo and San Diego can be expected to have differential impacts. They both may be positive relative to their region, but this success may be overwhelmed by the statistical noise in regional variations in the overall economy. In analyses ignoring locational differences or attempting to control for them with dummy variables for regions, significant explanatory power is lost.

The use of spatial statistical techniques to identify and control for these regional variations is discussed in Gaile and Willmott, 1984. Here we use trend surface analysis to measure the strength of regional variables and to create residuals to be used as independent nonautocorrelated variables in further analyses. More specifically, before analyzing the relationship of the new policy index and city growth features, it was necessary to develop residuals through trend surface analysis for each variable with strong regional patterns. The principal methodological argument is that regional trends exist in most of the variables typically used to identify policy effectiveness. The hypothesized strength of these regional trends is such that they would likely mask or misrepresent significant relationships between new policy efforts and city growth measures.

Essentially this is a straightforward modification of standard regression techniques; it uses regression to analyze a variable's locational attributes by disaggregating the variable into a *regional component* that is described by the regression "surface" and a *local component* that is described by the residuals from that surface. The validity of trend surface results can be judged by using the same appropriateness measures as in simple regression analyses. Thus F-tests are performed, and significance levels of the analyses are readily interpretable. Our trend surface analyses of the policy index showed no statistically significant regional trends. This finding suggests that these entrepreneurial, market-based policies have largely become employed on a national basis without strong regional emphases.

References

Barnekov, T., Boyle, R., & Rich, D. (1989). *Privatism and urban policy in Britain and the United States.* Oxford: Oxford University Press.

Bingham, R. D., Hill, E. W., & White, S. B. (Eds.). (1990). *Financing economic development.* Newbury Park, CA: Sage.

Bowman, A. O'M. (1987a). *Tools and targets: The mechanics of city economic development.* Washington, DC: National League of Cities.

Bowman, A. O'M. (1987b). *The visible hand.* Washington, DC: National League of Cities.

Brintnall, M. (1989, Winter). Future directions in federal urban policy. *Journal of Urban Affairs,* pp. 1-19.

Clarke, S. E., & Gaile, G. L. (1989). Moving towards entrepreneurial state and local economic development strategies: Opportunities and barriers. *Policy Studies Journal, 17,* 574-598.

Clarke, S. E., & Gaile, G. L. (1990). *Assessing the characteristics and effectiveness of market-based urban economic development strategies* (Grant #99-07-13709). Washington, DC: U.S. Department of Commerce, Economic Development Administration.

Clarke, S. E., & Kirby, A. M. (1990, March). The mysterious case of the local corpse. *Urban Affairs Quarterly*, pp. 389-412.

Committee for Economic Development. (1986). *Leadership for dynamic state economies*. New York: Author.

Dubnick, M. J., & Bardes, B. A. (1983). *Thinking about public policy*. New York: John Wiley.

Eisinger, P. (1988). *The rise of the entrepreneurial state*. Madison: University of Wisconsin Press.

Fosler, R. S. (Ed.). (1991). *Local economic strategy*. Washington, DC: International City Management Association.

Frieden, B. J., & Sagalyn, L. (1989). *Downtown, Inc.* Cambridge: MIT Press.

Gaile, G. L., & Willmott, C. J. (1984). *Spatial statistics and models*. Dordrecht, The Netherlands: D. Reidel.

Goetz, E. (1990). Type II linkage policies. *Urban Affairs Quarterly, 26,* 170-190.

Hatry, H., Fall, M., Singer, T. O., & Liner, E. B. (Eds.). (1990). *Monitoring the outcomes of economic development programs*. Washington, DC: Urban Institute.

Ladd, H. F., & Yinger, J. (1989). *America's ailing cities*. Baltimore: Johns Hopkins University Press.

Logan, J., & Molotch, H. (1988). *Urban fortunes: The political economy of place*. Berkeley: University of California Press.

Logan, J., & Swanstrom, T. (Eds.). (1990). *Beyond the city limits: Urban policy and economic restructuring in comparative perspective*. Philadelphia: Temple University Press.

Mayer, M. (1989, September). *Local politics: From administration to management*. Paper presented at the Cardiff Symposium on Regulation, Innovation, and Spatial Development, University of Wales.

Sternberg, E. (1987). A practitioner's classification of economic development policy instruments, with some inspiration from political economy. *Economic Development Quarterly, 1,* 149-161.

Vaughn, R. J., & Pollard, R. (1986). Small business and economic development. In N. Walzer & D. L. Chicoine (Eds.), *Financing economic development in the 1980s*. New York: Praeger.

Walters, J. (1991, April). The urban crisis, Act II. *Governing*, pp. 26-32.

— 11 —

Equity and Local Economic Development

Norman Krumholz

Local economic development is one of the hottest topics in America's cities. Snowbelt or Sunbelt, it is the topic that mayors talk about the most. Many of the announcements from the mayor's office and many of the headline stories in the local media deal with proposed and actual development projects that ostensibly promote the local economy, earn more taxes, and create more jobs for city workers. Public subsidies are used in increasingly ingenious ways to stimulate and "leverage" private investment. Yet, although politicians, economic development practitioners, and the media trumpet successes of local economic development in many cities, poverty deepens, unemployment rises, and neighborhoods deteriorate.

The effectiveness of local economic development in terms of net new jobs or taxes—its essential public purpose—is largely unknown. As Matthew Marlin (1990) recently has written:

> Despite billions of dollars and an ongoing controversy, practitioners and academics have generated surprisingly little empirical evidence regarding the effectiveness of economic development incentives or subsidies in promoting economic growth. Some have argued that they are an essential tool that stimulates local economic development, while others have argued that they are little more than a windfall subsidy for investment and would have occurred anyway. (p. 15)

In this chapter I look at local economic development from the perspective of a former Cleveland City Planning Director. I conclude (a) that the activities of local economic development practitioners are more promotional, marketing, and sales oriented than part of a rational planning process, (b) that they lack accepted record-keeping and accounting procedures, and—most importantly—(c) that they fail to negotiate equitably or aggressively on behalf of their increasingly distressed resident populations. To offset the latter deficiency, I offer a more equitable negotiating model.

The Ideal and Real Economic Development Process

Local economic development is a process by which local governments manage resources to stimulate private investment opportunities in order to generate new jobs and taxes. In the process, local government may try to execute its own plans and initiatives or, more likely, enter into partnerships with private enterprises or community-based groups. The core of locally based economic development is "the emphasis on 'endogenous development' policies using the potential of local human, institutional and physical resources" (Blakely, 1989, p. 59).

The local economic development process is divided into a number of tasks. Blakely (1989) lays these tasks out to include (a) gathering and analyzing data, (b) selecting local development goals, strategies, and criteria, (c) selecting local development projects, (d) building action plans and analyzing financial alternatives, (e) specifying feasibility and project details, and (f) preparing the overall development plan and scheduling its implementation (p. 74).

The process assumes rational planning and a broad range of technical studies—studies of labor markets and tax bases, interrelationships of land use and transportation, impact of employment growth on housing markets, and financial feasibility. Other technical studies might be the need for new or improved capital facilities, cost-benefit analyses, and industrial quotients to target various industries.

Regardless of the form of local economic development or the specific technical studies, local economic development has the primary goal of creating new jobs for local residents and providing a net tax increase to the local treasury. Indeed, it is these elements that provide the entire "public purpose" for the use of public subsidies.

The above description of the local economic development process implies a high degree of comprehensiveness and rationality. More limited approaches speak of "contingency" and "strategic" planning, but even these more limited efforts often serve as parts of a more comprehensive strategy (Malizia, 1985).

Local economic development practitioners are to be rational and systematic. It is a description reminiscent of the work of city planners operating in the "rational comprehensive model" (Banfield, 1973). The rational comprehensive model of city planning involves the defining the problem, gathering facts and analyzing them, selecting goals and optimal courses of action from among the many alternatives possible, developing detailed strategies and programs, implementing programs, analyzing their results, and feeding the results back into the model for corrective guidance. The rational comprehensive planning model is quite similar to textbook descriptions of local economic development.

Both formulations are, however, open to criticism. The rational comprehensive planning model is criticized on at least three levels. First, it is argued that the goal of optimization is impossible; the extensive data necessary to support "best" choices simply cannot be assembled within reasonable time spans. Most managers do not optimize with the best choices possible, but simply "satisfice"—that is, strive for merely adequate solutions within the time available for their decisions (Simon, 1955). This mode of decision making also has been described as "disjointed incrementalism" (Lindblom, 1959).

Second, value clarification is a problem. Different individuals and different groups in society have differing goals and objectives. Instead of a single, unitary public interest that a planner might clearly serve, there are many publics and multiple public interests. Under the circumstances, goal selection is, at best, a great deal more complex than it is made out to be; at worst, it is an exercise in elitism. Third, the model overlooks, or simplifies, the legal, political, and cultural constraints that inhibit the implementation of seemingly optimal solutions (Banfield, 1973).

The textbook model of local economic development seems also to be deeply flawed in practice. A recent survey finds that practitioners spend most of their time on public relations, marketing, advertising, and sales—not on research and analysis. Respondents to this survey described a reactive sales or marketing competition largely focused on downtown and relatively untouched by planning or targeting (Levy, 1990). Leadership is provided by private developers or real estate entrepreneurs seeking their own objectives, rather than by officials and citizens seeking public objectives. Other practitioners of local economic development have reported to me that their prime responsibility is simply "packaging the deals."

In the real world of municipal economic development, the local practitioner coordinates four components: land (or a building), public subsidies, political support, and private investment. The practitioner brings these four components together around a physical project usually initiated by the private sector. The official seeks such public subsidies as are available, "leverages" them with private funds, and supports the "deal" with confirmatory statements in order to justify the subsidy. At this level it is hard to distinguish the public official from the private developer.

In effect, the public economic development practitioner becomes an arm of the private developer; the success of the latter is a measure of the effectiveness of the former (for elaboration, see Rubin, 1988). Politically such closeness makes sense because it helps local government look good, strengthens the position of the economic development agency, and may even facilitate the practitioner's move to his or her next job with a private developer.[1] Closeness also builds a symbiotic relationship between the mayor's office (and economic development agency) and the business community—cemented, on the one hand, with public subsidies to private investors and, on the other hand, with private contributions to political campaigns.[2]

More significantly, economic development practitioners have been criticized for freely using various inducements with little idea of or interest in their effectiveness in achieving the overarching objectives—new jobs and net tax increases. A study of 21 subsidized projects in five cities (Birmingham, Hartford, Milwaukee, St. Louis, and Toledo) found that the projects had produced far fewer jobs and other benefits than were promised (Center for Community Change, 1991). This result should not be surprising, given that large corpora-

Table 11.1 Distribution of UDAG Dollars in Cleveland, 1981-1988

Location	Commercial/ Office	Industrial	Housing	Health/ Institutional	Total
Downtown[a]	72,221,495	0	0	351,750	72,573,245
University Circle	1,350,000	0	5,500,000	1,200,000	8,050,000
Neighborhoods	10,047,493	0	4,606,000	2,276,810	16,930,303
Other areas[b]	0	4,300,000	0	1,455,000	5,755,000
Total	83,618,988	4,300,000	10,106,000	5,283,560	103,308,548

SOURCE: City of Cleveland, Department of Economic Development.
NOTE: a. Includes central business district, Playhouse Square, Warehouse District, and the Flats.
b. Includes industrial and institutional areas outside of downtown, University Circle, and the city's residential neighborhoods.

tions get most of the subsidies and that small businesses produce most of the jobs (Redmond & Goldsmith, 1986). Thus public inducements are made available, economic development deals get made, and projects get built, but unemployment increases, poverty grows, and the city's neighborhoods become increasingly distressed. Let us look briefly at what the record seems to say about a few of these public economic development inducements.

Urban Development Action Grants (UDAGs)

On the heels of the demise of urban renewal in 1974, the Carter administration introduced UDAGs to "revitalize" American cities. Private investments across the cities were to be "leveraged" by federal grants and low-interest loan money. But for the UDAG investments, these projects allegedly could not go forward. New developments spurred by UDAGs would generate jobs for the disadvantaged, taxes for central city coffers, and affordable housing. Other entrepreneurs, encouraged by the UDAG-assisted projects, would invest in an ever-broadening cycle of redevelopment that would embrace the entire city.

As it turned out, relatively few UDAG dollars found their way into the neighborhoods; most of the money went to downtown hotels, office buildings, and trendy festival markets. Even with UDAGs, distressed neighborhoods were not attractive investment options. And sometimes UDAG grants actually followed a developer's announced intention to build (Gist, 1980).

As the UDAG program ended in 1988, it was estimated that nationally only 15% of the funds were spent on "neighborhood facilities" (Gist, 1980, p. 245). The overwhelming majority of the funding went to downtown projects—especially to hotel developments (Jacobs & Roistacher, 1980; Tabb, 1984). In Cleveland the distribution of UDAG funds (1981-1988) was close to the national average, with downtown seizing the lion's share, whereas the city's neighborhoods got little (Keating, Krumholz, & Metzger, 1989) (see Table 11.1).

Tax Abatements

Once confined mainly to Southern states, legislation enabling tax abatement had been made available in 19 states by 1988 (*Business Month,* 1988). Generally such legislation gives counties and municipalities the power to grant special reductions in property taxes in order to retain or attract new investment. Tax abatement trades off growth for greater inequality in the tax system. Property taxes generally are considered to be regressive (Pechman & Okner, 1974); tax abatement makes it more so by shifting the tax burden from "capital to consumers, from large corporations to small businesses and home owners" (Swanstrom, 1982). Tax abatement was to stimulate industrial investment in disinvested areas, but in many cities it has played into already strong real estate trends and has been used most extensively in downtowns.

Cleveland is an excellent example. Downtown has been booming, and that is where tax abatement has been used. Although tax abatement was to be limited to the most blighted and poorest areas of the city, developers got the enthusiastic support of Cleveland officials to designate the most valuable real estate in the city as "blighted." No challenge was raised as to the "need"; no questions were asked about who (city residents or suburbanites) would get the "new" jobs the development was supposed to create. Few questions were raised about the large and glowing projections for new jobs. In fact, the city's own studies made clear that the market for downtown office space was not growing but simply was shifting from older buildings into new ones.

The beneficiaries of tax abatement have been the most profitable businesses in the city, including the most profitable bank in the state of Ohio in 1977: BP America (formerly Standard Oil of Ohio, which later built without abatement) and developers who ranked among the 100 richest families in the United States. The absurdity of Cleveland, a city that defaulted on its fiscal obligations in 1978, and a public school system perpetually poised on the brink of fiscal disaster granting essential tax revenues to the richest corporations in the country seems lost on local officials. Cleveland's tax abatement proponents do not see this as a "giveaway" or a windfall profit or as tax reform. To them, tax abatement is simply another tool of economic development.

Would investment come without tax abatement? There is no way to prove or disprove this question. A large body of evidence, however, does support the conclusion that local taxes play a minor role in location decisions (Harrison & Kanter, 1976).

Most studies of location decisions show that tax concessions and other financial incentives are not important considerations in the decision to develop. That is true of recent studies (Hack, 1988; Wolkoff, 1985) and older studies (Schmenner, 1982; Vaughn, 1980). Despite these studies, tax concessions continue to be granted with an open hand. Local economic development practitioners collect excellent data on project costs, estimated numbers of new jobs created or old jobs retained, and tax projections. The record, however, is usually silent when fundamental questions like these arise: How and when should tax abatements be used? Where is the comprehensive city financing plan as the foundation for any proposed subsidy or off-the-budget allocation of resources? Who will consume what is produced? How many and what kinds of new jobs will be created and at what levels of pay? Who will get the jobs—city residents or suburbanites? How soon will the public investment in foregone revenues be recaptured?

Local and state governments are usually precise in recording, reporting, and monitoring all payments received and expenditures made. But a recent survey by the Government Finance Officer's Association reported that 37 of the 45 states returning the survey and 12 of the 14 largest cities returning the survey said they kept no records whatever on tax abatements (*Government Financial Review*, 1988). A 1986 survey of local government financial reporting practices found that only 25% of responding cities included information on tax abatements in their financial reports. The conclusion can be drawn that businesspersons and public officials alike do not know, and may not want to know, the costs versus the benefits of playing the economic development game. It would seem that, at a minimum, people entrusted to oversee the fiscal affairs of states and local governments need to know the amounts of revenue being lost to their governments—currently and cumulatively. Taxpayers have a right to have this information made public. Yet, as one cynical observer wrote: "One of the great arts of the tax game is to design revenue sources so that people will not know they are paying taxes. Taxes should not be seen or felt, only paid" (Meltsner, 1971, p. 38).

Tax Increment Financing (TIF)

Tax increment financing (TIF) dates back to 1952 in California; by the mid-1970s, 17 states had authorized this form of financing. TIF is used primarily to pay for public improvements to support a development project. When a project is built within a TIF district, the value of the land increases, and any increases in the property tax revenues are retained within the district to help pay for additional development or project costs.

The theory behind TIF is that because renewal pays for itself in higher tax revenues, cities should be able to use this "increment" in advance to fund renewal. Cities can declare an area "blighted," earmarking all net increases in property taxes for reinvestment in the area, and then acquire buildings, clear land, and subsidize development. Using increased taxes and rising assessments as a revenue stream, development agencies issue tax-exempt bonds and operate with great fiscal autonomy. The bonds are amortized by future tax-increment revenues from the project, but if that fails the bonds may be a charge against the full faith and credit of the community. There is often a vague hope that "leftover" money will be used to provide improved services or to finance low-income housing (Wolinsky, 1984).

Developers and other growth proponents use TIF for many reasons. The loans are tax free and below market interest, and no cumbersome federal application is necessary. Local controversy also is held to a minimum; once approved by the city council, the loans do not have to go on the ballot for local approval. Finally the city is the sole grantee, out of reach of the other taxing bodies even though these taxing bodies may have to provide uncompensated services.

Critics have raised a number of concerns. They observe that TIFs pay their own way only by making the public absorb any losses. Also, new property taxes, which might have accompanied unsubsidized development and been used to provide public services are instead used to subsidize still more development. Largely invisible, TIF financing is so seductive that a study in Minneapolis found that TIF funds committed for redevelopment costs grew from $437,000 in 1974 to $56 million in 1986 (Frieden & Sagalyn, 1989, p. 251).

Cities have also established TIF districts in areas that are not blighted (Klemanski, 1990). TIFs and tax abatements have been applied in healthy suburban land markets to draw investments out of weak central city markets. A few states have tried to restrain this intercommunity competition, but the problem persists. In Michigan, Public Act 281, passed in 1990, prohibits TIF authorities from offering TIF money or tax abatements to a business that relocates 50 or more jobs from one municipality to another, unless explicit permission is given by the affected community.

Equity or fairness is another issue. One careful analysis suggests that the larger, declining central cities are losers under TIF financing when compared with the wealthier suburbs (Huddleston, 1984).

Industrial Revenue Bonds (IRBs)

Industrial revenue bonds (IRBs) offer low-interest financing to privately built industrial plants and other facilities as diverse as fast-food restaurants and suburban health clubs. Industrial development agencies may acquire land and build structures that then may be sold or leased to private firms at cut-rate prices. The use of IRBs skyrocketed from $6.2 billion in 1975 to $44 billion in 1982. Because the bonds are tax exempt, the federal government provides the subsidy; in 1983 the federal cost was $7.4 billion ("Cap in Business," 1984). In a recent study, a positive relationship was found between the use of IRBs and economic growth, but it did not address the question of whether IRB benefits outweigh the costs of subsidy (Marlin, 1990). These costs include foregone federal tax revenues, higher municipal bond yields, and distortions in the allocation of capital resources. Perhaps it is just as well that the use of IRBs was virtually wiped out by the Tax Reform Act of 1986.

SUMMARY

To sum up, the mechanisms for generating public subsidy to induce private investment are many and are increasingly ingenious. UDAG grants and low-interest loans reduce the developer's need for equity and increase his or her return on investment; tax abatements reduce or eliminate an owner's property tax liability for a new building; TIFs are used to divert taxes from new development away from the general fund and instead reinvest them in still more development. Special assessment and tax districts operate on the same general principle, retaining property taxes generated in a development area. Other public inducements for local economic development include enterprise zones that offer waivers of regulations, tax abatements, and other subsidies within given areas (Mier, 1982); historic preservation subsidies that pay owners to let historic buildings survive (instead of simply forbidding their demolition); and publicly subsidized sports arenas and stadiums usually owned by the wealthiest people in town (see Johnson, 1984; Knack, 1986). Virtually all of these public inducements for local economic development have been used mostly in downtown areas where there has been strong market pressure for development anyway. The total public bill has varied. In 39 downtown projects studies between 1971 and 1985, the public share ranged up to 83% (Frieden & Sagalyn, 1989, p. 155).

The effect of this mix of financial and legal devices in the service of local economic development is to turn over governmental powers to private business. Because some governments see their aims as coinciding with those of business, public planning has taken on more and more of the aspects of private business (Fainstein, 1991). The policy ostensibly is justified because it produces investment and economic development, the beneficial effects of which are supposed to trickle down to the population at large. But there is little evidence that much of anything is trickling down. Indeed, although heavily subsidized massive construction projects reshape the skyline in many cities, the life situation of low- and moderate-income city residents continues to worsen.

Consider the following: During the 1970s and 1980s, many cities shifted from a manufacturing to a service economy, but most large cities of the Snowbelt lost manufacturing jobs faster than they gained white-collar jobs, leaving them with fewer jobs overall. Central cities in 1980 had higher unemployment rates than the rest of the country, and economic disparities between these central cities and their suburbs widened from 1970 to 1980. City families earned lower median incomes from 1969 to 1979, down 4% in constant dollars over that period.

Even in cities where local economic development efforts had been strongest, unemployment and poverty grew. As Frieden and Sagalyn observe: "The ten cities that led the nation in downtown office development from 1950-1984 had higher unemployment rates in 1982 than in 1970" (p. 288). One composite index using unemployment, poverty, and changes in real income between 1970 and 1980 showed four of the top cities in development slipping from better conditions in 1970 to join the 10 worst-off cities. Baltimore, hailed for its Inner Harbor project as a "Renaissance City," saw poverty in its black neighborhoods rise by 90% in the 1980s. From 1979 to 1985, central city poverty overall grew from 16% to 19% of total population. In addition to the increasingly troubled condition of their resident population, many of these cities are poised at the edge of bankruptcy.

Economic Development in Cleveland

Let us look more closely at how the local economic development process has played out in Cleveland. Cleveland is an older, industrial city subject to many of the problems of such cities. In 1940, two of every three workers had manufacturing jobs; the figure was down to one of three in the mid-1980s. In 1990, Cleveland had America's worst unemployment rate (20%) among black workers. Poverty has been increasing and growing deeper in Cleveland. About 40% of the city's population lives below the poverty line, and they are more deeply poor than they were 10 years ago (Coulton, Chow, & Pandley, 1990). In Cleveland's traditional poverty areas, the proportion of female-headed households jumped from 54% in 1970 to 77% in 1980 (Coulton et al., 1990). Two thirds of the children in Cleveland public schools are from families receiving Aid to Dependent Children. Half of the poor spend 70% or more of their incomes on their housing. At least 14 of the city's neighborhoods had poverty rates above 50% in 1990. Cleveland has one of the nation's largest disparities in income between center city and suburban areas. It also is ranked the second most racially segregated city in the nation and is one of the nine U.S. metropolitan areas that has been called "hypersegregated" (Massey, no date).

How is Cleveland dealing with its widespread poverty and inequality?—through "trickle-down" development projects that reinforce inequality. Two local banks have announced new headquarters buildings. They have been granted a 100%, 20-year tax abatement that will cost the taxpayers about $227 million. Of that, 60% is money that would have gone to the Cleveland public school system, which has been laying off teachers for lack of money. Some of it would have gone to the city, which is closing its health centers. Some of it would have gone to Cuyahoga County, which cannot pay for clothing for the children in its care.

The downtown Tower City development, an elegant shopping mall, has made ingenious use of UDAGs, TIFs, Urban Mass Transportation Administration funds, capital improvements, and tax abatements to the tune of over $100 million. Cleveland's next effort, a Rock 'n' Roll Hall of Fame, has the public in for more than $50 million so far, whereas private promoters have pledged less than $6 million.

Another recent economic development project is a new baseball stadium for the Indians and a basketball arena for the Cavaliers. These are to be financed mostly out of a local tax on tobacco and liquor. Because those products are more and more used by lower economic classes, and because sports teams are owned by the wealthiest class, this project represents a subsidy flowing from the very poorest to the very richest. Presumably some of the $350 million invested here might have been used for other, more humane purposes.

Cleveland's highest transit priority is a new 6-mile rail line to connect downtown with University Circle, two points already connected by excellent rail and bus service. Its cost: $600 million to $1 billion. Meanwhile, transit-dependent residents of Cleveland's poverty neighborhoods are further and further removed from suburban jobs.

These economic development projects are all promoted as fostering job creation. But much evidence suggests that business will build with its own money what it finds profitable to build, and public subsidies are largely public money substituted for entrepreneurial money. For example, the city of Vermillion, Ohio, recently sued to void a tax abatement granted by the city of Lorain to the Ford Motor Company. Vermillion won its suit, but the subpoenaed records of Ford's board meetings revealed that Ford had decided to build whether or not a tax abatement was granted.

There are other flagrant examples in other cities:

- New Orleans—Managed growth (involving heavy public subsidies with little direct benefit to city residents) "has altered the economic base of the city in such a way as to make life harder for low-income residents" (Smith & Keller, 1986, p. 100).
- St. Louis—Provides the power of eminent domain, a variety of public inducements, and up to 25 years of reduced taxes to private redevelopment corporations. After 25 years of local economic development "successes," St. Louis tops the nation in out-migration and has one of the nation's largest welfare roles.
- New York City—Since 1983 has granted $600 million in subsidies to six firms to keep them in town: Chase Manhattan Bank ($235 million); NBC ($98 million); Shearson Lehman ($74 million); Drexel Burnham ($85 million); and Dreyfus ($4 million a year) (Buskind, 1990).

Perhaps the most egregious example on what could be a very long list is the Poletown case in Detroit, in which local economic development destroyed 1,021 homes, 155 businesses, churches and a hospital, and obliterated a racially integrated community. Land acquisition, demolition, relocation, and new capital improvements equaled $200 million in city, state, and federal funds, plus some $170 million additional in tax abatements over 12 years. In return, General Motors was to retain 3,800 jobs and add 1,400, only about half of which existed in 1990. At the same time, Detroit's mayor suggested that citizens give cash gifts to the city and volunteer to maintain public parks and provide supplementary police and fire services. Unimpressed with this "do-it-yourself" government, Detroit's population has dropped steadily, with the 19% loss between 1980 and 1990 leading all cities; one third of all of its people are on welfare; unemployment is estimated at 18.6%, compared with 6.2% in the suburbs.

It seems clear that local economic development as currently practiced provides tangible benefits to developers, land owners, lawyers, construction union members, suppliers, politicians, and development officials. But tangible job benefits for the poor and unemployed residents of central cities cannot be demonstrated, and net benefits to the overall fiscal condition of these cities is questionable.

What To Do?

As we have seen, local economic development has been attempting to solve the problem of distressed cities and distressed people with general, trickle-down programs emphasizing new downtown construction. The results have been inefficient and inequitable; evidence is scant that the process works for those ends. Instead, cities, states, and the federal government should adopt policies that concentrate their resources directly on improving education, providing jobs, and improving the real income of the central city poor and working class.

Improved education is the single most important local economic development activity. Compared with improving education, tax incentives and other conventional development "tools" are insignificant. An educated labor force is competitive and productive. It can lift any city, region, or nation.

In upgrading education, the highest priority must be given to children from poor and minority families, many of whom are unprepared for an economy that demands constantly increasing levels of skill and knowledge. A key element in any educational policy would be an expanded K-3 Head Start program with extensive parental involvement. Reports from the Congressional Budget Office make clear that such a program offers enormous potential in reducing school failures.

In job development, it is time to realize that capital subsidies do not provide or upgrade employment in the ghetto. Labor subsidies would do much more to solve the problem. Rather than subsidize expensive (and not very efficient) filtering-down schemes of the sort we have described, a simple wage-subsidy program would be more logical and cost-effective. Employers hiring a worker who had been certified unemployed for a given period of time would receive an hourly cash subsidy, with the amount decreasing over time as the worker's productivity increased with training and experience. Public service employment also should be expanded because it can improve the urban environment and teach basic skills and work habits.

Full employment would do little to eliminate poverty among those millions of households with very low paid employment or without a member in the labor force. For those households, a national family assistance plan or income-maintenance program would be effective.

Continued state and federal participation is not enough. City governments also must do more directly for their troubled resident populations. Since the beginning of urban renewal, city officials have accepted the notion that the city and its people automatically would benefit from any development; therefore, public subsidies were appropriate. However, the use of federal and local subsidies to stimulate new development has not been carefully tied to the welfare of city residents—especially to those most in need. New development is not an end in itself. It is valuable only to the extent that it provides jobs for present or future city residents—especially city residents who are unemployed—and a net tax increase to city coffers.

Ideally, local economic development activities should be completely redirected. Instead of emphasizing tax inducements, sales, and marketing as is currently the case, they should emphasize the skills of the local labor force and its productivity, the quality of local schools, and the social and physical improvements completed and under way. Even if this approach fails to attract more investment, it will improve conditions for existing residents and businesses.

If subsidies are deemed absolutely essential, they should be programmed carefully to retain existing businesses and to start new businesses—particularly those that offer jobs to indigenous, low-skilled, unemployed workers. The city should not seek to stop development but should negotiate as an equal: "We'll give you this, if you give us that." As a prior condition for the use of public powers or subsidies, the city should seek credible guarantees on the number of new jobs generated, the number that will be filled by residents who have been unemployed for long periods, and the net revenue increase that will be generated. Performance guarantees should be written into all agreements, with sanctions provided if the guarantees are not fulfilled. Should the recipient company default on its agreements, it should be liable to reimburse the city for all benefits it received plus interest. The cities of San Francisco, Berkeley, Jersey City, and Boston provide ready examples of such hard bargaining.

Perhaps the most aggressive model thus far for such an approach is the *Chicago Development Plan, 1984* (see also Clavel & Wiewel, 1991; Mier, Moe, & Sherr, 1987). This plan focused narrowly on running the city for the benefit of its people. Its vision was driven by the notion that questions of economic equity (who gets what kind of jobs and the resultant income) were inextricably bound to the practice of public urban economic development planning.

The plan proposed to use the full weight of the city's leverage—tax incentives, public financing, city purchasing, and infrastructure improvements—to generate jobs for Chicago residents. City resources were seen as public investments with a targeted rate of return in the form of the number of jobs provided for Chicagoans. Specific hiring targets for minority and female employment were set; 60% of the city's purchasing was directed to Chicago businesses; and 25% of this spending was to be done with minority and women- owned firms.

The plan also sought to encourage balanced growth between downtown and city neighborhoods. Public support was offered to private developers building projects in "strong" market areas if they were willing to help neighborhood economic development projects in "weaker" market areas. The help could be provided through technical or legal assistance to local development corporations, joint venturing on neighborhood projects, or through contributions to a low-income housing fund.

As a final objective, the plan drew up a regional, state, and federal legislative agenda to advocate its interests. This was a reminder that the best laid and most equitable of city plans can be undercut by statehouse priorities and congressional budgets. It also reminded planners that forceful advocacy is a demand of their profession—and the only promise for the realization of their plans.

The *Chicago Development Plan, 1984* seems to be the strongest indication thus far that American cities are willing to try to harness local economic development for the benefit of their disadvantaged residents. If economic development practitioners in other cities modeled their efforts along the lines suggested here, their cities might have a more viable local economy, their poor and working class neighborhoods might be less disinvested, and their resident populations might be better employed, more responsible, and more self-respecting.

Notes

1. A flagrant example of this process was the 1986 resignation of New York City's Commissioner of Housing and Development, Anthony Gliedman, who promptly went to work for Donald Trump (Peirce, 1986). Also, in 1990 the former chief planner for Toronto, the former president of Battery Park City, and a former vice president of the New York City Public Development Corporation all worked for Olympia and York.

2. In Cleveland the developers of two new office towers proposed in 1989 got about $225 million in a 20-year tax abatement deal. They were the largest contributors to the mayor's war chest in his campaign for governor in 1990.

References

Banfield, E. C. (1973). Ends and means in planning. In A. Faludi (Ed.), *A reader in planning theory*. New York: Pergamon.

Blakely, E. J. (1989). *Planning local economic development*. Newbury Park, CA: Sage.

Business Month. (1988, June). Pp. 55-63.

Buskind, R. (1990, February). The giveaway game continues. *Planning*, p. 5.

Cap in business IRBs pushed for states, cities. (1984, March 15). *Public Administration Times*, pp. 1, 3.

Center for Community Change. (1991). *Bright promises, questionable results*. Washington, DC: Author.

Chicago Department of Economic Development. *1984 Chicago development plan*. Chicago: Author.

Clavel, P., & Wiewel, W. (Eds.). (1991). *Harold Washington and the neighborhoods: Progressive city government in Chicago, 1983-1987*. New Brunswick, NJ: Rutgers University Press.

Coulton, C. J., Chow, J., & Pandley, S. (1990, January). *An analysis of poverty and related conditions in Cleveland area neighborhoods over the past decade*. Cleveland: Case Western Reserve University, Center for Urban Poverty and Change.

Fainstein, S. S. (1991, Winter). Promoting economic development: Urban planning in the U.S. and Great Britain. *Journal of the American Planning Association*, pp. 22-33.

Frieden, B., & Sagalyn, L. (1989). *Downtown, Inc.* Cambridge: MIT Press.

Gist, J. R. (1980). Urban development action grants. In D. B. Rosenthal (Ed.), *Urban revitalization* (pp. 237-252). Beverly Hills, CA: Sage.

Government Financial Review. (1988, June). p. 3.

Hack, G. D. (1988, October). Location trends: 1958-1988. *Area Development*, p. 12.

Harrison, B., & Kanter, S. (1976, Spring). The great state robbery. *Working Papers for a New Society*.

Huddleston, J. R. (1984, April). Tax increment financing as a state development policy. *Growth and Change*, pp. 11-17.

Jacobs, S., & Roistacher, E. (1980). The urban impacts of HUD's Urban Development Action Grant Program. In N. Glickman (Ed.), *Urban impacts of federal policies*. Baltimore: Johns Hopkins University Press.

Johnson, A. T. (1984). *Economic and policy implications of hosting sports franchises: Lessons from Baltimore* (Working Paper 4, Maryland Institute for Policy Analysis and Research). Baltimore: University of Maryland.

Keating, D., Krumholz, N., & Metzger, J. (1989). Post populist public private partnerships. In G. Squires (Ed.), *Unequal partnerships* (pp. 121-141). New Brunswick, NJ: Rutgers University Press.

Klemanski, J. S. (1990). Using tax increment financing for urban redevelopment projects. *Economic Development Quarterly*, pp. 23-28.

Knack, R. E. (1986, September). Stadiums: The right game plan? *Planning, 52*, 6-11.

Levy, J. M. (1990, Spring). What local economic developers actually do. *Journal of the American Planning Association*, p. 121.

Lindblom, C. E. (1959, Spring). The science of muddling through. *Public Administration Review*, pp. 30-34.

Malizia, E. E. (1985). *Local economic development: A guide to practice*. New York: Praeger.

Marlin, M. R. (1990). The effectiveness of economic development subsidies. *Economic Development Quarterly, 4*, 15-22.

Massey, D. S. (no date). *Segregation and the underclass*. Unpublished manuscript, University of Chicago.

Meltsner, A. J. (1971). *The politics of city revenue*. Berkeley: University of California Press.

Mier, R. (1982, April). Enterprise zones, a long shot. *Planning, 48*, 10-14.

Mier, R., Moe, K. J., & Sherr, I. (1987). Strategic planning and the pursuit of reform: Economic development and equity. In H. A. Goldstein (Ed.), *The state and local industrial policy question* (pp. 161-175). Chicago: Planners Press.

Pechman, J., & Okner, B. (1974). *Who bears the tax burden?* Washington, DC: Brookings Institution.

Peirce, N. R. (1986, June 23). The ethics of revolving-doorism. *Boston Globe,* p. 16.

Redmond, T., & Goldsmith, D. (1986, April). The end of the high-rise jobs myth. *Planning,* pp. 18-21.

Rubin, H. J. (1988). Shoot anything that flies; Claim anything that falls. *Economic Development Quarterly,* 2, 236-251.

Schmenner, R. W. (1982). *Making business location decisions.* Englewood Cliffs, NJ: Prentice Hall.

Simon, H. (1955). *Administrative behavior.* New York: Macmillan.

Smith, M. P., & Keller, M. (1983). Managed growth and the politics of uneven development in th New Orleans. In S. Fainstein & N. Fainstein (Eds.), *Restructuring the city* (pp. 126-166). New York: Longman.

Swanstrom, T. (1982, Winter). Tax abatement in Cleveland. *Social Policy,* pp. 24-30.

Tabb, W. K. (1984). The failures of national urban policy. In W. K. Tabb & L. Sawers (Eds.), *Marxism and the metropolis* (pp. 363-382). New York: Oxford University Press.

Vaughn, R. J. (1980, January 8). How effective are they? *Commentary, National Council of Economic Development,* pp. 12-18.

Wolinsky, L. (1984, April 2). Cities fatten budgets in redevelopment law. *Los Angeles Times,* Sec. I, pp. 1, 3, 30.

Wolkoff, M. J. (1985). Chasing a dream: The use of tax abatements to spur urban economic development. *Urban Studies, 22,* 305-315.

The Politics of Economic Development Policy

Elaine B. Sharp

David R. Elkins

In the past 18 years, tax abatement has emerged as a favored policy for leveraging growth in communities. Its popularity stems from the perception of it as a relatively inexpensive and yet very effective means of enticing new businesses to locate in a community or as a means of providing incentives for existing businesses to expand. But tax abatement also has been characterized as inefficient, inequitable, and politically controversial (e.g., see Swanstrom, 1988; Wasylenko, 1981). The politically controversial nature of tax abatement may present one of the more formidable challenges for the use of this policy tool. In this chapter we explore the potential for political controversy over the use of tax abatement.

Such an exploration may appear contrary to the insights provided by *domain theory*—one of the more substantial developments in contemporary studies of urban politics. According to this theory, local politics consists of differing spheres or domains, each with its own unique political characteristics and policy issues. Especially important contrasts have been drawn between the sphere in which "developmental" politics is decided and the sphere in which "allocational" policies are handled.

The politics of allocation, which deals with typical distributive services such as snowplowing and garbage collection, is characterized as open, visible, competitive, controversial, and pluralistic. Peterson (1981, pp. 165-166) describes this domain as follows:

> There is no end to the politics of allocation. It is a continuing, thriving, potentially explosive political arena that . . . often subjects decisionmakers to intense political heat. . . . The widely held view that local politics is an arena of bargaining, compromise, cross-cutting cleavages, and changing political issues is not incorrect. On the contrary, such a view depicts and characterizes the most visible aspects of . . . the allocational arena.

Dye (1986, pp. 43-44) provides a similar characterization of decision making concerning allocational matters:

> The allocational policy arena is pluralist in character. . . . [Public officials] are responsive to the expressed demands of many varied and often competing groups within the community. Participation in decisionmaking is open. . . . Interest and activity rather than economic resources are the key to leadership in allocational policy.

By contrast, the domain of developmental politics, which is concerned with many of the policies of interest for economic development, is characterized as having a highly centralized, behind-the-scenes, consensual, business-elite-dominated form of politics. For example, consider Peterson's (1981, p. 132) description:

> [Developmental policies] are often promulgated through highly centralized decision-making processes involving prestigious businessmen and professionals. Conflict within the city tends to be minimal, decision-making processes tend to be closed until the project is about to be consummated, local support is broad and continuous, and, if any group objects, that group is unlikely to gain much support.

Similarly Dye (1986) argues that commercial elites are predominant in the domain of developmental politics. Adding a contingent perspective on the elitist-pluralist debate of community power studies, Dye argues that "reputations for power correlate with the 'reality' of power when the issues specified are developmental issues" (p. 41).

The strength of domain theory is that it focuses attention on the *contrasts* between economic development decision making and decision making about the basic housekeeping services of city government. By implication, it also emphasizes *uniformities* within each domain, across cities. However, the uniformities of domain theory may be overstated. Domain theory suggests that openness and public involvement is not typical—that instead the politics of development is relatively closed, quiet, and elite dominated. Case study evidence, however, suggests instances in which economic development decision making seems much more controversial and politicized, with the relatively high levels of citizen involvement that are supposedly characteristic of only the politics of allocation (e.g., see Jones & Bachelor, 1986). But do these cases constitute only visible but atypical episodes?

Drawing on a survey of economic development practices and decision-making arrangements in a national sample of cities, we suggest in this chapter that cities do differ substantially with respect to the openness or citizen involvement dimension of economic development decision making. Some have a more participatory approach, embracing a variety of mechanisms for popular input; others not.

But of what significance are such differences? Much of the literature that acknowledges differences in the character of economic development decision making focuses on the antecedents of one or the other style. Case study and other research in this vein includes an effort to identify contextual variables that pave the way for a more participatory, populist approach to economic development (e.g., see Clarke, 1987; Daykin, 1988; Swanstrom, 1988).

From both theoretical and policy perspectives, however, the consequences of differing modes of economic development decision making may be at least as significant as the contextual antecedents. For example, important case study evidence suggests that, in certain contexts, the populist, participatory approach leads to the delegitimation of tax abatement as an economic development policy. In his analysis of the Kucinich administration in Cleveland, Swanstrom (1988) acknowledges that tax abatement normally is a very desirable policy strategy for city officials because it requires no apparent financial commitment and appears to be cost free. In an economically declining city, however, tax breaks for well-to-do, downtown investors contrasted unfavorably with the tax burden on other Clevelanders. A variety of local organizations mobilized to oppose tax abatements in 1977, turning city council meetings into arenas for protest and confrontational tactics. This participatory fervor did not initially succeed in preventing tax abatements, but Kucinich's successful mayoral campaign incorporated the anti-tax-abatement fervor and, once in office, Kucinich succeeded in blocking many abatement efforts.

The foregoing characterization of anti-tax-abatement politics in Cleveland provides a stark contrast with the description of economic development decision making provided by domain theory. Rather than a quiet, consensual, elite-dominated setting, the Cleveland case presents a rowdy, conflictual, highly participative setting. That mode of decision making generated a rejection of tax abatement policy. But the case study leaves us with two important questions. The first is the generalizability question: Is Cleveland in the Kucinich era a unique case? Can one find systematic evidence that, at least where property taxes are problematic, citizen involvement in the politics of development tends to deter tax abatement? An initial purpose of this chapter is to transform the Cleveland case study findings into a working hypothesis and to test it on a national sample of cities.

Hypothesis 1: Where there is property tax stress, greater citizen involvement in development decision making diminishes the propensity to use property tax abatement.

As a corollary to this, we might expect that reliance on locally raised taxes to finance economic development activities, under those same circumstances, also would be heavily circumscribed. This generates the follow-up hypothesis.

Hypothesis 1a: Where there is property tax stress, greater citizen involvement in development decision making diminishes local government reliance on locally raised taxes for economic development projects.

The second important question is: What other policy consequences does a participatory mode of economic development decision making generate? Should the anti-tax-abatement case be logically extended to such a degree that participatory politics is presumed to generate opposition to any and all policies providing subsidies or other favorable treatment to businesses and potential investors?

For substantial reasons, one can expect quite the opposite. From two different theoretical perspectives, it should be hypothesized that higher levels of popular involvement actually enhance the adoption of many economic development policies, at least in fiscally stressed settings where there is a pressing need for development activity. Ironically these two theoretical perspectives, though generating similar hypotheses, are based on contrasting interpretations of government responses to a politicized, highly involved public.

Minimizing Apparent Costs

The first theoretical perspective argues that, as a strategic response to the pressures of citizen oversight and accountability control on traditional policy tools, government decision makers move toward alternative, less visible policies to achieve their goals (Bennett & DiLorenzo, 1983; Leonard, 1986). Traditional policy tools involve straightforward budgetary commitments hammered out in open legislative forums and paid for with tax revenues that are budgeted and subjected to accountability controls. Alternative policies are off budget, not subjected to normal appropriations processes, and generally less easily tracked by the public. Furthermore, many of these policies involve either resource commitments or financing strategies far removed from the traditional property tax and general obligation bonds, mainly to escape the citizen-oversight mechanisms with which property taxation and traditional debt instruments are encrusted.

Notable examples of newer, less visible policy tools include subsidized loans and loan guarantees, sale-leaseback arrangements, and tax-exempt bonds (Leonard, 1986, pp. 8, 183-184, 150-163). Similarly, local officials can use tax deferment as a tool for economic development. Unlike tax abatement, tax deferment is a more low key policy tool with the appearance of an extension of credit, rather than a tax break. In short, a variety of economic development policy tools are much less visible and presumably less politically sensitive than tax abatement.

If politicians cannot appear to create costless-development initiatives in these ways, they may be expected at least to adopt strategies that insulate general taxpayers from costs by shifting costs toward user-beneficiaries (Pagano, 1988). Thus strategies such as tax increment financing and special assessment districts, which isolate costs and tax impacts, may be almost as popular as the "quiet side" strategies outlined by Leonard, when taxpayer activism and economic distress confront the politician.[1] Finally, although they cannot be interpreted as particularly innovative financing tools, federal grants, if they can be obtained, permit local officials to pursue economic development goals without tax-cost implications for the local public.

From this first perspective, the adoption of innovative development policies is driven by the need to sidestep citizen oversight and, in particular, to diminish the apparent tax costs of development-policy initiatives. Presumably such sidestepping is most necessary in settings where citizen involvement is high; consequently the potential for controversy is high. By contrast, in settings where there is less citizen involvement or fewer avenues for citizen oversight of economic development activities, there should be less of a need for the adoption of policies that hide or minimize costs of development projects.[2] In short, innovative financing mechanisms can be viewed as devices to evade voter accountability and to hide the true scope of government activity, thus prompting the following hypothesis.

Hypothesis 2: Where property tax stress is high, citizen involvement is positively associated with the adoption of tax-cost avoidance strategies for financing economic development—that is, tax-exempt bonds, loan subsidies or guarantees, sale-leaseback arrangements, tax-increment financing, special assessment districts, and federal grants.

Maximizing Apparent Benefits

From a very different perspective, local officials need to claim credit for economic development activities. Drawing on Mayhew's (1974) description of the self-promotional activities of members of Congress, Feiock and Clingermayer (1988) argue that, particularly when the community is economically distressed, local politicians need to be able to show that they are "doing something" positive toward economic development (see also Swanstrom, 1988, pp. 103-105). From this perspective, hiding the costs of development initiatives is perhaps less important than publicizing the benefits.[3]

Four economic development strategies are assessed here from the credit-claiming point of view: (a) infrastructure improvements or construction designed mainly to encourage economic development purposes, (b) reform in building or zoning regulatory processes (e.g., the creation of an ombudsman, regulatory consolidation or streamlining, and modifications in zoning or building inspections), (c) reform of environmental regulations to facilitate economic development, and (d) promotional activities by community leaders, including field visits or meetings with managers of prospective businesses. In different ways, each of these development policies allows local politicians to show that they are taking action to foster investment.[4]

These considerations lead to another hypothesis:

Hypothesis 3: Where property tax stress is high, citizen involvement is positively associated with the adoption of projects that allow local officials to claim credit for economic development progress—that is, infrastructure improvements, promotional activities, reform of land-use regulations, and reform of environmental regulation.

Although distinctive, the cost minimization and the benefit maximization perspectives are complementary. One assumes that citizens are most worried about the tax costs of government activity and hence that officials are driven toward creative financing schemes that avoid the appearance of increased property tax costs for all. The other assumes that citizens are most worried about the need for visible progress on the economic development front and hence that govern-

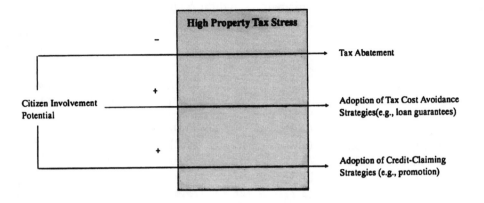

Figure 12.1. Summary of Hypothesized Relationships

ment officials are driven toward the appearance of action. Figure 12.1 is a summary of the hypothesized relationships to be examined.

Our emphasis on the contextual importance of property tax stress in shaping development policy responses to citizen involvement is consistent with previous studies (Feiock & Clingermayer, 1988; Rubin & Rubin, 1987) although it narrows the focus from economic distress more broadly to property tax stress specifically. The narrowing of focus to property tax stress is important for our purposes because it taps a highly salient and politicized aspect of fiscal stress. If fiscal stress in this sense provided fertile ground for the delegitimization of tax abatement (cf. Swanstrom, 1988, and our Hypothesis 1), it should likewise be expected to create especially strong pressures for the alternative, economic development policies at issue here.

Data Sources and Measurement Methods

The analysis is based on data from a 1984 survey of municipal chief administrative officers, focusing on economic development practices and policies. The survey, administered by the International City Management Association (ICMA), was targeted to municipalities of at least 10,000 population. The survey data are supplemented with data from the 1982 Census of Governments.

Because of a peculiarity in questionnaire construction, a subsample of the communities responding to ICMA's 1984 economic development survey constitutes the sample for this analysis. Crucial questions concerning citizen involvement in economic development decision making were asked only of administrators who had responded that the community has an economic development plan. Because those questions are essential for the measurement of our key independent variable, there is no choice but to restrict the analysis to that subset of municipalities (38% of the total) that had such a plan.

Table 12.1 Comparison of All Municipalities and the Sample Municipalities, by Population Size and Per Capita Property Taxes, 1982

Population Size	Distribution of Municipalities				Per Capita Property Taxes	
	N	All Percentage	N	Sample Percentage	All Dollars	Sample Dollars
300,000 or more	49	(2.2)	0	(0.0)	212.65	—
200,000-299,999	22	(1.0)	5	(1.2)	133.58	135.53
100,000-199,999	100	(4.5)	38	(8.9)	162.96	140.90
50,000-99,999	250	(11.3)	68	(15.9)	147.38	152.41
25,000-49,999	523	(23.7)	126	(29.4)	112.77	128.75
10,000-24,999	1,259	(57.1)	191	(44.6)	94.50	121.28

SOURCE: Data for all municipal governments are from the Bureau of the Census, *Finances of Municipal and Township Governments,* Washington, DC: Government Printing Office, p. 30. Data for sample cities are from the 1982 Census of Governments data tape for the set of cities available for analysis in the ICMA survey. A straight-line estimate of 1982 population from 1980 and 1985 figures is used in calculating 1982 per capita property taxes for the sample cities.

This narrowing of the sample introduces the possibility of bias. Several considerations, however, lead us to conclude that the restricted sample does not constitute a fatal flaw. First, the cities lost to analysis are predominantly small communities; 55% of them are under 25,000 population, and another 27% are between 25,000 and 50,000 population. Urban research using national samples typically draws from cities of at least 25,000 population and often is restricted to cities of at least 50,000. Consequently our subsample is not substantially different from the usual in terms of the range of population size that is functionally represented.

Second, this subsample appears to be a reasonable sampling of the universe of municipalities, with the exception of those at the extremes of population size. In Table 12.1, for example, is a comparison of our sample of 428 municipalities with the universe of municipalities in the United States, using data from the 1982 Census of Governments. The table shows that our sample cities approximate the shares of municipalities in the various population-size groupings, with the exception that the smallest municipalities are underrepresented,[5] and with the exception that the very largest cities are unrepresented. Although the percentage of sample cities over 100,000 population (10.1%) is similar to that for the population (7.9%), the sample distribution is skewed toward the smaller end of these largest cities. When sample cities are compared with all municipalities in the United States with respect to per capita property taxes—a key variable in this analysis—there is notable comparability, although by no means identical distributions.

Our measure of citizen involvement in economic development decision making is based on responses to four items asking whether or not the city used each of the following in its economic development planning: (a) appointed advisory committees representing the entire community, (b) open meetings or public hearings, (c) citizen surveys, and (d) elected neighborhood commissions.[6] Some of these devices imply greater citizen empowerment than others and therefore are used much more rarely than others. In particular, elected neighborhood commissions are used for economic development decision making by only 3.3% of the sample cities. By comparison, over half of the cities use public hearings (51.6%) or appointed communitywide advisory boards (56.3%). Given these important differences, a weighted index of citizen involvement was created by using all four of the citizen input methods mentioned here, but with the methods weighted according to their prevalence.[7] On this index, 23% of the sample cities have a score of 0, indicating that none of the four methods are used, and another 17% have a score of 1, indicating that only the most prevalent methods of citizen involvement are used. Combined, 40% of the sample cities constitute the "low" category on the citizen-involvement index. Roughly another two fifths of the cities (42%) score in the middle range (2-5) on the citizen-involvement index, and about one fifth (19%) score high on the index (6-10).

Two measures of property tax stress are used in the analysis: (a) an indicator of average property tax burden—that is, per capita property tax revenues in 1982; and (b) an indicator of the city's dependence on the property tax—that is, property tax revenue in 1982 as a percentage of total general revenue in that year. Data on property tax and general revenues are obtained from the 1982 Census of Governments data tape, and the 1982 population is a straight-line estimate from the 1980 and 1985 population figures for each community.

Measurement of each of the economic development activities and financing strategies is based on yes-no responses to individual ICMA survey items asking about each, with the exception of promotional visits or meetings, which is a simple, additive index of two items: one asking whether the community uses "field visits by city/community representatives," and the second asking whether "community leaders meet with managers of prospective new businesses." Each of the financing strategies asked about is linked specifically in the questionnaire wording to economic development purposes. For example, the items on tax abatement, deferred tax payments, loan guarantees, loan subsidies, and sale-leaseback arrangements are all specific probes following up a broader question on whether the city uses "financial incentives to encourage economic development activities by the private sector." Similarly items on tax-exempt bonds, tax-increment financing, special assessment districts, federal grants, and locally raised taxes were all presented in the survey as "revenue sources that might be used to fund a city's economic development activities." Respondents were asked to indicate those the city did use to fund economic development activities.

Although most of the financing strategies are broadly available to local decision makers, questions should be raised about whether tax abatement and the use of locally issued, tax-exempt bonds for private purposes are functionally a part of the possible repertoire of economic development approaches for all cities. As of 1984, only about half (26) of the 50 states fully authorized local property tax abatement for economic development purposes. In another 6 states, only a limited form of tax abatement is authorized (e.g., in New Hampshire, where tax abatement is authorized for new railroads or small-scale power facilities; in

Tennessee, where tax abatement is authorized for electric energy cooperatives and historical structures). In 18 of the 50 states, authorization for local property tax abatement was not reported as of 1984 (National Association of State Development Agencies [NASDA], 1984). Issuance of tax-exempt bonds such as industrial revenue bonds (IRBs) or industrial development bonds (IDBs) for private purposes is more widespread, but five states do not appear to have authorized these financing instruments (NASDA, 1984). In the analysis that follows, the hypothesis concerning linkage between citizen involvement and the use of tax abatement policies is tested only for cities in states that have at least limited tax-abatement authorization. Similarly the hypothesis concerning linkage between citizen involvement and use of tax-exempt bonds for private purposes is tested only for cities in states that allow for such revenue instruments.

As Table 12.2 shows, there is considerable variation in the extent to which cities use the various economic development tools and financing strategies considered here. The vast majority of cities use federal grants—when they can get them—to finance economic development initiatives, and most also must rely on local taxes for development activities as well. Promotional activities such as field visits by city representatives and meetings between city leaders and business prospects epitomize visible, symbolic activity in the name of economic development. Not surprisingly, city officials in the vast majority of cities engage in either one or both of these activities. Even in states in which they are authorized, only about half of the cities use tax-exempt bonds to spur private development, and only about one third of cities use tax abatement. Other activities, particularly those involving innovative tools for financing development projects, are even rarer. Tax-increment financing schemes and special assessment districts are used by fewer than 30% of the cities, and only a little over one fifth of the cities provide loan subsidies for the private sector. Fewer than 10% of all cities use loan guarantees, sale-leaseback arrangements, or deferred tax payments as development tools. Efforts to encourage development by relaxing environmental regulations are also quite rare (11.8%); reform of building or zoning regulations is more common (27.4%). Despite the importance of infrastructure for economic development, just over one fifth of the respondents reported that their city had undertaken construction or improvement of city facilities in order to encourage economic growth.

Table 12.2 also shows the results when the eight items listed as activities that minimize apparent tax costs are combined into an overall index of tax-cost minimization and when the four items considered as maximizing apparent benefits are combined into an overall index of benefit maximization. Nearly two fifths of the cities score 0 or 1 on the index of tax-cost minimization, and approximately another two fifths score 2 or 3. Only about one fifth of the cities score high on the index, indicating they used four or more of the economic development tactics that minimize apparent tax costs. About one fifth of the cities score only 0 or 1 on the index of benefit-maximizing activity, and roughly a third score 2. In over two fifths of the cities, respondents report either 3 or all 4 of the activities that we interpret as maximizing the apparent benefits of economic development. It is clear that cities vary considerably in the policy tools used to

Table 12.2 Frequency Distributions for Economic Development Activities

	Percentage Reporting Use	N
Activities with visible tax costs		
Tax abatement:		
All cities	23.6	428
In states authorizing tax abatement	33.8	284
Use of local taxes for economic development	61.7	428
Minimizing apparent tax costs		
Tax-exempt bonds:		
All cities	46.3	428
In states authorizing local IDBs	51.0	386
Loan guarantees	9.3	428
Loan subsidies	21.5	428
Sale-leaseback arrangements	7.5	428
Tax-increment financing	29.0	428
Special assessment district	27.1	428
Deferred tax payments	6.1	428
Federal grants	75.0	428
Index of cost-minimizing activity:		428
Low (0-1)	36.9	
Moderate (2-3)	43.5	
High (4-7)	19.6	
Maximizing apparent benefits		
Infrastructure improvement	21.3	408
Reform building/zoning regulations	27.4	424
Reform environmental regulations	11.8	422
Field visits/meeting business prospects:		
One or the other activity	30.6	428
Both activities	49.5	428
Index of benefit maximizing activity:		
Low (0-1)	22.0	
Moderate (2)	34.8	
High (3-4)	43.2	

further economic development. The key question driving our analysis is whether these differences in economic development practices are the result of differences in economic development politics—that is, levels of popular involvement in economic development decision making, conditioned by property tax sensitivities.

Analysis and Findings

A key component of each of our hypotheses is the stipulation that citizen involvement has effects on economic development policy, primarily in settings of property tax stress. By comparison, citizen involvement is expected to be of negligible importance in settings that are not experiencing property tax pressures. The hypotheses can be seen as positing an interaction effect, such that the combination of fiscal pressure and citizen-involvement drives local officials away from property tax abatement and toward the other economic development activities and financing methods outlined above. Our findings suggest that there are such interaction effects.

Before turning to them, however, it is useful to examine the bivariate relationships between property tax stress and adoption of the various economic development activities. These bivariate findings can be compared with findings from other research on the direct link between economic distress and economic development activity. For example, Rubin and Rubin (1987) test for and find some significant relationships between median family income, poverty, unemployment, property tax rate, and assessed valuation per capita on the one hand, and various economic development policies on the other hand. Their findings provide partial support for the general theory that economic need drives cities' economic development policy efforts. But, just as their research finds only mixed and relatively weak evidence of a direct link between economic distress and the use of various economic development policies, there is mixed evidence from our sample of a direct link between economic distress and the use of various economic development policies as well. As Table 12.3 shows, only four statistically significant relationships are found between per capita property taxation as an indicator of fiscal stress and the various economic development policies. All four are very weak, and one of the four is in a direction opposite that expected. When property tax dependence is introduced as an alternative indicator of fiscal stress, the results are not at all consistent with the thesis of a direct link between fiscal stress and economic development policy efforts. Four of the 14 possible relationships are significant, yet all but 1 of them are negative, suggesting that various strategies for the creative financing of economic development projects are more likely in cities that are less dependent on the property tax.

As Table 12.4 shows, the interactive model posited by the hypotheses in this chapter finds much better support in the data, at least when per capita property taxation is used as the indicator of fiscal stress. When property tax burdens are high, greater levels of citizen involvement are associated with greater propensities to use nearly all financing methods that minimize tax costs for the general citizenry and with greater propensity to engage in three of the four activities that provide credit-claiming opportunities for politicians. In stark contrast, but consistent with Hypothesis 1, that same set of contingencies leads to a diminished propensity to use tax abatement for economic development. The negative coefficient for tax abatement in cities with high property tax burdens is modest and, partly because of the smaller number of cases included in analysis of this item, barely under the .10 level of statistical significance. What is most notable, however, is the distinctive break in an overall pattern that this negative coefficient represents. Tax abatement is the only economic development policy that is less likely to be used when there are greater levels of popular involvement, and then only where property tax burdens are high.

Table 12.3 Bivariate Relationships Between Economic Development Activities and Alternative Measures of Property Tax Stress

	Property Tax Per Capita	Property Tax Dependence
Activities with visible tax costs		
Tax abatement[a]	—	—
Use of local taxes for economic development	—	—
Minimizing apparent tax costs		
Tax-exempt bonds[b]	—	—
Loan guarantees	+0.07 (.06)	—
Loan subsidies	+0.16 (.00)	+0.10 (.02)
Sale-leaseback arrangements	—	−0.07 (.07)
Tax-increment financing	—	−0.14 (.00)
Special assessment district	−0.07 (.06)	−0.10 (.01)
Deferred tax payments	—	—
Federal grants	+0.06 (.099)	—
Maximizing apparent benefits		
Infrastructure improvement	—	—
Reform building/zoning regulations	—	—
Reform environmental regulations	—	—
Promotional visits/meetings	—	—

NOTE: Figures presented are Kendall's Tau B coefficients and, in parentheses, significance levels. Only a "—" is shown if the significance level is greater than .10.
a. Includes only cities in states that authorize property tax abatement.
b. Includes only cities in states that authorize IDBs or IRBs.

Hypothesis 1a, which stipulates that any use of local taxes for economic development will be similarly constrained by the combination of property tax stress and citizen involvement, is not directly supported by the data. Where property tax burdens are high, no association is found between citizen involvement and use of local taxes for development projects. It is interesting to note, however, the evidence that citizen involvement is associated with greater use of local taxes for economic development in low property tax settings. In those settings, the property tax apparently is much less politically sensitive than it is in higher tax settings.

When we consider the cost-minimizing and benefit-maximizing strategies for economic development that are relevant to Hypothesis 2, we find that the linkage between heightened citizen involvement and the cities' use of these strategies is not the same in communities with low or moderate property taxes per capita as it is in communities with high property taxes per capita. In cities with moderate or low tax burdens, heightened citizen involvement typically is unrelated to the adoption of economic development strategies that minimize apparent costs or that maximize apparent benefits. There are some exceptions. Table 12.4 shows significant associations between citizen involvement and two

Table 12.4 Relationships Between Citizen Involvement and Economic Development Activities by Level of Per Capita Property Tax

	Per Capita Property Tax Revenue		
	High ($150 or More) N = 101	Moderate ($54-$149) N = 186	Low (0-$53) N = 141
Activities with visible tax costs			
Tax abatement[a]	–0.14 (.097)	—	—
Use of local taxes for economic development	—	—	+0.14 (.04)
Minimizing apparent tax costs			
Tax-exempt bonds[b]	+0.27 (.01)	—	—
Loan guarantees	+0.19 (.02)	—	—
Loan subsidies	—	—	—
Sale-leaseback arrangements	+0.26 (.00)	—	—
Tax-increment financing	+0.21 (.01)	—	—
Special assessment district	+0.14 (.07)	—	+0.14 (.04)
Deferred tax payments	+0.17 (.04)	—	—
Federal grants	—	—	+0.20 (.01)
Maximizing apparent benefits			
Infrastructure improvement	—	+0.17 (.01)	—
Reform building/zoning regulations	+0.21 (.01)	—	—
Reform environmental regulations	+0.13 (.09)	—	—
Promotional visits/meetings	+0.18 (.02)	—	+0.14 (.03)

NOTE: Figures presented are Kendall's Tau B coefficients and, in parentheses, significance levels. Only a "—" is shown if the significance level is greater than .10.
a. Includes only cities in states that authorize property tax abatement.
b. Includes only cities in states that authorize IDBs or IRBs.

of the seven cost-minimizing policies at issue, in communities with low tax burdens. Two exceptions also involve significant relationships between citizen involvement and benefit-maximizing activities in cities that have moderate or low property tax burdens.

Despite the exceptions, Table 12.4 presents a rather striking pattern. In cities with high property tax burdens, higher levels of citizen involvement appear to present relatively consistent pressure for the adoption of economic development policies that minimize apparent tax costs and that maximize apparent benefits. In cities with low or moderate property tax burdens, no such consistent linkage appears, only sporadic connections between enhanced citizen involvement and use of a few policy tools. In short, citizen activism in a high property tax context appears to yield nontrivial pressures against tax abatement and pressures for development policy alternatives.

Finally the set of hypotheses tested in Table 12.4 can be retested by using an alternative indicator of property tax stress—that is, the city's dependence on the property tax relative to other sources of general revenue. Table 12.5, which

Table 12.5 Relationships Between Citizen Involvement and Economic Development Activities by Level of Property Tax Dependence

	Property Tax Dependence		
	High (33% or More) N = 121	Moderate (16-32%) N = 161	Low (0-15%) N = 146
Activities with visible tax costs			
Tax abatement[a]	−0.15 (.06)	—	+0.16 (.06)
Use of local taxes for economic development	—	—	+0.13 (.05)
Minimizing apparent tax costs			
Tax-exempt bonds[b]	+0.14 (.08)	—	+0.14 (.04)
Loan guarantees	—	+0.10 (.09)	—
Loan subsidies	+0.16 (.04)	—	—
Sale-leaseback arrangements	+0.25 (.00)	+0.11 (.08)	—
Tax-increment financing	+0.13 (.06)	—	—
Special assessment district	—	—	+0.13 (.06)
Deferred tax payments	+0.12 (.09)	—	—
Federal grants	—	+0.14 (.03)	+0.11 (.09)
Maximizing apparent benefits			
Infrastructure improvement	—	—	—
Reform building/zoning regulations	—	—	—
Reform environmental regulations	+0.11 (.098)	—	—
Promotional visits/meetings	—	—	+0.12 (.06)

NOTE: Figures presented are Kendall's Tau B coefficients and, in parentheses, significance levels. Only a "—" is shown if the significance level is greater than .10.
a. Includes only cities in states that authorize property tax abatement.
b. Includes only cities in states that authorize IDBs or IRBs.

presents the results, shows that the evidence is more mixed when this alternative indicator of fiscal stress is used. Hypothesis 1, however, is not only reaffirmed but also is supported with even stronger evidence than in Table 12.4. Table 12.5 shows a striking reversal with respect to use of tax abatement. In low-tax settings, citizen involvement is positively associated with use of this economic development tools. By contrast, in high-tax settings, citizen involvement is associated with a diminished propensity to use tax abatement. Again, as in Table 12.4, the circumstances that diminish use of tax abatement do not appear to diminish use of taxes generally for economic development purposes. However, the positive relationship between popular involvement and use of local taxes for economic development disappears where property taxes are higher.

Hypotheses 3 and 4 are given more mixed evidence in Table 12.5 than in Table 12.4. Table 12.5 shows that, in settings with high property tax dependence, popular involvement is linked to most economic development activities that involve minimizing apparent tax costs. However, settings of high property tax dependence are not as distinctive in this regard as were settings of high property tax burdens. For several of the development tools (loan guarantees, tax-exempt bonds, sale-leaseback arrangements, special assessment districts, and federal grants), significant and positive relationships between citizen involvement and use of the development tool also appear for settings with lower levels of property tax dependence.

Hypothesis 4, involving activities that maximize apparent development benefits, is given even less support in Table 12.5. Only in the case of reforming environmental regulations do we find the expected pattern. For the other three policies in this category, citizen involvement is either unrelated to use of the policy, regardless of property tax dependence, or related to the policy only where property tax dependence is low.

Taken together, Tables 12.4 and 12.5 provide evidence of a pattern of local government use of economic development policies. There are some breaks in the pattern, and the pattern emerges much more clearly when property tax burden is at issue than when property tax dependence is the indicator of fiscal stress. Despite these flaws, a general pattern can be detected. It suggests that the combined pressures of popular involvement and property tax stress drive city officials away from the use of tax abatement and toward the use of alternative development policies that maximize apparent benefits and minimize apparent tax costs.

Discussion

Economic development policies in general, and tax abatement in particular, have been treated as largely symbolic policies that can appease voters with the vision of benefits for the community at large while meeting the desires of developmental elites (Swanstrom, 1988, p. 104). The symbolic policy approach assumes that voters will be sold on the collective benefits of tax abatement and distracted from the redistributional aspects of these subsidies. The research in this chapter suggests, however, that this is not always so. Where property tax burdens are problematic and citizens are more involved in the economic development sphere, local officials are constrained from property tax abatement. We refer to this as the "politics of delegitimation of tax abatement." Our evidence of this is cross-sectional analysis of national sample survey data, with all of its attendant limitations. Allowing for those limitations, we believe that we have evidence for a large group of cities that parallels Swanstrom's case study findings concerning the politics of tax abatement in Cleveland during the Kucinich administration.

Our analysis provides a number of partial challenges to contemporary assumptions that the realm of development policy is insulated from popular pressures. First, our analysis suggests that when local officials eschew tax abatement and adopt alternative policies that minimize tax costs and maximize benefits, they are responding to the public, or at least are constrained by the anticipated public reaction to tax abatement. Second, our results suggest a more populist face for development politics than is often recognized. Politicians concerned with electoral reprisals avoid tax policies that threaten their chances of reelection. Under the circumstances we have specified, tax abatement is one of those policies to be avoided.

Third, our analysis also suggests considerable variety in the politics of development. It is not just that cities have very different levels of citizen involvement in economic development decision making, although our analysis suggests that this is certainly the case. The impact of citizen involvement seems to differ, depending on the type of development policy at issue (tax abatement vs. alternative strategies) and depending on the community context (higher vs. lower property tax burdens). Public officials in tax-stressed communities who are wary of the potential consequences of popular mobilization may be relieved to find that heightened citizen involvement actually can enhance the climate for economic development policies other than tax abatement.

But it perhaps may be inappropriate to interpret the positive relationships in Tables 12.4 and 12.5 as instances of popular support for the various economic development strategies at issue. An alternative interpretation holds that the adoption of devices such as loan guarantees, sale-leaseback arrangements, and similar developmental strategies is a flight from popular accountability. That is, in high tax settings, such developmental strategies are positively associated with level of citizen involvement not because citizens support them, but because these strategies can be undertaken invisibly, without popular approval, and typically without public awareness of their existence. Ironically the pressures of popular sensitivities concerning property taxes drive politicians to adopt economic development strategies that evade popular scrutiny.

From this point of view, our results need not constitute a wholesale rejection of domain theory. Some cities may indeed have relatively elaborate mechanisms for citizen involvement in economic development decision making, as diverse as public hearings and elected neighborhood commissions that mobilize citizens around developmental issues. But these participatory devices may be empty exercises in popular involvement. "Real" decision making about economic development may be going on simultaneously in the insulated, centralized, elite-dominated format described by domain theorists. Many of the economic development tools considered in our list of those that minimize apparent tax costs could be used to sustain such an off-budget, relatively invisible form of developmental decision making.

Our analysis, however, suggests that the politics of a number of ostensibly different policy tools follows a common pattern—one influenced by citizen involvement and existing levels of property tax stress. The key distinction appears to be between economic development tools that have direct implications for the general property-tax payer and economic development tools that minimize apparent tax costs. This is not to say that there are no other important differences between the various policy tools considered here. For example, financing tools such as sale-leaseback arrangements and loan guarantees may be dependent on unique forms of financial expertise at city hall and the capacity for innovation, and the use of federal grants for development projects may entail a quite different set of requirements. Despite their other key differences, however, our analysis suggests that a great deal about cities' propensity to use many of these development policy tools can be understood from the perspective of the political pressures presented by high property taxes and a participatory approach to economic development decision making.

Notes

1. For a somewhat different interpretation of tax increment financing, see Rubin and Rubin, 1987, p. 41.

2. For example, Bennett and DiLorenzo (1983, p. 103) find that states in which 1970s taxpayer revolts generated populist controls on local taxing and spending show much greater growth in nonguaranteed debt issuances (revenue bond) than do states that did not experience such citizen mobilization.

3. See also a description of the "politics of announcement"—in which local politicians build favorable images of themselves by publicizing federal grants for redevelopment projects, development plans, promotional activities, and other signs that promise economic progress for the community, in Stone and Sanders, 1987, pp. 178-179.

4. Infrastructure improvements' openings, groundbreakings, and promotional activity provide excellent opportunities for politicians to be seen actively involved in community improvements. Regulatory reform is a somewhat more problematic strategy from a credit-claiming point of view because, on the one hand, it is less inherently visible and because questions can be raised about compromises of the public interest when regulations are relaxed. On the other hand, tackling allegedly outdated, inefficient, counterproductive regulations and slow-moving regulatory bureaucracies provides the occasion for local politicians to manipulate the potentially powerful symbols of bureaucratic reform and regulatory relief.

5. The very smallest municipalities in the United States—those under 10,000 population—were excluded totally from the study sample by virtue of the fact that questionnaires were sent only to cities of at least 10,000 population. Because the study sample, therefore, is not an acknowledged attempt to represent these numerous but extremely small jurisdictions, they are excluded from the comparisons in Table 12.1 as well.

6. Some caution must be reserved for this measure of citizen participation. The variables used to create this measure merely reflect the means by which an economic development plan is adopted in the city. The precise definition of "economic development plan" is left up to the survey respondent's interpretation. Consequently, what is entailed by economic development plan may bear no relevance to policies later adopted or to levels of citizen participation. The authors are indebted to the observation of an anonymous reviewer of the limitations of this measure.

7. Specifically, the citizen involvement index was a count of the number of citizen input methods reported to be used, with the most common device—appointed advisory committees representing the entire community—weighted once, the next most common device—open meetings or public hearings—weighted twice, the third most common device—citizen surveys—weighted three times, and the rarest device—elected neighborhood commissions—weighted four times.

References

Bennett, J., & DiLorenzo, T. (1983). *Underground government: The off-budget public sector.* Washington, DC: Cato Institute.

Clarke, S. (1987). More autonomous policy orientations: An analytic framework. In C. Stone & H. Sanders (Eds.), *The politics of economic development* (pp. 105-124). Lawrence: University of Kansas Press.

Daykin, D. (1988). The limits to neighborhood power: Progressive politics and local control in Santa Monica. In S. Cummings (Ed.), *Business elites and urban development* (pp. 357-387). Albany: State University of New York Press.

Dye, T. (1986). Community power and public policy. In R. J. Waste (Ed.), *Community power.* Beverly Hills, CA: Sage.

Feiock, R., & Clingermayer, J. (1988, November). *Development policy choice: Four explanations for city implementation of economic development policies.* Paper presented at the Annual Meetings of the Southern Political Science Association, Atlanta.

Jones, B., & Bachelor, L. (1986). *The sustaining hand.* Lawrence: University of Kansas Press.

Leonard, H. B. (1986). *Checks unbalanced: The quiet side of public spending.* New York: Basic Books.

Mayhew, D. (1974). *Congress: The electoral connection.* New Haven, CT: Yale University Press.

National Association of State Development Agencies (NASDA). (1984). *Directory of incentives for business investment and development in the United States.* Washington, DC: Urban Institute Press.

Pagano, M. (1988). Fiscal disruptions and city responses. *Urban Affairs Quarterly, 24,* 118-137.

Peterson, P. (1981). *City limits.* Chicago: University of Chicago Press.

Rubin, I., & Rubin, H. (1987). Economic development incentives: The poor (cities) pay more. *Urban Affairs Quarterly, 23,* 37-62.

Stone, C., & Sanders, H. (Eds.). (1987). Reexamining a classic case of development politics: New Haven, Connecticut. In *The politics of urban development* (pp. 178-179). Lawrence: University of Kansas Press.

Swanstrom, T. (1988). Semisovereign cities: The politics of urban development. *Polity, 21,* 83-110.

Wasylenko, M. (1981). The location of firms: The role of taxes and fiscal incentives. In R. Bahl (Ed.), *Urban government finance.* Beverly Hills, CA: Sage.

— 13 —

Women-Owned Businesses

DIMENSIONS AND POLICY ISSUES

Thomas A. Clark

Franklin J. James

Recent reports of substantial increases in the number and economic scale of women-owned businesses have stimulated commentary touting women's gains and the importance of ownership in their economic advancement. A report of the House Small Business Committee concludes:

> From less than five percent of the Nation's businesses prior to the 1970s, women have now come to own approximately thirty percent of American businesses. They are starting businesses at over twice the rate of men and could well own and operate fifty percent of the Nation's businesses by the year 2000. . . . Women-owned businesses may well provide the cutting edge—and the American advantage—in the worldwide economic competitiveness fast upon us. The loss to the Nation would be incalculable were public policymakers not to foster this development to the fullest extent possible. (U.S. House of Representatives, 1988, p. iii)

Gillian Rudd, president of the National Association of Women Business Owners (NAWBO), argues:

> Somewhere between 4 and 5 million women business owners have entered the economy in the last 15 years. They are the most vital and rapidly growing force in the U.S. economy—yet the government gives them minimal encouragement, (and) programs of technical or financial assistance are practically nonexistent. (NAWBO, 1988, p. 1)

Indeed, the rate of growth of women-owned businesses is remarkable. Evidence presented in this chapter shows that the number of women-owned businesses may have grown 6-fold, and their gross receipts 50-fold since 1970.

AUTHORS' NOTE: The authors wish to thank Nancy B. Leff for her comments and suggestions for this research, and Gary Berlin for his assistance in gathering data and materials. We also acknowledge the important comments of several anonymous reviewers.

In the face of such evidence, it is tempting to conclude that great strides have occurred, that a continuation of the trend is both likely and desirable, and that women's successes might help reformulate the flagging effort to promote minority business ownership (e.g., see James & Clark, 1987). It is tempting, as well, to conclude that business ownership is a significant contributor to economic progress for women in the labor force. A closer examination, however, calls into question such positive assessments of the gross trendline.

In this study we identify and assess the issues surrounding women's business ownership. First, we present the most up-to-date figures on trends in the scale of business ownership by women. These are 1987 figures from the business census of that year. The statistics verify that female ownership of business continues to grow rapidly. Second, and more importantly, we examine the economic payoffs to women from business ownership.

Our analysis suggests that few women have achieved high earnings from ownership. Indeed it appears that potential earnings for most women are significantly higher in management positions on payrolls than in self-employment, taking into account measures of experience and qualification that are available from the census.

Third, we consider the barriers to women's economic advancement through ownership. This analysis considers most women's limited experience in management, family and other constraints on the goals and objectives of women business owners, and women's access to financing. The evidence suggests that each of these may circumscribe the economic opportunities of many women. We conclude the chapter with a brief discussion of the policy implications of these findings.

The heart of our conclusions is that business ownership so far has conferred relatively few significant benefits on women. Most businesses owned by women are small and undercapitalized, and many offer only part-time work. Women lacking prior work or management experience (over half of all owners) are hard-pressed to find substantial opportunities either in self-employment or in the wage and salary sector. Women with continuing care responsibilities—most importantly for children, but also for the disabled and the elderly—often are constrained to work sites in or near the home and to work schedules that complement these allied activities. Many such women appear to seek to own businesses largely because they have no alternative, not because business ownership is so attractive.

At the same time, we acknowledge that some women, at least, prefer ownership and the personal autonomy it affords over work in other businesses in which invidious barriers limit advancement. Moreover, a number of positive factors may enhance women's success as business owners. In the concluding section we address the potential for positive developments in the returns to women from business ownership and briefly discuss their policy implications.

The Dimensions of Women's Business Ownership

Before beginning the analysis, it must be emphasized that available data on women-owned businesses are deficient, hamstringing some research undertakings. The bulk of the available statistics describes women's ownership of proprietorships, partnerships, and subchapter S corporations.[1] Most firms owned by women fall into these groups. No extant database, however, accurately tracks the ownership by women of nonsubchapter S corporations. This is an important omission because these corporations tend to be the largest in the economy, and they encompass a significant portion of the sales, employment, and profits of women entrepreneurs.

Moreover, women-owned businesses were inadvertently undercounted in the U.S. Economic Censuses prior to 1982, and industrial coverage was expanded after 1972, the first year of this data series.[2] Because the earlier counts were deflated and the later counts addressed an expanded base, growth rates derived from these data exceed the actual rates. Because of these data limitations, the research presented here is necessarily exploratory.

All available evidence points to a rapid, probably accelerating increase in the portion of the American economy owned by women (see U.S. Bureau of the Census, 1986a). The share of all businesses that was owned by women may have grown by as much as 552% from 1972 to 1987 (Table 13.1). In these same years, the share of gross receipts claimed by women-owned firms may have increased even more rapidly—by as much as 4,500%. By 1987, women owned almost one in three firms and oversaw over 14% of the nation's gross business receipts.

Women-owned sole proprietorships appear to have been growing in number especially rapidly since 1980. From 1980 to 1986, these rose by almost two thirds, whereas those of men and women combined increased by only 42%. Total (current dollar) receipts of women-owned nonfarm proprietorships nearly doubled in these same years (U.S. Small Business Administration [SBA], 1989, Tables 10 and 11).

The growth in women-owned businesses has been highly selective across industries. By 1982, women claimed the largest shares of all firms in trade (26.6%) and services (29.7%). Both industries, of course, are characterized by attenuated career tracks, lower levels of value-added per worker, lower wages, and lower capital requirements. It is significant, however, that in that same year, women also owned in excess of 15% of all firms in manufacturing and in finance, insurance, and real estate. Of the two, however, only in manufacturing did women's share of gross receipts (9.1%) approach the approximately 12% share secured in trade and in services. The current trendline, however, seems to favor goods production. Women-owned proprietorships in manufacturing, mining, and construction combined almost tripled (166%) in number between 1980 and 1987, those in trade grew by just 18%, and those in services grew by 94% (SBA, 1989, Chart 8).

Most women-owned firms are tiny. Indeed, most do not have any paid employees. The 3.5 million such small firms comprised 85% of all women-owned businesses in 1987. Almost all were sole proprietorships. In that same year,

Table 13.1 Women-Owned Firms[a] and the Value of Their Receipts as a Percentage of All Owned by Men and Women, by Industry, 1972 to 1987

| | Percentage of All Firms Owned by Men and Women | | | | | | | |
| | Firms | | | | Receipts[b] | | | |
Industry	1972	1977	1982	1987	1972	1977	1982	1987
All[c]	4.6	7.1	23.9	30.0	0.3	6.6	10.2	13.9
Construction	1.5	1.9	4.7	5.7	0.3	4.0	4.9	8.7
Manufacturing	1.8	6.6	15.8	21.7	NA	9.4	9.1	13.6
Transportation and public utilities	1.6	2.9	8.1	13.5	0.1	5.7	8.2	14.3
Trade	4.7	8.8	26.6	32.9	0.6	8.0	11.7	14.8
Finance, insurance, and real estate	2.8	4.7	15.5	35.6	0.2	3.2	5.4	14.4
Selected services	6.8	8.7	29.7	38.2	1.6	5.9	12.3	14.7

SOURCES: U.S. Bureau of the Census, 1976, "Women-Owned Businesses, 1972," *1972 Economic Censuses,* WB72, Washington, DC: Government Printing Office, Table A; U.S. Bureau of the Census, 1980, *"Women-Owned Businesses, 1977," 1977 Economic Census,* WB77-1, Washington, DC: Government Printing Office; U.S. Bureau of the Census, 1986b, "Women-Owned Businesses, 1982," *1982 Economic Censuses,* WB82-1, Washington, DC: Government Printing Office, Table E; and U.S. Bureau of the Census, 1990, "Women-Owned Businesses, 1987," *1987 Economic Censuses,* WB87-1, Washington, DC: Government Printing Office, Figure 4.
NOTE: NA = less than .05%.
a. Firms both with and without employees. The individual firm may include more than one business establishment.
b. Receipts include the gross annual value of all products sold and services rendered, plus other receipts, less returns and allowances.
c. Includes both unclassified and classified industries owned by women. Note that census procedures placed a larger share of women-owned firms in the "unclassified" category in 1982.

almost 300,000 women-owned firms had between 1 and 4 employees, and the number in each successively larger class decreased steadily (Table 13.2). Also, 115,000 firms had between 5 and 19 employees, and just over 25,000 had 20 or more workers. Nevertheless, in total, these 25,000 firms—less than 1% of the total firms—accounted for half of all employees in women-owned firms.

Whereas almost all women-owned firms lacking paid employees are individual or "sole" proprietorships (96%), just over 6 in 10 of those with paid employees are of this legal form (Table 13.3). Of firms with paid employees, about 1 in 10 is a partnership, and 3 in 10 are corporations. Corporations with paid employees, however, claim an inordinate share of both workers and receipts —64% of all paid workers employed by women-owned firms, and 73% of all of their gross receipts.[3]

Like other women-owned businesses, those having paid employees are highly concentrated in the trade and service industries (Table 13.4). In fact, almost 8 in 10 of all women-owned firms with paid employees were in these two industry sectors in 1987; and their employment was similarly concentrated there. But the average gross receipts per firm in retailing and services have been the lowest of all industries (excepting finance, insurance, and real estate) since the early 1970s. The fractions of all firms realizing gross receipts of over $500,000 have

Table 13.2 Firms Owned by Women, by Number of Employees, 1972 to 1987

Employment Size Intervals	Firms With Paid Employees				Number of Employees			
	1972	1977	1982	1987	1972	1977	1982	1987
1 to 4	39	119	207	297	79	199	387	586
5 to 9	10	29	45	79	61	186	288	509
10 to 19	4	14	18	36	46	181	236	471
20 to 49	1	6	7	17	32	168	211	500
50 to 99	(98)	1.2	1.5	5	7	84	99	319
100 or more	(21)	(437)	(668)	3	4.5	77	133	719

SOURCES: U.S. Bureau of the Census, 1976, "Women-Owned Businesses, 1972," *1972 Economic Censuses,* WB72, Washington, DC: Government Printing Office, Table 6; U.S. Bureau of the Census, 1980, "Women-Owned Businesses, 1977," *1977 Economic Census,* WB77-1, Washington, DC: Government Printing Office, Table 6; U.S. Bureau of the Census, 1986b, "Women-Owned Businesses, 1982," *1982 Economic Censuses,* WB82-1, Washington, DC: Government Printing Office, Table 8; and U.S. Bureau of the Census, 1990, "Women-Owned Businesses, 1987," *1987 Economic Censuses,* WB87-1, Washington, DC: Government Printing Office, Table 9.

Table 13.3 Characteristics of Women-Owned Firms With Paid Employees, by Legal Form of Organization[a],1977 to 1987

Legal Form	Firms (thousands)			Employees[d] (thousands)			Gross Receipts[e] (millions, $)		
	1977[b]	1982[c]	1987	1977[b]	1982	1987	1977	1982	1987
Number									
Sole proprietor[f]	64	181	377	187	500	727	8	21	38
Partnerships	56	64	69	291	332	392	17	16	32
Corporations	48	67	171	416	523	1,984	29	29	158
Percentage of all women-owned firms									
Sole proprietor	39	58	61	21	37	23	15	33	17
Partnerships	33	20	11	33	25	13	32	24	10
Corporations	28	22	28	46	38	64	53	43	73

SOURCES: U.S. Bureau of the Census, 1980, "Women-Owned Businesses, 1977," *1977 Economic Census,* WB77-1, Washington, DC: Government Printing Office, Table 8; U.S. Bureau of the Census, 1986b, "Women-Owned Businesses, 1982," *1982 Economic Censuses,* WB82-1, Washington, DC: Government Printing Office, Table 6; and U.S. Bureau of the Census, 1990, "Women-Owned Businesses, 1987," *1987 Economic Censuses,* WB87-1, Washington, DC: Government Printing Office, Table 7.
NOTE: a. Partnerships and small corporations were significantly undercounted in 1972.
b. Industrial coverage expanded in 1977 to cover industry groups 81, 804-1, 808, and 899, not covered elsewhere. All are services.
c. In 1982, for the first time, different businesses owned by the same individual proprietor were counted separately. Certain businesses owned by women in 1972 and 1977 were counted as men-owned, when co-owned by women.
d. Includes all paid employees, both full- and part-time.
e. In constant (1982) dollars, using the producer price index.
f. Identified as "individual partnerships" after 1977.

Table 13.4 Number, Employment, and Gross Receipts of Women-Owned Firms With Employees, by Industry, 1977 to 1987 (in thousands)

Industry	Number of Firms			Employment			Average Receipts per Firm[a]		
	1977	1982	1987	1977	1982	1987	1977	1982	1987
All firms	168	312	618	895	1,355	3,103	323	210	352
Construction	9	13	36	49	56	180	425	248	482
Manufacturing	8	10	27	94	93	364	635	465	1,079
Transportation and public utilities	5	8	19	30	39	106	376	297	483
Trade									
Wholesale	8	9	23	53	50	188	1,227	947	1,729
Retail	71	119	199	354	541	1,091	305	224	363
Finance, insurance, and real estate	12	16	37	37	49	109	177	172	247
Services	51	122	253	263	491	1,016	159	116	149

SOURCES: U.S. Bureau of the Census, 1980, "Women-Owned Businesses, 1977," *1972 Economic Censuses*, WB77-1, Washington, DC: Government Printing Office, Table 1; U.S. Bureau of the Census, 1986, "Women-Owned Businesses, 1982," *1982 Economic Censuses*, WB82-1, Washington, DC: Government Printing Office, Table 1; and U.S. Bureau of the Census, 1990, "Women-Owned Businesses, 1987," *1987 Economic Censuses*, WB87-1, Washington, DC: Government Printing Office, Table 1.
NOTE: a. In constant (1982) dollars.

been consistently higher in three sectors: manufacturing, transportation and public utilities, and wholesaling.

If women's propensity to own businesses depends, at least in part, on the health of regional economies, then we would expect to find geographical differences in the rate of formation of such businesses. The evidence at hand indicates little regional variation in the rate of women's business ownership, however. In the 1980s the largest absolute concentrations of women-owned firms having paid employees were in the Middle Atlantic, East North Central, South Atlantic, and Pacific census divisions (U.S. Bureau of the Census, 1976, Table 2; 1980b, Table 2; 1986b, Table 2). In those years, however, the number of women-owned firms with paid employees per capita was almost constant across the major census divisions. Regional variations—if they exist—apparently are obscured by such aggregate data as these.[4]

At the same time, women-owned firms with paid employees are somewhat underrepresented in the nation's larger metropolitan areas. In the early 1980s, almost 30% of the U.S. population resided in metropolitan areas that contained the 15 largest cities, but only one in five firms owned by women were located in those areas (U.S. Bureau of the Census, 1977, Table 5; 1980b, Table 5; 1986a). Within these areas, women-owned firms are distributed about equally between central cities and suburban rings.

Historically, small business ownership has been highest in the rural economies of nonmetropolitan areas. The relatively low rate of women's business ownership in large metropolitan areas may be a reflection of this basic pattern.

Business ownership by women has not yet played a large role in minority advancement. In the late 1970s, more than 9 in 10 women owners were non-Hispanic whites (U.S. Bureau of the Census, 1980b, Tables 2a and 2b). Blacks owned just 2% of all women-owned firms, and Hispanics just 2.6% in 1977. Most minority-owned firms were quite small; 96% of all black-owned firms were sole proprietorships, and 83% of all owned by Hispanic women were of this class.

Economic Opportunities Associated With Business Ownership for Women

Indicators of growth in numbers of women-owned businesses belie the continuing vulnerability of women-owned enterprises. A closer analysis of the data suggests that the increase in women's business ownership is largely a correlate of rising labor force participation by women. It is not the result of increasing rates of self-employment among women.[5] Between 1980 and 1987, the number of self-employed women rose by 728,000, or 36%. For purposes of comparison, numbers of self-employed men rose by 13% during the period (U.S. Bureau of Labor Statistics, 1981, 1988). For both men and women, however, the rate of self-employment fluctuated, rising from the mid-1970s to the early 1980s and then declining. Among women, the rate of unincorporated self-employment fell from 3.4% in 1967 to 2.7% in 1975, then rose to 3.9% in 1982, and finally declined to 3.3% in 1985 (SBA, 1989). There is thus no evidence of a significant rise in the rate of self-employment among women during the 1980s.

Available data suggest that self-employment has offered only modest economic opportunities for women. As a result of growing numbers of self-employed women, a burgeoning number of self-employed women were earning comparatively high incomes. In 1988, for example, about 13,000 self-employed women in executive, administrative, and management jobs earned more than $50,000. Controlling for inflation, there were only 3,000 such women in 1980 (U.S. Bureau of the Census, 1982, 1989).

Most self-employed women earned relatively little, however. In 1987 the median earnings from self-employment of all women reporting self-employment earnings (including those working part-time) was only $3,120. These women's median overall income was $8,345, including income from other sources (U.S. Bureau of the Census, 1989). For most self-employed women, it seems that self-employment provides a relatively small supplement to other sources of income.

It is not yet well recognized that the 1980s was a decade of marked economic progress for women. In general, the earnings of women increased during the period, relative to those of men. The median earnings of full-time, year-round women workers increased by 10% in real terms between 1980 and 1988, from $16,075 to $17,606 (in 1988 dollars). By contrast, the median earnings of full-time, year-round male workers actually diminished somewhat during the period, from $26,721 to $26,656. As a result, women's earnings reached 66% of men's in 1988, up from 60% at the start of the decade (U.S. Bureau of the

Table 13.5 Median Earnings of Full-Time Year-Round Workers, by Occupation
Employment Status and Sex (in 1988 dollars)

Occupation and Status	Women			Men		
	1982	1986	% Change 1982-86	1982	1986	% Change 1982-86
Executives, administrators, and managers						
Total	$21,242	$23,125	8.9	$35,333	$36,982	4.7
Self-employed	11,204	11,330	1.1	19,400	24,343	25.5
Total, all occupations	15,955	17,514	9.8	25,840	27,251	5.5

SOURCE: "Trends in Income, by Selected Characteristics: 1947 to 1988" by M. F. Henson, 1990, *Current Population Reports* (Series P-60, Report No. 167). Washington, DC: Government Printing Office.

Census, 1989). During the 1970s the relative earnings of women did not change significantly.

Between 1982 and 1988, the earnings of women rose relative to those of men in all major occupation groups, save one: the exception was service work, in which the real median earnings of men and women both grew by 5%. In executive, administrative, and management jobs, the real median earnings of women grew by 10%, whereas that of men increased by only 4%. In professional jobs the median earnings of women rose by 14%; those of men rose by 10% (U.S. Bureau of the Census, 1989).

Self-employment has not been a major factor boosting the economic relative status of women in the labor force. Strikingly, the earnings of self-employed female managers and executives have not kept pace with those of self-employed men or with those of women on payrolls. Table 13.5 lists the earnings of self-employed women working full-time, year-round. As can be seen, the median earnings of full-time, year-round, self-employed women executives, administrators, and managers rose more slowly (in constant 1988 dollars) between 1982 and 1986 than did the earnings of women in such jobs on payrolls or than did the earnings of women in total. Moreover, the gap between the increase in the median earnings of men and women was largest among the self-employed, compared with other groups identified in the table, and was the only one to favor men rather than women. The median earnings of self-employed women executives, administrators, and managers were lower by half than were those of all women executives as a group, even though the comparison is limited to full-time, year-round workers.

It thus appears that self-employment has not been an effective route to economic opportunity for many women, although it has been a well-traveled one. The evidence that self-employment has not proven a highly effective route to economic opportunity for many women is stronger for earlier periods. Multivariate analysis of the determinants of the annual earnings of self-employed persons and of persons on payrolls provides useful insight into the economic payoffs to women entrepreneurs of self-employment in 1970 and 1980.

For purposes of this analysis, a regression model of annual earnings was developed and applied to samples of men and women in the labor force by using unpublished public use data from the 1970 and 1980 censuses. Women were included in the analysis only if they were of working age (18 to 65) and had worked the bulk of the year prior to the decennial census.[6] Separate models were estimated for blacks, Hispanics, and non-Hispanic whites in order to control for the effects of race and ethnicity on potential labor market earnings. The statistical model can be summarized as follows:

$$E = f(\text{CLS, EX, ED, DIS, MIG, CIT, IMM, REG, MET}),$$

where

       E = The natural logarithm of annual earnings of workers

       CLS = Class of worker (dummy variables denoting self-employed, federal government worker, state and local government worker, private payroll employee)

       EX = Indicators of potential work experience (age minus years of schooling plus five)

       ED = Years of schooling (dummy variables denoting high school graduates, some college, and college graduates)

       DIS = Indicators of partial or total work disabilities

       MIG = Indicators of interstate migration within 5 years of the census

       CIT = Indicators of national citizenship, for Hispanics only

       IMM = Indicators of international immigration and country of origin, for Hispanics only

       REG = Region of residence

       MET = Metropolitan residence

Exactly the same model was applied to both 1970 and 1980 data.

A comprehensive report of the results is beyond the scope of this chapter (for a complete presentation of the analysis, see James ,1985). Table 13.6 presents the portion of the findings pertaining to three cohorts of workers. Coefficients in the table are interpretable as proportional differences in annual earnings to be expected if a woman were employed in the identified class of job, compared to the earnings she could expect on private payrolls, holding other factors including race/ethnicity constant. Strikingly, expected annual earnings from self-employment, relative to earnings on private payrolls, fell markedly for every group of women between 1970 and 1980. In 1970 the expected annual earnings of self-employed non-Hispanic white women, for example, were 24% lower than their expected earnings on private payrolls. By 1980 this gap had risen to 65%. The gap in expected earnings rose for Hispanic women from 15% in 1970 to 55% in 1980. For black women the gap grew from 14% to 45%. In contrast, the gap in annual earnings between government work and the private sector closed for each female cohort in the same period.

Table 13.6 Percentage Differences in Women's Annual Earnings Among Various Types of Jobs, Compared to Expected Earnings on Private Payrolls: 1970 and 1980

	1970	1980
A. Non-Hispanic whites		
Self-employment	−.2428**	−.6468***
Federal government	.2464***	.1644
State and local government	.0371	.0228
B. Hispanics		
Self-employment	−.1452	−.5516***
Federal government	.2780***	.2079**
State and local government	.1338**	.0139
C. Blacks		
Self-employment	−.1412	−.4501*
Federal government	.4866***	.2236***
State and local government	.1773***	−.0414

NOTE: These estimates are based on a multivariate regression model of the natural logarithms of individual annual earnings reported to the decennial census. Only persons of working age, with earnings during the year prior to the census, and who were in the labor force at the time of the census were included. Persons were also required to have worked for the bulk of the year prior to the census.
*Statistically significant at the .10 level, one-tail test.
**Statistically significant at the .05 level, one-tail test.
***Statistically significant at the .01 level, one-tail test.

Declining economic returns to self-employment explain why *rates* of self-employment did not rise significantly during the period. As the relative potential payoff to self-employment over payroll employment has declined, the incentive to become self-employed has diminished. Clearly more research needs to be done on the factors producing such a marked decline in the economic payoffs to self-employment.

Constraints on Women's Success in Business Ownership

Several factors appear to have significantly circumscribed the potential productivity and profitability of women-owned businesses. First, as we have noted, many women seek to own their own businesses in order to accommodate parallel responsibilities for the care of children, the disabled, and the elderly. As a result, they are frequently constrained to work sites in or near the home, and require some measure of flexibility in work hours. When few potential employers are geographically accessible, or when employers are unwilling to accommodate these special needs or to provide compensation that is sufficient to offset home care expenses, women find self-employment to be the only option.

Second, many women pursuing business ownership have little capital. As a result they frequently enter industries such as trade and personal services in which they can gain a foothold with a relatively small front-end investment.

The growing dispersal of production within so-called post-Fordist production regimes is clearly congruent with the development of such enterprises (see Lovering, 1990; Scott, 1988). Unfortunately, small suppliers serving large corporations tend to trim labor costs to remain competitive. As a result, this arrangement denies workers the full compensation they might have received in the direct employment of the corporation itself. Operating out of the home and without many, if any, employees further cuts initial capital requirements. Such labor-intensive businesses are consequently poorly positioned to reap substantial financial gains. These factors curtail the size and potential returns to women's entrepreneurial efforts. They are largely systemic in character—products of conditions and barriers.

A number of analysts claim that a third factor—namely, women's management aspirations and approach—serves to dampen the financial rewards of business ownership. As we argue subsequently, however, these are more properly understood as derivatives of experience, life circumstance, and invidious barriers, not as intrinsic features of women's business involvement. Nevertheless, we later argue that women—as well as men—who have no prior management experience can indeed benefit from technical assistance, especially during the start-up phase. Each of the three is considered next.

Characteristics of Women Business Owners

As many as half of the firms owned by women are located in the home, and this rate may be even higher among those without employees (U.S. Bureau of the Census, 1980a, 1980b). Moreover, women-owned businesses typically provide only part-time work for their owners. The median time worked per week by women owning the largest firms was 22 hours, just ahead of those owning the smallest firms. Women entrepreneurs devote fewer hours to their businesses than do men. Fully 63% of women business owners report working part-time—that is, fewer than 40 hours per week. Only 43% of men owners work less than full-time (SBA, 1988).

Many women, as we have noted, desire only part-time involvement in order to accommodate other responsibilities. Part-time work in the home is often seen as a practical complement to parallel work or family responsibilities; 60% of women owners realized less than 75% of their total income from the business they owned in 1977 (U.S. Bureau of the Census, 1980b, Table 7a). In-home communications and production technologies have made this practice more and more feasible in recent years (Brophy, 1989).

Strikingly, 60% of women owners report no paid years of managerial experience prior to owning their own business; only 45% of men had no such experience (SBA, 1988). In 1982, 60% of women-owned businesses were less than 7 years old; the figure for men-owned businesses was 52%. Most women owners have had at least some education beyond high school: Half had 4 or more years of college, most in business and education—but most had no prior ownership experience (U.S. Bureau of the Census, 1980b, Tables 31, 3b, and 4b).

Lack of prior involvement in management occupations places women at some disadvantage as entry is sought: "She enters a traditionally male-dominated arena without many of the tools generally developed by males. These include experience in the relevant industry; general business training and education; and a vital network of business, industry, government, and other contacts" (Ford, 1989, p. 79).

Other evidence suggests that female entrepreneurs tend to have had more experience over a longer period of time *in the business area from which they generate their venture* than is the case with male entrepreneurs (see also Ronstadt, 1984). Brophy (1989, p. 59) notes, "Most studies report females have more relevant direct experience before starting their venture." Such experience may not be in management positions, however.

Compared with men business owners, women business owners more frequently represent "opportunistic entrepreneurs" characterized by broad-based, nontechnical educations, less technical specialization, high social awareness and communication skills, and longer planning horizons. In contrast, men may be more "craft-oriented," in the sense that their enterprises have grown out of technical educations and previous work experiences (Smith, McCain, & Warren, 1982). Indeed, such a conclusion is consistent with the concentration of women in retail trade and services and with the domination of men in goods-production. Surveys of women business owners have found that women in "nontraditional" economic sectors—sectors other than trade and services—report more problems than those in trade or the services (Hisrich & O'Brien, 1981). Women business owners are reported to consider themselves adept at idea generation, product development, and dealing with people, but weak in financing business development (Brush & Hisrich, 1985).

Overall, Ronstadt (1984) concludes that, to this point at least, women have been most successful in owning businesses that require less than a full-time commitment, afford ease of financing, and are less likely to grow beyond a manageable size. Such characterizations as these, of course, emerge from a panoply of employment experiences constrained by prevailing norms having a systemic origin.

Capitalization of Women's Businesses

There is no direct evidence that women business owners are discriminated against, relative to men business owners, in the search for business financing. Such discrimination, however, frequently is alleged. A national survey of women business owners estimated that "68 percent believe they have been discriminated against in business loan applications, and 29 percent of those who received loans believed they were offered on discriminatory terms" (U.S. House of Representatives, 1988, p. 14). The causes of women's problems in getting business capital "can include the general difficulty faced by all business in getting capital, the shortcomings of her own business, the lack of experience and training that may result from being a woman with a traditional education and occupation, or outright discrimination" (p. 16).

There is evidence that the capital base of women's businesses is very fragile. Of all women owners in 1977, most—70%—were the original founders, while one fifth purchased the business. Just 8% inherited the business. Women starting their own businesses started the smallest firms. Of the very largest firms—those having over $1 million in annual receipts—only 40% of women owners were founders. One quarter were purchasers, and 36% inherited the business. At the low end of the size spectrum, proportionately far more were founders, and far fewer inherited the business (U.S. Bureau of the Census, 1980b, Table 7b).

The small size of firms started by women is not surprising. Surveys of business owners have shown that women-owned businesses generally are started with less capital from fewer sources than is true for male-owned enterprises. Fully 89% of women's businesses are started or acquired for less than $10,000. Only 78% of men's businesses require so little capital (SBA, 1988). Men and women business owners who borrowed money used much the same sources: Most borrowed from either banks (over 60%) or families (about one fourth).

Access to capital appears to be a particular problem for women who are single. Married women are more likely to own larger firms, and of these, more are in the most profitable industry sectors such as wholesaling, manufacturing, construction, and transportation (SBA, 1988, p. 5). Single women tend to have less capital to invest, irrespective of their marital or work history. Married women frequently have access to capital through home equity and husbands, which single women do not have. In 1977 three quarters of women business owners were not currently married. Over half of all owners were widowed, divorced, or separated (U.S. Bureau of the Census, 1980b). Many middle-aged women owners start their businesses on the breakup of their marriage or the death of a spouse.

Women's Ownership Goals and Management Styles

A number of analysts point to a third factor—ownership goals and management style—that they assert may further impede women's progress in business ownership. We report these observations with some misgivings, inasmuch as they may tend to posit behaviors and values without acknowledging their underlying, systemic origin. Although these characterizations fall short of an outright "blaming of the victim," they do tend to call into question certain value orientations—such as ones associated with parenting and caregiving—which we believe have intrinsic legitimacy. The fact is that women are as diverse as men and that no simple characterization of inclinations, capabilities, and motives can suffice for all.

Many women apparently manage their business to yield a profit, but not to grow substantially. Small enterprises offer income, independence, autonomy, and control of their owners, yet can be consistent with other family or career responsibilities. Rapid growth, in particular, is seen by some women to be disruptive and to jeopardize family and other commitments (Brophy, 1989).

Women who eschew business with high potential for rapid growth may be reducing the viability of their businesses. National business failure rates, for example, tend to be higher among nongrowing firms than among growing ones (SBA, 1989). Efforts by women business owners to mute the growth of their businesses—a reasonable objective for many who seek to balance work and other commitments—also can undercut the access of their firms to some potential institutional sources of financing, especially venture or equity capital, that are so important for new businesses.

> Professional venture capitalists manage the funds of financial institutions in pursuit of unusually high rates of return. As the investee firm approaches maturity, the incremental return on investment declines with the risk and the venture capitalist wishes to "harvest" the initial investment by selling the company in a private transaction or public offering of securities (Brophy, 1989, p. 67).

Such a harvest clearly requires significant firm growth and development, particularly if stock is to be offered publicly. On the basis of such observations as these, some assert that more female business owners should somehow redirect their energies. Donald Sexton (1989b, p. 185), for example, argues: "The key to the economic importance of women-owned businesses is to move beyond the part-time or life-style businesses and into the emerging or entrepreneurial business organizations that create the vast majority of new jobs."

Surveys of men and women entrepreneurs have found significant differences in the value orientations of persons aiming to create a growing, dynamic enterprise, compared with businesspersons having more modest goals. In particular, growth-oriented businesspersons are said to be highly motivated, emotionally aloof, challenged by risk, willing to subvert the status quo, and less in need of social confirmation or reassurance.[7]

If indeed women profess more modest goals for business growth, then perhaps fewer women owners now share these characteristics. But in time, and with experience, there is every reason to suppose that many will develop an orientation toward ownership that fosters success on a larger scale. It seems likely that more experience in business management will generate a greater willingness to undertake risk or to subvert the status quo. Such experience might result also in a lessening of the need for social confirmation, which is said by some to dampen initiative.

Federal Policy and Women Business Enterprise

The recent rapid growth in the number and economic impact of women-owned businesses is not attributable in any significant degree to public policy initiatives. Rather, public policy, until recently, has been largely indifferent to women's business ownership. Generally women are not included among disadvantaged groups receiving special set-asides or preference in federal financing or procurement programs. Indeed there is considerable opposition to including women in these programs because such a policy could increase the competition for already limited resources (U.S. House of Representatives, 1988).

Initial national concern with women-owned business occurred during the 1970s. In 1977 President Carter appointed an Interagency Task Force on Women Business Owners. This appointment was followed in 1979 by Carter Executive Order 12,138, which created a National Women's Business Enterprise Policy calling for affirmative action by federal agencies to foster such business and to create programs for women. The Interagency Committee on Women's Business Enterprise was established that persists as part of the federal bureaucracy.

The Carter executive order remains in effect. Evidence is scant that it has been effective in stimulating federal programs responsive to the needs of women's businesses. These early actions, however, may have had some benefits. The U.S. Small Business Administration (SBA) established an Office of Women Business Ownership, which has undertaken limited initiatives in research: It was a cosponsor of the 1982 Characteristics of Business Owners Survey (Interagency Committee on Women's Business Enterprise, 1988, p. 29).

The authority in the Carter executive order for preferential treatment of women-owned businesses in federal procurement has never been implemented. Nevertheless, growth has occurred in procurement contracts with women-owned businesses. In fiscal year 1979, for example, federal government prime contracts awarded to women-owned businesses amounted to only $181 million; by fiscal year 1988, this figure had risen to $1.9 billion, about 1% of federal prime contract actions (SBA, 1989, Table C.12; undated). This increase, in part, may be attributable to efforts by the SBA's Office of Women Business Ownership to get women business owners listed on the SBA's "PASS" Program. This is a computerized list of small businesses interested in selling to the federal government (Interagency Committee on Women's Business Enterprise, 1988).

In addition, since 1980, federal agencies have been required to set procurement goals for prime contracts with women-owned businesses, and Federal Acquisition Regulations require larger prime contracts to state "the national policy to utilize women-owned businesses" and to require "that the prime contractors use 'best efforts' to give such businesses the maximum practicable opportunity to participate in subcontracts" (U.S. House of Representatives, 1988, p. 20). These policies may have sensitized federal agencies and federal contractors to the potential contributions and needs of women-owned businesses.

There is a broad consensus that more management and technical assistance could increase the effectiveness of many women entrepreneurs. This consensus has not been matched by sustained action by the federal government. SBA Small Business Development Centers offer some training and assistance programs for women. The SBA and other federal agencies also offer some assistance for women's firms interested in export activity or in selling to the federal government. Generally, however, federal government assistance has been episodic and unreliable. Private sector organizations such as the National Association of Women Business Owners (NAWBO) and the American Women's Economic Development Corporation are active in this area of service.

With respect to financing assistance, federal action largely has been lacking until recently. As has been suggested, women face significant barriers to getting business credit. Federally guaranteed loans for women-owned businesses under standard SBA programs—those available to all businesspersons—are reported to be diminishing. In 1984, for example, the SBA guaranteed 2,103 loans for women, 10.7% of the total number of SBA guaranteed loans. By 1987 these figures had declined to 1,565 and 10.1%, respectively (U.S. House of Representatives, 1988). Only 5 of 224 loans made by the Business and Industrial Loan Program of the Farmers' Home Administration during a recent 4-year period were made to women entrepreneurs. No federal financing or guarantee programs are limited to women-owned businesses or give women preference for loans.

In 1988 the Congress passed into law the Women's Business Ownership Act, the only law devoted to women-owned business. Strikingly this act offered distinctly limited benefits for women. It authorized $10 million over 3 years to finance "demonstration" projects offering women training and counseling in financing, managing, and marketing businesses. That they are demonstration projects, rather than new permanent programs, would appear to guarantee unstable government funding. The act also called for a continued investment in research on women-owned business and established a new "high level" National Women's Business Council to develop a long-run federal strategy for aiding women-owned businesses. Finally the act offered limited additional help regarding financing. First, it clarified that the Consumer Credit Protection Act bars discrimination against women in lending for business or commercial loans. Second, it set up a new "small loan" program offering SBA guarantees for business loans up to $50,000. As in previous SBA programs, women are not given preferences for getting such loans, but the SBA suggests that such small loans will be particularly attractive to businesses in service industries in which women entrepreneurs remain highly overrepresented.

Conclusions

The 1980s clearly witnessed a significant evolution in the roles of women in the economy. During the decade, large numbers entered the workforce, concurrent with the rapid growth in the service industries. Growth in women's share of the workforce is likely to continue for some time. Between 1985 and 2,000, women are projected to contribute 59% of net growth to the national labor force (Johnston & Packer, 1987). Moreover, women realized a significant income gain in recent decades. The median earnings of full-time, year-round female workers grew markedly relative to those of males in the 1980s, although a full convergence of the income profiles of men and women has yet to be attained.

The bulk of the economic strides of women, however, were made on payrolls. Although large in numbers, women-owned businesses are generally small and remain concentrated in the retail and service sectors. Their overall economic output and employment remain relatively limited. The payoffs to women from entrepreneurial activity are modest at present, compared to opportunities available on payrolls.

To be sure, women able to acquire or develop corporations can realize appreciable gains. But few women who begin as sole proprietors appear to advance to the more expansive and lucrative corporate sector. Small sole proprietorships are likely to be targeted for exploitation by the larger firms with which they subcontract work and are highly vulnerable to economic setbacks. For many women these enterprises are a pursuit of last resort when other employment will not pay a wage sufficient to offset child care and other expenses or afford the flexibility that many parents (single or not) require.

In the long run, women's earnings gains will be shaped by trends in both the payroll and the ownership sectors. We are optimistic concerning women's future economic opportunities. We believe that payroll enterprises will make progress in accommodating the particular needs of female workers. To attract and retain qualified female workers in the face of impending shortages of qualified workers, employers already have begun to incorporate flextime, accommodate part-time workers, tailor benefits, and assist in child care. The result will be more alternatives to self-employment for women. The "control" available through self-employment will become less critical for many women.

We are guardedly optimistic about the future of self-employment among women. The current, large scale of women's business ownership is producing a generation of women with business experience in all major economic sectors of the U.S. economy. As women's experience grows and as successful role models emerge of women owning powerful economic enterprises, we believe that the economic goals of many women owners will expand. Longer track records of success in business ownership, as well as recent modest federal financing initiatives, also may increase the access of women to business capital. The current problems of women business owners may signal the early stages of a developmental process that will culminate in much greater economic achievements in the future.

What are appropriate public policy goals in such an uncertain and fluid environment? There are two potentially potent rationales for public sector encouragement of women's business ownership. The first is that such ownership might proffer external benefits for others in the economy. The prevalent assertion that small businesses are fecund generators of new jobs and innovation has been used as the basis for such arguments. The conclusion by the House Small Business Committee that "women-owned businesses may well provide the cutting edge in the worldwide economic competitiveness" is another version of this argument (U.S. House of Representatives, 1988, p. iii). Unfortunately, evidence is scant that women-owned businesses offer such significant positive externalities at present. In particular, Birch's evidence for rapid job generation by small business has been largely discredited by more in-depth analysis (Armington & Odle, 1982). The continued concentration of women entrepreneurs in trade and service industries deeply undercuts the belief that they will enhance the economic competitiveness of the United States.

The second possible argument for public assistance to female entrepreneurs is that such aid would, at a minimum, increase the economic opportunity of the businesswomen who receive the aid. The rationale for such a limited goal would be to redress the effects of past economic discrimination against women. The evidence in our chapter clearly weakens this argument. In general, payrolls appear to offer more opportunity for most women. If true, this implies that public civil rights activities might better focus on existing businesses, rather than on entrepreneurial activity.

We are led to conclude that there is little economic rationale for intense, new public efforts to stimulate or support women-owned businesses. It seems highly appropriate for the federal government to provide general advocacy of and research on business ownership by women. Such advocacy and research can help eliminate or restrict unnecessary barriers to female business owners. The Office of Women's Business Enterprise of the SBA currently is playing this role, as are other interagency groups. In parallel the federal government must continue to promote equal opportunity for the advancement of both women and minorities within businesses at large.

We believe there is reason to support somewhat more direct aids to women business owners as well. If we are witnessing the early stages of a developmental process for women, then the growing experience of women in business ownership will help erode certain barriers that currently impede success in self-employment. This optimistic scenario argues for public efforts to facilitate a transformation in the nature of women-owned businesses. The new SBA small loan program may alleviate some of the credit problems of new business start-ups. We also would suggest that technical assistance needs of women-owned businesses should be examined carefully. Technical assistance programs that focus on narrow technical skills may have a payoff, but the larger objective may be to foster networking among women business owners. This networking could generate a greater awareness of the opportunities available through ownership and surely would lead some women to pursue a more ambitious ownership agenda. At base, we concur with the modest scale and general directions of federal policy taken by the Congress with respect to women business ownership.

The national economy is fast reshaping itself in response to competition from abroad and to structural change in the domestic economy and its labor force. In this fluid setting, entrepreneurs can find an abundance of new opportunities. For women and minorities alike, these conditions may hold particular, as yet unrealized, promise.

Notes

1. *Sole proprietorships* are unincorporated businesses owned by an individual. *Partnerships* are unincorporated businesses owned by two or more persons, each of whom shares in its ownership. A *subchapter S corporation* is defined by the Internal Revenue Service to be a legally incorporated business having 35 or fewer shareholders. These shareholders elect—because of tax advantages—to be taxed as individual shareholders, rather than as corporations. In prior years the size limit was 10 or fewer shareholders, so the nonsubchapter S corporations' share probably has increased. A firm is considered *women-owned* if the sole owner or one half or more of the partners were women. A corporation fell in this class if half or more of its shares were owned by women.

2. Partnerships and small corporations were substantially undercounted in 1972, thereby deflating estimates in the earlier years. Prior to 1982 no single proprietor was credited with the ownership of more than a single firm, even when several were owned simultaneously. In addition, census procedures erroneously attributed ownership to the first-named person—generally the husband—in joint tax returns before 1982. At

the same time, industrial coverage was expanded slightly in the services during 1977 and thereafter. This increased estimates in later years.

3. Employeeless sole or "individual" proprietorships are important in their own right as a source of income for many women, but they also appear to be a seedbed for larger enterprises. Some inevitably grow and require employees and so move into the category documented here. The process is borne out by a survey of women-owned businesses in the late 1970s. Almost 60% of all women-owned firms that had more than $500,000 in receipts were founded prior to 1960. Just under a quarter of those having less than $5,000 in receipts at that same time were founded before 1960 (U.S. Bureau of the Census, 1980a, 1980b).

4. These rates, we must caution, are spatially uniform at the scale examined here. Other dimensions of comparison may yield a different conclusion. We do have some evidence that the characteristics of women-owned businesses may vary geographically. Average receipts per firm differ significantly across census divisions. Part of this variation is certainly attributable to interregional differences in industrial composition and economic performance. Part also may reflect differences in the distribution of ownership across industries within regions themselves. In 1982, average receipts per woman-owned firm exceeded the national average in three divisions: Middle Atlantic (121% of the U.S. equivalent), West South Central (117%), and East North Central (104%). Remarkably, both the New England (88%) and Pacific (90%) divisions lagged well behind the national norm.

5. Here the rate of self-employment is simply the number of self-employed females expressed as a fraction of total female employment.

6. Specifically, women were included only if they were of working age (18 to 65), not enrolled in school, and received wage or salary income in the year prior to the census. In addition, persons were included only if they worked at least 35 weeks in the year prior to the census.

7. Growth-oriented women have been found to differ somewhat from growth-oriented men entrepreneurs. Compared with growth-oriented men, growth-oriented women show higher autonomy and liking for change; they also show lower energy and willingness to take risks. See Sexton, 1989a, p. 143.

References

Armington, C., & Odle, M. (1982, Winter). Small business: How many jobs? *The Brookings Review*, pp. 14-17.

Brophy, D. J. (1989). Financing women-owned entrepreneurial firms. In O. Hagan, C. Rivchun, & D. Sexton (Eds.), *Women-owned businesses*. New York: Praeger.

Brush, C. G., & Hisrich, R. D. (1985). Women and minority entrepreneurs: A comparative analysis. In *Frontiers of entrepreneurship research—1985* (pp. 566-587). Wellesley, MA: Babson College.

Ford, R. H. (1989). The board of directors: A tool for the future. In O. Hagan, C. Rivchun, & D. Sexton (Eds.), *Women-owned businesses*. New York: Praeger.

Henson, M. F. (1990). Trends in income, by selected characteristics: 1947 to 1988. *Current Population Reports* (Series P-60, Report No. 167). Washington, DC: Government Printing Office.

Hisrich, R. D., & O'Brien, M. (1981). The woman entrepreneur as a reflection of the type of business. In *Frontiers of entrepreneurial research—1981* (pp. 54-67). Wellesley, MA: Babson College.

Interagency Committee on Women's Business Enterprise. (1988). *1988 annual report*. Washington, DC: U.S. Small Business Administration.

James, F. J. (1985). *The economic status of Hispanic females in the United States: 1970-1980*. Unpublished manuscript, University of Colorado, Graduate School of Public Affairs, Denver.

James, F. J., & Clark, T. A. (1987). Minority business in urban economies. *Urban Studies, 24*, 489-502.

Johnston, W. B., & Packer, A. H. (1987). *Workforce 2000: Work and workers for the 21st century*. Indianapolis: Hudson Institute.

Lovering, J. (1990). Fordism's unknown successor: A comment on Scott's theory of flexible accumulation and the re-emergence of regional economies. *International Journal of Urban and Regional Research, 14*, 159-174.

National Association of Women Business Owners (NAWBO). (1988). *Survey reveals startling new data on U.S. women business owners*. Chicago: Author.

Ronstadt, R. C. (1984). *Entrepreneurship: Text cases and notes*. Dover, MA: Lord.

Scott, A. J. (1988). Flexible production systems and regional development: The rise of new industrial spaces in North America and Western Europe. *International Journal of Urban and Regional Research, 12*, 171-185.

Sexton, D. L. (1989a). Growth decisions and growth patterns of women enterprises. In O. Hagan, C. Rivchun, & D. Sexton (Eds.), *Women-owned businesses.* New York: Praeger.

Sexton, D. L. (1989b). Research on women-owned businesses: Current status and future directions. In O. Hagan, C. Rivchun, & D. Sexton (Eds.), *Women-owned businesses.* New York: Praeger.

Smith, N. R., McCain, G., & Warren, A. (1982). Women entrepreneurs really are different: A comparison of constructed ideal types of male and female entrepreneurs. In *Frontiers of entrepreneurship research—1982* (pp. 68-77). Wellesley, MA: Babson College.

U.S. Bureau of Labor Statistics. (1981). *Employment and earnings.* Washington, DC: Government Printing Office.

U.S. Bureau of Labor Statistics. (1988). *Employment and earnings.* Washington, DC: Government Printing Office.

U.S. Bureau of the Census. (1976). Women-owned businesses, 1972. *1972 economic census* (WB72). Washington, DC: Government Printing Office.

U.S. Bureau of the Census. (1980a). Selected characteristics of women-owned businesses, 1977. *1977 economic census* (WB77-2). Washington, DC: Government Printing Office.

U.S. Bureau of the Census. (1980b). Women-owned businesses, 1977. *1977 economic census* (WB77-1). Washington, DC: Government Printing Office.

U.S. Bureau of the Census. (1982). Money income of households, families, and persons in the United States: 1980. *Current population reports* (Series P-60, No. 132). Washington, DC: Government Printing Office.

U.S. Bureau of the Census. (1986a). Women in the American economy. *Current population reports* (Special Studies, Series P-23, No. 146). Washington, DC: Government Printing Office.

U.S. Bureau of the Census. (1986b). Women-owned businesses, 1982. *1982 economic censuses* (WB82-1). Washington, DC: Government Printing Office.

U.S. Bureau of the Census. (1989). Money income of households, families, and persons in the United States: 1987. *Current population reports* (Series P-60, No. 162). Washington, DC: Government Printing Office.

U.S. Bureau of the Census. (1990). Women-owned businesses, 1987. *1987 economic censuses* (WB87-1). Washington, DC: Government Printing Office.

U.S. House of Representatives. (1988). *New economic realities: The rise of women entrepreneurs: A report of the Committee on Small Business.* Washington, DC: Government Printing Office.

U.S. Small Business Administration (SBA). (1988). *Small business in the American economy.* Washington, DC: Government Printing Office.

U.S. Small Business Administration (SBA). (1989). *The state of small business: A report to the president.* Washington, DC: Government Printing Office.

U.S. Small Business Administration (SBA). (undated). *Facts and figures.* Washington, DC: Author.

PART FIVE

Community Services and Infrastructure

Introduction to Community Services
and Infrastructure

Robert Warren

Any large population center must have an infrastructure that can deliver basic urban services if the biological, socioeconomic, and cultural needs of its residents are to be met (Herman & Ausubel, 1988). Without a water supply, a means to remove human and industrial waste, and transportation and communication systems, places as different as Denver, Cleveland, Tulsa, and New York would be unlivable in terms of modern standards of health, safety, and economic and social exchange. These traditional infrastructure elements are necessary but not sufficient for providing an adequate quality of urban life. In addition to police and fire protection, most citizens also expect parks, schools, public health facilities, libraries, and a variety of other services in their communities (Herman, Ardekani, Govind, & Dona, 1988).

For much of this century, there was little question that the primary responsibility for the urban infrastructure and production and distribution of services should be located in the public sector. As a research and policy focus, the field of urban services was technical, predictable, and dull. To the degree that conflict existed, it was over whether services in metropolitan areas should be produced by a multitude of municipalities, districts, and authorities or by a single, regional-scale unit of government. Today, however, what urban services should be provided; whether they should be produced by the public, private, or third sectors (voluntary and nonprofit organizations); who should receive them; and how they should affect the environment have become problematic and highly politicized matters.

Debates have been going on for more than two decades over whether public works, garbage collection, and even law enforcement should be produced by public agencies or private firms. Groups from totally opposed ideological camps are supporting a larger role for the third sector in meeting urban service needs (Wolch, 1990, pp. 4-7). Rapidly evolving telecommunications and information technology are changing the options available for dealing with infrastructure and urban services needs, but these innovations have not been fully integrated into policy-making discussions (Brooks, 1988).

Questions of fairness and ecological values have also become an important part of the discourse. Attention has ebbed and flowed since the 1960s but never disappeared on the issue of whether urban services are equitably distributed and accessible to all municipal residents (Hero, 1986; Warren, Rosentraub, & Weschler, 1992). Demand has been growing during this same period that greater weight be given to environmental priorities in the design and operation of urban infrastructure and service systems. It has come from traditional groups concerned with ecological viability (Brown & Jacobson, 1987) and a newer environmental justice movement (Ong & Blumenberg, 1990).

Effectively dealing with community infrastructure and service needs and the policy dilemmas related to these issues presents a major challenge to urban governance. Success in doing so depends, to an important degree, on adopting new ways of thinking about them. We must be able to enlarge the boundaries of analysis so that infrastructure and services are not viewed as independent and separate from one another, but as interrelated subsystems that have both individual and collective social, economic, and environmental impacts on urban communities. Equally important, analysis and policy making cannot take the status quo as a given. Automatically assuming that service inadequacies can be remedied by providing more of the same service is counterproductive. For example, reducing the conditions that give rise to the service needs or source control may be the better solution. Using a different technology may offer greater advantages in other instances. Similarly the assumption that a service is always more efficiently produced in the private or the public sector is equally dysfunctional.

A consideration of these technological, environmental, and sectoral dimensions requires a more complex way of thinking about community services and infrastructure. But it also offers a means of expanding the range of options for making informed choices in the future about repairing, building new, or reducing the needs for infrastructure, and using the public, private, and third sectors in community service provision and production in ways that account for ecological and equity values. These perspectives are explored in more detail in the sections that follow.

Infrastructure and Telecommunications and Information Technology

The incorporation of telecommunications and information technology into the planning, design, and operation of the transportation infrastructure of a city provides an opportunity to explore ways in which developments unimagined a few decades ago can become part of the urban infrastructure in theory and in fact. The nature of America's infrastructure crisis, which Kaplan (chap. 15) and Nathan (chap. 16) discuss in Part 5, was defined in a series of reports in the 1980s. In the case of transportation, they point to the neglect of maintenance of roads, bridges, and highways and a failure to invest in new facilities expected to be necessary to meet needs based on population and economic growth projections. Although recognizing that funding problems exist, there is agreement that major investments will be needed at the local level. At the same time, the almost exclusive focus on these traditional elements of the transportation infrastructure constitutes a problem itself.

The growing, if not already dominant, view at the national level is that the rapid creation of an "information superhighway" is the key infrastructure need for America's economic and social future. The building of a national information infrastructure has become a priority of the Clinton administration. Paradoxically discussions about the urban infrastructure per se have been very slow in recognizing that, in the future, miles of fiber-optic cable will be as critical to an urban area as miles of freeways, water mains, and sewers. The importance of building the fiber-optic/coaxial cable and satellite-linked communication systems within urban areas has remained outside the mainstream dialogue of infrastructure needs that has focused on the crisis of maintaining and building roads, highways, bridges, and sewerage systems. In the 21st century, cities with local area networks (LANs), teleports, wide area networks, and integrated services digital networks (ISDNs) that provide interactive audio, video, and data linkages among and between businesses, homes, and public institutions such as schools, libraries, and hospitals will have pronounced advantages as centers of economic activity and in quality of life over those that do not (Graham, 1992; Hepworth, 1990).

Considering transportation specifically, telecommunication and information technology can contribute in a number of ways to the future development of a region's infrastructure. Intelligent vehicle highway systems (IVHSs) can reduce the need to build new highways. By using computer, communications, and sensing technologies, IVHSs can "transform collections of passive thoroughfares into a smart, responsive system adept enough" to control commuter patterns, make public transportation more efficient and attractive, and help motorists drive better (Arlook & Jones, 1993).

There is also growing use of global positioning systems (GPSs), through a network of satellites, to track continuously the location of and communicate with mobile equipment. Autos, trucks, and buses are equipped to emit radio signals that allow their locations to be instantly determined and road data in map form to be transmitted to a computer screen in the vehicle. GPS is being applied to such diverse things as the real-time rerouting of buses for greater efficiency and demand responsiveness and to combat car theft and freeway crimes against tourists (Wald, 1994).

Substitution of communication for travel through the use of telecommunications is used increasingly to reduce traffic congestion and vehicle-related air pollution. Not only telecommuting and teleconferencing but also telebanking, teleshopping, and telemedicine are functional areas in which this substitution can occur. Telecommuters, using computers and modems or LANs, can work at home or decentralized suburban locations, rather than travel to their firm's main office in city centers (Moss & Carey, 1993), with the result of assisting firms in meeting guidelines of the Clear Air Act of 1990 and increasing employee productivity (*Telecommuting Times,* 1993).

The reasons for broadening the definition of urban infrastructure to include telecommunications go well beyond transportation benefits. The efficiency of other basic services can be affected positively by new technology in several ways. Water and electric utilities in a number of areas, for example, are building broadband fiber-optic/coaxial cable networks into the homes and businesses that they serve to gain greater efficiency in billing and to achieve reductions in energy consumption (Dawson, 1994). A utility, in cooperation with customers, can engage in demand-side management by controlling the flow of energy to each end user to meet individual need patterns (Joyce, 1993). The results can reduce the need for increased production capacity and the building of additional fossil fuel or nuclear plants.

Interactive electronic information kiosks provide a way for municipalities to deliver multiple services through a single channel (Speed, 1993). Kiosks are being located in government office lobbies, civic centers, and even shopping malls by public agencies to provide citizens with information and even direct services such as job referrals, payment of traffic fines, listing of public meeting schedules, and vehicle registration. A number of cities are experimenting with providing citizens access to municipal databases and electronic communication with public officials from neighborhood libraries linked to city hall by LANs or home computers and modems (Guthrie & Dutton, 1992; Young, 1990).

The promise of the infrastructure and service enhancements possible through the information superhighway carries with it a caution that a deep division can be created in urban centers between those who have and do not have access to the benefits of the emerging information society (Gillette, 1988, p. 243; Murdock & Golding, 1989).

The Ecology of Cities

Another critical element that has not been a consistent part of the mainstream discussion on infrastructure and service needs is their environmental dimensions. A massive body of literature exists concerning locational and environmental problems with landfills, refuse incineration, and toxic waste dumps, but these issues seldom are reflected adequately in writings that focus on urban services and infrastructure. Similarly such things as the effect of highways on wildlife, of high-voltage power lines on human health, and of drawing water for urban populations from distant wilderness sources are also absent. It is only through understanding the metabolism of a city that these and related issues can be taken into account in infrastructure and service policy making.

Great amounts of raw material, fuels, agricultural and forest products, water and minerals, and manufactured goods flow into and move through cities and must be eliminated. How this is done has consequences for the ecological systems of which cities are part. As Girardet (1990) argues, however, the current "linear" model of production, consumption, and disposal that the infrastructure is designed to serve is not the least concerned with the ecological viability of cities. These diverse inputs and outputs, which may be in toxic forms, are not linked in infrastructure planning. The dominant response to increasing amounts of refuse, for example, is to expand the city's disposal capacity into land, water, or air, rather than develop policies to reduce the input of materials that become the refuse. Such policies, however, are no longer uncontested.

A growing grassroots-based ecological consciousness has entered a second stage in its evolution. Initially it took the form of a Not In My Backyard (NIMBY) response by neighborhoods to such things as landfills, waste incinerators, and, most recently, high-voltage power lines. The response of mainstream policy-makers has been to try to overcome NIMBY behavior by siting facilities in areas where citizens will generate less resistance (Szasz, 1994). A growing body of opinion, however, is taking the view that NIMBY behavior is rational and justified but must be transformed into a policy of Not In Anybody's Backyard that involves "up-front" controls and source reduction. In reference to siting hazardous and solid waste facilities, Heiman (1990, p. 359) explains that this policy calls for "no new facilities until industry and government are committed to a level of source reduction that goes well beyond the process, recycling, and waste stream modifications" now being pushed by public agencies. Similarly Lake (1993, p. 88) points out that building hazardous waste incinerators is only one of a number of possible responses and one that "concentrates the costs on host communities, as compared to the alternative strategy of restructuring production so as to produce less waste."

Long-term infrastructure cost reductions for energy through demand management mentioned above can be applied also to water supply. New York City was able to drop plans in the mid-1980s to spend $6.2 billion to increase its water supply by reducing demand "by installing residential water meters, charging higher prices, and encouraging conservation" (Brown & Jacobson, 1987, p. 43). When public officials are not willing to substitute up-front controls and demand management for new infrastructure facilities, grassroots movements have been able to oppose successfully the construction of waste incinerators, hazardous waste sites, and high-voltage power lines (Heiman, 1990; Marks, 1994). Increasing and effective opposition to public facilities is being mounted by minority and low-income communities, in coalition with national environmental organizations, under the banner of environmental justice (Ong & Blumenberg, 1990).

The protection of urban wildlife and biodiversity has become an issue in infrastructure development as well. Soule (1991) argues that to maintain wildlife and ecosystem viability in population centers, it is necessary to preserve and link habitat areas through habitat corridors in order to minimize habitat fragmentation. To pursue such a policy, urban land-use planners must have the necessary authority and knowledge to deliberately incorporate conservation biology principles into the design of such infrastructure elements as highway shoulders, utility rights of way, and culverts and underpasses to create habitat linking corridors.

Multisectoral Production
of Community Services

The frequent and often ideologically laden debates over whether community services should be produced publicly, privately, or by nonprofit organizations tend to obscure a number of points important to understanding the actual options available to citizens. First, these are not either/or choices. As Warren, Rosentraub, and Weschler propose in Chapter 14 "A Community Services Budget," the provision, funding, and production of urban services in any complex community are commonly achieved through a multisectoral mixture of public, private, and third-sector agencies acting individually and conjointly. The structure of services is dynamic, rather than static. It will change over time through a process influenced by experimentation, social learning, and ideology.

Current revisionist thinking about community services is most evident in relation to the benefits of using private firms for collective goods and services. Advocates of "privatization" have tended to dominate much of the discussion concerning urban administration since the mid-1970s (Poole, 1980; Savas, 1982). Some have called for the contracting out of municipal services to gain cost-reducing efficiencies for cities. Others, motivated by a desire to reduce the role of government in society, have called for "shedding" services from the public to the private or third sectors. By the early 1990s, however, a growing body of experience and reflection indicated that the benefits of substituting private for public production are less obvious and more difficult to measure than assumed.

In the United States, localities have been using private firms for services as diverse as road construction and social welfare since the 19th century (Kettl, 1993, p. 155). Even with this long history of awareness of the options, however, there has never been a significant shift of major responsibilities for community services from the public to the private sector.

Surveys of city and county governments conducted by the International City Management Association in 1982 and 1988 found that, with very few exceptions, no specific service had been totally dropped by more than 4% of the local governments reporting (Morley, 1989, p. 35). Contracting is used more widely. Between 23% and 50% indicated they used contracting for at least 1 of a set of 21 services that included such things as solid-waste collection, street repair, tree trimming and planting, ambulance service, drug and alcohol treatment, operation of homeless shelters, building and ground maintenance, legal services, and labor relations (Morley, 1989, p. 36).

Contracting does not erode the role of municipal government as would shedding services, because full responsibility for their provision remains in the public sector. Even so, evidence suggests that both the full costs of contracting out and the efficiencies of public agencies have been underestimated. Kettl (1993, p. 193), for example, concluded that the more a government relies on external contractors, the greater its overhead management costs. A need is created for the city to become a "smart buyer" of contract services. This requires that a city have personnel with the expertise and time to (a) define their agency's programmatic goals independently of what private vendors offer, (b) actively identify firms that can sell them what they want, rather than passively rely on bids, and (c) monitor and evaluate the contractors' performance (Kettl, 1993, p. 180). Too often, however, local governments do not invest the resources needed to ensure accountability by the private or third-sector producers (Prager, 1994). In such cases a second type of cost can be incurred in the form of failure to meet service standards contracted for, lack of responsiveness to citizen complaints, and low levels of production efficiency (Mead, 1994). These, in turn, can undermine public trust and confidence in government itself (Kettl, 1993, p. 194).

Even if contracting out reduces the amount a city spends for a service, the total cost for the community as a whole may still be the same or greater. This fluctuation can happen when larger private or voluntary expenditures are required to maintain the same level of service that the city had provided because of higher charges imposed by the private producer or the need for businesses and residents to augment the contractor's service because of reductions in its amount or quality. When this increase occurs, equity problems arise as well. Low-income residents and minority neighborhoods are the least able to afford such costs and are likely to have fewer resources to create volunteer organizations to supplement the reduced level of public services.

All of this being said, there is no question that contracting by public agencies can be beneficial when all costs are identified, adequate monitoring is done, and social equity is maintained. The more important point, however, is that it is the introduction of competition into what had been bureaucratic service monopolies, rather than contracting out, that results in greater efficiencies. When existing municipal departments compete with private and third-sector agencies for city contracts to produce services, they can tender winning bids. Phoenix is the most frequently cited example. Hatry (1983) notes that city agencies won in 10 of 22 cases. City workers won back Phoenix's large contract for refuse collection in 1884 (Moore, 1987, p. 64). In 1987 the Kansas City, Missouri, city council decided to switch back to municipal production of fire and emergency services at the city's airport after contracting privately for the services for 7 years (Sharp, 1990, p. 117).

The Larger Picture

An adequate discussion of all aspects of community services and infrastructure that could benefit from reassessment in theoretical and applied terms would require far greater space and more varied perspectives than can be provided here. Similarly, rethinking a selected set of issues relies on but does not review the extensive basic body of literature that describes and analyzes the general state of the field. What I hope this discussion has been able to do is to elaborate on some of the topics raised in the readings that follow, suggest issues that deserve more attention, and underscore the necessity of asking questions even though no obvious answers are available.

There is little doubt that telecommunications and information technology will have an increasing, if not revolutionary, role in the form of community services and infrastructure in the foreseeable future. There is equal certainty of the importance of understanding how to incorporate these changes while protecting and enhancing the ecological viability of cities, social equity for their citizens, and biodiversity. To do so will require creating and using a dynamic and adaptive mix of public, private, and third-sector resources and organizations that are guided by democratically determined policies designed to meet the collective needs of urban communities.

References

Arlook, J., & Jones, R. (1993, November/December). Tracking IVHS. *Geo Info Systems, 3*, 38-47.

Brooks, H. (1988). Reflections on the telecommunications infrastructure. In J. H. Ausubel & R. Herman (Eds.), *Cities and their vital systems* (pp. 249-257). Washington, DC: National Academy Press.

Brown, L. R., & Jacobson, J. L. (1987). *The future of urbanization: Facing the ecological and economic constraints* (Worldwatch Paper 77). Washington, DC: Worldwatch Institute.

Dawson, F. (1994, January 24). Arkansas utility will build broadband networks. *Multichannel News, 15*, 39.

Gillette, D. (1988). Combining communications and computing: Telematics infrastructure. In J. H. Ausubel & R. Herman (Eds.), *Cities and their vital systems* (pp. 233-248). Washington, DC: National Academy Press.

Girardet, H. (1990). The metabolism of cities. In D. Cadman & G. Payne (Eds.), *The living city* (pp. 170-180). London: Routledge.

Graham, S.D.N. (1992). Electronic infrastructures and the city: Some emerging municipal policy roles in the U.K. *Urban Studies, 29*(5), 755-781.

Guthrie, K. K., & Dutton, W. H. (1992). The politics of citizen access to technology: The development of public information utilities in four cities. *Policy Studies Journal, 20*(4), 574-597.

Hatry, H. (1983). *A review of private approaches for delivery of public services*. Washington, DC: Urban Institute.

Heiman, M. (1990, Summer). From "Not in My Backyard!" to "Not in Anybody's Backyard." *Journal of the American Planning Association, 56*, 359-362.

Hepworth, M. (1990). *Geography of the information economy*. New York: Guilford.

Herman, R., Ardekani, S. A., Govind, S., & Dona, E. (1988). The dynamic characterization of cities. In J. H. Ausubel & R. Herman (Eds.), *Cities and their vital systems* (pp. 22-71). Washington, DC: National Academy Press.

Herman, R., & Ausubel, J. H. (1988). Cities and infrastructure: Synthesis and perspectives. In J. H. Ausubel & R. Herman (Eds.), *Cities and their vital systems* (pp. 1-21). Washington, DC: National Academy Press.

Hero, R. E. (1986, Summer). The urban service delivery literature: Some questions and considerations. *Polity, 18*, 659-677.

Joyce, F. A. (1993, Winter). Telephone and cable TV: Two industries or one. *Convergence '93, 3*, 34.

Kettl, D. F. (1993). *Sharing power*. Washington, DC: Brookings Institution.

Lake, R. W. (1993, Winter). Rethinking NIMBY. *Journal of the American Planning Association, 59*, 87-93.

Marks, P. (1994, January 6). Electric fields create nebulous peril but real fear on L.I. *The New York Times*, pp. B1, B7.

Mead, T. D. (1994, April 1). Public administrator debunks some myths of privatization. *PA Times, 17*, pp. 1, 15.

Moore, S. (1987). Contracting out: A painless alternative to the budget cutter's knife. Cited in S. Hanke (Ed.), *Prospects for privatization: Proceedings of the Academy of Political Science, 36*(3), 60-73.

Morley, E. (1989). Patterns in the use of alternative serviced delivery approaches. In *Municipal Year Book 1989* (pp. 33-44). Washington, DC: International City Management Association.

Moss, M. L., & Carey, J. (1993). *The New York Telecommuting Project*. New York: New York Metropolitan Transportation Council.

Murdock, G., & Golding, P. (1989, Summer). Information poverty and political inequality: Citizenship in the age of privatized communications. *Journal of Communications, 39*, 180-194.

Ong, P. M., & Blumenberg, E. (1990). *Race and environmentalism* (D905). Los Angeles: University of California, Graduate School of Architecture and Urban Planning.

Poole, R. (1980). *Cutting back city hall*. New York: Universe.

Prager, J. (1994, March/April). Contracting out government services: Lessons from the private sector. *Public Administration Review, 54,* 176-184.

Savas, E. S. (1982). *Privatizing the public sector: How to shrink government.* Chatham, NJ: Chatham House.

Sharp, E. B. (1990). *Urban politics and administration.* New York: Longman.

Soule, M. E. (1991, Summer). Land-use planning and wildlife management. *Journal of the American Planning Association, 57,* 313-323.

Speed, V. (1993, September). The kiosks are coming. *Government Technology, 6,* 40.

Szasz, A. (1994). *EcoPopulism.* Minneapolis: University of Minnesota Press.

Telecommuting Times. (1993, Fall). [Pilot issue]

Wald, M. L. (1994, March 3). For the map-challenged, technology to the rescue. *The New York Times,* p. B4.

Warren, R., Rosentraub, M. S., & Weschler, L. F. (1992). Building urban governance: An agenda for the 1990s. *Journal of Urban Affairs, 14*(3/4), 399-422.

Wolch, J. R. (1990). *The shadow state: Government and voluntary sector in transition.* New York: Foundation Center.

Young, M. A. (1990). Local government use of communication and information technologies for citizen participation. *Municipal year book 1990* (pp. 8-14). Washington, DC: International City Management Association.

Suggestions for Further Reading

Brudney, J. L., & Warren, R. (1990, Spring). Multiple forms of volunteer activity: Functional, structural, and policy dimensions. *Nonprofit and Voluntary Sector Quarterly, 19,* 47-58.

Davis, C. E., & Lester, J. P. (Eds.). (1988). *Dimension of hazardous waste politics and policy.* Westport, CT: Greenwood.

Farr, C. A. (1989). *Service delivery in the '90s: Alternative approaches for local governments.* Washington, DC: International City Management Association.

Giuliano, G. (1992, Summer). Transportation demand management: Promise or panacea? *Journal of the American Planning Association, 58,* 327-335.

Government Technology: A Tool for State and Local Government in the Information Age. Current issues.

Hilke, J. C. (1992). *Competition in government financed services.* New York: Quorum.

Krol, E. (1992). *The whole internet.* Sebastopol, CA: O'Reilly & Associates.

Mayo, A. L. (1990, Spring). A 300-year water supply requirement: One country's approach. *Journal of the American Planning Association, 56,* 197-208.

McFall, M. (1993, Winter). The internet: route 66 and soul of the electronic infrastructure. *Convergence '93,* pp. 36-41.

Netzer, D. (1992, Spring). Do we really need a national infrastructure policy? *Journal of the American Planning Association, 58,* 139-142.

Newman, P.W.G., & Kenworthy, J. R. (1992, Summer). Is there a role for physical planners? *Journal of the American Planning Association, 58,* 353-362.

Organization for Economic Cooperation and Development. (1987). *Managing and financing urban services.* Paris, France: Author.

Robinson, S. G. (1990). *Building together: Investing in community infrastructure.* Washington, DC: National Association of Counties.

Tarr, J. A. (1984). The evolution of the urban infrastructure in the nineteenth and twentieth centuries. In R. Hanson (Ed.), *Perspectives on urban infrastructure* (pp. 4-66). Washington, DC: National Academy Press.

Toulmin, L. M. (1988, March). Equity as a decision rule in determining the distribution of urban services. *Urban Affairs Quarterly, 23,* 389-413.

14

A Community Services Budget

PUBLIC, PRIVATE, AND
THIRD-SECTOR ROLES IN URBAN SERVICES

Robert Warren

Mark S. Rosentraub

Louis F. Weschler

Assessment of Community-Level Service

How to achieve efficiency, adequacy, and equity in the provision of public goods and services within cities have been long-term issues of concern to urban scholars. To address these questions, social scientists commonly have studied the allocative policies and outputs of municipal agencies. A growing body of evidence suggests, however, that a municipal budget, even including services contracted for, accounts for far less than the total range of urban goods and services that citizens consume in a community. Many of the same goods and services that cities make available are provided simultaneously through the private sector and the *third sector.* The latter term, as used here, is meant to include nonprofit and voluntary organizations and individual volunteer activities.[1]

Answers to both normative and analytic questions concerning the efficiency, adequacy, and distributional equity of urban goods and services in a community must be grounded in data that go beyond the information contained in a municipal budget and the performance of public agencies. Consequently the purpose of this discussion is to explore the utility for policy analysis of taking into account the private and third-sector, as well as municipal, provision of urban goods and services within a city through the construction of a *community services budget.*

One intent of creating an information base of this type is to facilitate research on questions concerning distributional equity in the provision of urban services that have been partially or wholly shifted from direct municipal to private-

or third-sector production. Further, we argue that analysis of the production efficiency of municipal agencies is incomplete without data relating to private- and third-sector production and distribution of similar services.

The use of such a nonconventional term as *community services budget* necessarily invites problems. Clearly *budget* is not being used in its normal meaning. Here it accounts for a variety of nonmonetary, as well as monetary, inputs to the production of municipal-type goods and services. Furthermore, the term is designed as a framework for integrating data related to a number of agencies in three sectors, rather than to one agency in a single sector. Perhaps "gross urban services product" or "multisector urban services analytic matrix" would be more descriptive, but they would invite even greater problems of meaning. *Community services budget* seems a more appropriate term to use in attempting to develop a way of reconceiving the overall state of affairs in a city in relation to the production and distribution, from whatever sector, of all goods and services normally provided publicly at the community level.

In the following sections, the rationale and logic of this framework are considered in more detail. The framework's applicability for dealing with normative and analytic aspects of questions relating to the efficiency, adequacy, and equity of urban goods and services also are explored. The discussion provides a way for empirical research to span the three sectors with theoretical and applied benefits.

The Complex Nature
of Community-Level Service Production

Virtually every good or service produced by local public agencies can be provided or produced by other governments, nonprofit organizations, voluntary associations, private firms, or citizens themselves. If a city government is only one of several providers of collectively consumed goods and services in a community, the traditional boundaries between the public, private, and third sectors should be relaxed for some analytic purposes. The inputs and outputs of the private and third sectors, for example, should become part of the database in studies of local urban services that seek to evaluate the performance of the local public sector and to ask questions about the adequacy, equity, and efficiency of such services in the community as a whole.

The importance of the private and third sectors as providers of collective services is not a new idea (Douglas, 1983; Ginsberg, Heistad, & Reubens, 1965; Savas, 1982; Smith, 1975; Terrell, 1979; Weisbrod, 1977; White, 1981). Douglas (1983, pp. 51-52) notes that:

> A very wide range of goods and services could, in pure theory, be allocated to any of the three sectors. . . . Highways, police services, street lighting, fire services, drainage, parks, libraries, and so on—could be, have been and, in some cases, still are found in any of the three sectors.

Similarly, Weisbrod (1977, p. 171) reports, on the basis of his research with nonprofit hospital, educational, library, and employment services, that "all lend support to the view that many nonprofit organizations are like governments in their orientation toward the provision of collective goods."

Looking at the consequences of this blurring of sector boundaries from a policy-making perspective, Kamerman (1983, p. 8) states:

> Social policy, now as in the past, is being made and carried out in the public sector, in the voluntary sector, and in the marketplace. . . . [T]he mutual interpenetration of the public and private sectors has reached such a level that it is often impossible to distinguish one from the other.

Other writers have pointed out that provision, production, and funding of a particular service are separable and can be mixed, not only among the public, private, and third sectors, but also in terms of local, state, regional, and national scales when multiple organizations are involved (Advisory Commission on Intergovernmental Relations, 1985; Galaskiewicz, 1979; Kirlin, 1984; Ostrom, Tiebout, & Warren, 1961; Warren, 1972; Warren & Weschler, 1975; Wirt, 1974). A substantial body of writing has developed on the role of citizens as volunteer coproducers of public services (Brudney & England, 1983; Levine, 1984; Parks et al., 1981; Rich, 1979; Warren, Rosentraub, & Harlow, 1982).

The possibilities for urban governments to contract out or totally transfer the responsibility for the production of selected urban services to the third and private sectors have received attention (Hatry, 1983; Poole, 1980; Savas, 1982). Research has been initiated by the Urban Institute, under the impetus of cutbacks in federal programs in the public sector, to identify the magnitude of services provided by nonprofit organizations in urban centers and the extent to which they are dependent on government funding (Gutowski, Salamon, & Pittman, 1984; Rosentraub, Musslewhite, & Salamon, 1985; Salamon, 1984; Salamon & Abramson, 1982).

The Need for a Conceptual Reformulation

A variety of authors, then, recognize that firms and nonprofit organizations can produce urban goods and services similar to those provided by municipalities, that there is sectoral interpenetration in the provision of these services at the community level, and that volunteerism is widespread. Despite the substantial research and commentaries that exist, however, few attempts have been made in the study of urban services to include all three sectors as the basis of analysis at the municipal level. As a result, research on urban services tends to be based on partial data, and questions concerning the effects of the provision of services in the private and third sectors on the performance of the public sector seldom are asked. Further, looking at one sector limits our ability to describe what benefits are derived from the aggregate services output in a community and how and to whom they are distributed (Kamerman, 1981, p. 11).

The logic of this situation suggests that we devise and apply a conceptual framework that takes into account all three sectors if we wish (a) to understand fully who receives and benefits from such things as protection of person and property, recreation, education, and other collective goods and services available in a community and (b) to determine the costs of services and how the costs are distributed.

To accomplish this goal, a shift in thinking about urban services is needed. Analysts must recognize that the overall situation of individuals and subgroups within a community is the result of the total set of public-, private-, and third-sector urban goods and services they are able to consume or use. The concept of a community services budget offers one way of proceeding with this task.

Community Services Budget

As an analytic tool, the concept of a *community services budget* has two critical dimensions, First, as noted, it encompasses all urban goods and services normally provided publicly in a community, including those resulting from the private and third sectors. Second, the concept carries the assumption of three distinct and separate functions in the process by which goods and services are made available in a community: provision, financing, and production. *Provision* refers to the decision that determines what goods and services will be provided and to whom. *Financing* concerns how fiscal resources are generated and how the costs of the goods and services are distributed. *Production* involves the actual creation of the goods and services. Each of these can be undertaken by a different entity that may be internal or external to the community in which the goods and services are consumed. Further, as elaborated below, two or more entities may engage in conjoint behavior by mutually agreeing to participate in the provision, financing, or production of a particular good or service.

If one looks only at the public sector, all of these functions for a given good or service could be handled by a municipality if it (a) has authority to provide the good or service and chooses to do so, (b) supports it from local taxes, and (c) uses municipal personnel to produce the service. More complex arrangements are common. A city council could vote to provide a special recreational program for handicapped children, with most of the funds coming from the national government and the remainder from the city. In addition, citizen volunteers could, along with city personnel, participate in the production of the service.

In the latter case, conjoint or mutually agreed behavior occurs in each of the three functions. Although the city voted to provide the program for the disabled, it would not have done so if the national government had not made the decision that such a service was desirable at the community level and provided financial incentives. Thus the city and national governments can be viewed as coproviders. Cofinancing also is involved because federal and local monies are used, as are the inputs of citizen volunteers. Coproduction is present, too, with both citizens and municipal employees staffing the program.

Sector	Function		
	Provision	Financial	Production
Public			
Private			
Third			
Conjoint			

Figure 14.1. Community Services Budget Matrix

Similar combinations can exist in the private and third sectors. For example, a local nonprofit agency such as the YMCA or a youth club may have the provision decision for some of its programs made by its national headquarters. The local unit could receive funds from the national level, other local nonprofit groups (e.g., the United Way, which aggregates the private contributions of individuals), and the city government. Volunteers also could be involved in the production function, along with the agency's staff. Finally a good or service in a community may be provided only in the private or third sector, as in the case of a hospital.

For a specified set of urban goods and services, the community services budget serves as a means to frame data in terms of who makes provision decisions, how costs are distributed, and who are the producers. As Figure 14.1 indicates, a wide range of options is possible.

The sectoral and functional configuration of actors can differ from one good or service to another within a community, as well as differ for the same type of good or service among communities. Also involved can be inter- and intrasectoral arrangements with two or more entities. It is quite prevalent, for example, for several public agencies to participate jointly in the provision, funding, or production of a good or service.

Adequacy, Efficiency, and Equity in the Distribution of Urban Services

A number of long-standing issues in urban policy assume different dimensions when this framework is used. The question of equity, for example, now can be formed in terms of how the total supply of an urban service, not just that portion provided by the city, is distributed within a community. Similarly studies of the adequacy of service levels can take into account all outputs accessible to local residents, including those provided in the private and third sectors. Finally the assessment of the efficiency of a municipal department can incorporate all costs associated with the production of a service, including those not recorded in the city budget, such as the time contributed by citizen volunteers and private investments.

The matrix in Figure 14.1 sets out the variety of intra- and intersectoral combinations possible in the provision, financing, and production of urban goods and services. An aspect of recreation—swimming—can be used as a more specific example of the arrangements that could be available within a community. An inventory of the swimming options in a city could find that:

1. All swimming facilities are within the public sector.
   a. All are owned and operated by the municipal government.
   b. All are owned and operated by other governmental units, such as a special district or a county, where the territory is less than, the same as, or more than the area of the city.
2. Swimming facilities are completely within the private and third sectors.
   a. All are owned by individual property owners.
   b. All are owned by private entrepreneurs.
   c. All are owned by homeowners' associations and apartment complexes.
   d. All are owned by voluntary and nonprofit organizations.
3. A mixed ownership system exists with any combination of the above.
4. Conjoint arrangements occur among any of the above, such as a city contracting with a private firm to operate a municipal pool or to subsidize a youth club's swimming program.

After the actual arrangements within a community are identified, the possible policy ramifications can be analyzed. If, for instance, most neighborhoods of a city except the poorest have pools that belong to homeowners' associations or apartment complexes, little support could be expected among the middle- and upper income groups for allocating municipal resources for public swimming facilities (Rich, 1979; Weisbrod, 1977). Yet the absence of a municipal pool does not mean that access to swimming is necessarily unavailable in the poorer neighborhoods. One or more nonprofit organizations may provide access to residents of lower income areas at a comparable quality and cost to public facilities in cities of similar size and characteristics. If a subsidy is required to operate and maintain the pool, it is absorbed by the third, rather than the public, sector.

The assessment of the adequacy of a service can be affected also by a single- or multisector analysis. Assume that a medium-sized city has a municipal library and several college libraries. Further assume that the budget and book-collection decisions of the public library take into account the reference and technical resources of the colleges' collections because the latter make their facilities accessible to community residents. By nationally recommended professional standards, the city library may be inadequate because of its small budget, limited reference and technical material, and low book circulation per capita. Yet if the total set of library services available and the use patterns are considered, the community may exceed professional norms on all of these measures.

In each of these cases, the combined output of the public and third sectors may be as equitable, adequate, or efficient, or more so, as would or could be obtained from the public sector alone. The matrix in Figure 14.1 indicates that a great range of possible sectoral and scale mixes could result in positive public policy outcomes. Although the options will differ from community to community, a good deal can be foregone if this is not recognized. Negative public effects, however, also may occur from the way the three sectors participate in the provision and distribution of a particular service in a community.

Measuring the economic efficiency of a municipal agency only in terms of the city budget, for example, may be misleading. Accounting for private- and third-sector providers in a community may result in a different assessment of public-sector efficiency. To illustrate, consider two cities of similar size, demographic characteristics, and crime rates. Public investment in the police department is lower in City A on a per capita basis than in City B. Using public expenditures and crime rates as indices, the data would show that the police are more efficient in City A. By looking at all providers of security within each city, however, this assessment can change.

City A, in fact, has much higher investment in private patrols, industrial police, home alarm systems, and gun purchases, as well as extensive voluntary citizen participation in crime-prevention activities at the neighborhood level. If all of these expenditures of money and time are taken into account and used as the input measure in calculating efficiency, the total amount collectively paid in City A is considerably more than in City B to produce the same outcome in crime level.

The variety of security services available in the private sector makes the possibility of this scenario quite real. A 1971 study reported that 36% of all security personnel in the United States were employed in the private sector (Fisk, Keisling, & Muller, 1978, p. 32).

Kakalik and Wilhorn (1971, p. 2) indicate that private contract agencies and in-house security forces provide, among other things, roving patrols, armored-car escorts, central station alarm service, and various investigative functions. They go on to say: "Both types of security personnel are utilized by a wide variety of consumers, including individual citizens, banks, retail establishments, insurance companies and other financial institutions, and apartment houses, and, at recreational events."

In discussing private patrols, Poole (1980, p. 39) describes 62 police beats in San Francisco that are " 'owned' by private police officers who are paid by their customers—the businesses, apartment owners, and home owners. The 'Patrol Specials,' as the officers are called, receive a complete police academy training, carry guns, and have full arrest powers."

Privatization of security services between 1975 and 1985 has continued to increase. The number of private security guards in the United States rose during this period by more than 50%. Currently, for example, they outnumber the 96,000 state and local public law enforcement officers in Connecticut, New Jersey, and New York by more than two to one (Tolchin, 1985).

This discussion of police services has another implication beyond efficiency. These examples also suggest that the actual amount of protection
available to citizens can differ considerably within a community even though
the level provided by the public sector is equal. When public security services
are supplemented by inputs from the private and voluntary sectors, there is no
reason to believe that they will be distributed equally or by need. If, as would
be expected, a positive relationship exists between the amount of a family's income
and its willingness and ability to make investments in the private and third
sectors for security services (Warren, Rosentraub, & Harlow, 1984), questions
of distributional equity necessarily arise.

This is particularly true in cases in which public investment is low and an
effort is made by the city to use citizen volunteer activity or coproduction to
augment municipal services at the neighborhood level. For example, in areas
where residents have marginal incomes, the probability is higher that they will
have a relatively greater need for security services, have limited resources to
command private services, and be less likely to participate in coproduction.
Further, equity questions of this type are not limited to the security of person
and property. The actual absolute and relative distribution of any service can be
significantly different among neighborhoods from what a municipality provides
when there is nonpublic involvement in its provision, financing, and production.

Citizen Volunteers and Conjoint Behavior

Using volunteer citizens in the production of urban services traditionally
has been relied on by nonprofit agencies and is cited frequently as a strategy for
cities to consider in responding to fiscal stress (Brudney, 1984; Levine, 1984).
This use is in contrast to a city's simply "shedding" the service or shifting its
funding to user charges. The most obvious benefit of coproduction (the direct
involvement of citizens with public personnel in the production of municipally
provided goods or services) is the ability to maintain a service in the public
sector while reducing the costs that are directly borne by the city budget. This
maintenance is achieved by transferring a portion of the production costs from
the municipal budget to participating citizens. Some authors have noted that
coproduction also can enhance the quality of a service and result in improved
citizen-bureaucratic relations and greater general support for the public sector
(Sharp, 1980; Whitaker, 1980; Wilson, 1981).

However, the nature and implications of citizen participation related to
public goods and services now is recognized to be more complex than initially
conceived (Brudney & England, 1983; Ferris, 1984; Warren et al., 1982). The
possibility of citizen involvement is not limited to the production function but
can include coprovision and cofinancing as well. A municipal soccer league with
volunteer coaches illustrates this point. It can be assumed that a city would not
have started such a league without citizen interest and willingness to voluntarily
staff the program. If the city would not have created and would not continue
to support the league without citizen demand and participation, the provision

decision is a joint one, requiring the agreement, even though implicit, of both the city and the residents.

Cofinancing is involved in the example as well. The true cost for the league includes the overhead, facilities, and equipment provided by the city's recreation department and the value of the time, services, and equipment contributed by citizen volunteers. Usually, however, the latter will not be included in the municipal budget. Thus neither the magnitude of the citizen input nor the full cost of this service is known and is not formally recorded.

In contrast, the cost is incorporated into the municipal budget when the conjoint behavior involves a market relationship. The case of a city engaging in coproduction with a private firm to clear the snow from its streets provides an example. A contract is entered under which a private firm agrees to provide and operate its snowplows to augment the city's during heavy snowstorms. The city pays a stipulated sum of money, and the cost incurred is included in the budget of the public works department. Quantifying the value of voluntary citizen inputs and devising methods of formally including them in a municipal budget are far more formidable tasks. To ignore them, however, will perpetuate significant gaps in the knowledge and understanding of the service infrastructure of the urban community.

Education reflects a particularly complex set of analytic issues and raises a number of policy questions when a community services budget framework is used. In most urban areas, private- and third-sector schools are substitutes for public schools. Approximately 10% of all students at the elementary and secondary levels in the United States in 1977-1978 were enrolled in nonpublic schools (Poole, 1980, p. 184). The ratio, however, can vary considerably. For example, in the New Castle County portion of the Wilmington, Delaware, metropolitan area, 26% of all students were enrolled in nonpublic schools in 1982 (Boozer, 1982). Thus a quarter of the production of education in the area occurs outside the public sector. The number of students in a community who are in public and nonpublic schools can be an important policy fact for several reasons. If a substantial percentage of children are in schools operated by private and nonprofit organizations, parents, as voters, may resist the allocation of money to the public schools. Further, how well the public school students score on national tests or the number who go on to college would not be a true indicator of the overall performance of the community in educating its children.

Similarly the economic efficiency of the public schools in such a community could not be determined without the ability to control for demographic and other variables in measuring the cost of producing a specified level of student achievement in public, private, and nonprofit schools. Measures of efficiency would be distorted if either a disproportionate number of high or low achievers in the community's student population were in nonpublic schools and this was not taken into account in the analysis of public schools.

Even if public-sector schools were the only focus of concern, the requirement of a community services budget that citizen participation be accounted for directs attention to additional policy questions. Citizens can participate jointly with public schools in a number of ways. First, parents may work with their children or purchase goods and services to improve their academic achievement levels and competencies. Second, parents or other citizens may donate their time in a school as teachers' aides, tutors, monitors, or chaperons. Third, parents, other citizens, and private and third-sector organizations can make donations to schools in the form of cash, equipment, or special services. Each of these types of input has the potential for substantially changing the quality and distribution of the education produced within a specific school or the district as a whole.

Conjoint behavior, then, can affect how much is expended for public education and the adequacy and efficiency of the product. In some schools within a district, for example, parents are expected to make contributions to the learning process by providing a supportive environment and tutoring, if necessary. In other schools the opposite may be true. In each case, the expected parental behavior will affect how administrators and teachers allocate their resources, as well as the *total* resources expended for the education of students at each school.

Citizens can become involved also in cofinancing through direct monetary inputs to a school. If these are identified, what appears to be an equitable distribution of resources among the schools in their formal public budgets may be inequitable in terms of the actual resources available in various schools. It is not uncommon for parent support groups and local merchants to raise and donate money to a particular school for such things as band uniforms and travel, athletic equipment, and discretionary funds for teachers. Many Parent-Teacher Association groups also collect money specifically to buy and then donate such things as microcomputers to the schools with which they are affiliated.

There is little question that middle- and upper income area schools are more likely to have their resource base increased by this process than those in lower income neighborhoods. There also is little question that such cofinancing becomes more important as school districts are forced to cut back or eliminate various programs because of fiscal problems. Thus, unless a means is available to identify and compare the direct monetary and service inputs that are outside the formal budget, the extent and policy implications of the resulting inequities will seldom be noted and, if they are, will be difficult to identify, analyze, and, particularly, redress.

Advantages of Applying
a Community Services Budget Matrix

The quality of life for people in a community depends, in part, on the range and quality of urban services that are accessible to them. As has been argued in this discussion, the adequacy, efficiency, and equity of these services in a city are the joint product of the policies and actions of various levels of government, nonprofit and voluntary agencies, private firms, and individuals. Yet most policy studies concerned with these issues have focused exclusively on the performance of public agencies.

As has been shown, there can be considerable differences in what is found as the result of research on distributional equity or production efficiency for an urban service, depending on whether the analysis is based on what is publicly produced or the total output of all sectors. This analytic problem is compounded unless it also is recognized that the decisions and processes concerning the provision, financing, and production of urban services are separable and can be mixed among actors within and between the three sectors at the local, regional, and national levels. Thus to assess adequately the actual state of affairs in a given community or to make a comparison among cities, it is necessary to account for the inputs, linkages, and distributional effects of the public, private, and third sectors.

A number of problems are to be faced in applying a community services trisectoral matrix to the study of urban services. Obtaining the necessary data, if they are available, to measure the service outputs of individuals, private firms, and nonprofit and volunteer organizations and their distribution will be a difficult task.[2] At the same time, perceiving a community through the prism of a community services budget can expand the range of policy and programmatic opportunities that officials, citizen organizations, and researchers may find useful.

The utility of the construct is not that it "discovers" new phenomena. The role of private- and third-sector activities in a city are normally well known. It does, however, provide a means of aggregating and thinking about these elements in a more systematic and usable way. Further, to be of benefit, it is not necessary to produce a fully developed community services budget of the type discussed here. Simply perceiving the urban service output of a community as the product of a multisector network directs the attention of public-sector officials to an expanded resource base. Further, the concept can be applied to a single service, rather than to all services. The possibility of obtaining the data required for creating a community services budget for education or recreation may be feasible and strategically important for community groups concerned with distributional equity. In all of these cases, going beyond the information available in public-sector budgets is the key factor.

During a period of municipal austerity in which local officials are placing a priority on cost-reducing policies, a community services budget framework is particularly relevant from two perspectives. One relates to municipal management. It provides officials and planners with a systematic way of identifying the extent to which comparable or compatible urban goods and services are being provided in the private and third sectors. This knowledge, in turn, offers the opportunity to undertake initiatives to maximize conjoint relationships among the sectors that are perceived to be in the public interest and to obtain a more valid database for assessing the production efficiency of municipal agencies.

At the same time, questions of distributional equity tend to be both more complex and of less interest to municipal officials when policies are considered involving such things as transferring the production of a municipal service from the city to a private firm by contracting, augmenting reduced departmental service levels through the use of citizen volunteers, or shedding all or part of a function with the expectation that nonprofit and voluntary organizations will assume responsibility for its provision at an adequate level. Just as a community services budget enlarges the database available so that options such as these can be readily identified, it also directs attention to the possibility that such arrangements may result in inequitable distributions of an urban good or service for which the municipality has a responsibility or is considering shedding.

It should be underscored, however, that the useful application of this framework as a tool of urban policy analysis is heightened by, rather than limited to, a time of fiscal crisis. Unless the analysis of the adequacy, efficiency, and equity of urban services as a normal practice goes beyond the public sector, we will continue to produce partial pictures of reality and will base policies on incomplete data and potentially misleading premises.

Notes

1. It is useful to clarify briefly what is meant by the private and third sectors. *Private-sector* actors are those producing goods and services for profits and engaging in market transactions. Firms offering security services, garbage and trash collection, and a water sports recreational park are examples. It is now widely recognized, however, that the traditional public/private dichotomy does not adequately capture significant differences among nonpublic entities. Several labels have been used to identify entities that are neither public bodies nor engaged in profit-making activities—labels including *nonprofit sector, independent sector,* and *third sector.* The term *third sector* is used because it is a more inclusive term that, for the purposes of this discussion, can include individual volunteer actions, as well as nonprofit and voluntary organizations. Commenting on the difficulty in precisely defining the third sector, Douglas (1983, p. 16) suggests that the merit of the term "is that it draws attention to what the organizations constituting it *are not*"—commercial firms and governmental agencies.

Salamon and Abramson (1982, pp. 9, 10) provide a useful characterization of nonprofit organizations as private in structure and public in purpose and as constituting the primary mechanism for converting "private charitable resources into the solution of community problems." They distinguish four types of nonprofit organizations: funding agencies or fund-raising intermediaries, including foundations (e.g., Blue Cross, Blue Shield, United Way), organizations that provide goods and services primarily to their members (e.g., unions, professional associations), organizations with service and welfare goals that primarily serve others (e.g., hospitals, social welfare agencies, cultural and educational institutions), and religious bodies. This set of organizations and individual volunteering represent third-sector phenomena of basic relevance to the discussion.

2. Some local governments have citizen volunteer banks that can provide records of such activities. Surveys have asked citizens about involvement in coproduction, particularly in the area of security of person and property. Municipal building, license, and permit records can establish starting points for identifying such things as nonpublic schools and recreational facilities. Umbrella organizations such as the United Way have records showing the type and magnitude of the activities of many nonprofit organizations. Saying this does not eliminate the real possibility that comprehensive data may not exist in a given community, that those with the information may not be willing to make it available, or that the resources necessary for acquiring, formatting, updating, and maintaining such data may not be easily obtainable. It does indicate, however, that where there is interest, a start can be made.

References

Advisory Commission on Intergovernmental Relations. (1985). *Intergovernmental service arrangements for delivering local public services.* Washington, DC: Government Printing Office.

Boozer, R. F. (1982). *Summary of report on nonpublic schools in Delaware 1982-83.* Dover: Delaware Department of Public Instruction.

Brudney, J. L. (1984, June). Local coproduction of services and the analysis of municipal productivity. *Urban Affairs Quarterly, 19,* 465-484.

Brudney, J. L., & England, R. E. (1983, January/February). Toward a definition of the coproduction concept. *Public Administration Review, 43,* 59-65.

Douglas, J. (1983). *Why charity?* Beverly Hills, CA: Sage.

Ferris, J. M. (1984). Coprovision: Citizen time and money donations in public service provision. *Public Administration Review, 44,* 324-333.

Fisk, D., Keisling, H., & Muller, T. (1978). *Private provision of public services: An overview.* Washington, DC: Urban Institute.

Galaskiewicz, J. (1979). *Exchange networks and community politics.* Beverly Hills, CA: Sage.

Ginsberg, E., Heistad, D., & Reubens, B. (1965). *The pluralist economy.* New York: McGraw-Hill.

Gutowski, M., Salamon, L. M., & Pittman, K. (1984). *The Pittsburgh nonprofit sector in a time of government retrenchment.* Washington, DC: Urban Institute.

Hatry, H. P. (1983). *A review of private approaches for delivery of public services.* Washington, DC: Urban Institute.

Kakalik, J., & Wilhorn, S. (1971). *The private police industry: Its nature and extent.* Santa Monica, CA: RAND.

Kamerman, S. B. (1981, November). *The public and the private intertwined: The new mixed economy of welfare.* Paper presented at the Seventh National Association of Social Workers, Professional Symposium, Philadelphia.

Kamerman, S. B. (1983, January/February). The new mixed economy of welfare: Public and private. *Journal of the National Association of Social Workers, 28,* 5-10.

Kirlin, J. J. (1984). A political perspective. In T. C. Miller (Ed.), *Public sector performance* (pp. 161-192). Baltimore: Johns Hopkins University Press.

Levine, C. H. (1984, March). Citizenship and service delivery: The promise of coproduction. *Public Administration Review, 44,* 178-187.

Ostrom, V., Tiebout, C. M., & Warren, R. (1961, December). The organization of government in metropolitan areas: A theoretical inquiry. *American Political Science Review, LV,* 831-842.

Parks, R. B., Baker, P. C., Kiser, L., Oakerson, R., Ostrom, E., Ostrom, V., Percy, S. L., Vandivort, M. B., Whitaker, G. P., & Wilson, R. (1981, Summer). Consumers as producers of public services: Some economic and institutional considerations. *Policy Studies Journal, 9,* 1001-1011.

Poole, R. W., Jr. (1980). *Cutting back city hall.* New York: Universe.

Rich, R. C. (1979). Neglected issues in the study of urban service distributions: A research agenda. *Urban Studies, 16,* 143-156.

Rosentraub, M. S., Musslewhite, J. C., Jr., & Salamon, L. M. (1985). *Government spending and the non-profit sector in Dallas.* Washington, DC: Urban Institute.

Salamon, L. M. (1984). Nonprofit organizations: The lost opportunity. In J. L. Palmer & I. B. Sawhills (Eds.), *The Reagan record* (pp. 261-285). Cambridge, MA: Ballinger.

Salamon, L. M., & Abramson, A. J. (1982). *The federal budget and the non-profit sector.* Washington, DC: Urban Institute.

Savas, E. S. (1982). *Privatizing the public sector.* Chatham, NJ: Chatham House.

Sharp, E. B. (1980, June). Toward a new understanding of urban services and citizen participation: The coproduction concept. *Midwest Review of Public Administration, 14,* 105-118.

Smith, B.R.L. (Ed.). (1975). *The new political economy: The public uses of the private sector.* New York: Macmillan.

Terrell, P. (1979, March). Private alternatives to public human services administration. *Social Service Review, 53,* 56-74.

Tolchin, M. (1985, November 29). Private guards get new role in public law enforcement. *The New York Times,* p. 1.

Warren, R. (1972). *The community in America.* Chicago: Rand McNally.

Warren, R., Rosentraub, M. S., & Harlow, K. S. (1982, March). Citizen participation in the production of services: Methodological and policy issues in coproduction research. *Southwestern Review of Management and Economics, 2,* 41-55.

Warren, R., Rosentraub, M. S., & Harlow, K. S. (1984, June). Coproduction, equity, and the distribution of safety. *Urban Affairs Quarterly, 19,* 447-464.

Warren, R., & Weschler, L. F. (1975, Spring). Governing urban space: Multi-boundary politics. *Policy Studies Journal, 3,* 240-247.

Weisbrod, B. A. (1977). *The voluntary nonprofit sector: An economic analysis.* Lexington, MA: D. C. Heath.

Whitaker, G. (1980, May/June). Coproduction: Citizen participation in service delivery. *Public Administration Review, 40,* 240-246.

White, M. J. (Ed.). (1981). *Non-profit firms in a three sector economy.* Washington, DC: Urban Institute.

Wilson, R. K. (1981, Fall). Citizen coproduction as a mode of participation: Conjectures and modes. *Journal of Urban Affairs, 3,* 37-50.

Wirt, F. M. (1974). *Power in the city.* Berkeley: University of California Press.

15

Infrastructure Policy

REPETITIVE STUDIES,

UNEVEN RESPONSE, NEXT STEPS

Marshall Kaplan

During the last 16 years or so, the word *infrastructure* has become familiar to scholars who focus on the problems of America's regions and its urban areas. From time to time, the term has traveled from the halls of ivy to the halls of Congress and to the hallways of federal, state, and local government bureaucracies. Indeed, periodically, often after a bridge collapses and loss of life occurs or after a large pothole appears on a major interstate highway, causing a multicar accident, media attention grants infrastructure a temporary seat at the nation's public policy table.

The Studies: Themes and Approaches

Since the late 1970s, several assessments of national infrastructure needs have been completed by reputable national panels and/or individual analysts. Surprisingly, given their different genesis, their varied client groups, and their diverse methodologies, most of the studies contain similar kinds of general conclusions. Among them are the following:

1. The United States has failed to invest sufficient funds in its roads, bridges, transit systems, airports, water systems, and the like. Investment, generally, has not kept up with growth of the gross national product (GNP). For example, total public spending on infrastructure was 3.6% of the GNP in 1960 and only 2.6% in 1985. Expenditures for infrastructure have not kept pace with expenditures for other domestic purposes. The relative share of public works spending decreased from nearly 20% of total expenditures in 1950 to less than 7% in 1984. During this same period, government spending for welfare and education increased from 10% to more than 40% of total expenditures (National Council on Public Works Improvement, 1988, p. 8).

2. Although state and local governments have picked up some of the slack in total infrastructure investment, the pattern between and among states has been uneven. More relevant, perhaps, very few state governments and even fewer local governments have developed rational investment plans and strategies. Meaningful capital improvement programs and/or capital budgets premised on precise needs assessment, benefit-cost, or rate of return analyses are rare commodities.

3. The gap between current and anticipated infrastructure investment by the public sector and assumed needs is a relatively large but manageable one—if the nation can mount a solid multiyear commitment.

4. Efforts to initiate user fees, exactments, urban services fees, tax increment financing, public/private sector partnerships, and other unique or innovative forms of financing are noteworthy and growing. From a public policy perspective, however, many such techniques raise serious equity and efficiency questions.

5. Without a national infrastructure policy and an integrated approach to public sector capital investment, significant negative externalities will occur between and among regions, states, and cities. The nation's water, sewer, and transportation systems are inextricably linked. Failure to attend to problems in one area of one system increases the burdens and problems, or so the argument goes, in other areas of the system. For example, the inability or unwillingness of a jurisdiction to develop strategies to ensure water quality will negatively affect the water supply in other contiguous jurisdictions; the inability or unwillingness of a state or city government to improve key roads and bridges within its boundaries will negatively affect the safety and efficiency of transportation in other nearby states or cities; the inability or unwillingness of regions, states, or cities to invest in undersized or deteriorated airports will impede the ability of other airports to function effectively. And so on and so forth. We are one nation, indivisible. The nation must respond to its primary infrastructure needs in a coordinated manner or else it will face serious consequences.

6. Although different regions of the country have different kinds of infrastructure needs, interregional differences often seem less important than differences internal to each region. For example, although differences exist between the infrastructure problems of the Northeast and the South or the Midwest and the Rocky Mountain region, equally great or greater differences apparently exist between rural and urban areas in each state and between states in each region. Overall and over time, regional differences, particularly with respect to the dollar needs involved, seem to be lessening.

7. Severe infrastructure problems are being encountered in cities, particularly in central cities. They result from the absence of maintenance and preservation investments. They occur because, almost by definition, water, sewer, and transportation facilities in central cities are older and in some cases in need of replacement. They have outlived their "life cycle."

Most authors of the studies warn their readers of the consequences of the failure of public and private sectors to act on the nation's infrastructure problems.

The intensity of their individual forebodings varies considerably. Although no authors reach the apocalyptic tone that Choate and Walter (1981) did in their popular book *America in Ruins* in their predictions of doom, almost all who offer national needs estimates project difficult times ahead, if their perceptions of policy paralysis continue among federal, state, and local governments. For example:

From *Fragile Foundations:*

The quality of a nation's infrastructure is a critical index of its economic vitality. Reliable transportation, clean water, and safe disposal of wastes are basic elements of a civilized society and a productive economy. Their absence or failure introduces an intolerable dimension of risk and hardship to everyday life, and a major obstacle to growth and competitiveness. . . . Unless we dramatically enhance the capacity and performance of the nation's public works, our own generation will forfeit its place in the American tradition of commitment to the future. Without such an effort, our legacy will be modest at best. At worst, we will default on our obligation to the future, and succeeding generations will have to compensate for our failures. (National Council on Public Works Improvement, 1988, p. 1)

From *Hard Choices:*

Yet, the inability to adequately address the infrastructure needs of the nation constitutes a major problem that seriously threatens our economic and physical well being. . . . With virtually every level of government facing either major revenue shortfalls or significant unmet needs, the temptation to further postpone addressing this situation is great. It is time to measure the need, to understand that investment in our capital stock must be a priority to help keep our economy competitive and sustain our quality of life. . . . There's a bill coming due. We should start to pay now, or we'll have to pay a lot more later. (National Infrastructure Advisory Committee, 1984, p. 3)

From *A Consensus in Rebuilding America's Vital Public Facilities:*

Difficult political decisions are necessary. Citizen resistance may be strong, and improved government management and organization will not come about easily, quickly, or without resistance. However, the time for action is now. An informed citizenry will be the safeguard that will ensure that our vital facilities and in turn our nation stay healthy and safe for generations to come. (Labor Management Group, 1983, p. 12)

From *America in Ruins:*

America's public facilities are wearing out faster than they are being replaced. The deteriorated condition of the basic public facilities that underpin the economy presents a major structural barrier to the renewal of our national economy. In hundreds of communities, deteriorated public facilities threaten the continuation of basic community services such as fire protection, public transportation, water supplies, secure prisons and flood protection. . . . The United States is seriously underinvesting in public infrastructure. (Choate & Walter, 1981, p. 1)

From *Capital Budgeting and Infrastructure in American Cities:*

Whether it is Wheeling, West Virginia, struggling with inadequate roads, or Albuquerque, New Mexico, grappling with 70 million dollars worth of mandated wastewater treatment upgrading, many of America's communities face substantial backlogs of needed repair work. . . . No region, population size category, or type of city, however, is free of pressing infrastructure problems. (U.S. Conference of Mayors and National League of Cities, 1983, p. iii)

Although researchers provide their own distinct menu of policy options, certain important commonalities appear among them as follows:

1. They call for more federal dollars and more flexible federal administration of federal aid initiatives.
2. They call for state and local governments to allocate increased funds for infrastructure purposes.
3. They call for better planning and management of infrastructure investment in all levels of government.
4. They suggest that budget realities require the development of innovative new financing techniques and public-private sector partnerships.
5. They hold out varying degrees of hope that new technology concerning infrastructure development may reduce the dollars needed to finance future needs.
6. They indicate the need to allocate more budget funds to maintenance, as opposed to new construction of infrastructure.
7. They indicate the need to improve information about infrastructure conditions and to develop better analytical tools to determine priorities among infrastructure projects.

Differences do exist among and between published analyses. In reports completed toward the end of the 1970s and in the early 1980s, in effect before the extent of the deficit had become imprinted in the minds of the general public, authors tended to emphasize requirements for major public expenditures to cure infrastructure problems and, more often than not, granted emphasis to the related role of the federal government in providing significant new monies.

In *America in Ruins* (Choate & Walter, 1981), the first comprehensive study of capital facilities, the authors placed a good deal of blame for the deteriorated condition of basic public facilities on the cutbacks in public-sector dollars available for infrastructure maintenance and development. Clearly, according to these authors, the road to infrastructure recovery requires collaborative efforts among all levels of government and between public and private sectors. But although they were vague on specifics, they suggested that the road to recovery could not be traveled without the development of comprehensive federal policies and the provision of increased federal financial support. Choate articulated the need for $2.5 trillion to $3 trillion to maintain current levels of infrastructure services. (Choate's estimate is contained in his special report provided to the House of Representatives' "Wednesday Group" in May 1982.)

The Associated General Contractors of America (AGCA) (1983) followed and to some extent parroted the tone and content of *America in Ruins*. Based on a collage of data secured from federal, state, and industry sources, *America's Infrastructure: A Plan to Rebuild* identified just over $3 trillion in infrastructure investment needs or $140 billion per year through the turn of the century. The AGCA called on the federal government to play a major role in securing a necessary national response. In a follow-up report, the group indicated:

> Judicial pruning of federal spending should be undertaken where the maximum savings can be achieved while sacrificing the least in terms of the country's future potential. Sacrificing infrastructure investment for short term gains can only serve to saddle future generations with a low growth economy and unfairly penalize them for our fiscal imprudence. (AGCA, 1983, p. 27)

And that

> The long-range impact on the nation of not going forward with vital infrastructure projects would be disastrous. (p. 3)

Toward the end of President Reagan's first term, the administration's efforts to contend with seemingly intractable federal budget and national economic problems, combined with its attempt to shift key domestic initiatives to states, made some infrastructure analysts wary about highlighting requests for large amounts of grant money and new federal roles. For example, in the joint survey and study completed by the U.S. Conference of Mayors and the National League of Cities (1983), the national nature of the infrastructure problem was noted. But the suggestion was made that it was a "manageable problem." Steady, strategic investments over a number of years could and would enable communities to start work on the "capital assets ranked as highest priorities" (p. 4).

In the Joint Economic Committee of Congress's (JEC) report *Hard Choices* (National Infrastructure Advisory Committee, 1984, pp. 13, 17), advocacy of a national infrastructure fund (a bank) that would lend money to states for infrastructure purposes was coupled with a plea to initiate efforts to reconsider standards governing the design and development of infrastructure. The former approach, ostensibly, would lessen the demand for "on budget" funds, and the latter effort, if it led to the lowering of standards, could reduce future investment costs. As noted in the JEC study, the "bottom line is that the problem is solvable, that the solutions are manageable, and that the time to start is now" (p. 4). The total investment need was listed at $1.157 trillion, or $64.3 billion annually through the turn of the century (p. 5).

The U.S. Congressional Budget Office (CBO) (1983) published a report titled *Public Works Infrastructures: Policy Considerations of the 1980s* in which it was concluded that from $48.8 billion to $54.8 billion is required annually to meet infrastructure needs. It was suggested that costs could be controlled by avoiding undercharging users and by eliminating the bias in federal aid programs toward capital-intensive projects. The infrastructure problem again was presented as significant but manageable.

The U.S. Office of Management and Budget (OMB) (1985) followed the CBO's analysis with a review of infrastructure needs particularly related to federal investment levels. Contrary to most national needs studies, the OMB's report, *Supplement to Special Analysis D,* reflecting administration budget policy, indicated that no compelling justification existed for increased federal expenditures. "For the eleven categories of public civilian investment reviewed (including highways, bridges, wastewater facilities, and federal buildings . . .) here, this report finds that current federal investment levels are generally sufficient to maintain investment considered of national importance" (p. 1). OMB's arguments for maintenance of effort stemmed, in part, from opportunity costing and, in part, from an effort to limit federal exposure.

Although calling for up to a 100% increase in the amount of capital the nation spends each year on public works and for a related continued and vital federal role in infrastructure financing, the National Council on Public Works Improvement's recently published report *Fragile Foundations* (1988) highlights the need to improve infrastructure management at state and local government levels. The need to secure more financing from state and local governments is emphasized, as well as the need to develop innovative financing techniques between public and private sectors.

CBO (1988) has had the last word concerning national infrastructure needs. In its report *New Directions for the Nation's Public Works,* published in September 1988, CBO severely faults the proposal of the National Council on Public Works Improvement contained in *Fragile Foundations* that national infrastructure expenditures must almost double. CBO's criticism ranges far and wide. Although directed at *Fragile Foundations,* it fits most national needs estimates and studies. Among its most salient points are the following:

1. The assertions in *Fragile Foundations* that national public works investment is inadequate to sustain future economic growth is not based on hard data. "The relation between infrastructure investment and GNP is poorly understood; economic theory provides no indication of the optimal level of infrastructure investment relative to GNP" (p. 130).

2. The recommendation in *Fragile Foundations* to increase significantly infrastructure spending because economic growth requires a constant proportion of GNP to be allocated to infrastructure investment is open to question. The efficiency of infrastructure use is not considered, nor are changes in the structure of the economy (services may require less infrastructure for each dollar of GNP). Further, the country may not need as much infrastructure as it once did. The nation may be moving into an era of infrastructure management, as opposed to construction.

3. The need and use estimates in *Fragile Foundations* may be overinflated. They reflect engineering standards, and not benefit-cost analyses. The council's needs studies ignore savings from more productive use of existing capital. The council

Table 15.1 Estimates of Annual Capital Investment Needs: Comparison of Studies

	AGCA	CBO	Labor/ Management	JEC	Other[h]
Highways and bridges	$85.4	$27.2	$24.0	$40.0	N/A
Other transportation	7.8[a]	9.0	N/A	9.9[e]	N/A
Wastewater systems	35.6[b]	6.6[g]	6.6[g]	9.1	N/A
Water systems	11.7[c]	15.0[d]	7.2	5.3[f]	N/A
Total	$140.5	$57.8	$37.8	$64.3	N/A

SOURCE: Data concerning AGCA, CBO, Labor/Management Group, and JEC reports from analyses by Peggy Cuciti, University of Colorado's Graduate School of Public Affairs, prepared for National Council on Public Works Improvement.
NOTES: a. Includes airports, mass transit, waterways, locks, and ports.
b. Includes wastewater treatment, drainage, minor flood control.
c. Includes potable water (urban) and dams/reservoirs.
d. This figure is an extrapolation to all systems, based on 756 systems included in a federal report.
e. Excludes water transportation and the air traffic control system.
f. Excludes facilities owned or operated by federal agencies.
g. Only includes spending eligible for federal assistance.
h. Choate, coauthor of *America in Ruins*, estimates that nearly $3 trillion is required to maintain current capital facilities (Peterson et al., 1986). The National Council on Public Works Improvement (1988, p. 2) recommends up to a doubling of current annual investment each year.

inflates required spending in both needs and use studies by assuming that both technology and existing pricing policies will remain unchanged (pp. 131-133).

4. No priorities are set in *Fragile Foundations*. All infrastructure expenditures are treated as equal. This, according to CBO, is a mistake. The council should have considered estimating rates of return from different infrastructure investments (pp. 134-136).

Table 15.1 illustrates the dollar estimates of national needs described in many of the key studies. Although the numbers vary, understandably, in light of different methodological approaches, different time periods covered, and different categories of infrastructure included, the same general picture is reflected in the totals of each study. Simply put, according to each of the groups, the nation's bill to fix up and respond to its infrastructure problems will be a relatively large one.

The Policy Response

Despite the number of analyses and their coincidence with respect to general findings, no ground swell has emerged among the American public for new infrastructure spending. Arguably the public has a better understanding of infrastructure problems and the parameters associated with a required response to them. Congress has passed legislation mandating the OMB to project federal capital outlays and to link spending projections to needs. The gas tax has been increased and, as a result, additional funds for highway repair and development are being generated. Some state and local governments have initiated needs assessments and have begun to develop improved capital budget procedures. Yet the overall response, in the words of one respected individual associated with two major studies, has been "underwhelming."

No popular continuous outcry among public or private sector leaders for action has occurred to stimulate sustained congressional or administration concern. Congress has not mounted the infrastructure barricades; the White House and federal agencies have been relatively quiet.

The pattern has varied among states and municipalities. Each national report has generated episodic newspaper and television coverage. Each report has led to supportive quotations from governors, legislators, and mayors. In some areas, aggregate levels of infrastructure investment have increased. New and innovative forms of taxes and fees have been used to finance infrastructure in several areas of the country. No consistent fiscal response, however, has emerged across state boundaries or among cities.

Why, given the number of reports and their respected authorship, has it been so difficult to develop consensus concerning federal, state, and local government infrastructure policy? Put another way, what variables have muted the ability of the nation to develop a seemingly efficient and equitable response to its infrastructure needs? Several factors apparently are at work. Among them are the following:

1. *The numbers are not believable.* Big numbers, particularly if they describe future needs and/or calamities, are unreal to most Americans. Choate and Walter's (1981) or the AGCA's (1983) $3 trillion estimate of needs provides good copy for *Time* and *Newsweek*. But it is unfathomable to most Americans and to most in Congress. Although the JEC, OMB, CBO, and National Council on Public Works Improvement studies portray more manageable figures, their scale is still bigger than any one level of government could contemplate politically in light of uncertain economic times and many unmet needs. Proposals to shift funds from one or more federal or state budget items to infrastructure defy political and economic logic. They do not reflect the competing claims for use of marginal public dollars made by other worthy petitioners. Neither do they acknowledge budget realities.

2. *Household opportunity costing concerning infrastructure does not favor collective community action.* Language concerning a general infrastructure crisis— current or anticipated—makes for good after-dinner speeches to professional groups linked to infrastructure development, urban scholars, and sometimes interested government officials. More often than not, however, it falls on the deaf ears of most citizens. Polls consistently show that Americans care about potholes, particularly in their neighborhoods, that they are bothered by streets in their area that are accident prone, and that they get terribly upset by sewers that back up into their houses or the houses of their friends. Bond elections devoted to providing funds to meet specific problems in local communities face reasonable chances of success. But to jump from perceived manageable local infrastructure ills to a national infrastructure malaise seems to most Americans too high a leap.

The infrastructure problem perceived by most Americans is a very personal one; the effort to classify it as a national crisis requires difficult inductive reasoning on the part of nonprofessionals. Airports, for the most part, are functioning, water systems generally work, and interstate highways seem to take people reasonably efficiently where they want to go. Protestations to the contrary, Americans just do not believe things are falling apart.

Further, the "tragedy of the commons" (Hardin, 1968) is more relevant than the theory of community beneficence, guilt, or the public interest when dealing with perceptions or lack of perceptions concerning the national infrastructure crisis. Citizens will continue to use the "commons"—read highways, water facili-ties—as long as it meets their individual personal needs; they will be unwilling to move easily to respond to community, state, or national problems that they believe do not generally affect them or, if they affect them, do not do so in a manner that is personally very costly. In this context, individual marginal dollars are priced high. They will be expended to meet valued household needs. They will not be used, without political tensions, to respond to often-devalued public needs.

3. *The methodology used in many of the studies appears flawed.* Put two scholars in a room, and you generally will get three or four answers to most questions. Put two infrastructure analysts in a room, and you may never get agreement on the questions to ask and/or if you gain consensus on the questions, you still may have a rough time gaining agreement on the answers.

Most, if not all, efforts to define aggregate national infrastructure needs rely on secondary data. Some premise their need and investment figures on anec-dotes from state and local governments or on interviews with respected public- and private-sector leaders. Some rest their case concerning national infrastruc-ture ills on case studies either of states or cities or both. Most support their analyses of infrastructure expenditure shortfalls on the assumed wisdom re-flected in either the historical relationship between federal or federal-state and local budgets and GNP or federal expenditures for infrastructure to total federal budget expenditures. Some analyses place great weight on the historical com-parison between federal infrastructure expenditures and federal spending for other key domestic priorities. Consistency, if and when it appears among scholars, generally results from data secured from federal agency-prepared functional studies—studies often premised on reasonably sophisticated state reporting systems and common measurements of need.

As noted earlier, different national analysts produce different national esti-mates of needs. The same approach is rarely used, nor is an easily replicable approach. The whole is often more than the sum of the parts. The studies taken together seem to present a convincing case concerning current and future infra-structure problems. Yet the studies looked at individually often cannot hold together under methodological criticism.

Anecdotes, professional or bureaucratic wish lists, case studies, comparisons of infrastructure expenditures over time with expenditures for other public goods, with GNP, or with industrial investment generates only circumstantial evidence about needs. In a similar vein, without common standards defining need, it is difficult to secure national totals that make definitive sense. Failure to use benefit-cost or rate-of-return techniques in needs analyses limits their usefulness in determining priorities and in gaining consensus concerning invest-ment choices.

Although they often are cited, state and local government studies of infra-structure needs have been of little help to the national investigators. Generally "stories" are provided to lend support to the authors' claims that there are prob-lems in their states. But, with some exceptions, the authors neglect or cannot offer credible comparable data to extend or fine-tune national analyses. Standards defining problems and needs often vary from state to state and city to city. Very few state and even fewer city governments have initiated or completed compre-hensive needs assessments. Studies of current and future needs that exist appear premised on assumptions concerning use and demand; assumptions that almost always rest on uncertain population and economic projections and the continu-ation of present federal aid formulas, present pricing of infrastructure services, and present infrastructure costs. No two infrastructure accounting systems seem alike. Treatment of depreciation, maintenance costs, and revenue varies consid-erably. Rarely do state governments weigh the opportunity costs or risks associated with alternative investments.

4. *Infrastructure lacks a strong, unified constituency.* Infrastructure needs are important to many groups—some professional and self-interest groups such as the American General Contractors, and some public but still special-interest focused, such as those representing state and local governments. Infrastructure investment is important to many private-sector groups such as the chambers of commerce and municipal bond associations. Infrastructure proposals frequently appear in professional and lay journals, as well as in daily newspapers and weekly magazines.

Despite the variety of organizations expressing infrastructure concerns, many key political figures read the constituency as wide but not very deep. "They couldn't rally the public. . . . They speak only for themselves and probably only for their officers." Similarly political leaders view the support for specific new infrastructure initiatives as predictable, sometimes self-serving or serving the interest represented, and rarely representative of the public. Finally the dispa-rate nature of the groups and their failure to select similar priorities from among a common policy wish list often weaken their political clout. Many organiza-tions marching to a different drumbeat make it difficult for political leaders to read a single sheet of music or music with the same notes and sounds.

5. *Federalism combined with electoral politics makes it difficult to develop agreement concerning needs.* More than 80,000 jurisdictions now exist in America. Strategic intergovernmental efforts to engage in cooperative planning and policy development are difficult, if not impossible, in this institutional environ-ment. Interests, as well as role, mission, and access to resources, are different. Federalism and jurisdictional pluralism may have many benefits, but these do not include ease-of-action or agreement concerning shared objectives, policies, and problems.

Responsibility for infrastructure development, maintenance, repair, and management is divided among several layers of government and among several kinds of government entities. In some areas of the nation, it is difficult to determine who is responsible for infrastructure and/or what the responsibilities for infrastructure are among diverse governmental entities. Absence of specific assignments and the clear-cut attribution of benefits and costs among diverse public entities mutes their willingness to own up to infrastructure needs and/or supposed crises. The crisis sometimes is perceived as just over the county line, in the other cities, at the state level, or at the federal level. In this environment it is difficult to secure an agreed-upon definition of the national infrastructure interest—or even state and local government infrastructure interests.

In a similar vein, crisis estimates premised on future occurrences and/or future neglect bother political leaders elected for relatively short terms. Projected crises, if something is not done, fit bureaucrats better than politicians, except perhaps when the politicians are campaigning. Future crises require present money that could be used to meet present needs; blame for inaction often is suggested without the promise of praise for action. Response to anticipated crises, particularly if they are complex and expensive, like infrastructure, must occur over time and sustain bipartisan support and bipartisan willingness to take the heat if something goes wrong. Political figures do not like to be responsible for expenditures or their surrogate—taxes—that do not lead to visible immediate public goods and/or services. They do not want to assume any fault or blame for crises that cannot be solved easily and during their term of office. Their willingness to tackle long-term problems buys them ownership of the problem without ownership of the solution. Political behavior, in this context, resembles business management behavior. Short-term profits are often preferred over long-term profits. Investments to gain long-term profits are often frowned on by current corporate managers or their accountants.

Long-term commitments have been made by federal, state, and local governments. For example, the Congress and successive administrations have adhered to and, indeed, over time, amplified the social security program. It has become part of the culture. Woe to the congressperson or even president who proposes significantly cutting or restructuring social security. Similarly national and state park systems have secured bipartisan support over the years. Proposals to tamper with public parks often bring down howls of protest and a quick retreat from proponents.

Infrastructure also has had its share of winners, such as the interstate highway system and the cleaning up of the nation's waterways, lakes, and rivers. Contrary to the present situation with respect to infrastructure, however, in almost every instance, when long-term commitments have been made and sustained by federal, state, and local governments, the perception of the problem was widespread among the public, the benefits to relatively large numbers of people appeared real, consensus concerning solutions was apparent, and vigorous lobbying groups or groups with perceived widespread support existed in Washington, DC, and/or the affected areas. Over time, "something like a theology has enveloped programs like social security and national/state parks." To be against them is to be against the fundamental principles of righteousness, goodness, and the like. Infrastructure needs assessments have not yet granted infrastructure expenditures for specific projects or purposes of a religious status.

Next Steps

It will be difficult for the federal government and most states to initiate major new sustained infrastructure spending programs. At the federal level, the economy and the deficit remain a challenge; at the state level, budgets remain uncertain, and surpluses existing in some states are not matched in other states.

No additional nationwide infrastructure needs assessments would likely grant a more favorable environment for infrastructure advocates than what exists at the present time. As reported by the Urban Land Institute:

> It is now well understood that aggregate national investment "need," when computed as the annual investment required to bring public capital facilities to an adequate standard, exceeds by a wide margin aggregate investment in civilian public works. The magnitude of this gap is inherently uncertain. Further data collection will not help close it, because the principal uncertainties now concern the standards against which need should be measured. (Peterson et al., 1986, pp. 1-2)

Or as was suggested in a recent National Academy of Science report:

> The result of the traditional needs approach generally has been estimates of capital investment requirements far in excess of available resources. Long range needs estimates in the abstract and independent of short range budget decisions are difficult to understand and often simply are not very useful. Ultimately, tough priority decisions have to be made about how to spend available resources and too often needs studies have provided no real guidance on how to separate desirable improvements from critically important investments required to maintain essential service levels. (O'Day & Neumann, 1986)

During the next few years, less talk about the need for a comprehensive infrastructure policy and more talk about strategic initiatives is appropriate and would reflect political and economic realities. Similarly, fewer admonitions to acknowledge global needs and more urgings to define priorities within acknowledged national, state, and local needs and budget constraints would ensure more progress toward responding to specific infrastructure problems. In this context the following proposals are in order:

1. *Refinement of needs assessment and information system techniques at state and local government level.* Further efforts to develop national needs estimates will have minimal payoffs. Sufficient information is available to indicate even to the most obdurate that the nation has many infrastructure needs. Federal support of a number of pilot state and local government efforts to develop

refined needs estimates and information system techniques, however, would be worthwhile. If governed by comparable methodologies and if premised on consistent standards or definitions of need, they will help participating state and local governments determine strategic priorities concerning investment and investment purposes. They also will provide capacity-building lessons to all public jurisdictions about improved infrastructure management techniques. Matched with a set of "what if" questions concerning benefits and costs of proposed projects and used in conjunction with alternative revenue projections, they can maximize the advantages of capital-budgeting and capital-improvement programs. They can help prove to the public that public officials know what they are doing when they recommend infrastructure investments.

2. *Development of cost-benefit/rate-of-return methodologies.* Cost-benefit/rate-of-return analyses are sometimes difficult to apply to infrastructure projects. Clearly, analytical problems exist in determining direct and indirect benefits and costs of many kinds of infrastructure projects. Just as clearly, reaching consensus on discount rates and rates of return is sometimes difficult. Yet as a recent CBO study indicated, cost-benefit/rate-of-return studies have come a long way. Paraphrasing the CBO, many federal agencies collect data sufficient to estimate rates of return, rate-of-return studies now take into account the effect of public capital on the productivity of private sector capital, and rate-of-return studies have accommodated theories related to the measurement or value of future public benefits.

Capacity now exists to carry out meaningful cost-benefit/rate-of-return studies. The federal government should encourage all levels of government increasingly to use such studies to help determine infrastructure priorities. Studies, once concluded, can be amended or fine-tuned to consider often nonquantifiable variables. Put another way, qualitative judgments concerning environmental and social factors—judgments often difficult to convert to financial indices—can be used to extend or change the outcomes of cost-benefit/rate-of-return analyses.

3. *Development of increased consensus concerning appropriate federal, state, and local government roles in infrastructure investment.* Allocation of infrastructure responsibilities among and between various levels of government is a difficult task. Simple textbook formulas related to the attribution of benefits to well-defined jurisdictions or geographic areas as the precursor to the allocation of costs and/or administrative functions do not work. Infrastructure impacts are often difficult to discern. Use of infrastructure facilities and services often generates spillover effects—some positive and some negative—on jurisdictions, people, and geography many miles away. The definition of a national or state interest is tough to agree on. Overlapping interests—some federal, some state, and some local—are more often than not associated with infrastructure use, development, and maintenance. Separate roles and a return to "dual federalism," with the federal and state governments each maintaining well-defined infrastructure missions, is highly unlikely in the light of the interconnectedness of America's key infrastructure elements.

Efforts should be made to clarify or fine-tune federal, state, and local government infrastructure roles and to define them in a manner that proves supportive, rather than disruptive, to collaborative intergovernmental relationships. Although it will be difficult to determine precise national infrastructure objectives and interests, the task, if undertaken, will assist Congress to sort out issues related to federal mandates and funding of infrastructure systems. Initiatives that will allow states and cities to capture benefits from their infrastructure investments that accrue to other areas and to avoid infrastructure costs from actions or nonactions by jurisdictions outside their boundaries deserve encouragement. Similarly initiatives that will test more efficient and equitable ways to allocate the benefits and costs of infrastructure investment to different governmental levels, though methodologically tough, deserve support. Even their partial success would help foster the development of more rational roles and funding approaches among and between different levels of government.

4. *Development of more flexibility concerning existing federal tax and financial-assistance programs to reflect state and local government infrastructure priorities and problems.* The federal government's role in planning, creating standards, and providing financial support for major elements of the nation's infrastructure has been and remains crucial. Federal actions, though generally helpful, at times have restricted state and local government infrastructure choices in sometimes unanticipated ways. For example, recent federal tax reforms have narrowed opportunities to create public-private-sector partnerships (the tax law restricts the growth of tax-exempt bonds and, because of rate reductions, reduces incentives to invest in tax-exempt bonds). In a similar manner, statutory and administrative criteria governing existing grant programs often skew state and local government infrastructure activities toward, assumably, federal objectives, rather than local needs (e.g., new construction rather than maintenance of infrastructure). Finally federal parsimony concerning use of unspent balances in earmarked trust funds—such as highways, transit, and aviation—denies state and local governments the options to respond to their respective infrastructure problems.

Future partnerships among all levels of government will necessitate flexible administration of existing support programs by the federal government and a willingness to acknowledge and, when possible, accommodate the negative impact of noninfrastructure policies and programs. To the extent possible, federal dollar assistance should be distributed in the form of block grants—that is, either consolidated categorical or new grants with broad objectives concerning infrastructure development and maintenance. As an alternative, jurisdictions eligible for federal categorical grants should be allowed to transfer funds from one infrastructure grant account to another. They should be permitted to use funds in locally relevant ways as long as objectives remain consistent with broadly stated legislative purposes. Succinctly, recipients should be allowed maximum choice concerning use of available infrastructure grant funds.

Increasingly the federal government should direct its aid programs at state governments to extend or induce state infrastructure involvement in planning and to expand state infrastructure investments. Crucial or key federal policy objectives should be secured through use of incentives over and above block grant or categorical grant threshold levels and by allowing states increased authority to commingle different grant monies. Federal trust funds created by Congress to meet specific infrastructure needs should be allocated by administering agencies for relevant infrastructure improvements, consistent with proper accounting procedures. Although the need to maintain working balances is an appropriate reason for holding back full use of trust funds, the assumed desire on the part of the administration to use trust funds to lessen budget deficits is not.

5. *Inventory, review, and dissemination of case studies concerning collaborative public sector and public/private sector infrastructure roles.* Many interesting and potentially far-reaching replicable infrastructure partnerships and/or collaborative approaches involving two or more governmental entities and governments and the private sector have occurred and continue to occur. Some have shown significant promise concerning the financing, production, and maintenance of infrastructure. They range from the by-now-conventional agreements between cities and developers concerning user fees and exactments to government agreements with private firms to turnkey and sometimes lease/manage infrastructure developments. They include the relatively common use of urban renewal powers to secure land and improvements for new infrastructure developments to the relatively uncommon but growing use of profit-sharing arrangements between public and private sectors to stimulate infrastructure development and operation. They involve different equity and debt arrangements concerning financing, and they involve innovative use of municipal bond insurance and other credit-enhancement measures.

Information concerning what works and what does not work, if made fairly readily available, would make it easier for both public and private sectors to act creatively and efficiently in responding to difficult infrastructure problems. The federal government and/or consortiums of state governments should take the lead in developing and distributing data and case studies concerning innovative infrastructure projects. On-line computer capacity concerning "success stories" would provide a unique benefit to state and local governments, as well as to their potential private-sector partners.

6. *Development of more efficient and equitable means to finance infrastructure, including more efficient and equitable means to link financing to individuals and groups who benefit from infrastructure use.* Users now pay for three quarters of all public investment in infrastructure. Yet "only 50% of spending on operations and maintenance" comes from users (National Council on Public Works Improvements, 1988, p. 11). Increasingly all levels of government must find ways to price adequately and accurately the costs of development, management, maintenance, and repair of water supply, transportation, waste-water treatment facilities, and services and, to the extent possible, to pass on such costs to beneficiaries—public and private. Fees in the form of exactments, tolls, special taxes, and service charges should be examined and, when feasible, used

increasingly to capture direct and indirect benefits from use. Conventional and historical patterns of behavior combined with political facts of life will make the process of moving to coincidence between costs and benefits sometimes difficult. Problems that inure with respect to fairness and equity and/or that occur with respect to limiting household and jurisdictional choices can be accommodated by innovative funding techniques and/or direct subsidies.

Infrastructure: Testing and Amending the Federal System

Needs estimates prepared during the last 10 years have helped make Americans aware of the dimensions of the difficulties we face. However, they have not enervated leadership able to make hard choices among competing demands for public services. Neither have they established the wisdom of significant increases in infrastructure expenditures. Purported crises, described in most of the current or recent infrastructure reports, often are perceived as such only by limited numbers of academics and professionals. To secure attention to infrastructure will require Americans to understand far better than they do today their personal stake in a healthy infrastructure. Understanding alone, however, will not convert readily into policies and commitments. In the light of resource, knowledge, and political constraints, infrastructure policy in the 1990s likely will be incremental rather than far-reaching, and strategic rather than comprehensive. The nation will have an opportunity to evaluate and test alternative infrastructure approaches. It will be a time to combine reflection on where we have come from with carefully defined actions concerning where we want to go. It will disappoint those who look to Washington for major new and expensive programs; it will excite those who sense that America's infrastructure ills lend themselves to resolution by and through unique, still-to-be-defined amendments to the federal system and still-to-be-developed public-private-sector partnerships.

References

Associated General Contractors of America (AGCA). (1983). *America's infrastructure: A plan to rebuild.* Washington, DC: Author.

Choate, P., & Walter, S. (1981). *America in ruins: Beyond the public works pork barrel.* Washington, DC: Council of State Planning Agencies.

Hardin, G. (1968). *The tragedy of the commons* (162 American Association for the Advancement of Science 1243).

Labor-Management Group. (1983). *A consensus in rebuilding America's vital public facilities.* Washington, DC: Author.

National Council on Public Works Improvement. (1988). *Fragile foundations: A report on America's public works.* Washington, DC: Government Printing Office.

National Infrastructure Advisory Committee. (1984). *Hard choices: A report on the increasing gap between America's infrastructure needs and our ability to pay for them. A summary report of the National Infrastructure Study.* Washington, DC: Joint Economic Committee of Congress.

O'Day, D. K., & Neumann, L. A. (1986). Assessing infrastructure need: The state of the art. In R. Hansen (Ed.), *Perspectives on urban infrastructure*. Washington, DC: National Academy of Science.

Peterson, G., et al. (1986). *Infrastructure needs studies: A critique*. Washington, DC: Urban Institute.

U.S. Conference of Mayors and National League of Cities. (1983). *Capital budgeting and infrastructure in American cities: An initial assessment*. Washington, DC: Author.

U.S. Congressional Budget Office (CBO). (1983). *Public works infrastructures: Policy considerations of the 1980s*. Washington, DC: Government Printing Office.

U.S. Congressional Budget Office (CBO). (1988) *New directions for the nation's public works*. Washington, DC: Government Printing Office.

U.S. Office of Management and Budget (OMB). (1985). *Supplement to Special Analysis D*. Washington, DC: Government Printing Office.

16

Needed: A Marshall Plan for Ourselves

Richard P. Nathan

Grid•lock *n.* (1980) **1:** a traffic jam in which a grid of intersecting streets is so completely congested that no vehicular movement is possible.
2: a situation resembling a gridlock [as in congestion or lack of movement].
Source: *Webster's Tenth New Collegiate Dictionary,* 1993.

By the time "gridlock" traffic jams got into *Webster's* in 1980, we had become all too familiar with their meaning for America's cities. Visit almost any big city today, and you can learn about blocked roads, jammed traffic, closed bridges, flooded streets, and blown sewers. Gridlock also has a second definition in *Webster's,* as quoted above, that can be political: "lack of movement." The two kinds of gridlock—physical and political—go together. Until recently, when Congress passed the Intermodal Surface Transportation Efficiency Act of 1991 (quickly dubbed "iced tea"), the national government and, for that matter, many state and local governments have had a congested policy backlog with little movement on measures to keep up and build up the nation's public physical plant.

The United States has been strong in demanding that other countries adopt policies to stimulate economic growth. We need to take some of our own medicine. The national government cannot keep hiding behind the budget deficit; it must adopt policies to stimulate the investment side of the economic equation.

The time has come for a shift in U.S. economic policy to favor investment, including state and local infrastructure investment. A special fund is needed to rebuild the nation's roads, bridges, water and sewer systems, waste treatment systems, and other public facilities in order to spur the economy. What we need now is a Marshall Plan for ourselves.

AUTHOR'S NOTE: I gratefully acknowledge help from Robert M. Solow, Steven D. Gold, Paul Page, and our competent editor, Laurie Norris, in writing this chapter. The ideas here, however, are in the usual way my responsibility.

The 1992 presidential campaign heightened concern about the sluggishness of the national economy. With the beginning of the presidential term, we can expect (or at least hope for) attention to the nation's domestic agenda. One would anticipate the focus to be on the economy. The key is likely to be *investment* to increase economic growth and enhance the nation's competitiveness.

The 1991 Surface Transportation Act authorizes spending of $155 billion over 6 years, gives states added flexibility, and increases the national commitment to urban mass transit. Federal matching grants would be provided on an 80/20 cost sharing basis, including:

$121 billion for highways

$31.5 billion for mass transit

$3 billion for safety and research programs (U.S. House of Representatives, 1991, pp. 4-8)

Recognition is growing that further action is needed to improve our public capital stock beyond the 1991 Surface Transportation Act. We need investments that will enhance the movement of goods and provide transportation and other public services essential to the efficient working of markets and the smooth flow of goods and the factors of production.

Going back over a decade, a series of studies has documented public infrastructure needs and urged action:

- The *Hard Choices* report prepared for the Joint Economic Committee of the U.S. Congress (National Infrastructure Advisory Committee, 1984) proposed a partnership between federal, state, and local governments in the form of a National Infrastructure Fund, with the planning and implementation of aided projects assigned to state and local governments.
- The National Council on Public Works Improvement in *Fragile Foundations* (1988) argued that the quality of America's infrastructure was barely adequate to fulfill current requirements and insufficient to meet the demands of future economic growth and development.
- In *Rebuilding the Foundations* (1990), the Office of Technology Assessment said that it is time for the federal and state governments to create a coherent, supportive, management framework for infrastructure projects that includes adequate financing.

A growing body of public opinion and analysis argues that government spending for infrastructure is deficient and should be increased. Many experts are urging us to take action now:

- Robert Reich in *The Work of Nations* (1988) sees the federal withdrawal from infrastructure spending as precipitous and recommends that savings from the defense budget be applied to public investment in infrastructure, education, and training (pp. 254-260).

- Felix Rohatyn (1991, p. 8) recently called for a vast national public investment program both to meet the long-term needs of the country and to provide a countercyclical boost to the weak economy.
- Economist Robert M. Solow (1991) believes it is essential for the United States to get serious now about private and public investment to make our economy competitive. He sees the Unites States as looking feckless and drifting in the eyes of other industrial countries.

Historically, federal funds have been an indispensable part of the financing for public capital for ports, highways, bridges, transit systems, airports, wastewater treatment, and water supply plants and systems. During the past decade, however, the federal government has withdrawn much of this support for state and local infrastructure. The Congressional Budget Office's latest report on federal capital spending shows a steady decline in infrastructure spending as a percentage of all federal outlays, hitting a low of 2.5% in 1990. It was 4.5% a decade earlier in 1980, and throughout the 1970s this ratio was over 4%. The downward trend of infrastructure spending is depicted in Figure 16.1, which shows that the amount of nondefense public physical capital public investment —on a net basis—has changed dramatically since 1970. Measured in constant dollars, net investment financed by federal grants-in-aid declined by 60% from 1970.

The reduction of federal assistance in the 1980s has been wrenching and painful for state and local governments. Both construction and maintenance expenditures for public infrastructure have been the victim of fiscal stress. During the last decade, cities and states across the United States have been caught among demands for better services, rising costs, the inability to raise and accumulate revenues, and decreased funding from the federal government. Recent data show a decline in total state and local government capital spending during the last 15 years (Figure 16.2). As a percentage of state and local government expenditure, capital spending from all sources declined from 16.6% in 1975 to 10.8% in 1989. Cities and states are unable to raise enough revenue to keep up and fix aging roads, outdated water supply systems, overcrowded airports, and other public facilities.

America in Ruins

Thirteen years ago Pat Choate and Susan Walter published the report *America in Ruins: Beyond the Public Works Pork Barrel* (1981), which dramatized the case for rebuilding our public capital infrastructures. The message of this report is just as alive today: "A large and growing number of communities are now hamstrung in their economic revitalization efforts because their basic public facilities—their streets, roads, water systems, and sewerage treatment plants— are either too limited, obsolete, or worn out to sustain a modernized industrial economy" (p. 15).

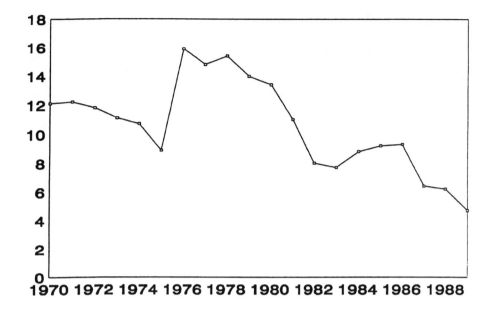

Figure 16.1. Federally Financed Net Investment in Nondefense Public Physical Capital (constant 1982 billion dollars)

SOURCE: Executive Office of the President. Office of Management and Budget. *Special Analyses: Budget of the U.S. Government, 1989.* Washington, DC: Government Printing Office.

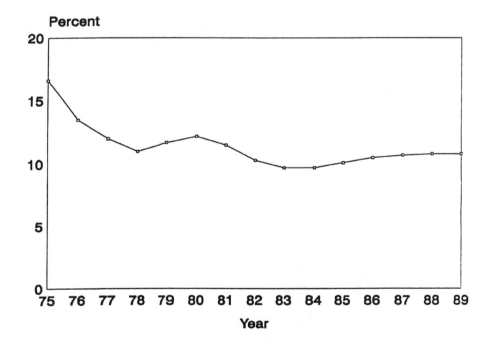

Figure 16.2. Real Capital Expenditure as a Percentage of Total Expenditures (constant 1982 billion dollars)

SOURCE: U.S. Bureau of the Census, Government Finances, selected years.

Choate and Walter reviewed the major past efforts to use infrastructure investment for countercyclical policy purposes. The most recent such effort was the $6 billion Local Public Works (LPW) Impact Program in 1976-1977. Choate and Walter's description of this program showed how good intentions can go awry. Like its predecessors, the LPW was unable to meet its objective of stimulating employment in distressed areas during an economic downturn. Choate and Walter wrote: "The temporary countercyclical Local Public Works program of 1976-1977 did nothing to relieve the 1974-1977 recession until late 1976. Over 80 percent of the direct employment generated by the LPW projects did not occur until the recovery phase of the cycle had begun" (p. 24). Choate and Walter found the LPW to be inadequate primarily because of legislative and executive incapacity, rather than failures with the program concept itself. "Lags occurred because of delays in securing passage of legislation, Presidential approval, appropriation of funds, selection of projects, and construction" (p. 24).

Infrastructure and the Economy

The condition of our public infrastructure is bound up intrinsically in our economic renewal. Evidence suggests that infrastructure expenditures have been a key ingredient to the robust performance of the economy in the 1950s and 1960s (Aschauer, 1990, p. 48). Carefully selected public investment in infrastructure can improve national productivity and output. The building of the interstate highway system was a major contributor to the rise in national productivity during the 1960s (Schultze, 1990).

The argument made here that the United States has underinvested in public capital and that public capital has a positive impact on economic activity does not mean that the United States should suddenly and indiscriminately pump up the amount of money it spends on public capital, nor does it mean that careful cost-benefit analyses are no longer needed for individual projects (Munnell, 1990, pp. 94-95). Wisely planned infrastructure investments, coupled with efficient pricing, can improve federal and state finances and national economic conditions now and over the long run (Winston, 1990, p. 199).

Hard Choices Report

Three years after the Choate-Walter report was published in 1981, the National Infrastructure Study, prepared by the University of Colorado under the direction of the National Infrastructure Advisory Committee, produced the report called Hard Choices. This study was done for a subcommittee of the Joint Economic Committee of the U.S. Congress. The study group was chaired by Henry S. Reuss, former chair of the Joint Economic Committee. The two vice chairs were Peter C. Goldmark, Jr., Executive Director of the Port Authority of New York and New Jersey; and Lee White of Smith Barney.

Hard Choices analyzed national infrastructure repair needs and policy options in 23 states and evaluated projected repair needs and financial resources for highways and bridges, transportation facilities, water supply and distribution, and wastewater collection and treatment. The report highlighted the increasing gap between America's infrastructure needs and the ability to pay for them. The authors concluded that infrastructure deterioration had reached the point where it limits the ability of the United States to achieve a satisfactory rate of economic growth. Other conclusions of the National Infrastructure Advisory Committee were as follows:

- The problem is national in scope, and it is not limited to a region or to any state.
- The problem is manageable if the public sector's capacity to define needs and budget scarce resources is increased.
- The problem requires that state and local governments assume basic management and funding responsibilities.
- The problem requires a predictable and long-term response by the federal government. (p. 113)

The examination of infrastructure needs and revenues in this report indicated:

For the 23 states studied, total infrastructure needs (highways, other transportation, water sewerage) for the 1983 to 2000 period are projected to be about $750 billion in 1982 dollars. Revenue to meet these needs is to be about $460 billion resulting in a revenue shortfall of $290 billion.

For the country as a whole, infrastructure needs for the four categories addressed in this study are estimated to be $1,160 billion in 1982 dollars. Revenue to meet these needs is projected to be $710 billion leaving a financing gap of $450 billion. (p. 8)

The authors found that the single most dominant need across the country is for additional spending on highways and bridges—an infrastructure category traditionally financed in large measure by the federal government. This study also looked at waste treatment and the needs of the water supply.

The authors proposed that a new infrastructure-financing mechanism be established, capitalized with long-term federal debt issued over a period of 10 years. Interest on this debt would be borne by the federal government. The report stated that the National Infrastructure Fund would provide state and local governments with a means to increase available capital at reasonable rates. It would avoid the numerous legal, financial, and institutional problems that 50 state governments and more than 10,000 local governments would encounter in trying to raise the same amount of capital through many individual debt issues.

1984 National Council

In 1984 the National Council on Public Works Improvement was established by Congress to analyze the nation's public works improvements, including their age and condition, changes in their condition from preceding years, their maintenance needs and projected expenditures, and their capacity to sustain economic development and to support an expanding economy.[1]

This congressionally established council concluded that the quality of America's infrastructure—highways, mass transit, aviation, water resources, water supply, waste-water, solid waste, hazardous waste handling, and so on—is barely adequate to fill current requirements and is insufficient to meet the demands of future economic growth and development (National Council on Public Works Improvement, 1988, p. 10). It recommended a national commitment shared by all levels of government to upgrade the infrastructure that could require an annual national capital investment increase of up to 100% in new and existing public works.

The Case of New York State

New York offers a good example of the problems in state and local capital investment. According to the U.S. Census, capital investment by all levels of government in New York State declined in 1982 dollars throughout the 1970s and early 1980s. This was especially true of highway-related expenditures. As a result of this decline, New York State now faces the serious consequences that come from inadequate maintenance and aging facilities. A report by a national group of government organizations described the problems facing New York's public facilities and concluded:

> Over the next five years, 78 projects must be built to bring New York's municipalities into compliance with water quality standards mandated by the Clean Water Act, at a cost of $1.7 billion. To remain in compliance with these standards, in the next five years 547 additional projects must be funded at a cost of $5.6 billion.
>
> Of New York's 17,313 bridges, 11,808 (63 percent) are either structurally deficient or functionally obsolete.
>
> Three percent, or 798 miles, of paved rural highway miles in the state are rated as deficient under the Federal Highway Administration pavement conditions rating system; that is, the pavement is in "poor" condition and is therefore in need of resurfacing, rehabilitation, or reconstruction. (Rebuild America Coalition, 1991, p. 163)

The numbers are high and may be overestimated, but the general assessment is on target.

New York now is trying to take up the challenge and reverse the decline in its capital stock. Governor Mario M. Cuomo has proposed an $800 million infrastructure bond act to prime the economy and lay a foundation for future economic growth. The money from the Jobs Bond Act would be disbursed among local governments to finance improvements of public facilities and to create private-sector jobs.

As noted above, state and local government spending has declined over the past 18 years, and cities and states are unable to keep up. No state can entirely fill the vacuum created by the reductions in federal public works support because right now they are under the gun to meet many burdens, for example, of Medicaid, education, and corrections.

Federal support of state and local capital investment should be more than a jobs program. It should go beyond the 1991 Surface Transportation Act and address a range of needs intrinsic to the growing national concern about America's economic future. According to the Office of Technology Assessment, a strong case can be made for a dedicated federal fund to assist states and local governments in the construction and repair of deteriorating public facilities.

A Capital Investment Block Grant

The history of federal grant-in-aid programs in the United States suggests a number of lessons for thinking about how the federal government should operate in this field. What is needed now is a Capital Investment Block Grant. Such a fund, administered by the U.S. Treasury, would be used to expand and modernize capital facilities. Aid should be provided on a flexible basis to avoid interference with state and local practices. It should be allocated according to a redistributive formula, favoring communities and regions with the greatest capital deficiencies and economic needs. States should be free to use these grant funds for a wide range of capital investments, including roads, bridges, other transportation facilities, waste treatment, water systems, mass transit, schools, and other public facilities. The Capital Investment Block Grant should be a multiyear program.

1. Grants of $25 billion per year should be made annually to the states in predictable annual installments.
2. States should be required to use 60% of the total funding received for areas that are eligible for aid under a similar program—the Community Development Block Grant—that has withstood the test of time and that aids urban development. This requirement could be waived by the secretary of the treasury for sparsely settled and smaller states.
3. Funds should not be used as a substitute for planned or existing state and local public works projects.
4. The secretary of the treasury should be required to issue an annual inventory statement on infrastructure needs, an evaluation of basic infrastructure conditions in the states, and a report of the uses of the grant funds provided under this program.
5. An automatically triggered 50% countercyclical add-on should be provided when there is a one quarter decline in GNP or the classification of a given period as a recession by the National Bureau of Economic Research, whichever comes sooner.

The Capital Investment Block Grant would be an instrument of both long-term and countercyclical economic policy. Choate and Walter, in *America in Ruins* (1981), analyzed the ways capital investments can be used to stabilize the economy. The basic problem limiting the impact of public works investment in the past, they said, has been the tendency to increase expenditures during expansionary phases and decrease them during contractionary periods.

Under the Capital Investment Block Grant Fund proposed here, states and local governments would be required to maintain a list of projects that would be ready to go as the economy begins to slide into recession. A number of benefits accrue from the fund having a permanent countercyclical add-on:

- It would reduce the adverse consequences caused by procyclical investment patterns of these investments.
- A permanent, more systematic countercyclical public works policy would reduce the "crisis" management atmosphere that surrounds temporary public works programs.
- It would reduce the lag between the beginning of a recession and the time when benefits flow.
- It would facilitate better targeted investments in order to address regional and sectoral variations in the economic cycle.
- It would improve fiscal stability to state and local finance. (Choate & Walter, 1981, pp. 25-26).

At the state level, I favor a requirement under this block grant for the establishment by the states of panels such as the U.S. military base-closing commissions. The panels would be charged to recommend a program of priority projects for development that the governor and the legislature would have to consider en bloc. They could send it back, but not modify the list.

A critical operational question for such a program is, What is its impact? One needs to think through the substitution issue carefully. Our earlier evaluation research on revenue sharing and block grants suggests that, for capital projects, this is a harder issue to deal with than for operating programs. For the latter, recipient jurisdictions of revenue sharing and block grant funds tend to avoid substitution, not wanting to lock in new operational commitments that would be hard to cut off if the federal government—often unpredictable and fickle—cuts their money or changes the rules.[2]

Substitution is not always bad. The issue is complicated. One good way to handle it would be to empower the secretary of the treasury to establish oversight committees in each state to report on the use of Capital Investment Block Grant funds and on the substitution issue, assuming the law requires "maintenance of effort," which I believe it should. In any event, Congress is likely to want to do this. These state committees, in turn, would be charged to report publicly on an annual basis, would have access to state and local records, and would use a common analytical framework and structure for their reports. The secretary of the treasury, on a longer term, follow-up basis, would be required to compile these reports and to submit an overall report to the Congress on the use of these block grant funds and compliance with maintenance of effort.

In the long run, my Rockefeller Institute colleague Steven D. Gold and I favor special borrowing through a federal capital budget for these types of investment purposes. In the near term, an increase in the gasoline tax is also a possible and logical funding source for the Capital Investment Block Grant. Another funding option for such a program is to adopt the approach of the "National Infrastructure Fund" proposed in the 1984 *Hard Choices* report (cited above), though on balance I think the block grant approach suggested above is a better way to proceed.

The start of a new presidential term in January 1993 served as the impetus for adopting new economic and domestic policies, including a public capital program—on a substantial scale—what I called earlier, "a Marshall Plan for ourselves." The program described here could generate upwards of half a trillion dollars in investments in public facilities (planned, committed, or under way) between now and the millennial year, 2000.

The economy of the United States has been a beacon for industrial development throughout the world. Its hallmark is a high standard of personal consumption for all of its citizens. But we got lost in the 1980s. We overdid it. We ate our seed corn. Now we need to adopt such a Marshall Plan for ourselves to stimulate the investment side of the American economic equation. This would be sound economics, good for growth, and good for the American future.

Notes

1. The chairman of this National Council was Joseph M. Giglio, Managing Director, Bear, Stearns and Company.

2. Earlier research is discussed in Nathan, Manvel, Calkins, and Associates (1975); Nathan, Adams, and Associates (1977); Nathan, Dommel, Liebschutz, Morris, and Associates (1977); and Nathan, Cook, Rawlins, and Associates (1981).

References

Aschauer, D. A. (1990, June). Why is infrastructure important? In A. H. Munnell (Ed.), *Is there a shortfall in public infrastructure?* (Conference proceedings). Harwich Port, MA: Federal Reserve Bank of Boston.

Choate, P., & Walter, S. (1981). *American in ruins: Beyond the public works pork barrel.* Washington, DC: Council of State Planning Agencies.

Munnell, A. H. (1990, June). Is there a shortfall in public capital investment? In A. H. Munnell (Ed.), *Is there a shortfall in public infrastructure?* (Conference proceedings). Harwich Port, MA: Federal Reserve Bank of Boston.

Nathan, R. P., Adams, C. F., Jr., & Associates. (1977). *Revenue sharing: The second round.* Washington, DC: Brookings Institution.

Nathan, R. P., Cook, R. F., Rawlins, V. L., & Associates. (1981). *Public service employment: A field evaluation.* Washington, DC: Brookings Institution.

Nathan, R. P., Dommel, P. R., Liebschutz, S. F., Morris, M. D., & Associates. (1977, January). *Block grants for community development.* Washington, DC: U.S. Department of Housing and Urban Development.

Nathan, R. P., Manvel, A. D., Calkins, S. E., & Associates. (1975). *Monitoring revenue sharing.* Washington, DC: Brookings Institution.

National Council on Public Works Improvement. (1988). *Fragile foundations: A report on America's public works.* Washington DC: Government Printing Office.

National Infrastructure Advisory Committee. (1984). *Hard choices: A report on the increasing gap between America's infrastructure needs and our ability to pay for them. A summary report of the National Infrastructure Study.* Washington, DC: Joint Economic Committee of Congress.

Office of Technology Assessment. (1990). *Rebuilding the foundations: Public works technologies, management, and financing.* Washington, DC: Government Printing Office.

Rebuild America Coalition. (1991). America's infrastructure: Preserving our quality of life. In U.S. Congress, Committee on Public Works and Transportation, *To examine the future of our nation's infrastructure needs.* Washington, DC: Government Printing Office.

Reich, R. (1992). *The work of nations.* New York: Vintage.

Rohatyn, F. (1991, November 21). The new domestic order? *New York Review of Books,* p. 8.

Schultze, C. L. (1990). The federal budget and the nation's economic health. In H. J. Aaron (Ed.), *Setting national priorities: Policy for the nineties.* Washington, DC: Brookings Institution.

Solow, R. M. (1991, May 26). Address on the occasion of his receiving an honorary degree from the State University of New York at Albany.

U.S. House of Representatives. (1991). *Intermodal Surface Transportation Efficiency Act of 1991.* Washington, DC: Government Printing Office.

Winston, C. M. (1990, June). How efficient is current infrastructure spending and pricing? In A. H. Munnell (Ed.), *Is there a shortfall in public infrastructure?* (Conference proceedings). Harwich Port, MA: Federal Reserve Bank of Boston.

PART SIX

Community Development

Introduction to Community Development

Margaret Wilder

The Concept of Community Development

At first glance, the meaning of *community development* seems self-evident: to develop a community. Once we begin to explore the broad literature on this topic, however, the assumed simplicity of the term disappears. Various discussions of community development imply that it is alternately thought of as a process and an outcome. No simple definition can encompass all of the perspectives found in the community development literature. For our purposes, however, a working definition can be constructed by examining the very definition of the terms *community* and *development*. According to *Webster's New World Dictionary* (1986):

> **community 1:** *a*) all the people living in a particular district, city, etc. *b*) the district, city, etc. where they live **2:** a group of people living together as a smaller social unit within a larger one, and having interests, work, etc. in common. (p. 288)

> **development 1:** a developing or being developed. (p. 386)

> **develop 1:** to build up or expand **2:** to make stronger or more effective. (p. 386)

As these notions suggest, a *community* can be thought of as a specific group of people who share some common link or as a specific place where people live. This place is considered a subpart of a larger social unit. *Development* refers to an act of changing some entity so as to strengthen or extend its capacity. Thus *community development* can be thought of as strengthening or extending certain attributes of a specific group of people or of the place in which they reside.

In the U.S. context, communities typically are identified as individual cities or neighborhoods. These communities are differentiated by attributes such as their location, population characteristics, economic activities, physical condition, and political structures. Although, in theory, *community development* could refer to any process whereby the attributes of a city or neighborhood are improved, the term has come to have a more limited set of connotations with respect to U.S. communities.

Community development in the U.S. context has had a long history. That history reflects both community-based grassroots efforts, as well as more formal government-initiated programs. Early efforts to assist rural and small town areas led to the establishment of Cooperative Extension Programs in land-grant state colleges during the 1920s (Phifer, List, & Faulkner, 1980). In the post-World War II era, the concept of community development broadened to include urban efforts at community organizing by activists such as Saul Alinsky of Chicago, as well as more formal urban development programs funded by the federal government (e.g., public housing, urban renewal). A major thrust to community development was provided in the late 1960s under the massive War on Poverty programs of the Johnson administration. The multibillion dollar programs were coordinated by the Office of Economic Opportunity (OEO) and operational-ized through the establishment of a national network of "community action agencies" (Phifer et al., 1980). These agencies were formed at the neighborhood level to provide opportunities for direct involvement by residents in the improve-ment of their communities. In subsequent years, funding for these initiatives dissipated, leaving a skeletal web of community agencies, most struggling to survive more modest funding from the Community Services Administration (OEO's successor agency).

Beginning in the 1970s, community development became increasingly asso-ciated with the activities of neighborhood-based organizations. The earlier govern-ment initiatives were retained through local public housing and renewal agencies. A newer set of activities evolved, however, that sought to address changes in the economic character of U.S. communities. These new initiatives became identified as "economic development." The distinction between community and economic development is continually blurred. Some analysts associate commu-nity development with the provision of social services (e.g., welfare, housing, day care) and housing. Economic development is associated with activities that focus on business development, commercial/industrial development, and infra-structure. This dichotomy frequently is ignored by writers who choose to view all local improvement efforts as community economic development. Although the terminology of community development is often ambiguous, the objectives and issues associated with this phenomenon are much more distinct.

The Objectives of Community Development

Communities in the United States, particularly those in urbanized areas, have experienced a unique set of transitions during the post-World War II era. The three most important changes are (a) an increase in urban poverty, (b) the suburbanization of the middle class, particularly among white residents, and (c) the transition from an industrially based economy to a service-based economy. These changes have brought about an increasing concern with *community revi-talization* as communities seek to adapt to changing sociodemographic, physi-cal, and economic realities. Not surprisingly, most of the emphasis in community development is on the following:

- the physical development of land (e.g., residential, commercial, industrial development)
- the rehabilitation of physical structures (particularly residential structures)
- the provision of forms of social assistance (e.g., job training, child care, youth programs)
- the mobilization of interests to affect political outcomes
- the use of resources to create employment opportunities

The basic goal of community development would seem to be one that all citizens could agree on. However, much conflict and controversy surrounds community development. Differences in perspective arise for several reasons. First, community development usually requires the use of limited resources (e.g., land, money, natural resources). Thus decisions regarding the use of these resources generate competing interests. Second, most development affects both the immediate community and neighboring areas. Opposition may develop due to assumed or anticipated side effects. Third, social differences (particularly those based on race, ethnicity, and income class) within and between communities tends to create intolerance and a desire for separation. Fourth, differences in the value orientations of communities lead to distinct priorities. Some communities place great importance on economic growth; others are more concerned with aesthetics and environmental quality. Still others seek to meet basic needs in housing and employment training.

The conflicts that arise over community development are played out as communities interact with various agents of change within communities. These agents are the *public sector* (federal, state, and local government), the *private sector* (privately owned businesses), and the *nonprofit sector* (community-based organizations, neighborhood associations, etc.). In recent years the locus of activity in community development has shifted toward localized efforts, particularly those of the nonprofit sector.

Current Issues in Community Development

The current issues in community development revolve around the three basic themes of *what, where,* and *who.* The first theme raises the issue of what forms of community development should be given priority. The old War on Poverty and urban development programs have been largely discredited. Moreover, funding of new initiatives is severely restricted in the current era of limited resources. Thus there is considerable concern that community development efforts be both effective and cost-efficient. Significant debates rage over the most effective strategies. Some analysts argue that "human capital" programs (e.g., education, job training), which invest in developing people, are more effective than "private capital" programs, which assist private businesses.

A second set of issues revolves around the theme of where community development should occur. Given multiple-need contexts and limited resources, there is constant competition and conflict over community development. Within cities, neighborhoods frequently become pitted against one another in an effort to garner resources or unique opportunities (e.g., new parks, schools). Alternatively neighborhoods may resist certain development activities that are disruptive (e.g., stadium construction projects, landfills).

Finally issues arise surrounding the theme of who should plan and carry out community development. The three most obvious candidates are the public sector, the private sector, and nonprofits. Each of these entities has a different set of priorities and strategies for community development. The public sector has multiple interests in community development. Local government as well as state and federal agencies often are motivated by broad social and economic goals. Most recently, economic restructuring in cities has created a strong interest on the part of local governments in revitalizing communities. This revitalization typically is conceptualized as new business development, construction of commercial or industrial sites, and development of middle-class housing. The perspective of the private sector often is dominated by concerns of profitability and competitiveness. Community development is seen as an opportunity to promote these concerns. Nonprofits hold a somewhat different perspective in which concerns over the development of human capital and neighborhood resources is paramount. Each potential agent of community development is subject to different levels of political control and pressure. In general, residents of communities have the greatest measure of potential influence over nonprofit organizations. Some level of influence may be exercised over local government through elected representatives or organized protest, but the private sector primarily operates as a free agent.

In summary, the issues that surround community development are complex and varied. Moreover, the goals, strategies, and agents of community development are subjects of debate and conflict.

The Readings in Part 6

The readings in Part 6 address two major themes surrounding community development: (a) strategies and programmatic approaches; and (b) the role of private, public, and community organizations. A brief overview of each reading reveals both unique and related aspects of the authors' perspectives.

Charles Bartsch, Chapter 17

Bartsch introduces Chapter 17 "Government and Neighborhoods" by discussing briefly the changing economic realities of American communities. As he suggests, the basic shift from an industrial to a postindustrial economy has meant lost jobs and businesses, which has undermined the basic stability of American communities. Bartsch examines the nature of efforts by communities to stimulate local development or redevelopment in the face of this economic transformation. The primary purpose of these efforts is to generate new employment and investment. To reach this basic goal, Bartsch argues, communities must create and/or maintain a climate that is supportive of new development.

Bartsch provides an overview of major federal and state strategies designed to stimulate development activity. These strategies include both *direct* subsidies for new development through various grant and loan programs, as well as *indirect* subsidies provided by tax incentives and the streamlining of governmental regulations. Although the federal government provides an important set of programs, the funding associated with these initiatives has declined in recent years. As federal support has diminished, state governments have taken on a decidedly larger role.

Bartsch provides a more detailed discussion of the various approaches used by state governments. These approaches include both *business development* initiatives, such as incubator programs, and *financial assistance* programs, which provide capital for new business start-ups, as well as expansion of existing enterprises. The federal and state programs are used by local governments, businesses, and neighborhood-based organizations to stimulate new economic activity at the community level.

Although most economic development programs are accessible to any community or business within a state, some programs are *targeted* at specific groups of businesses, people, or communities (e.g., minority hiring/contract programs, enterprise zones). Overall, government-sponsored economic development programs have provided a basis for improving the economic climate of communities and for encouraging partnerships between the public and private sectors. Although much development has been generated by these initiatives, their ultimate success in revitalizing communities is still being assessed.

Louise Jezierski, Chapter 18

The partnerships between the public and private sectors that Bartsch regards as positive vehicles for stimulating development in communities are viewed in Chapter 18 "Neighborhoods and Public-Private Partnerships in Pittsburgh" from a decidedly different perspective by Louise Jezierski. Through her examination of community development politics in Pittsburgh, Jezierski argues that the assumed neutrality and representativeness of such partnerships are misleading. Although these partnerships are highly efficient in organizing development interests, expertise, and resources, they tend to exclude the interests of such groups as labor unions, small businesses, and neighborhood groups.

Public-private partnerships are by nature quasi-governmental; that is, they are given certain powers similar to those exercised by formal government (e.g., issuing tax-exempt bonds, exercising eminent domain), but they are not typically accountable to the local electorate. As such, public-private partnerships represent a type of "governing structure" that is more difficult for neighborhood interests to access and influence. This inaccessibility is problematic for neighborhood groups that often represent people and areas directly affected by the decisions of the partnerships.

In Pittsburgh the alliance between the public and private sectors has sought to facilitate the transformation of this former "steel city" into a postindustrial center for high-tech and corporate service industries. Throughout this process, neighborhood groups have fought to have their concerns incorporated into the local development agenda. The relative effectiveness of neighborhood groups has varied over time. From 1945 to 1969, local government worked in conjunction with the Allegheny Conference on Community Development (a group of local corporate executive officers) to clean up the air, redevelop the deteriorating downtown, and construct major freeway routes.

The general public supported these initiatives. Certain neighborhoods that were negatively affected by renewal projects, however, began to protest the development actions. These protests laid the groundwork for mobilization of neighborhood interests that resulted in the election in 1969 of a mayor who supported grassroots community interests over those of the traditional development-oriented partnership. Private business interests became alienated from local government, and redevelopment stalled. Neighborhood organizations flourished and consolidated power under an umbrella organization called the Pittsburgh Neighborhood Alliance (PNA).

The impasse between neighborhoods and development interests was resolved when a new compromise-oriented mayor—Richard Caliguiri—was elected in 1977. From that point until 1988, local government followed a dual strategy of partnership with the private sector and support for neighborhood-based planning and community development. This political transformation occurred at the same time that local corporations were undergoing extensive restructuring (e.g., mergers, head office relocations, plant closings). New industries in high technology and services have emerged that also have broadened and diversified the traditional partnership. In recent years neighborhood organizations in Pittsburgh have increased their technical expertise and become more bureaucratic in structure in order to forge direct working relations with the private sector, as well as to act as development entities in their own right.

The Pittsburgh experience shows that neighborhood groups can influence local public-private partnership structures and activities through strategies as diverse as opposition in the form of protest, confrontation, and political/electoral mobilization to cooperation in joint ventures and independent development. Jezierski believes that neighborhood coalitions must acquire the same level of political power and access to resources as business associations if they are to influence directly the form and extent of development within their communities.

John R. Logan and Gordana Rabrenovic, Chapter 19

The capacity of neighborhood organizations to rise to the challenge identified by Jezierski's analysis is examined in Chapter 19 "Neighborhood Associations," on neighborhood associations in the Albany, New York, area. The authors are concerned with understanding the role of neighborhood associations in the context of local politics. Specifically they examine the degree to which these groups challenge the traditional *growth machine* (the coalition of private and public interests that tend to dominate development decision making in urban areas).

Logan and Rabrenovic provide an in-depth examination of (a) the evolution of neighborhood associations in the Albany area, (b) the issues around which they were formed and continue to focus, and (c) their perceived opponents and allies. Furthermore they examine differences in the character of neighborhood associations found in the central city versus their suburban counterparts.

In the Albany area, the authors found that most city-based associations were established between 1975 and 1979, while suburban associations developed primarily during the 1980-1985 period. Neighborhood associations in the central city evolved as part of grassroots resistance to massive state office development and the increased availability of federal community development funds in the 1970s. Suburban organizations seemed to evolve as their communities became more completely developed.

Overall, city and suburban associations shared concerns over traffic, parking, land use, and development. City neighborhood associations, however, were more concerned over issues of cleanliness, streets/sidewalks, recreation areas, and public safety. The organizations had mixed perceptions about their opposition and sources of support. Local government was seen as a supportive agency on safety and quality of life issues but as a common opponent over land development. Business and land developers, however, were perceived overwhelmingly as opposing interests. Regardless of location, the basic aim of neighborhood associations appeared to be the protection of the "residential environment." But because city residents are less likely to own the land in their neighborhoods, their struggles are more extensive.

In the city of Albany, neighborhood associations have formed a key component of political power that has eroded the city's long-time political machine. As Logan and Rabrenovic suggest, however, the old-style political machine has been replaced by the growth machine. Thus neighborhood associations must play a direct political role in promoting the interests of residents.

Concluding Comments

As a group, these readings reveal the range of issues confronting communities as they engage in the processes of social reorganization and physical/economic revitalization. Conflicts generated in the context of community development between public and private sectors, as well as between different "communities" within urban areas, must be managed and resolved. Community organizers, urban planners, and policymakers play crucial roles in the resolution of these conflicts. The ability of these professionals to make equitable and effective policy decisions will either strengthen or undermine efforts at community renewal. As these readings demonstrate, community development is a crucial process, particularly in the context of urban areas.

References

Phifer, B. M., List, E. F., & Faulkner, B. (1980). History of community development in America. In J. Christenson & J. W. Robinson, Jr. (Eds.), *Community development in America*. Ames: Iowa State University Press.

Webster's New World Dictionary of the American Language (2nd coll. ed.). (1986). New York: Prentice Hall.

Suggestions for Further Reading

Bruyn, S., & Meehan, J. (1987). *Beyond the market and the state: New directions in community development.* Philadelphia: Temple University Press.

Community Development Journal: An International Forum. (1993, October).

Darden, J. R., Hill, J. T., & Thomas, R. (1987). *Race and uneven development.* Philadelphia: Temple University Press.

Davis, J. E. (1991). *Contested ground.* Ithaca, NY: Cornell University Press.

Fasenfest, D. (1993). *Community economic development.* New York: St. Martin's.

Logan, J. R., & Molotch, H. L. (1987). *Urban fortunes.* Berkeley: University of California Press.

Melvin, P. M. (1986). *American community organizations.* Westport, CT: Greenwood.

Mier, R. et al. (1993). *Social justice and local development policy.* Newbury Park, CA: Sage.

Nyden, P. W., & Wiewel, W. (1991). *Challenging uneven development.* New Brunswick, NJ: Rutgers University Press.

Pierce, N., & Steinbach, C. F. (1990). *Enterprising communities.* Washington, DC: Council for Community-Based Development.

Squires, G. (Ed.). (1989). *Unequal partnerships.* New Brunswick, NJ: Rutgers University Press.

Twelvetrees, A. (1989). *Organizing for neighborhood development.* Brookfield, VT: Avebury/Gower.

Vidal, A. (1991). *The community economic development assessment: A national study of urban community development corporations.* New York: Community Development Research Center, The New School for Social Research.

Williams, M. R. (1985). *Neighborhood organizations.* Westport, CT: Greenwood.

17

Government and Neighborhoods

PROGRAMS PROMOTING

COMMUNITY DEVELOPMENT

Charles Bartsch

The term *economic development* has been defined in many ways over the years. Basically it is the process by which individuals and organizations decide to, and then do, invest capital in an area—central business district, neighborhood, or exurban site. The results are new, expanded, or retained industrial, commercial, or service enterprises, and new or retained jobs. Economic development activities are carried out in areas of all types and sizes, ranging from corner stores in inner-city neighborhoods to full-scale high-technology industrial centers at the outskirts of suburbia. Across the country, governments at all levels, as well as local development corporations, community-based organizations, and neighborhood groups, are becoming more involved in meeting the challenges posed by the economic development process.

A transformed economy is the most critical issue facing communities of all sizes in many states, keenly affecting community life in neighborhoods hit by the tangible impacts of economic change—closed plants, lost jobs and reduced wages, and plummeting demand for local goods and services. Adapting to a changing economic base requires a coordinated effort on the part of government, business, neighborhood and nonprofit organizations, universities, and labor. New companies and industries have to be generated, capital sources uncovered, and new jobs created. Viable older firms often need to be revitalized as well, so that they can adapt and grow.

Accordingly, promoting job creation and economic development activities has become a top priority of mayors and community leaders. Their efforts often require a shift from an outdated economic base to a new, more productive one with the financial, physical, and labor-force capability to sustain growth. But in many ways, spurring the economic development process at the neighborhood level is more an art than a science and is neither exact nor fully predictable.

This chapter is intended as a snapshot of current public-sector efforts to spur neighborhood development activities. I examine a number of federal and state policies and programs that have substantial impact on neighborhood-level economic development activities and describe emerging development tools being used in neighborhoods to stimulate new economic activity. Because municipal programs are so widely varied and often tailored to unique needs in specific cities, an overview is not offered here.

The Neighborhood Economic Development Climate

Neighborhood economic development, in the context of this chapter, consists of generating new investment activity (from owners/entrepreneurs, as well as from third-party investors) and jobs. Human and social services such as housing, day care, and health care play an important role in a community's well-being and its ability to maintain and attract economic activity, but they are not the focus of this discussion. For economic development to occur—in a neighborhood, as well as in a city, state, or region—the climate must be conducive to investment. The components of this climate have been variously described and much debated by practitioners and scholars; they do include several critical factors:

- availability of suitable land or space at an affordable price
- accessible, affordable capital
- adequate infrastructure and other support services, including adequate transport for products
- research and development capacity that can support innovation and entrepreneurship
- a workforce that is appropriately educated, skilled, and suitably trained
- volume and flexibility of regulations directed at business and investment activities
- a tax system that enhances, rather than inhibits, development efforts
- a public-sector capacity to support development

The match between these physical, financial, technical/research, and labor components and the needs of investors and entrepreneurs makes a neighborhood area attractive for development. In practice, they vary by location, project, and opportunity; different players emphasize different factors of the economic development climate, but all elements are represented in some way. The significance of any one factor relative to the others depends on the economic development game plan pursued by a given community and its economic situation and advantages. Neighborhoods unable to attract investment or facing disinvestment lack (or are perceived to lack) at least one of these elements. Development officials must identify these needs or gaps and work to fill them. Moreover, they now must assume greater responsibilities in the development process because of the federal government's de-emphasized role.

During the last decade, public-sector leaders have devised many tools and techniques to stimulate economic development activities. They have become much more sophisticated and entrepreneurial in their approach to attracting development, no longer simply abating taxes or spending more money. Now programs and incentives offered at all levels of government may be very different in terms of targeted clientele, size, and anticipated outcomes. These efforts fall into two broad categories: financial help and nonfinancial assistance.

Financial assistance programs are shaped to meet several objectives, such as reducing the lender's risk, reducing the borrower's cost of financing or easing repayment requirements, improving business cash flow, or providing equity capital. They include grants, loans (primary or subordinated), loan guarantees, interest rate subsidies, bond financing (taxable as well as tax-exempt), equity financing (usually through venture or seed capital initiatives), royalty agreements, and tax credits, deductions, or abatements.

Nonfinancial assistance reduces the cost of doing business and can have as much impact on development as do financial incentives. Most nonfinancial aid is aimed at improving business or development opportunities by providing skills or information to entrepreneurs or investors. Common types of help include management assistance, such as seminars, counseling, or tailored consulting services; referral or liaison services (often with potential clients or vendors); loan packaging; market identification services; site selection; regulatory relief; training; and research and development.

Both financial and nonfinancial programs are shaped by different levels of government, and their implementation is influenced or directed by neighborhood leaders. They are the building blocks of community development programs.

Federal Programs and Policies
Affecting Neighborhood Development

In the 1980s, federal community development policies mirrored the free-market approach that crosscut most of the Reagan domestic agenda. The administration argued that it is the responsibility of states, local governments, and the private sector to provide resources for community development and revitalization. The rhetoric in the president's congressionally mandated biennial urban policy reports had mellowed since its first draft statement in 1982; that report chided the nation's mayors as "wily stalkers" of federal largesse. The underlying policy tenets of subsequent reports, however, remained the same.

First, the administration continued to promote expanded state and local flexibility in the use of federal resources and curbed federal involvement, while it shifted more funding responsibilities to those units of government—the so-called devolution policy. Second, there was increased reliance on public-private partnerships to both reduce federal spending and tap the expertise of the business community in carrying out public/social objectives. Prime examples are community development partnerships initiated with CDBG funds and the

Private Industry Councils authorized under the Job Training Partnership Act (JTPA) that design and implement training programs.

Third, the administration continued to push for consumer choice in publicly supported services to make greater use of market forces in the provision of public services and assistance. These choices included tuition tax credit and compensatory education vouchers, as well as housing assistance vouchers.

As part of its devolution philosophy, the Reagan administration continuously called for steep reductions in spending for community development programs. The administration maintained that it is the duty of the states, local governments, community organizations, and the private sector to provide the resources and expertise needed for neighborhood development and revitalization activities. Administration officials contended that community development programs such as UDAG- and SBA-direct loans provide unwarranted subsidies to some firms at the expense of others and lead to no net national increase in jobs but merely relocate them from one community to another.

Community development program advocates countered these outlooks by noting that federal assistance is often the first and most critical help that distressed neighborhoods receive and is pivotal in community economic revitalization. They contended that these programs provide tangible benefits to neighborhoods and their residents by creating jobs, stimulating private business investment, and reversing blighted conditions in declining neighborhoods.

Major federal community development programs commonly used in neighborhood economic development efforts include the following:

- Community Development Block Grants (CDBG), funded at $3 billion in fiscal 1987. Recipients must use at least 60% of their funds to benefit low- and moderate-income persons. In 1987, communities earmarked $335 million for economic development activities and $940 million for public works projects, amounts consistent with prior years.
- Urban Development Action Grants (UDAG), funded at $225 million in fiscal 1987. About 17.6% of project awards went to HUD-defined "neighborhood" applications. In the first 10 years of UDAG life, $920 million in program funds helped finance 731 neighborhood projects.
- Economic Development Administration (EDA) programs traditionally devote most of their resources to smaller cities or rural areas; $182 million in fiscal 1987 funds helped construct public works facilities, capitalize revolving funds, and combat the effects of plant closings.
- Small Business Administration (SBA) programs had a loan, guarantee, and technical assistance service commitment level of $2.8 billion in fiscal 1987. Many SBA programs are used by neighborhood businesses and bankers to promote start-ups and expansions, SBA-chartered local development corporations, investment companies receive SBA financial backing and are active in many neighborhoods, and universities and community-based organizations help provide technical and informational assistance.

These programs have borne a disproportionate share of budget reductions since 1981 (see Table 17.1). From fiscal 1981 to fiscal 1987, the level of appropriated funds for community and neighborhood development declined by 19.4%,

Table 17.1 U.S. Budget Outlays in Selected Categories in Current Dollars,
Fiscal Years 1980, 1981, and 1985 (fiscal year 1980 = 100)

	1980	1981	1985
Total U.S. budget outlays (excluding off-budget outlays)	100.0	114.0	162.4
Selected grants to state and local governments (by function):			
Transportation	100.0	103.1	131.3
Community/regional development	100.0	93.8	78.5
Education, training, employment	100.0	98.2	87.6
General purpose fiscal assistance	100.0	78.8	· 71.8

SOURCE: Congressional Research Service, Library of Congress.

from $4.68 billion to $3.77 billion. This drop is even more striking when fiscal 1987 dollars are adjusted for inflation and converted into constant fiscal 1981 dollars. According to Congressional Research Service staff calculations, the actual program buying power of these development programs plunged 36.5% since fiscal 1981, to the equivalent of $2.976 billion (Boyd, 1987, p. 3).

State Economic Development Programs
Affecting Neighborhoods

State programs are as diverse as local economic conditions and development needs. Many types of state programs are not targeted solely to neighborhood economic development but directly affect the climate in which neighborhood-based projects and programs are devised and implemented. Those cited next are based on broad patterns of state initiatives that became apparent during a mail and telephone survey of economic development officials in all 50 states undertaken by Northeast-Midwest Institute staff in late 1987 and early 1988 (see Bartsch, 1988, for complete information on state programs surveyed; also Table 17.2 provides a summary of the state program survey findings).

Business Development

Implementing business development initiatives with their job creation benefits is a top priority in many cities and communities—the driving force defining the role of the public sector in the economic development process. Many efforts were inspired by research findings indicating that a substantial portion of the nation's new job creation and technological innovation takes place in the small business sector; according to some experts, as many as one

Table 17.2 State Economic Development Programs: Summary of Selected Initiatives
(number of states available)

Business Financial and Technical Assistance Programs
  States Guarantees for Building Loans—22 states
  One-Stop Licensing Centers—16 states
  State Training Finance Mechanisms—23 states
  Taxable Bond Programs—8 states
  Venture Capital—24 states
  Private Development Credit Corporation—30 states
  State R&D Promotion Program—44 states
  State-Owned Industrial Parks—11 states
Programs Targeted to Distressed Areas
  Neighborhood Improvement Programs—25 states
  Customized Job Training—37 states
  Industrial/Commercial Site Development—23 states
  Capital Improvements—25 states
  State Enterprise Zones—34 states
  "Mini-UDAG" Programs—14 states
  Incentives for Locating Plants in Distressed Areas—30 states

---

SOURCE: Northeast-Midwest Institute 1987-1988 survey of state development programs.

job in five is attributable to an entrepreneur starting his or her own operation
(Armington & Odle, 1982; Birch, 1978; Vaughn, 1988).

Every state has adopted a series of business development programs that
community-based organizations and local governments often implement in neighborhood commercial and industrial districts. Crafted to take advantage of a
locality's specific economic assets, these programs play an integral part in
community economic revitalization and growth strategies. Although they vary
widely in scope and level of effort, Northeast-Midwest Institute survey information from the 50 states indicated that programs having a significant impact
on neighborhood-based economic development efforts focus on one or more
of the following four broad categories or characteristics:

- an individual industry or sector, as diverse as textiles or high technology, that may
  be used to promote specific neighborhood development activities with clearly
  defined job and investment outcomes
- the entrepreneur or small business owner/operator and his or her most critical
  development needs, such as access to capital, management assistance, and marketing
  advice, that can be used in conjunction with community efforts to encourage
  new business formations in the neighborhood area
- existing businesses, in terms of both expansion and retention strategies—often
  the key to maintaining the economic base necessary to attract additional activity
  to the neighborhood
- filling development climate gaps—providing space or sites, support services, finance, and other help necessary to launch an enterprise or get it through start-up
  or transition phases; business incubators and Small Business Development Centers are popular entities of this type.

Business development programs are innovative and diverse. Those with the greatest impact on city and neighborhood efforts are those that support business development by increasing the new business birthrate; by nurturing budding small firms that help essentially viable companies enhance their probability of survival during their early, volatile phase; by easing the expansion process for firms already in place; and by aggressively addressing business retention issues to prevent closings and relocations. They include management and technical assistance efforts, such as marketing and financial planning seminars, counseling, training, and advocacy. According to institute survey data, 29 states support some type of incubator facility to encourage new business formations and to combat their high failure rate. The number of incubators sponsored by states, cities, and nonprofit organizations surpassed 200 in mid-1987. Many are housed in underused or abandoned manufacturing complexes in inner-city neighborhoods.

On a different tack, states increasingly are trying to reduce bureaucracy to improve business climate. Many are streamlining license and permit procedures and reducing regulatory paperwork requirements that have proven particularly onerous to new and small enterprises. In all, 16 states have established "one-stop" centers, whose services include referrals to and liaisons with other agencies for financial packaging and other business resources. Nearly a dozen have designated a business ombudsman or added toll-free hot lines for business information—and granted them enough authority and access across bureaucratic levels to effectively perform their jobs. Many of these programs feature sophisticated computer operations to aid efficiency. These types of state initiatives can be used skillfully by neighborhood leaders or economic development organizations in a liaison or advocacy role with local enterprises. Such efforts can reduce business operation and compliance costs, an important factor for new or struggling neighborhood-based firms with little cash to pay professional help.

Program Examples

Noteworthy state business development programs include New Jersey's Business Retention and Expansion Program, which offers an avenue for companies to relay their concerns to municipal government: Problems can be identified and addressed early and potential business relocations or shutdowns avoided. Program staff recruit knowledgeable volunteers from the community to interview the chief executive officers of local firms. The information gained from this process can have important impacts on neighborhood economies; the program helped save 100 jobs in an older, distressed area in Elizabeth. The New Jersey program has served as a model for city-sponsored programs around the nation.

New York's Center for Employee Ownership and Participation, located in New York City, promotes worker ownership by coordinating state educational and outreach efforts and financial and technical assistance; such a program could prove essential to keeping small, viable firms operating when their owners retire or die. As of mid-1987, NYCEOP had helped 25 firms. Special projects currently underway include the organization of local ownership development corporations in Albany and Buffalo. This initiative grew out of the U.S. Catholic Bishops' Pastoral Letter on Economic Justice, which endorsed employee ownership as a tool for job creation in distressed communities. The center is acting as an organizer-coordinator on the local level, helping to secure funds (about $90,000 in seed money to date) and providing technical assistance to local development corporations in their role as employee-ownership advocates.

Ohio's Small Business Enterprise Centers are sources of free advice and assistance for all types of small businesses. By working in cooperation with community-based organizations, local labor and trade associations, chambers of commerce, and other organizations, the centers have access to resources that allow them to offer incubator services, financial help, legal services, and a variety of technical assistance efforts.

Pennsylvania offers the nation's largest state-assisted small business incubator program. The state provides financial and technical assistance to eligible incubator developers and managers, including municipalities, economic development organizations, and private nonprofit groups. Many of the 28 incubators already under way are housed in renovated industrial facilities in distressed communities such as Scranton, Johnstown, and Bethlehem. By the end of 1987, more than 1,600 jobs had been created by the 220 businesses located in these incubators.

Financial Assistance Programs

Financial support is often the critical missing piece of the economic development puzzle, especially in business and commercial districts outside the central city. Small, neighborhood-based businesses face significant problems in securing capital for both long-term investments in plant and equipment and short-term inventory finance. They often rely on the public sector for help. Some states face constitutional and statutory barriers that limit the types of programs they can sponsor—and these restrictions also can affect local governments and quasi-public development organizations—but all states offer creative and increasingly sophisticated financial assistance to promote business and community development activities. They are structuring programs to meet a variety of fixed asset and working-capital needs, investing in start-ups as well as in existing firms.

Moreover, as their financial assistance programs evolve, many states are starting to address capital availability, as well as its cost. As a result, states are beginning to provide equity financing to encourage new or additional business activity, in addition to low-interest direct loans or rate-reducing guarantees of private-sector loans. In a 1987 study, the Economic Policy Institute noted that "this strategy of filling capital gaps encourages new players to enter the game, rather than merely subsidizing those who are already playing. In particular, it improves the odds for new and small firms—precisely the kind that make bankers nervous" (Osborne, 1987, p. 61).

State-assisted business finance efforts occupy an increasingly prominent place in local economic development strategies because of shrinking federal resources. With the relative reduction in CDBG resources and the pending demise of UDAG, state initiatives take on growing importance in neighborhood economic development activities. The 1987 reductions in permitted levels of tax-exempt bonds and eligible IDB activities had an adverse effect on most states. This, in turn, affected many communities, which were forced to compete for economic development bond authority with other jurisdictions and other local needs, such as infrastructure and hospitals.

In the face of these constraints, states are devising ways to replace lost federal development grants, loans, tax incentives, and other assistance. Although tremendously varied, most state programs fall into one of the following categories:

- financial assistance grants (offered by 45 states), usually to local governments or nonprofit organizations such as local development corporations, most commonly as "deal clinchers" in response to specific business development situations
- direct lending, often through a revolving-fund mechanism, frequently in conjunction with local governments, development corporations, or community-based nonprofit organizations. According to institute data, 42 states have loan programs that focus on small business capital needs, 32 provide loans to help finance buildings, and 34 offer loans for equipment and machinery
- loan guarantees, made by 22 states
- equity financing programs, most often venture capital programs (in 25 states) and pension fund initiatives (12 states), usually offered through separate agencies or entities; 30 states have chartered development credit corporations, most relatively new (only three predate 1979)

In addition, state governments are turning to a variety of creative bond finance mechanisms to supplement traditional bond issues. These new methods include variable-rate obligations, zero coupon bonds, put option bonds, warrants, and short-term borrowing. Taxable bonds have been developed as another way of providing affordable financing for economic development projects.

Nearly every state provides some type of loan packaging services, often in conjunction with local development corporations or similar community-based organizations. These services usually include liaison with private financing institutions and public or private technical assistance. Some of these finance programs promote high-technology or high-growth industries. Others are designed to shore up existing sectors, such as manufacturing, that will continue to play an important economic role in specific communities or cities. They all can provide a path to state and local economic diversification and lead to a more favorable climate in which entrepreneurship can flourish—making it possible, for example, to introduce new products and new services to strengthen the community economic base.

Program Examples

With nearly $300 million invested, Michigan operates the nation's largest publicly capitalized venture capital fund, providing access to capital resources to companies with significant growth potential. Unlike other state-sponsored pools, it serves as a limited partner in other venture funds; this has spurred greater venture capital flow into the state. In the 4 years ending in May 1987, the program is credited with creating or saving 3,180 jobs even though companies assisted are largely in their formative stages. Michigan's venture fund has encouraged commercialization of industrial automation and robotics technologies—industries directly associated with the revitalization of the state's older industrial cities.

Massachusetts's Economic Stabilization Trust (EST) provides flexible financing to troubled but viable firms in mature industries. EST help is offered in various forms of debt and equity financing, structured to meet the repayment ability of the borrowing firms. Eligible businesses, usually in the steel or textile industries, may use the resources to reorganize under existing ownership or to bring in new ownership or arrange an employee buyout.

New Jersey's Urban Development Corporation offers financial help to firms and individuals undertaking redevelopment projects in distressed older communities; it can make loans, provide construction funds, invest in joint ventures, create subsidiaries, and buy and sell stock in those subsidiaries. A related initiative, the New Jersey Local Development Financing Fund, helps finance commercial and industrial projects in older neighborhoods where sufficient conventional financing has been difficult to obtain. Local projects must be sponsored by a public entity such as a municipality or a local development corporation. These efforts have leveraged nearly $129 million in private investment since early 1985, creating or saving more than 4,000 jobs.

Florida's Community Contribution Tax Credit provides businesses with a 50% credit (up to $200,000 annually) for contributions to community development projects in distressed neighborhoods. Eligible projects include construction or rehabilitation of commercial and industrial properties, housing, and public facilities, and activities to improve entrepreneurial and job development opportunities. Projects must be sponsored by community development corporations, revitalization agencies, or action programs.

Missouri's Neighborhood Assistance Program provides tax incentives to businesses underwriting projects designed and operated by neighborhood groups. Companies may earn credit on their state tax liability of up to $250,000 or 50% of the value of the contribution, which can take the form of cash, services, staff salaries, or donated space. During the 10-year period ending in mid-1987, more than $20.5 million in credits had been granted under this program, representing nearly $41 million in donations to community projects. Credits have been awarded to a range of businesses as diverse as neighborhood corner stores and Hallmark Cards' world headquarters in Kansas City. Hallmark received credit for numerous efforts, including a "donated labor" project. During periods of low work in the plant, Hallmark recruits personnel to work with neighborhood betterment projects on company time and on the company payroll.

Targeted Development Initiatives

Public programs exist to provide assistance to people, places, or firms with special needs. Some states direct considerable resources to older communities and industries needing special help to make the transition to a new set of economic and market circumstances. More than 40 states have adopted targeted development programs to get maximum return on public resources. Although the programs have diverse frameworks, all have a common goal: to channel development activities to and achieve economic benefits for locations, sectors, or groups that are at a perceived disadvantage in the marketplace. They aim to improve the economic climate by identifying the obstacles to investment and attempting to overcome them.

Most targeted development programs are designed to influence private investment decisions by providing a cushion against the risk of investing or offering incentives to counter negative factors associated with the target area or group. They fall primarily into three categories:

1. enterprise zones and other geographically defined areas (usually distressed communities) that offer a variety of tax, training, financial, and technical help and incentives
2. assistance to specific clienteles, such as entrepreneurs, women, minorities, and the economically disadvantaged
3. targeted industry programs, either to prospective high-growth industries with good potential to expand in a community or region or to existing mature industries facing a significant transition period

According to Northeast-Midwest Institute survey data, 25 states have initiated programs to spur neighborhood improvement in economically distressed cities. In all, 23 states have adopted industrial or commercial site programs to encourage investment, and 30 have adopted special incentives to encourage companies to locate new or branch plants in distressed areas. According to a recent survey by the National Governors' Association, 20 states are earmarking state resources to invest in emerging or high-growth industries such as biotechnology and telecommunications, which were chosen after careful analysis of the state and local economic advantages (Clarke, 1986, p. 3). Some 34 states now designate enterprise zones, which feature geographically targeted investment incentives. According to a HUD analysis, $6 billion in new capital investment has been made at more than 3,000 new or expanded facilities in 1,400 state enterprise zones nationwide. More than 100,000 jobs have been created or retained (U.S. Department of Housing and Urban Development, 1986, pp. ii-iii).

Program Examples

A number of targeted state programs concentrate on using public resources to fill one critical shortcoming in the economic development climate that affects the targeted community, industry, or constituency. For example, Wisconsin concentrates on providing investment capital to small-town areas, and Maryland does the same for inner-city, minority-owned firms. Massachusetts provides technological and marketing help to the metalworking industry, and New York to the garment industry. Minnesota's Tourism Loan Program provides matching loans and shared credit risks for projects to upgrade facilities for visitors.

States also may focus on one particular element or subset of a larger targeting strategy. California has implemented Employment Incentive Areas to stimulate job creation among the unemployed and economically disadvantaged. An innovative feature of the Tennessee enterprise zone program with neighborhood implications is an excise tax credit for corporate contributions to public schools within the zone.

The state-operated Main Street Program, sponsored by the National Trust for Historic Preservation, helps rejuvenate commercial districts by attracting new businesses to them and improving their physical appearance. By increasing community involvement in economic development strategy making, this program promotes the restoration and renovation of historic commercial cores of business districts in cities of all sizes. Currently 27 states operate Main Street Programs.

Florida's Community Development Corporation Support and Assistance Program provides loans and grants for business development projects to community development corporations in distressed communities. The loan program is divided into two parts: a loan guarantee program and a direct loan program. One of the most active CDCs under this program is in Pensacola. In 1987, five loans totaling $115,000 yielded 36 new jobs.

The Maryland Equity Participation Investment Program helps socially, physically, or economically disadvantaged persons finance the purchase of franchise businesses. Franchise companies—such as fast-food chains, auto service centers, and convenience stores—are attractive business-development options because they offer training, support, and a readily marketable product to relatively inexperienced entrepreneurs. Franchise businesses can add important value to neighborhood economies. More importantly, they have a higher success rate than other kinds of start-up companies, lessening the risk to owners and financiers. The program complements other sources of start-up financing by providing up to $100,000 or 45% of the total initial investment. A summary of state economic development programs is given in Table 17.2.

Community Development Corporations

Community development corporations (CDCs) are private, usually nonprofit organizations established and controlled by the residents of specific neighborhood areas. They were first started as part of the War on Poverty in the 1960s, but their biggest growth has taken place in the last decade in response to cutbacks in federal programs earmarked for neighborhood improvement.

CDCs are similar in mission to local development corporations (LDCs), but unlike them, are not certified or regulated by the Small Business Administration.

CDCs serve an important economic development role in many neighborhood areas. They undertake a wide range of activities, including technical assistance to and investment in small business, commercial revitalization, and development planning. CDCs are community-based organizations that set their own priorities. Some choose to serve as a catalyst to development (e.g., coordinating public and private projects in their area); others become full-fledged economic development entities in their own right, operating nonprofit technical assistance centers and for-profit ventures themselves.

CDCs are eligible for a variety of federal and state economic development programs, including all EDA programs and many HUD programs. CDCs also have structured projects by using SBA development company and venture capital programs. Many foundations also support nonprofit CDCs, notably the Ford Foundation through its Local Initiatives Support Corporation (LISC).

When they serve as private community developers, CDCs can attract private sector financing. The most common sources of private financing for CDC projects are bank loans, syndications, and investments by project partners. Private loans for CDC activities often are guaranteed by state or federal programs, such as those operated by SBA. More sophisticated CDCs can use government equity grants to leverage private debt financing for specific community development projects. This kind of financing flexibility permits the corporations to undertake a broad range of projects.

In the past few years, states have broadened their use of CDCs in state economic development policies. Several have started their own programs. Massachusetts helps with organizational development, project planning, and corporation management, and provides equity for specific community projects. Minnesota provides CDCs with small grants for planning and venture investment. Wisconsin was the first state, in the late 1970s, to start a state office for CDC technical assistance. Through its CDC support program, Florida provides administrative grants to defray CDC operating and staffing costs.

Private-Sector Initiatives
to Promote Neighborhood Economic Development

Although the purpose of this chapter is to examine public-sector programs that support neighborhood economic development activities, a number of emerging community development initiatives stem from private-sector efforts to improve the neighborhoods in which they are based. They assume considerable importance at a time when federal resources have dissipated and state funds face increasing and competing pressures. These private sector strategies illustrate that investing in community improvement projects outside the office building or plant and making company profits are not mutually exclusive goals; in other words, a business can have a conscience and fund it too. In addition, many company executives have recognized that continuing to be competitive relies on an economic development climate—including factors such as technology, education, and supportive service businesses—that in many ways is structured beyond their front doors. They realize that maintaining these portions of the economic base should not be totally a public-sector responsibility.

Community development approaches used by businesses depend on such factors as the size of the company, the skill level and home neighborhood of its employees, its facilities and the services they require, and the types of goods and services the company supplies to the marketplace. Firms may make monetary contributions to neighborhood development efforts, lend their staff or facilities (e.g., computer systems, libraries), directly invest in community development projects, or provide their reputation or clout in community endeavors before government agencies or leaders. Other private sector efforts include the following:

- first-source agreements, in which companies agree to hire first from individuals or groups referred by local officials (usually, first-source clients are neighborhood residents meeting minimum company job qualifications or graduates from employer-approved training programs)
- participation in technology transfer efforts with other (often small or start-up) firms
- participation in capital pools, venture capital clubs, and similar business finance projects

In some cases, of course, business makes the contributions because they must in order to gain permission to undertake other activities. For example, the "linked development" policies adopted for economically desirable areas such as downtown San Francisco require developer contributions (cash or in-kind) for housing or related development in distressed neighborhoods away from the downtown site.

Conclusion

The government role in devising and promoting economic development policies and programs will continue to grow in both scope and sophistication. Some 15 years ago, few government agencies at any level had the technical ability or the will to implement leveraged financial assistance programs or package economic development deals with private companies or local development corporations.

This has changed significantly since 1980, for two reasons. First, the U.S. economic engine emerged from the most recent recession in far different shape than it did from the previous downturn. Changes in the nation's business, employment, and technology bases became increasingly pronounced, and the economic downturn that began in 1980 became an economic shakeout. Second, unlike many previous slumps, the last recession ran its course in the face of reduced federal assistance programs. As economic conditions worsened and unemployment rose to record levels in many areas, cities and states came under increasing pressure to do something. Faced with mounting budget pressures themselves, they responded by structuring new economic development programs that would have the most impact for the least cost. "Public-private partnerships" became the direction and the theme in the public agencies and local development organizations that emerged to carry out this new role.

City and state economic development efforts have multiplied and will continue to do so. Many of the highly touted programs now in place are less than 9 years old, but they have tallied impressive achievements in terms of investment leveraged and jobs created. Development officials must plan carefully to ensure that deserving initiatives become institutionalized within their sponsoring governments. They must continue once the initial enthusiasm fades. Programs built on the presence of one administration or department are frail; they must attract and maintain strong and broad-based support. Moreover, the resilience of many of these new economic development efforts has yet to be tested. They have grown up in the face of improving economic conditions, but they cannot be judged as mature until they confront—and survive—an economic slump as severe as the one that inspired their creation.

References

Armington, C., & Odle, M. (1982, Fall). Sources of job growth: A new look at the small business role. *Economic Development Commentary,* 3-7.

Bartsch, C. (1988, August). *The guide to state and federal resources for economic development.* Washington, DC: Northeast-Midwest Institute.

Birch, D. (1978). *The job generation process.* Cambridge: MIT Program on Neighborhoods and Regional Change.

Boyd, E. P. (1987). *Community and neighborhood development: Issues in the 100th Congress.* Washington, DC: Congressional Research Service.

Clarke, M. K. (1986). *Revitalizing state economies.* Washington, DC: National Governors' Association.

Osborne, D. (1987). *Economic competitiveness: The states take the lead.* Washington, DC: Economic Policy Institute.

U.S. Department of Housing and Urban Development. (1986, August). *State-designated enterprise zones: Ten case studies.* Washington, DC: Office of Community Planning and Development.

Vaughn, R. (1988, April). *Linking education and economic development: Providence, Rhode Island.* Paper delivered at a conference sponsored by Congresswoman Claudine Schneider (R-RI).

— 18 —

Neighborhoods and
Public-Private Partnerships in Pittsburgh

Louise Jezierski

Public-private partnerships have become increasingly popular for local development policy and planning, especially to declining industrial cities like Pittsburgh that require enhancement of public and private resources and investment
to undertake restructuring and renew growth (Brooks, Liebman, & Schelling,
1984; Fosler & Berger, 1982). The public-private partnership is supposed to
provide a rational, flexible, voluntary, and cooperative alternative decision-
making structure to augment the local state and the market—to "rehabilitate
the civic tradition" (Committee for Economic Development, 1982). Liberal assumptions that partnerships provide for equal representation of fundamental,
stakeholder interests, pluralist decision making through negotiation, and a
neutral arena for rational planning are implicit in the concept (see Adams, 1983;
Brooks et al., 1984; Committee for Economic Development, 1982; Fosler &
Berger, 1982).[1]

Yet, although public-private partnerships can be efficient at organizing
growth coalitions that enhance expertise and funding resources, their impact
as governing structures over the long term has not been fully explored. The aim
to make government decision making more "efficient" leads to the organization
of expertise in an exclusive, technocratic, and paternalistic way that can preclude
a wide representation of interests and hinder legitimacy. Most partnerships are
arrangements between business and government (Clarke, 1986; Fleischmann &
Feagin, 1987).[2] The incorporation of such other interests as neighborhood groups,
labor, or small businesses will appear under certain conditions. As a locus of quasi-
governmental power, a partnership can become as politically contested as any
formal state structure (Silver & Burton, 1986; Squires, 1989). The durability of
partnerships for initiating and coordinating urban social change requires constant efforts to institutionalize conflicting interests and construct legitimacy for
development policy and for the partnership itself.

AUTHOR'S NOTE: This research was supported with dissertation grants from the University of California
at Berkeley. I would like to thank Barry Goetz, Sam Kaplan, Brian Powers, the anonymous reviewers, and the
editors for their helpful comments on earlier versions of this chapter.

In most analyses partnerships are viewed as static structures or only short-term projects are evaluated, and therefore conflict is underestimated (Adams, 1983; Brooks et al., 1984). In this analysis I emphasize the *dynamic* aspects of partnerships as attempts are made to integrate interest groups *variously* within the development agenda. I focus on neighborhood organizations as one interest group with a fundamental stake in the restructuring of the city (Castells, 1983; Cunningham, 1983; Fainstein & Fainstein, 1985) and show that their interests in community and self-management are often in contradiction with the interests of the local state and large corporations organized in partnerships. Neighborhood interests must be included in a partnership in order for this form of governance to remain legitimate over the long term. Decision-making power can become concentrated within partnership structures, and neighborhood groups, as an organized form of the electorate, are important because they wield an essential veto power over local government through voting and protest. Especially within the United States, neighborhood organizations represent working-class interests as unions shrink, ethnic associations fade, and political parties become de-aligned (Castells, 1983; Katznelson, 1981). Partnerships must respond to these challenges.

The city of Pittsburgh provides examples of both public-private partnerships and effectively organized neighborhoods. This is not coincidental, because the organization of one has encouraged the shaping of the other. The longevity of the partnership in Pittsburgh and the renown of voluntary associations and programs developed there make it a useful case for examination.[3] Moreover, the greater organizational capacity found in Pittsburgh helps explain the success of economic restructuring in Pittsburgh, compared with other industrial cities such as Cleveland (Jezierski, 1988a, 1988b; Shearer, 1989; Swanstrom, 1985). The partnership has been striving for the past 49 years to generate the *Pittsburgh Renaissance,* the economic transformation of Pittsburgh from Steel City to a postindustrial city. Its success is evident in the growth of high-technology and service industries and in Rand McNally's designation of Pittsburgh as the nation's "most livable city" in 1985 (Jacobson, 1987). This revitalization has come about because of a reconstitution of the local political system. Here I examine the process through which neighborhood groups organized to challenge the restructuring of their community life, political institutions, and jobs.

This study of community struggles indicates inherent limitations to the legitimacy of partnership structures, and I suggest that forms of neighborhood mobilization and organization develop as a response to the partnership form. Mobilization strategies of protest, coalitions, and electoral power are useful to delegitimize the authority vested in partnerships, and incorporation requires consensual, technical, and rational organizations. The context of urban decline and restructuring provides a unique set of constraints and opportunities to local political bargaining that may contribute to both activism and new organizational forms. The consent of an electorate to a partnership is achieved more easily under conditions of economic growth, whereas decline and restructuring will bring about more challenges to economic and political elites.

Neighborhood Access to Partnerships

The capacity for neighborhood groups to participate in partnership structures can be informed by two analytical approaches: the theory of incorporation and the theory of corporatism. Different understandings of the political bargaining inherent in partnerships and different strategies for gaining access to them are suggested in these two approaches. Incorporation research is not focused specifically on partnerships, but it indicates that if neighborhood groups can gain local-state power, they automatically will exert some influence within public-private partnerships. Incorporation theorists assume that local politics is a competitive arena in which interest groups can capture state power by mobilizing resources to overcome organizational and structural barriers. Neighborhood groups, however, lack financial, organizational, and ideological resources that allow them to compete effectively with business interests (Gittell, 1980; Henig, 1982; Stone, 1986). Research focused on the incorporation of racial and ethnic minorities and neighborhood interests into the local state shows that a strategy of protest may be useful in the short term, but incorporation is better ensured through the formation of broader based coalitions and decentralized government structures (Browning & Marshall, 1986; Browning, Marshall, & Tabb, 1984). Centralized political structures such as party machines and at-large-council or city-manager arrangements restrict access, whereas decentralized bureaucratic administrations and district elections facilitate incorporation (Clarke, 1986; Mollenkopf, 1986).

Incorporation theory is useful for explaining how neighborhood interests might gain representation in the local state, but it is less useful for predicting whether the state will adequately represent their interests as part of a public-private partnership. Incorporation theorists tend to view state power as an object that needs to be captured and fail to view the state as an actor with its own interests. The Pittsburgh case shows that conditions necessary for incorporation within the state may be different from conditions for participation in partnerships.

Although state incorporation strategies require the formation of broad-based coalitions, partnerships are strategically constructed to promote efficiency by bypassing them and narrowing representation. Piven and Friedland (1984) suggested that the incorporation of too many interests within the state can cause hyper-demand and fiscal crisis. A dual-state structure consisting of an electoral sphere and a quasi-state sphere typical of partnerships allows the state to avoid the constraints of the electoral system. Clarke and Meyer (1986) argued that because grassroots interests in the United States do not have the benefit of alternative parties, interests are more likely to be channeled into bureaucratic, quasi-state organizations such as partnerships.

Partnerships can be alternative mechanisms for interest intermediation, by which interests can be independently organized and negotiated outside the state or in quasi-public agencies, facilitating greater autonomy and maneuverability for state bureaucracies (Offe, 1985). Partnerships transcend the more usual, elite relationship of influence between government and business because they rely on formal agreements and legal institutional arrangements such as task forces, authorities, and economic development corporations (Clarke & Rich, 1982; Goldstein & Bergman, 1986). Whereas incorporation theorists focus on politics within legal city or county boundaries, partnerships afford the local state more autonomy as it gains access to private and state- and federal-level resources and extends its authority to metropolitan or regional jurisdictions.

In corporatist literature, authors go beyond examining the construction of coalitions and the distribution of political influence to examining the process of how competitive politics is transformed to noncompetitive and rational politics by putting decision making above mobilization of demand. Coleman (1983) argued that partnership forms are "corporatist-like" because the mediation of interests is singular, noncompetitive, hierarchically ordered, functionally differentiated, and technocratic (Schmitter, 1979).[4] Most important in corporatist theory, the state is recognized as a separate interest.

An assumption in incorporation theory is that voter coalitions can influence state policy through the process of majority rule. With a partnership form, however, the state can bypass this process and recognize minority interests that lack voting power, such as business organizations. This detour strengthens the political position of groups legally outside state jurisdiction (Offe, 1985; Schmitter, 1983). Only private interests that are normatively recognized to have "public status" are assigned places within the partnership (Offe, 1985). According to Goldstein and Bergman (1986, p. 266), "key actors are persons (or organizations) who directly or indirectly control resources or possess the expertise needed to make things happen." Businesses may represent universal interests because their private interests (profit) may seem to coincide with the public interest in economic growth and jobs (Habermas, 1975; Przeworski, 1980). They can contribute technical expertise, money, and equipment that the local state lacks, and these are especially important in cities suffering from economic decline and fiscal stress (Spiegel, 1981). Participation in a partnership also requires that interest groups are represented as autonomous organizations so that government may enter into formal negotiation (Goldstein & Bergman, 1986; Schmitter, 1983). The political participation of business interests is facilitated by their prior organization into firms and business associations (Offe, 1985).

Theories of corporatism advance our understanding of partnerships because they reveal the dilemmas of increased state intervention and privatism, which have deleterious effects on sovereignty and the legitimacy of policy (Clarke, 1986; Habermas, 1975; Langton, 1983; Offe, 1985). Indeed, Schmitter (1983, p. 927) warned that corporatist arrangements may be considered "vicarious democracy." In corporatist approaches, however, the ability of local actors to achieve consensus and compromise is overestimated, and the inequalities among interest groups are underestimated. For example, the state is too often a weak partner (Spiegel, 1981),[5] and better theories of the local state (and its limits) must be developed (Gottdiener, 1987). Furthermore, one cannot assume that neighborhood interests (or any minority interest) can be institutionalized within a partnership structure (Sbragia, 1989). In reality they are limited to playing a consensual and consulting role on issues of housing and neighborhood economic development (Clarke & Rich, 1982; Lurcott & Downing, 1987). They are not recognized for possessing any key resources that could enhance regional economic growth (Weiss & Metzger, 1987). As one skeptical business representative of the Pittsburgh Partnership asked, "What do these neighborhood groups have to bring to the table?" (Jezierski, 1988a, p. 185). The answer to this question, as revealed here, lies in the extent to which neighborhood interests are recognized as general community interests that can contribute to development and provide a legitimation function.

The theories of incorporation and corporatism suggest opportunities and limitations for the organization and mobilization of neighborhood interests and for the consequences of partnership membership. Incorporation theorists would suggest that neighborhood groups can successfully use political powers of election of representatives to promote their agendas inside the state and that collective powers of protest and association can check an exclusive power bloc contained in the public-private partnership (Schmitter, 1983). Corporatist theorists explain that a partnership form presents an entirely different political process based on consensus and requires a mobilization of resources that can be brought to the partnership rather than a mobilization of demand.

Taken together, these theories suggest that neighborhood groups will have an ambivalent relation to partnerships and that they have to create new and different types of organizations—some that enable them to become partners, and some that challenge the partnership. Qualitative differences between their goals and mode of organization and those subscribed by the partnership form will make their participation problematic. Corporatist theorists point out that the partnership mode requires a rationalized organization process; yet community groups organize around nonrational ties such as traditional authority (Castells, 1983; Schwartz, 1979). Because class-based, racial, or territorial interests help define groups and create solidarity, their parochial concerns are counter to the focus on the "public good" (Goering, 1979). No separate neighborhood leadership may exist to coordinate these separate interests besides a mayor, and this pivotal position may not always be reliable. As incorporation theorists recognize, adversarial tactics are useful for those groups that have few resources because they enhance solidarity. Protest has often served as an effective veto strategy (Lipsky, 1968; Mollenkopf, 1983). The partnership's reliance on consensus, however, keeps them wary of neighborhood groups dependent on conflict, and the strategy of protest may have to be abandoned to create the trust and certainty required to maintain negotiations and coalitions of partnerships.

Neighborhood groups seeking incorporation in the partnership must adopt businesslike organizational modes, develop technical expertise, turn from advocacy to economic development, and apply for funding—sometimes from business foundations (Lurcott & Downing, 1987; Thomas, 1983). Thus neighborhood groups must engage simultaneously in contradictory activities.

The research presented here indicates that thresholds of participation for efficiency and legitimacy are built into partnership structures.[6] Some of these are as follows:

- Neighborhood incorporation may require different strategies that are useful at different junctures, targeted to specific forms of political structure.
- A partnership may be organized to promote exclusivity and efficiency, but it is not so autonomous a structure that it can remain above local mobilizations of protest and electoral power.
- The partnership must be considered legitimate to be effective.
- Rather than being based on consensus, which implies equal access to decision making, the politics of partnerships is more accurately understood as *consent* (Przeworski, 1980) because neighborhood groups, as well as the state, are dependent on private resources to some extent for investment.[7]
- Neighborhood groups will consent to an exclusive partnership, even at the risk of losing some sovereignty, only if certain conditions are met—primarily that private interests will invest in future jobs and neighborhoods.
- Political negotiation is achieved through mutual exchange of threat: On the one side, neighborhoods can withdraw consent and deem partnerships illegitimate if their interests are not addressed; on the other side, the private sector can withhold participation and investment, and the local state can withhold public investment.

In this work I attempt to specify thresholds of consent and protest and to explain how exclusive partnerships are constructed and challenged.

The Pittsburgh Partnership and the Organization of Neighborhoods

Pittsburgh's public-private partnership was established primarily as a business-government relationship, but political struggles to expand this arrangement to include neighborhood groups have developed during its 40-year history. The partnership was formed by the cooperative relationship between the administrations of Mayor David Lawrence (who served from 1945 to 1961) and Mayor Joseph Barr (from 1961 to 1969) and the Allegheny Conference on Community Development (ACCD), an organization of chief executive officers (CEOs) of the major local corporations that was created in 1943 with the backing of Richard King Mellon of Mellon Bank. The coordination of these forces was brought about by the creation of a number of specialized organizations for planning and implementation. The partnership's aims were to facilitate development in response to economic and environmental decline, as well as to a significant period of strike activity (Lubove, 1969; "Pittsburgh's New Powers," 1947).

The *Pittsburgh Renaissance I* development strategy (from 1945 to 1969) included a number of successful projects such as air and industrial smoke abatement and flood controls, traffic and bridge reconstruction, and downtown renewal projects such as the Gateway Center—a complex of hotels and office buildings, Point State Park, and Three Rivers Stadium—that replaced warehouses, port facilities, factories and workshops, and rail lines of the dense industrial entrepôt located at the intersection of Pittsburgh's three rivers. This area became known as the *Golden Triangle* (see Figure 18.1) (Lorant, 1980; Lubove, 1969).

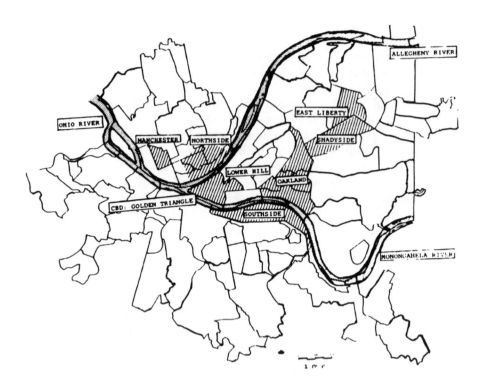

Figure 18.1. Pittsburgh Neighborhoods
SOURCE: Pittsburgh, PA, Department of City Planning (1980).

A second development strategy, *Renaissance II* (from 1977 to 1987), included construction of new hotels, riverfront development, airport development, construction of a convention center, and the installation of a new subway system and the support of business services, light industry, and high-technology industries in the areas of computer, electrical, and optics engineering and medicine (Ahlbrandt & Weaver, 1987; Stewman & Tarr, 1982).

Although the partnership largely excluded union, neighborhood, minority, and small business participation, the improvement program was supported widely because it represented growth and had the backing of the popular Democratic party machine. Tangible results were evident in building and environmental quality (Lorant, 1980; Lubove, 1969; Robin, 1972; Stewman & Tarr, 1982). A long-standing domination of local corporations fostered an ideology of paternalism, and new organizational capacities also aided the construction of a stable political climate that enabled restructuring to take place. The durability of these relationships can be attributed to the establishment of new public commissions and authorities and the institutionalization of associations and personnel in a centralized and interlocking network of staff positions (Stewman & Tarr, 1982). These new legal organizational structures channeled representation and mediated conflict.

The formation of new organizations that specialized in identifying and organizing business interests was a necessary condition for the creation of the partnership. The ACCD consolidated fractions of corporate interests within the private sector and provided them with an identity of general interest in the community that allowed the public sector to negotiate with them collectively. The ACCD also helped create spin-off specialty organizations such as the Regional Industrial Development Corporation (RIDC) in 1955 to develop industrial parks, Penn's Southwest Association in 1971 to market the Pittsburgh area, and the High Technology Council in 1981 (see Table 18.1) (Ahlbrandt & Weaver, 1987; Stewman & Tarr, 1982). These civic associations, in conjunction with the representation subsumed by the Democratic party, provide the basis for characterizing the partnership as corporatist (Ahlbrandt & Weaver, 1987; Coleman, 1983; Weaver & Dennert, 1987).[8]

The coordination of the public sector was constituted by a strong-mayor form of government, a nine-member, at-large council, and a reliable Democratic party machine. Mayoral administrations had continuity and autonomy: Only four mayors presided over the city between 1945 and 1987. This structure provided electoral control and a relative consensus within city government. A semi-autonomous Urban Redevelopment Authority (URA) was the first instrument created by the partnership through state legislation. This five-member commission has served as the legal arm for amassing land and capital, as the crucible of development expertise, and as the broker of demands, channeling conflict away from the mayor and the ACCD. The URA was created in 1948, and Mayor Lawrence served as its first chair; two former URA directors moved on to become the executive director of the ACCD (Coleman, 1983; Stewman & Tarr, 1982).

Economic decline and social transformation in Pittsburgh changed the balance of interest organization and created new opportunities for neighborhood mobilization. Traditional means of interest organization became less effective with industrial restructuring and the collapse of the domestic steel industry. Local class structure has changed because union membership has declined precipitously, and the turn toward service and high-tech industries has bolstered a middle-class labor force that promotes middle-class culture and real estate. Manufacturing employment in the city dropped from 27% to 15% of the total workforce, and employment in services rose from 23% to 39% between 1940 and 1980. By 1980, 23% of the workforce could be described as professionals and managers, whereas only 15% were operators and laborers (U.S. Bureau of the Census, 1980). This altered the social class of some neighborhoods, leading to gentrification.

In the midst of restructuring, Pittsburgh's neighborhoods provided a secure quality of life for its residents but needed further investment. A well-developed sense of place and commitment in Pittsburgh (Ahlbrandt, 1986) has supported neighborhood organizations and networks substantially. Pittsburgh has 88 neighborhoods that contain from 400 to 16,000 residents, that are territorially and socially well defined, and that are recognized by city hall (Lurcott & Downing, 1987). Although all but seven neighborhoods lost population during the 1970s, Pittsburghers remain attached to their neighborhoods. Some neighborhoods

Table 18.1 Community Organizations in the Pittsburgh Partnership

Organization	Date Founded	Membership	Primary Functions
Private Sector			
Allegheny Conference for Community Development (ACCD)	1943	CEOs of large corporations	planning and research, regional development
Regional Industrial Development Corporation (RIDC)	1955	corporate executives, public officials, labor	industrial park development
Penn's Southwest Association	1972	executives of large and small corporations and public officials	marketing the Pittsburgh region nationally and internationally
High Technology Council	1983	executives of high-technology corporations, Penn's Southwest, RIDC, universities, investment banking	marketing and coordination of capital and informational resources
ACTION-Housing, Inc.	1957	civic agency; board includes private housing industry, business, and public officials	loan fund, planning, housing development, and organizational service
Allegheny Housing Rehabilitation Corporation (AHRCO)	1968	19 executive committee members from neighborhoods, public officials, and businesses	housing rehabilitation
Public Sector			
Urban Redevelopment Authority (URA)	1946	5 members appointed by the mayor plus an executive director	land acquisition, planning, development, urban renewal, grant management
Department of City Planning	1911	technical staff	research, legal management of land use
Neighborhoods for Living Center	1979	mayoral appointed staff; URA and neighborhood group representation	neighborhood fair, residence referral
Steel Valley Authority (SVA)	1985	appointed board; unions and municipal and county officials	land and capital acquisition, industrial redevelopment

Community Groups

Organization	Year	Structure / Membership	Focus
Neighborhood Housing Service (NHS)	1968	15-member board; citizen groups, mayor's office, bank officers, and private foundations	housing loan fund
People's Oakland	1970	neighborhood residents and staff	advocacy
Shadyside Action Coalition	1970	residents	party reform
Pittsburgh Neighborhood Association (PNA)	1970	coalition of neighborhood groups	advocacy
Manchester Community Corporation (MCC)	1979	neighborhood residents and staff	advocacy
Save Nabisco/Our Neighborhoods Action Coalition (SNAC)	1982-1986	residents, unions, and public officials	job retention
Pittsburgh Partnership for Neighborhood Development	1983	staff; city, neighborhood, and foundations	coordination of funding and research for development
Community Technical Assistance Corporation (CTAC)	1980	staff; city and neighborhood representation	organizational development
The Working Group	1985	informal; neighborhood groups	neighborhood policy strategy, advocacy
Community Reinvestment Group	1988	neighborhood groups	bank investment, consumer advocacy

have been intact for 80 years. Home ownership and neighborhood identification are grounded in the kin networks and the place of work for particular ethnic groups (Bodnar, Simon, & Weber, 1982). In a survey, Ahlbrandt (1984) found that 44% of households have kin in the neighborhood, 75% of households have at least one friend in the neighborhood, and a majority (55%) have their primary social contact in the neighborhood. More than half of the residents engage in significant amounts of neighboring. Of those surveyed, 63% reported they were strongly attached to their neighborhood, and 72% rated their neighborhood a good or excellent place to live.

The partnership's response to decline, however, was to raze slum areas, build highways for access to suburbs, and develop the downtown area. The public and private sectors did not address the needs of neighborhoods until some countermobilization was organized, and they responded by committing to an agenda of community development in order to gain greater legitimacy and contain conflict, acting on the assumption that "worthy living environments are necessary for economic growth" (Lubove, 1969, pp. 136-142). Negotiation through partnership channels, however, required that neighborhood interests become organized first. Even the ACCD recognized this by the early 1960s and proceeded to aid in this effort (Ahlbrandt & Brophy, 1975; Cunningham, 1983; Cunningham & Kotler, 1983). Coleman (1983, p. 27) has argued that as party structures broke down, business interests were concerned about the lack of mediating structures and lacked confidence that local government could meet their needs. Local foundations were recruited to enhance this organizational effort and remain an important source for organizational seed money and even for operating expenses.

Neighborhood groups began to achieve some autonomy once they were able to develop a cadre of experienced leaders and to acquire real estate development skills that matched the planning capacities of the associations and task forces in the business and local-state sectors. Some of this autonomy was gained with the help of federal programs and national movements but was threatened with federal cutbacks (Mollenkopf, 1983). This promotes a greater dependency on foundation and development money and therefore demands more compromises from neighborhoods.

In the following sections, I examine three empirical examples of how neighborhood interests in Pittsburgh have mobilized to challenge the development policies and structure of the Pittsburgh public-private partnership. The analysis is limited to the years 1943, when the partnership began, through 1988, the end of Mayor Caliguiri's administration. The examples are presented in historical order, but they also represent a continuum of mobilization efforts from the most defensive and reactive to the most proactive strategies that these neighborhood groups have developed over the years. These examples include (a) antirenewal and defense of neighborhoods, (b) transformation of political institutions, including the Democratic party, the political machine, and the URA, and (c) defense of jobs and alternative economic development policy.

Neighborhood Mobilizations

Neighborhood Defense and Antirenewal

The aims of the Renaissance I strategy (from 1945 to 1969) to revitalize the downtown, improve the environment, and build a highway system were supported by Pittsburgh residents, in part, because the downtown area was perceived as the locale of the business community. The legitimacy of the partnership's plans, however, generally was threatened when the plans involved neighborhoods. A major exception was the plan to remove 728 families from the Southside, Hazelwood, and Scotchbottom neighborhoods to redevelop and expand the Jones & Laughlin steel plants: This negotiation was uncontested because it was undertaken through the United Steelworkers and was basically a trade of homes for jobs (Lubove, 1969; Robin, 1972).[9]

Consent broke down with the clearance of the Lower Hill, a low-income, primarily black neighborhood adjacent to the downtown area. Demolition of 95 acres began in 1956, causing the displacement of 1,551 families and 413 businesses (Lubove, 1969). The purpose of the renewal was to anchor the civic-arena development to the downtown, but the policy blatantly supported downtown development at the expense of the neighborhood. No plans had been made for rehousing or resettling the families and businesses. Protests were mobilized in the midst of the project, which was decried as "Negro removal." The efforts of residents to stop renewal were supported by the momentum of the civil rights movements of the late 1950s. These renewal efforts made other black neighborhoods, such as Manchester and Homewood, more precarious because those displaced from the Hill crowded into these more stable black neighborhoods. Renewal plans for the Upper Hill were dropped when an impasse developed between the black community and the partnership.

The strategy of protest was useful to neighborhoods for transforming their plight into a general issue concerning quality of life. The failure of renewal efforts was quickly realized by the partnership and prompted the ACCD to respond with programs (Hritz, 1968). ACTION-Housing, Inc.—a civic agency with business support that extended the partnership to represent a coalition of the housing industry, government agencies, and top civic leadership—was formed in 1957 (Coleman, 1983). ACTION-Housing focused on neighborhood reinvestment and was careful to coordinate neighborhood interests, even assisting with the development of community organizations. Further support for community groups came with the Department of City Planning's Community Renewal Plan of 1961 and an innovative Neighborhood Urban Extension Policy that provided for the employment of social workers to organize neighborhoods for participation and compliance in neighborhood plans. The partnership also introduced the Allegheny Housing Rehabilitation Corporation (AHRCO) to rehabilitate, rather than clear, housing. Despite the partnership's efforts, the success of these programs was limited to a mostly middle-class clientele (Ahlbrandt & Brophy, 1975).

Economic and population decline continued in Pittsburgh during the 1960s. The city's population decreased by 84,000, or 14%. The city's housing stock was reduced by 3% during this period; yet the vacancy rate increased by 6% (Ahlbrandt & Brophy, 1975). Neighborhoods were in flux because of displacement. Economic and housing crises increased racial strife. These conditions prompted citizens to take control of their own neighborhoods with support from an emerging ideology that called for racial equality and citizen participation. But tensions were created between the partnership and community groups, stemming from the dual directives of top-down renewal policy and neighborhood demands for self-management (Coleman, 1983; Lubove, 1969). An autonomous community movement emerged that was counter to partnership efforts, as well as to the Democratic party machine and to the Barr administration (from 1959 to 1969; Coleman, 1983).

The most important mobilization toward self-management occurred in Manchester, a black, middle-class neighborhood. The Manchester Community Corporation (MCC) secured federal historic preservation protection for many of the blocks that were constructed during the 19th century. Historic preservation designation halted renewal, took redevelopment plans away from the public-private partnership, and kept control in the hands of the neighborhood. One neighborhood activist (see Jezierski, 1988a, p. 458) described these events as follows:

> How did Manchester happen? The black power movement helped, luck too, in that you had the right people at the right time, and people committed to the public benefit rather than for private benefit. The initiative was ours to do rehab instead of renewal. Everyone thought we were crazy, including the URA, including HUD. But we thought it was crucial tactically to take it away from white people, from the URA. We made historical preservation history. We did our homework, traveled to other cities, and discovered the potential for rehabilitation. To stop renewal, we realized the possibility of protection by registering as an historic district. We became a nationally registered district six years before the city recognized this.

The MCC was successful because it was able to establish some autonomy based on Manchester's access to special-redevelopment status and federal funding as a historic district. Manchester is a unique case, however, because of its architecturally significant housing stock and the MCC's ingenious organizing capacity around this resource.

Throughout the 1960s, protest served as the prime strategy of neighborhood groups. They demanded protection from renewal and won rehabilitation and redevelopment. They demanded, through the usual client relationship, better state intervention and enforcement of established regulations and statutes. Protest threatened the redevelopment agenda of the partnership, so a compromise was instituted as programs to address community needs were added to the agenda. New leadership in the ACCD, under Robert Pease, turned its attention from physical development to programs concerned with social development. Minority scholarships, business loans, and community-service grants were developed in response to the black community and neighborhood groups (Stewman & Tarr, 1982). Control over these programs, however, was retained by the business-government elite. If neighborhood groups were to represent themselves in negotiation and administer programs, they had to develop more sophisticated organizations.

In the city's North Side, Neighborhood Housing Service (NHS) was created in 1968 as a response to direct appeals from citizens to Mayor Barr's administration, which answered with stepped-up code enforcement and public works projects. Barr also persuaded the private sector to provide technical assistance and set up a foundation fund for loans to nonbankable owners in the neighborhood (Coleman, 1983). The NHS's success was replicated later in cities across the country. The significance of the NHS was that for the first time, residents had some voice in the investment practices of private financial institutions. Subsequent evaluations of the program, however, found that bank reinvestment in the area was limited to the four banks represented on the NHS board and that the bulk of loan money generated reinvestment in only one part of the neighborhood—the part in which some gentrification had occurred. Moreover, 34% of the loans went to developing rental property (Ahlbrandt & Brophy, 1975; Stewman & Tarr, 1982).

By the end of the 1960s, the neighborhood groups' strategy of protest proved successful for halting the partnership's renewal policy of slum clearance and substituting housing rehabilitation. It also opened the development agenda to include social programs. Protest alone, however, was insufficient to challenge the basic power structure of the partnership or to halt the basic agenda of downtown development. Priority went to expensive physical projects, such as a new stadium and a sky bus, and neighborhood demands for reinvestment seemed to be ignored. In the 1970s, demands for social welfare and tactics of neighborhood defense escalated to attempts to control the public-private partnership itself by wielding electoral power over it. The Barr administration had lost support in the wake of strikes, a fiscal crisis, and rebellion in black neighborhoods (Coleman, 1983). Political restructuring during this period marked the "end of oligarchy" in Pittsburgh, including a breakdown in its political machine, the emergence of federal money and guidelines for urban redevelopment, and the activism of neighborhood associations (Coleman, 1983). This restructuring ushered in a new era that changed the partnership fundamentally. Federal urban policy to fund neighborhood-based nonprofit organizations allowed community groups to establish independent political bases, so these groups turned their attention to city hall and launched incorporation strategies (Mollenkopf, 1983).

The Transformation of Local-State and Party Organization

In the following discussion, I examine four examples of how neighborhood groups challenged the public-private partnership by changing the public sector: (a) targeting the mayoral administration, (b) targeting the city-planning structure and URA, (c) targeting the Democratic party machine, and (d) creating neighborhood coalitions as a check to city council.

Pittsburghers rejected the downtown agenda of the public-private partnership when, in 1969, they elected Pete Flaherty to serve as mayor. He ran on a platform of being "nobody's boy" and "no business as usual," without the support of the Democratic party leadership or the business community (Stewman & Tarr, 1982). Unlike mayors before him, Flaherty actively supported neighborhoods instead of merely responding to their demands. He promised that federal development money would be split 50-50 between downtown and neighborhood projects. As a result, the trust and consensus of the past between the mayor's office and the ACCD broke down. Stewman and Tarr (1982) characterized the period as "the Interlude," but its significance was a grassroots rejection of the partnership through elections. Flaherty instituted populist reforms that cut the size of the administration and machine-patronage jobs, thus severing traditional linkages between the Democratic party and businesses. He generally antagonized the corporate community. Flaherty would not provide the leadership, organization, or support for many of the physical development plans put forth by the ACCD; however, his populist agenda did allow fulfillment of plans to build a convention center and a light-rail transit system.

The Flaherty administration made key changes in development policy that widened neighborhood participation. The mayor replaced the directors of many city departments and severed staff links between the ACCD, the planning department, and the URA. Flaherty created a community-planning program that divided the city into districts and assigned city planners to work with neighborhood organizations. Progress reports for each neighborhood were published annually. Planners and residents worked together to coordinate neighborhood-improvement strategies and to review plans for housing and economic development and the city budget (Ahlbrandt, 1986). The reorganization of the planning arm of the local state under Flaherty opened up the planning process. He weakened the power of the relatively autonomous URA by removing city-planning functions from the URA and placing the director of the Department of City Planning under the mayor's office so that the director would report directly to the mayor, rather than to the appointed URA board. This changed the linkages between the URA and the ACCD and increased neighborhood input.

The best example of the new community planning effort was the Oakland Plan, produced by People's Oakland, a community group that organized in 1970 to oppose university and hospital expansion and the URA's master plan for the neighborhood. The master plan was abandoned after People's Oakland negotiated with the city and the major educational and medical institutions located in the Oakland neighborhood to draw up an alternative plan that was highlighted by a grassroots research effort to assess residents' needs. The new plan was developed with a labor intensity that would be hard to replicate—the group met every week for 4 years. The Oakland example is a lesson that neighborhood input is not spontaneous—a tremendous effort is required to transform residents into planners (see interview with Sandra Phillips of People's Oakland in Jezierski, 1988a, p. 463; also see Weiss & Metzger, 1987).

The proliferation of neighborhood organizations during the Flaherty years in the 1970s contributed to the undermining of the machine ward system that had supported the government-business partnership under Mayors Lawrence and Barr, thus allowing a new partnership between the local state and neighborhoods to emerge. More specific grassroots efforts were undertaken to overturn the Democratic party political machine and to sever its legitimacy function. The best example of these efforts is provided in the case of the Shadyside neighborhood (Cunningham & Kotler, 1983, pp. 131-140). In 1970 a small group of independent Democrats organized the Shadyside Action Coalition (SAC) to transform its 7th Ward Committee, capturing 14 ward committee seats (out of 26). The new committee stressed neighborhood-centered issues over party-centered issues and backed reforms such as merit-based employment to replace party patronage and candidate endorsement based on open, neighborhoodwide reviews to replace party loyalty. The committee took advantage of the 1974 changes in the city's charter to elect a neighborhood government for the ward, the Shadyside advisory board. The new 7th ward was in the forefront of the efforts to create neighborhood-based government. The neighborhood groups were trying to carve out a direct and separate authority from both the Democratic party and the local government that presumably represented them.

Another example of growing neighborhood power was the formation of the Pittsburgh Neighborhood Alliance (PNA) in 1970, which at its peak in the late 1970s served as a coalition for 80 neighborhood groups. A neighborhood leader explained its importance: "The Alliance was organized originally for political reform, to set up an alternative to the ward system, and create coalitions across party lines. The City Council did not realize its significance" (see interview with PNA activist in Jezierski, 1988a, p. 464). These interneighborhood coalitions provided autonomous alternatives to the council and the party machine and served as pressure groups for neighborhood interests.

The incorporation of neighborhood interests into the local state, coupled with the loss of the machine, undermined the business-government relationship of the partnership. The increasing activism of neighborhood groups and the hostility of Mayor Flaherty to the growth agenda of the ACCD contributed to the impression that the local state was a less reliable partner in redevelopment negotiations, and the business community withdrew from the partnership. Redevelopment plans were halted, and a feeling of stagnation emerged, exacerbating decline. If plans for growth were to continue, a compromise had to be introduced into local politics.

The momentum of neighborhood power elected another independent Democrat to the mayor's office, Richard Caliguiri, who served from 1977 to 1988. Caliguiri had support from neighborhood groups, and he recruited activists for positions in his administration, but he also sought to reestablish ties with the corporate sector. Caliguiri reemphasized consensus and repudiated the conflict style of Flaherty and of the activist neighborhood organizations that alienated the private sector. He introduced a new agenda, Renaissance II, which was a dualistic approach that retained the participation of citizen organizations in the formation of neighborhood policy but also resurrected the privilege of business to develop the downtown. Caliguiri also embraced the public-private partnership concept and the ACCD because he recognized the importance of private sector resources of capital and information to sponsor development. The contradiction of Caliguiri's dualistic approach, however, is illustrated by a new city agency that his administration established, the Neighborhoods for Living Center. The center was created to promote neighborhoods as communities and to target services to them. Caliguiri appointed the former head of the activist PNA as director. Although the center provided identities and political visibility for individual neighborhoods, it also served the city's promotional effort to attract capital in a real estate function.

The dualism of the Caliguiri approach heightened contradictions within local development policy and within neighborhood strategies. Structural change in the economy and the loss of federal funds to support neighborhood groups have brought about some retrenchment of the incorporation strategy. Neighborhood groups have supported reforms to create district elections in the city council in reaction to a crisis over adequate minority representation. The at-large council structure, instituted in a business-backed municipal reform movement in 1911 (Lubove, 1969), failed to elect a black to the council in 1985 for the first time in more than 20 years. This event, in conjunction with other neighborhood concerns, led to a grassroots political reform movement to return to a neighborhood-based, district-council organization. Caliguiri opposed district elections and offered a compromise of a mixed system, but a coalition of civil rights groups and neighborhood organizations took the issue to court. The lawsuit was rendered moot when a referendum for representation by district passed in 1987 and district council elections were held in 1988 (Sbragia, 1989). This change in the structure of representation will alter considerably the balance of power by weakening the mayor's power, strengthening the input of neighborhood interests, and making policy planning and implementation subject to much more political scrutiny. This weakening of mayoral autonomy could disrupt the partnership's basic operating principle of negotiation by consensus.

The successful incorporation of neighborhood interests depended on new coalitions among neighborhood groups (Browning et al., 1984), but it also required the restructuring of political institutions that blocked neighborhood influence, including opening up the URA and planning department to neighborhood participation, decentralizing the city council, and dismantling the Democratic machine that previously had maintained an elite relation with the ACCD and had supported downtown development. The Flaherty years, however, provided a hard lesson for political actors in Pittsburgh. Incorporation strategies worked almost too well for neighborhood interests because the populism of Flaherty alienated business interests. Seizing local-state power prompted disengagement of business interests, so consent had to be reestablished with Caliguiri's dualistic approach. The local state has its own interest in maintaining ties with business for growth.

Neighborhoods Take on Economic Development

Some political restructuring has occurred in response to economic restructuring in the Pittsburgh area. Mergers, loss of headquarters, less familial control of local corporations, loss of manufacturing—especially steel—and the emergence of new service and high-technology companies within the Pittsburgh economy have fragmented and extended the partnership. New organizations, such as the High Technology Council, have joined such traditional participants as the ACCD, and the University of Pittsburgh and Carnegie-Mellon University have increasing influence. The inclusion of these new corporations has broadened the composition of Pittsburgh's private sector and changed its development agenda, requiring more economic expertise and further compromise. Neighborhood representatives recognize that extracting concessions from the partnership may be harder than before and less predictable. One neighborhood activist (see Jezierski, 1988a, p. 246) said:

> The paranoia about the ACCD isn't as strong as it used to be. Pittsburgh hasn't learned to do business any differently than they did forty years ago, but we live with that. However, the power structure is changing: the High Tech Council is becoming more important, it's more rough and tumble. It's still a highly centralized town— seven phone calls is all that is needed. And when the mayor, a popular one that mends its fences with the neighborhoods, gets together with them, it's hard to think about how you can win with confrontation. How do you argue with "the number one city in America"? People just write us off as kooks.

This new partnership climate has forced neighborhoods to become less adversarial and increasingly technical and bureaucratic in order to participate. Caliguiri encouraged neighborhood groups to work more directly with the private sector on development as they found themselves more dependent because federal programs were cut. Some have begun to forge new linkages and to work directly with the private sector, providing they can acquire the sophistication to undertake development on their own. Although neighborhood investment was not forthcoming through usual business channels, neighborhood groups in Pittsburgh, like those elsewhere, found more success when they turned to intermediaries—primarily local foundations—willing to take on risks that banks would not. Foundations and the ACCD responded with development programs in an effort to support neighborhood self-sufficiency.

The ACCD and private foundations have been involved in communitywide efforts to improve schools, notably with the creation of the Public Education Fund. In 1983 the ACCD sponsored the creation of the Pittsburgh Partnership for Neighborhood Development (PPND), which was designed to provide operating funds for some community development organizations and to coordinate technical assistance. This group administered funding and technical support for the Ford Foundation's Local Initiatives Support Corporation (LISC) program. The successful projects, as judged by the private sector, have a track record of development and satisfy criteria such as "strong leadership, community support, good ties with local government, experience in program management and the potential to increase its local public and private sector backing" (LISC, 1984).

As a result of this new era of *entrepreneurialism* among neighborhood groups, contradictory goals of advocacy and economic development have occurred. As neighborhood groups assume the responsibility for economic development, private businesses can abdicate this local arena. Unequal access to organizational resources puts those groups most in need at a disadvantage because they also are more dependent on confrontation as a strategy of solidarity (Lurcott & Downing, 1987; Sbragia, 1989). Support for advocacy groups again has been encouraged through the Partnership for Neighborhood Development, the Neighborhood Fund, and the Community Technical Assistance Corporation (CTAC)—all supported by foundations (Lurcott & Downing, 1987). Yet neighborhoods often find themselves in competition with each other for funding, which undermines solidarity. An overall strategy for neighborhood groups is necessary for them to relate to the private sector in a collective and unified way. Communitywide resources had to be organized and made available in order to support a grassroots base for a corporatist form.

Two groups provide ideological and organizational support linkages that represent the interests of all neighborhood groups: the voluntary activist PNA, and the Working Group. The Working Group is composed of community group leaders and provides strategy. The PNA's mission is to provide leadership on neighborhood issues. The PNA has had to resist institutionalization; for example, in 1985 the group decided to sacrifice a city contract worth $320,000 because its advocacy role was becoming co-opted. Coordinating these efforts is difficult; no neighborhood organization can match the planning and technical expertise and financial power of such private-sector associations as the ACCD.

Those groups that have developed the individual capacity to work with the private sector have tried to assert their independence from the local state. Historically the unwieldiness of the URA, entrenched as "middleman" for development deals, has monopolized planning expertise. One neighborhood activist (see Jezierski, 1988a, p. 470) perceived that a change has occurred in the dynamic of the neighborhood-URA relationship: "Now, the intellectual movement is not there. Now more frustration is felt on an operational level and there are bureaucratic complaints. Some of the staff at the URA are good at the technical level but are not good at process. They are too elitist."

The URA, as the organizational mechanism through which negotiation takes place, accumulates contradictions and risks delegitimation because it is constructed in a way to fend off accessibility and to enhance its autonomy. Ironically, on the one hand, neighborhood groups complain that the local state is not fulfilling its social welfare functions, and on the other hand, they complain that the local state serves as an impediment to the neighborhoods' own efforts at market activities. These complaints show that some neighborhood organizations do not recognize the partnership as necessary as they develop their own ties with the private sector and that they simply demand more direct neighborhood investment. Yet no matter how substantially neighborhood interests are incorporated into local economic development, they have not gained enough public status to be included in citywide or regional development (Weiss & Metzger, 1987).

The emerging partnership between the ACCD, the High Technology Council, the universities, and the local state reasserts the fundamental interest in economic growth and developing a new Pittsburgh (Ahlbrandt & Weaver, 1987) without consultation of neighborhood interests. A recent comprehensive and long-term development policy report known as *Strategy 21* (City of Pittsburgh, 1985) was prepared by the ACCD at the request of state legislators. In it, the ACCD provided a consolidated package of capital budget requests for large capital improvements. The report was controversial because neighborhood groups were not consulted and it did not include any of their projects. More recently, the Working Group provided an education campaign to point out that airport expansion plans may divert development to the suburbs, rather than to the city of Pittsburgh.

The explicit economic policy to create new industrial parks and to pursue high-technology and service industries was developed in the midst of an economic recession and deep-cutting retrenchment in manufacturing jobs, especially steel. The unions involved were incapacitated to the extent that they could not provide much leadership, nor could they save these jobs themselves. The political impetus for job development has been spearheaded by neighborhood groups that have realized they have a direct interest in participating in citywide economic policy forums that affect their livelihoods. Their new forays into economic development began with protest and increasingly became institutionalized (Plotkin & Scheuerman, 1987). They have created new alliances to lobby for a different economic development agenda from the one being considered by the ACCD and the High Technology Council. The new alliances are no longer content with previous concessions to the partnership that such groups be involved in the parochial sphere of neighborhood development alone. The first lobbying attempts involved labor unions, churches, and neighborhood groups organized into protest movements in the steel-valley towns around Pittsburgh (Plotkin & Scheuerman, 1987).[10] This activist coalition demanded small business development, eminent domain to take over abandoned factories, and plant-closing-notification laws.

A new wave of mobilization for jobs was begun with an ad hoc coalition called the Save Nabisco Action Coalition (SNAC), which successfully halted the closing of a Nabisco bakery in Pittsburgh's East Liberty neighborhood in late 1982. SNAC received broad-based community support; it consisted of about 30 organizations, including the bakery labor union, supporting unions, neighborhood groups, churches, and leftist activists. SNAC pressured Nabisco by organizing a postcard campaign to support a boycott of company products and to threaten to withdraw funds from Equibank, a Pittsburgh bank with interlocking directorates, if the company closed. Well-attended public rallies provided a forum for group demands that the URA take over the plant through eminent domain and won the backing of the local media and the city council. Mayor Caliguiri, however, was more conciliatory to Nabisco, offering the company low-interest loans and free or low-cost land as incentives to keep the plant open. Only 2 months after the company's announced closing and the SNAC mobilizations, the company decided to stay in Pittsburgh, negotiating with the union and the city to reorganize production. All 650 jobs were saved.

The quick success with Nabisco provided SNAC with a moral authority that allowed the coalition to enlarge, changing its name to Save Our Neighborhoods Coalition, and to broaden its interests in other arenas of economic development. SNAC became a strong political force, though a controversial one:

> SNAC is nothing if not ubiquitous. Nabisco plans to close. SNAC springs up. Budget Laundry plans to close. SNAC's there again. Coca-Cola says it's moving. Here comes SNAC. The state closes an office. SNAC sues. Equitable Gas Co. workers strike. SNAC boos the company. Council approves the plant-closing bill. SNAC cheers. Caliguiri vetoes the bill. SNAC jeers. (Fontana, 1983)

The momentum of SNAC's successful mobilization led to the advocacy of a plant-closing-notification ordinance that the Pittsburgh city council passed. The mayor vetoed the bill, however, supporting business groups organized in opposition, afraid that the ordinance would scare away business. The ordinance was contested and found to violate the city's home-rule charter, which prevents the city from regulating business.

The plant-closing bill had upped the stakes in that it was a direct attempt to control private property and accumulation in the city. Although the media had supported SNAC in the Nabisco event, they did not back the plant-closing bill. Because Nabisco was not a local firm, it was easier to target, but the proposed legislation fundamentally challenged the local private sector. SNAC's position was labeled "confrontational" to delegitimize it. The SNAC coalition finally lost steam after 2 years, partly because the broadness of its mission weakened its organizational solidarity. The protest movement did succeed, however, in electing some of its leaders to the city council.

The SNAC movement represents a case of "protest is not enough." SNAC was demanding equal participation in citywide economic development policy, but limits to its incorporation into local-state policy were bound by legal prohibitions to seizing the private property of corporations. Protest affirmed the right of neighborhoods to defend themselves, and this stance was institutionalized in 1988 in a new organization, the Pittsburgh Community Reinvestment Group, to watch-dog bank investment. Other efforts were channeled into linkage relations, demanding job training and hiring in exchange for hospital and university expansion plans. The partnership's response was to continue to encourage development corporations for individual neighborhoods but to pursue regional development on their own, emphasizing high technology and services.

An alternative economic agenda has been pursued by the Tri-State Conference, a regional nonprofit group of activists intent on saving steel and manufacturing plants. They lobbied unions and Pittsburgh-area municipal and county administrations for support and created a quasi-public legal organization, the Steel-Valley Authority (SVA), to gain state control over industrial property and business development. Its significance was that theoretically it could parallel the power vested in the URA and the ACCD, instituting an alternative partnership. The composition of the SVA board included the interests of workers, union representatives, municipal and state officials, and industry representatives. The SVA represents the closest to a corporatist ideal yet envisioned in any city. This SVA plan appealed to all in the Pittsburgh community who were concerned about keeping manufacturing jobs, including Mayor Caliguiri, who was a key supporter, as well as to neighborhood activists and unions, to the officials in nine of the most economically discouraged municipal and county jurisdictions, and to state legislators. The SVA's budget was based on financial support from the city of Pittsburgh, Allegheny County, the URA, and other sources. As of early 1990, the SVA was negotiating to purchase land and capital to set up a new community-and-employee-owned steel mill that would generate income. The SVA's negotiation for the property, however, had been limited by a lack of commitment by private investors who were wary of the use of eminent domain to secure industrial property and of the difficulty of deal making in the public sphere where other civic groups were involved.

Neighborhood groups have had to begin a new cycle of protest, incorporation, and association to tackle a new area of concern over jobs. Corporatist relations within the public-private partnership had maintained the private sector's traditional control of city and regional economic development, while neighborhood expertise had been limited to housing and neighborhood commercial development. Meanwhile, neighborhood groups were demanding more participation in private-sector decision making. Broadening access to job-development policy first required that they create new coalitions and protest to delegitimize the corporate agenda. The partnership had failed to maintain working-class jobs and to provide growth, given industrial restructuring. The participation of neighborhood and worker interests required the formation of new organizations to enhance expertise and legitimacy.

Building Community Consent to Partnerships

The Pittsburgh case illustrates the limits to legitimacy inherent in public-private partnerships. Truly inclusive, corporatist partnerships in which neighborhood interests are equally and legally represented with business and state interests are difficult to construct because these groups often have fundamentally different and contradictory demands. The longevity of the Pittsburgh partnership has not relied on consensus, but on a constant renegotiation of consent built on a dynamic of mutual exchange of threat of disinvestment and delegitimation among neighborhood groups, business interests, and quasi-governmental entities.

Partnerships alter the governing structure of a community by limiting participation to enhance efficiency. Therefore, they become an object of political struggle. When a public-private partnership is introduced into local politics, two separate spheres of political bargaining are created, each with its own conditions. On the one hand, partnerships rely on quasi-public legal authorities and commissions that have their own rules for participation based on public status, exclusivity, and technical expertise. Electoral politics, on the other hand, rely on the mobilization of voting resources to enhance decentralized, partisan interests within a competitive forum.

Neighborhood groups must maneuver between these two spheres, often pursuing contradictory strategies. Strategies of protest, coalition, and electoral participation are powerful tools through which neighborhood groups assert sovereignty and legitimacy. They are useful for bringing visibility to neighborhood issues, to open up negotiation, and to serve as a veto power over unpopular partnership policies. More institutionalized organizational forms are required, however, for long-term negotiation. These dual requirements can hinder neighborhood partnership incorporation but do not affect business interests because they do not represent an elective constituency.

Theories of incorporation (Browning & Marshall, 1986; Browning et al., 1984) indicate that the inclusion of neighborhood interests into urban policy must get beyond protest and require electoral power. Incorporation theorists would suggest that for the case of the public-private partnership, neighborhood groups must capture state power to advance their interests. In Pittsburgh, neighborhood groups were successful at gaining incorporation into state structures through traditional forms such as protest, coalitions and parties, and electoral power only after dismantling the Democratic machine and reorganizing election procedures. They were able to wield some influence in the partnership indirectly because they controlled public sector input. This option was limited, however, by the local state, as an actor with its own interests, to pursue downtown and regional development, to maintain private investment, and to extend its own security and resources. The law also prohibited the state from interfering with private (business) property. Indeed the state was able to protect itself from incorporation strategies by forming exclusive quasi-legal structures to limit participation and to maximize flexibility and efficiency. Local-state policy resulted in a contradiction as a dual approach to development was attempted.

The cases presented here show a threshold to incorporation: If working-class, neighborhood, or minority interests are too entrenched in the state, business also can withdraw from political and economic engagement because the state becomes an unruly partner. Incorporation theory is limited because it does not explain the actions of the state or the significance of the dual-state structure. Theories of corporatism (Clarke, 1986; Clarke & Meyer, 1986; Coleman, 1983; Schmitter, 1983) remind us that the organizational form of the state is a crucial limiting factor and that the state must be considered an actor in its own right.

The Pittsburgh case also indicates that protest remains a useful strategy for neighborhood groups to delegitimize partnership policy when they find themselves excluded. It also enhances the solidarity and visibility of advocacy groups. When taken too far, however, these groups may become viewed as too irrational or parochial and therefore unsuitable for negotiation in partnerships. Thus partnership involvement required that neighborhood groups change their strategy and become more institutional by constructing new political organizations in which resources could be coordinated and representation could be stabilized for negotiation. Some groups in Pittsburgh were successful at acquiring expertise in economic development and planning and gained the special public status required to be included in neighborhood development policy. This would seem to conform with the corporatist model that is based on the assumption that each interest will participate equally and contribute in its own area of expertise. One key qualifier is that neighborhood groups were organized in response to and with the aid of prior organizational forms representing opposite interests such as the ACCD, business foundations, and the local state. It may be argued that this higher degree of organizational capacity in Pittsburgh—at many levels—has allowed a more successful transformation of the local economy.

This corporatist-consensus model is limited, however, even in the Pittsburgh case of greater organizational capacity. The participation of neighborhood groups was limited to social services and neighborhood economic development because, on the one hand, the partnership did not recognize them as having the expertise and authority to make decisions about regional economic investment. Business interests, on the other hand, were recognized as having the public authority to direct housing policy, neighborhood development, and educational issues. Moreover, the internal contradictions of pursuing neighborhood advocacy and an institutionalized form of development corporations weaken neighborhood organization. These groups were viewed as separate interests, and they found themselves in competition with each other for partnership recognition and resources. Traditional forms of protest and electoral power remain important tools for advancing neighborhood interests despite partnership requirements. A more decentralized and controllable form of local state may be a better alternative than expending the resources required by corporatist forms of organization and the possible co-optation that may result.

Neighborhood participation will remain limited until individual interests are represented in a collective association that is parallel in scope and power to business associations such as the ACCD. The most advanced organizations in Pittsburgh are the PNA (for advocacy), the Pittsburgh Partnership for Neighborhood Development (for technical help), and the SVA (which incorporates labor and other community-based groups in economic development on a regional scale). Yet these organizations are limited by the lack of legal and economic resources, autonomy, and veto power that business groups maintain.

This discussion leads to the consideration of whether partnerships are a viable and legitimate means to address urban decline. The Pittsburgh case provides evidence that a complex political struggle has emerged in response to economic restructuring. On the one hand, residents have mobilized to save their homes, to maintain sovereignty, and to save traditional working-class jobs. They often do accept, on the other hand, control of downtown development by a corporate elite because this is the domain of business. When exclusive business-government elites run partnerships under growth conditions, such as in Sunbelt cities (Clarke & Rich, 1982; Fleischmann & Feagin, 1987), residents are content to accept the arrangement under the condition of promised growth in jobs and other amenities.

The legitimacy of an exclusive partnership rests in an interpretation by the electorate that city government is bargaining successfully with their interests in mind. What already exists is only enhanced, including expanding a community with new neighborhoods. But even growth has its limits when quantitative change affects quality of life (Logan & Molotch, 1987; Vogel & Swanson, 1989). Under conditions of economic decline, restructuring prompts skepticism of business and government just when residents should be most willing to accept corporate directives because they are desperate for business involvement. Consent will be withdrawn when restructuring threatens traditional forms of community and political participation; thus exclusive partnerships can be delegitimized. The instability inherent in restructuring of community opens new challenges over development agendas, and stability will be reasserted only when legitimacy is achieved through the inclusion of neighborhood groups in partnerships.

Notes

1. Ahlbrandt and Weaver (1987, p. 457) define the partnership as a "purposive institutional form that uses a corporative transactive model based upon normative compliance, mutual learning, bargaining, and negotiation." Ideally partnerships could enable broader representation and progressive planning (Clarke, 1986; Schmitter, 1983).

2. Clarke (1986, p. 46) identified four types of partnership sponsors in a survey of Urban Development Action Grant (UDAG) packages: business and government partnerships accounted for 49%; neighborhood groups or other nonprofits 23%; other governmental units 17%; and combination 19%.

3. Pittsburgh organizations and programs, including the Neighborhood Housing Service, the Public Education Fund, and the Allegheny Conference on Community Development provided national models (Ahlbrandt, 1986; Ahlbrandt & Brophy, 1975; Fosler & Berger, 1982; Stewman & Tarr, 1982; Weaver & Dennert, 1987).

4. Because there are no labor parties in the United States, there is some debate whether corporatism applies to the U.S. case (Salisbury, 1979; Wilson, 1982). Corporatism could exist, however, at the local level under the decentralized federalism of the United States and may be applicable under specific projects in partnerships (Clarke, 1986; Clarke & Rich, 1982; Coleman, 1983; Schmitter, 1983). Clarke (1986, pp. 29-40) argued that partnerships, which are corporatist-like, are basically nonpluralist and that the state works not as a broker but as an active agent, an interested participant on its own. Whether local interest-intermediation happens through pluralist or corporatist dynamics may be a matter for empirical determination.

5. The public sector is usually reactive—limited to the role of referee and coordinator, rather than initiator—because the process often is not legally binding. Ahlbrandt and Weaver (1987) argued that the public sector had neither the expertise nor the investment funds to undertake the massive redevelopment achieved in Pittsburgh.

6. The analysis is based on a case study of Pittsburgh that included archival data and 35 in-depth interviews with representatives from business, labor, community groups, voluntary associations, and local government (see Jezierski, 1988a, for sources and methodology).

7. Przeworski (1980, pp. 31-33) quoted Gramsci to explain the reproduction of capitalist society: Wage earners consent to capitalist organization of society when they act as if they could improve their material conditions within the confines of capitalism. . . . Since the capacity of any group to satisfy material needs in the future depends fundamentally upon decisions of capitalists concerning the volume and direction of investment, democracy is a system through which these decisions can be influenced by any one *qua* citizen. The application of the concept of consent to neighborhood and urban investment is developed elsewhere (Jezierski, 1988a).

8. Coleman (1983, pp. 62-63) defined Pittsburgh's corporatist-like power structure as follows: In Pittsburgh the dominant value of urban development was to achieve efficient private growth. The political/economic structure of the city with its powerful corporate community and its strong mayor form of government provided a hierarchical base for urban development decision making. Pittsburgh's urban development process includes many of the characteristics of corporatism, including pre-emption of issues, permanent institutionalization of access, control of technocratic planning and allocation, extensive development of para-state agencies, and a political culture stressing consensus and continued bargaining.

9. This information was derived from an interview with John P. Robin, Pittsburgh, PA, in 1986 (see Jezierski, 1988a).

10. The radical nature of the movement was exemplified by the Denominational Ministry Strategy (DMS), which received national media coverage for its extreme strategies of throwing skunk oil and blood on the homes of USX (formerly United States Steel) corporation executives.

References

Adams, S. (1983). Public/private sector initiatives: Principles and action strategies. *National Civic Review, 72,* 83-98.

Ahlbrandt, R. S., Jr. (1984). *Neighborhoods, people, and community.* New York: Plenum.

Ahlbrandt, R. S., Jr. (1986). Public-private partnerships for neighborhood renewal. *Annals of the American Academy of Political and Social Science, 488,* 120-134.

Ahlbrandt, R. S., Jr., & Brophy, P. (1975). *Neighborhood revitalization: Theory and practice.* Lexington, MA: Lexington Books.

Ahlbrandt, R. S., Jr., & Weaver, C. (1987). Public-private institutions and advanced technology development in Southwestern Pennsylvania. *Journal of the American Planning Association, 53,* 449-458.

Bodnar, J., Simon, R., & Weber, M. P. (1982). *Lives of their own: Blacks, Italians, and Poles in Pittsburgh, 1900-1960.* Urbana: University of Illinois Press.

Brooks, H., Liebman, L., & Schelling, C. (1984). *Public-private partnerships.* Cambridge, MA: Ballinger.

Browning, R., & Marshall, D. R. (1986). Forum: Protest is not enough. *PS, 19,* 573-581.

Browning, R., Marshall, D. R., & Tabb, D. (1984). *Protest is not enough.* Berkeley: University of California Press.

Castells, M. (1983). *The city and the grassroots.* Berkeley: University of California Press.

City of Pittsburgh, PA, Department of City Planning. (1980). Map of Pittsburgh's neighborhoods.

City of Pittsburgh, PA, Office of the Mayor. (1985). *Strategy 21: Pittsburgh/Allegheny economic development strategy to begin the 21st century: A proposal to the Commonwealth of Pennsylvania.* Pittsburgh: Author.

Clarke, S. E. (1986). Urban America, Inc. In E. M. Bergman (Ed.), *Local economies in transition* (pp. 37-58). Durham, NC: Duke University Press.

Clarke, S. E., & Meyer, M. (1986). Responding to grassroots discontent: Germany and the United States. *International Journal of Urban and Regional Research, 10,* 401-417.

Clarke, S. E., & Rich, M. J. (1982). Partnerships for economic development: The UDAG experience. *Community Action, 1,* 51-56.

Coleman, M. (1983). *Interest intermediation and local urban development.* Unpublished doctoral dissertation, University of Pittsburgh.

Committee for Economic Development. (1982). *Public-private partnership: An opportunity for urban communities.* New York: Research and Policy Committee.

Cunningham, J. V. (1983). Power, participation, and local government: The communal struggle for parity. *Journal of Urban Affairs, 5,* 257-266.

Cunningham, J. V., & Kotler, M. (1983). *Building neighborhood organizations.* Notre Dame, IN: University of Notre Dame Press.

Fainstein, S. S., & Fainstein, N. I. (1985). Economic restructuring and the rise of urban social movements. *Urban Affairs Quarterly, 21,* 187-206.

Fleischmann, A., & Feagin, J. (1987). The politics of growth-oriented urban alliances: Comparing old industrial and new Sunbelt cities. *Urban Affairs Quarterly, 23,* 207-232.

Fontana, R. F. (1983, August 10). SNAC becomes a fighting advocacy group. *Pittsburgh Post-Gazette.*

Fosler, R. S., & Berger, R. A. (Eds.). (1982). *Public-private partnerships in American cities.* Lexington, MA: Lexington Books.

Gittell, M. (1980). *Limits to citizen participation: The decline of community organizations.* Beverly Hills, CA: Sage.

Goering, J. M. (1979). The national neighborhood movement: A preliminary analysis and critique. *Journal of the American Planning Association, 45,* 506-514.

Goldstein, H. A., & Bergman, E. M. (1986). Institutional arrangements for state and local industrial policy. *Journal of the American Planning Association, 52,* 265-277.

Gottdiener, M. (1987). *The decline of urban politics: Political theory and the crisis of the local state.* Newbury Park, CA: Sage.

Habermas, J. (1975). *Legitimation crisis.* Boston: Beacon.

Henig, J. (1982). *Neighborhood mobilization: Redevelopment and response.* New Brunswick, NJ: Rutgers University Press.

Hritz, T. M. (1968, June 26). New goals set for better city: Conference turns emphasis toward social reforms. *Pittsburgh Post-Gazette*, p. 1.

Jacobson, L. (1987). Labor mobility and structural change in Pittsburgh. *Journal of the American Planning Association, 53*, 438-448.

Jezierski, L. (1988a). *The politics of urban decline: Accumulation and community in Cleveland and Pittsburgh.* Unpublished doctoral dissertation, University of California, Berkeley.

Jezierski, L. (1988b). Political limits to development in two declining cities: Cleveland and Pittsburgh. In M. Wallace & J. Rothschild (Eds.), *Research in politics and society* (Vol. 3, pp. 173-189). Greenwich, CT: JAI.

Katznelson, I. (1981). *City trenches: Urban politics and the patterning of class in the U.S.* New York: Pantheon.

Langton, S. (1983). Public-private partnerships: Hope or hoax? *National Civic Review, 72*, 256-261.

Lipsky, M. (1968). Protest as a political strategy. *American Political Science Review, 62*, 1144-1158.

Local Initiatives Support Coalition (LISC). (1984). *A two-year report.* Washington, DC: Ford Foundation.

Logan, J., & Molotch, H. (1987). *Urban fortunes.* Berkeley: University of California Press.

Lorant, S. (1980). *Pittsburgh: The story of an American city.* Lenox, MA: Author's Edition.

Lubove, R. (1969). *Twentieth century Pittsburgh.* New York: John Wiley.

Lurcott, R. H., & Downing, J. A. (1987). A public-private support system for community-based organizations in Pittsburgh. *Journal of the American Planning Association, 53*, 459-468.

Mollenkopf, J. (1983). *The contested city.* Princeton, NJ: Princeton University Press.

Mollenkopf, J. (1986). New York: The great anomaly. *PS, 19*, 591-597.

Offe, C. (1985). *Disorganized capitalism.* Cambridge: MIT Press.

Pittsburgh's new powers. (1947). *Fortune, 35*, 69-77.

Piven, F., & Friedland, R. (1984). Public choice and private power. In A. Kirby, P. Knox, & S. Pinch (Eds.), *Public service provision and urban development* (pp. 391-420). New York: St. Martin's.

Plotkin, A., & Scheuerman, W. (1987, March). *Two roads left: Strategies of resistance to steel plant closings in the Monongahela Valley.* Paper presented at the Annual Meeting of the Southwestern Social Science Association, Dallas, TX.

Przeworski, A. (1980). Material bases of consent: Economics and politics in a hegemonic system. In M. Zeitlin (Ed.), *Political power and social theory* (pp. 21-66). Greenwich, CT: JAI.

Robin, J. (1972). *The Pittsburgh renaissance project.* Staunton Balfour Oral History Collection, Archives of Industrial History, University of Pittsburgh.

Salisbury, R. (1979). Why no corporatism in America? In P. Schmitter & G. Lembruch (Eds.), *Trends toward corporatist intermediation* (pp. 213-230). Beverly Hills, CA: Sage.

Sbragia, A. (1989). The Pittsburgh model of economic development: Partnership, responsiveness, and indifference. In G. Squires (Ed.), *Unequal partnerships: The political economy of urban redevelopment in the post-war era* (pp. 103-120). New Brunswick, NJ: Rutgers University Press.

Schmitter, P. C. (1979). Modes of interest intermediation and models of societal change in Western Europe. In P. Schmitter & G. Lembruch (Eds.), *Trends toward corporatist intermediation* (pp. 63-94). Beverly Hills, CA: Sage.

Schmitter, P. C. (1983). Democratic theory and neocorporatist practice. *Social Research, 50*, 885-928.

Schwartz, E. (1979). Neighborhoodism: A conflict in values. *Social Policy, 5*, 9-14.

Shearer, D. (1989). In search of equal partnerships: Prospects for progressive urban policy in the 1990s. In G. Squires (Ed.), *Unequal partnerships: The political economy of urban redevelopment in the post-war era* (pp. 289-307). New Brunswick, NJ: Rutgers University Press.

Silver, H., & Burton, R. (1986). The politics of state-level industrial policy: Lessons from Rhode Island's Greenhouse Compact. *Journal of the American Planning Association, 52*, 277-289.

Spiegel, H.B.C. (1981). The neighborhood partnership: Who's in it? Why? *National Civic Review, 70*, 513-520.

Squires, G. (Ed.). (1989). *Unequal partnerships: The political economy of urban redevelopment in the post-war era.* New Brunswick, NJ: Rutgers University Press.

Stewman, S., & Tarr, J. A. (1982). Four decades of public-private partnerships in Pittsburgh. In R. S. Fosler & R. A. Berger (Eds.), *Public-private partnerships in American cities* (pp. 59-127). Lexington, MA: Lexington Books.

Stone, C. (1986). Atlanta: Protest and elections are not enough. *PS, 19*, 618-625.

Swanstrom, T. (1985). *The crisis of growth politics.* Philadelphia: Temple University Press.

Thomas, J. C. (1983). Citizen participation and urban administration: From enemies to allies? *Journal of Urban Affairs, 5*, 175-183.

U.S. Bureau of the Census. (1980). *Census of population, industry, and employment* (Table 122). Washington, DC: Government Printing Office.

Vogel, R. K., & Swanson, B. E. (1989). The growth machine versus the antigrowth coalition: The battle for our communities. *Urban Affairs Quarterly, 25,* 63-84.

Weaver, C., & Dennert, M. (1987). Economic development and the public-private partnership. *Journal of the American Planning Association, 53,* 430-437.

Weiss, M. A., & Metzger, J. T. (1987). Technology development, neighborhood planning, and negotiated partnerships. *Journal of the American Planning Association, 53,* 469-477.

Wilson, G. K. (1982). Why is there no corporatism in the States? In G. Lembruch & P. Schmitter (Eds.), *Patterns of corporatist policy-making* (pp. 219-236). Beverly Hills, CA: Sage.

— 19 —

Neighborhood Associations

THEIR ISSUES, THEIR ALLIES,

AND THEIR OPPONENTS

John R. Logan

Gordana Rabrenovic

More than 35 years ago, counter to then-current fashion, two social scientists—Form (1954) and Long (1958)—argued that urban researchers should focus more directly on the various actors whose intentions, strategies, and interactions affect community development. As urban theory has veered away from the economic determinism and "invisible hand" metaphors of human ecology, consensus has grown among researchers regarding the need to understand land developers and realtors, financial institutions, associations of homeowners and tenant unions, and other political players. Although new theories are being based on assumptions about the interests and activities of such players, great gaps exist in the research base.

In our study we examine one type of actor: the neighborhood association. For our purposes a *neighborhood association* is defined as a civic organization oriented toward maintaining or improving the quality of life in a geographically delimited residential area. We are interested in the neighborhood association, not simply as a form of local voluntary organization, but more precisely as a distinctive form in which the common interests of residents of a bounded community area are expressed in American cities. The neighborhood association is not the only possible form; political party organizations, churches, union locals, ethnic clubs, and chapters of environmental groups can, and sometimes do, play this role. Nevertheless in this country the neighborhood association is commonly the vehicle through which neighbors learn about problems, formulate opinions, and seek to intervene in the political process to protect their local

AUTHORS' NOTE: This research was supported by a grant from the National Science Foundation (SES 8606586). We appreciate the assistance of Carolyn Horowitz and Gaye Reinhold in the interviewing of neighborhood association leaders.

interests. Further, although at one time these organizations may have had largely social functions, the current tendency seems to be that instrumental political activities predominate (Crenson, 1983; Goering, 1979; Guest, 1985). In this study, then, we examine how and why these associations come into being, what issues become the focus of their attention, and how they understand the political context in which they act.

Up Against the Growth Machine

The most vivid account of the politics of urbanization is Molotch's (1976) model of the city as a growth machine (see also Logan & Molotch, 1987). In this model the central issue in local politics, the issue that captures the essence of locality, is the pursuit of ever more intensive growth. Political control is attributed to a coalition of progrowth entrepreneurs (aiming for profit from rent intensification) against whom the interests of city residents (desiring price stability and security of their daily activities in the neighborhood) seldom prevail.

Mollenkopf (1983) portrayed growth politics in similar fashion, although in his view, neighborhood mobilization has stalled the growth machine in a number of major cities. The neighborhood association is often the vehicle for such mobilization. Elkin (1985) and Fainstein and Fainstein (1983) also have emphasized the centrality of progrowth coalitions in urban politics. In their accounts, however, the greater oppositional role is played by lower income and minority residents, and middle-class residents are viewed as important constituents whose interests often are protected by local regimes.

A parallel view of neighborhood associations has emerged from the geographic literature on locational conflicts (see especially Cox, 1981). Janelle (1977) emphasized transportation, high-density residential development, and commercial expansion as specific issues about which business and residents may have opposing views. Boyte (1980) stressed urban renewal and highway programs as the bases for urban confrontations in the 1960s.

Types of Issues

These researchers gave prominent attention to land-development questions. Others have given equal weight to issues of public service delivery—schools, crime prevention, traffic and parking regulation, and garbage collection (Burnett, 1983; Davidson, 1979; Lowe, 1977; Thomas, 1986). In one of the best empirical studies to date, Guest and Oropesa (1984) found that Seattle residents were most concerned with services and facilities (mentioned by 29.7% of respondents) and physical characteristics of the community environment, including streets and sewers (27.7%). Land-use change and zoning were mentioned by only 18.2%, which is a substantial percentage but is lower than might be expected on the basis of the growth-politics literature. We must note, however, that these data refer to the concerns of individual residents, not to the activities of neighborhood associations.

Thus one question that we must address is the relative importance of land-development issues and other sorts of issues for neighborhood associations. We take two approaches. One is to explore the conditions in which associations were formed: when, in what political context, and on what issue or issues. But, as Arnold (1979, p. 15) pointed out, many associations are founded to address a single issue but their concerns are broadened as they become permanent organizations. Therefore we also examine the current issues addressed by neighborhood associations.

The Balance of Power

A second question is how neighborhood associations fit into the local balance of forces. In the growth machine model, residents are portrayed as solo players with few allies against a powerful alliance of business and real estate developers. This image also has been disputed. Arnold (1979) believed that the key protagonist of neighborhood associations is municipal government and that their relationship to each other is becoming more cooperative over time as municipal officials realize that neighborhood associations are valuable for information and advice.

Similarly Taub and Surgeon (1977) argued that neighborhood associations increasingly are created or supported by government agencies that need them as channels of communication, sources of legitimation, vehicles of social control, and a means to organize and direct resources. This need has grown as the ideology emphasizing local participation and control has become more widespread. Of course, cooperation may tend toward cooptation, and some researchers and community activists are concerned that "the supposed opening of government is only a technique for seducing citizen groups" (Thomas, 1986, p. 5). But others (Ley & Mercer, 1980) emphasize that local government itself is not a monolithic institution and that citizen groups may find supporters in some quarters while battling other city officials.

The relationship between neighborhood associations and land developers also may be complex. Logan and Molotch (1987, pp. 139-146) pointed out that in gentrifying city neighborhoods, neighborhood associations representing more affluent residents may well play a double role in growth politics. On the one hand, they protect their turf against nonresidential development. On the other hand, they promote the continued conversion of the neighborhood. For example, Mollenkopf (1983, p. 175) reported close links between the Boston Redevelopment Authority and the leadership of the South End neighborhood in Boston. In another permutation of this pattern, the University of Chicago established and funded the Southeast Chicago Commission as a supposed community organization as part of its efforts to redevelop the Hyde Park-Kenwood neighborhood (Rossi & Dentler, 1961).

Certainly these visions of local political alignments depend on the kinds of issues that analysts have in mind. Some, intent on exploring land-development questions, emphasize business and developers as opponents (or sometimes as manipulators) of neighborhood associations; others, thinking of the demand for public services, emphasize local government. A major research question for us is how neighborhood associations themselves define their issues and, in turn, whom they perceive to be either their key allies or opponents.

It is unnecessary to posit a unitary pattern across time and space. Neighborhood associations may vary according to the logic of their own evolution or the character of their environment or members (Fisher, 1984; Henig, 1982; Thomas, 1986). We look specifically for differences between neighborhood associations in central-city and suburban locations. Suburbs are distinctive in the composition of their populations and housing stock. Furthermore, a much narrower range of public services is provided in the suburbs than is provided in central cities (Cox, 1973, pp. 30-48; Dye, 1970). Also, according to some political analysts, a particularly high level of community identity is enjoyed in the suburbs, and the means to enhance and perpetuate that identity are provided through public policy (Greer, 1962, p. 109; R. Wood, 1958). The continuing "close contact" of people to government in the suburbs makes the public's response to political decisions predictable and forces (or enables) local officials to adhere to public opinion (Crain & Rosenthal, 1967). Therefore one might expect to find relatively little emphasis on issues of service and public infrastructure in suburban neighborhoods and higher levels of cooperation with local government, compared with that in city neighborhoods.

In sum, our purpose here is to examine the causes of the founding of neighborhood associations, the kinds of issues on which they focus their attention, and their understanding of the forces that support and oppose their goals. The analysis is as much historical and descriptive—generating hypotheses for future work—as it is statistical, although some specific hypotheses are tested. We adopt an implicit causal model here in which some objective characteristics of the larger political environment and of neighborhoods themselves are expected to affect creation of neighborhood associations and the relative salience of different types of issues, which, in turn, affect perceptions of cooperation and conflict.

Research Design

This study is based on survey research in 1986 among neighborhood associations in New York's Capital District (encompassing Albany, Schenectady, Troy, and surrounding suburbs). A list of 99 existing neighborhood associations was compiled in the following manner. First, we contacted every local (municipal or county) planning office to obtain names of organizations. Second, in the city of Albany, we obtained a list of member organizations of the Council of Albany Neighborhood Associations. Third, we reviewed relevant newspaper articles from the *Albany Times Union,* which is indexed by the library staff at the State University of New York at Albany. Finally, in every interview, we asked respondents for names of other neighborhood associations in the area. Because this last method provided no new names, we are confident that our list is comprehensive.

Most of these organizations explicitly refer to themselves as neighborhood associations; a small number refer to themselves as homeowner, home improvement, tenant, resident, and civic associations. The majority are formally incorporated. On average, they have a membership of 156 persons and an annual budget of under $1,000. Only four have any paid staff. Most are financed entirely by dues paid by members.

We acknowledge some ambiguity in the definition of *neighborhood association* as we use it here. As noted earlier, our emphasis is on the representation of collective interests of residents in a bounded community area. During interviews with local informants, we learned that in one city neighborhood, this function was for many years handled by lay organizations of a local church. Of course, the church drew parishioners from a wider area than the neighborhood, and its principal purpose was religious worship, but our research would be incomplete if we ignored the activities of the organization in this church prior to about 1970. Another illustration of definitional ambiguity is provided by merchants' associations. We know about one case in which residents of a neighborhood who also own small businesses in the same area are the dominant voice in what is formally designated as a neighborhood association, and this association is included in our sample. In another neighborhood, business proprietors along a single street have formed a merchants' association that often acts in concert with local residents to regulate new development; this merchants' association is not included in our sample.

A total of 74 personal interviews were completed; thus the response rate was about 74%. Of these, 72 included sufficient information to be used in this analysis. In fact, most refusals were from organizations that existed in name only, having ceased to function years before. In every case, we attempted to interview the current top elected representative of the association. In some organizations with an informal leadership structure, we were referred from one person to another until an appropriate and willing respondent was found. Interviewing leaders is a strategic choice, in the light of their high levels of involvement and information. As Barber (1965) noted, "There exists in any given association, an active minority and inactive majority among the members" (see also Mansbridge, 1980; J. Wood, 1981).

In the following analysis, comparisons are made between neighborhood associations in central-city and suburban areas (39 in cities, 33 in suburbs). Three central cities are in this metropolitan area, each in a separate county, and for some purposes it might be interesting to distinguish among them. We do not distinguish among them here, however, because there are few cases in the smaller cities of Schenectady (9) and Troy (7) and because it is not meaningful to categorize suburbs in this area according to the city to which they are more closely connected. The city of Albany and Albany County contain the majority of neighborhood associations in this sample, and at some points we give special attention to these cases.

Table 19.1 Neighborhood Associations by Year of Establishment and City/Suburban Location

Year of Establishment	City	Suburb	Total
Before 1960	7.7%	9.1%	8.3%
1961-1970	10.3%	18.2%	13.9%
1971-1974	12.8%	0.0%	6.9%
1975-1979	46.2%	21.2%	34.7%
1980-1985	23.1%	51.5%	36.1%
Total	39	33	72

The Formation of Neighborhood Associations

The Capital District's neighborhood association movement is relatively young (compare to Fisher, 1984). Of the neighborhood associations in our sample, only 6 were founded before 1960. Another 10 date from the 1960s. By contrast, in the 6 years between 1980 and 1985, 26 neighborhood associations were established, and since we completed our interviews in mid-1986, several new neighborhood associations have come into existence. We searched carefully for evidence of associations from earlier periods that did not survive into the present. We found only 4, all in the city of Albany. More common in earlier years were settlement houses and other social service organizations oriented to specific neighborhoods.

The peak period of establishment differs slightly between city and suburban neighborhoods, as shown in Table 19.1. The high point for new neighborhood associations in the city neighborhoods was 1975-1979, and the greatest number of new neighborhood associations in the suburbs were established between 1980 and 1985. It is tempting to interpret this difference as a demonstration effect; that is, perhaps this type of political activity in the central cities paved the way for such activity in the suburbs. Before adopting this view, however, one must look for the distinct sources of neighborhood association formation in both parts of the metropolitan area.

Regarding the city neighborhoods, the surprise is that Albany, Schenectady, and Troy were affected so little by the explosion of grassroots organizing that occurred elsewhere in the country during the 1960s and early 1970s primarily as a result of the civil rights movement and federal support of community organizations (Schoenberg & Rosenbaum, 1980). Our discussion focuses on the Albany case because nearly two thirds of the city neighborhood associations are in Albany. In this case the failure to organize more neighborhood associations may be attributable, in part, to the endurance of the county's Democratic political machine. Following 30 years of Republican party rule in Albany, the Democratic party, under Dan O'Connell, took control in 1921. O'Connell maintained a political monopoly until his death in 1977, when power passed to his hand-picked mayor, Erastus Corning, who held office from 1941 until he died in 1983. This machine enjoyed the longest period of domination of any party apparatus in the country (Brown, 1986, pp. 69-70). Like many others, it operated through precinct leaders who were responsible for securing the Democratic vote and for keeping an eye on potential local issues. Mayor Corning maintained a reputation for responsiveness, and he actually encouraged people to approach him directly with their problems. This combination of political machine and political boss left little space for independent voluntary associations.

Nor did federal antipoverty programs of the 1960s have much lasting effect on community organizing in Albany County. As the novelist William Kennedy noted in his history of Albany, the party machine opposed federal antipoverty programs as "interference" in the local scene (Kennedy, 1983, p. 278). Like New Orleans, where rejection of outsiders is notorious (Smith & Keller, 1983), Albany was among the last major cities to take advantage of the federal programs. The Albany County Urban Renewal Agency was not established until 1960, and ground breaking for the first housing project did not take place until 1967. Curiously, then, it was the Republicans, not the Democrats, who promoted greater participation in the War on Poverty (an issue in the 1965 mayoral election). The first application for antipoverty funds for the city was denied because local officials failed to sign it; the next application was denied because the city's three-person "community representative" advisory board was understood even by the federal bureaucracy to be unrepresentative of the community.

In several city neighborhoods, independent community organizers backed by voluntary social service agencies attempted to promote local action groups in the mid-1960s. The most telling example is Albany's South End neighborhood, where a church-backed settlement house—the Trinity Institution—encouraged the formation of four small community action organizations (the Progressive Community Betterment Association, the Catherine Street Civic Association, the Grand Community Association, and the Better Homes and Community Organization). But precisely at the time that these groups began to agitate vocally for improved city services, the city government retaliated by cutting off funding of the Trinity Institution (from a level of about $13,000 per year; see Kennedy, 1983, p. 341). None of the four action groups survived beyond the 1960s.

By the late 1960s, however, the machine was beginning to weaken. From 1966 until 1970, a Republican represented the area in Congress; and in 1968, voters elected a Republican state senator, two Republicans to the state assembly, and a Republican to the critical position (for operation of a political machine) of County District Attorney (McEneny, 1981). Although redistricting prevented the reelection of these Republican officials, a new level of competitiveness had been introduced into local politics (Swanstrom & Ward, 1987). Then, in the early 1970s, large-scale redevelopment associated with the construction of a new state office complex (a $2 billion project completed in 1973) removed much of the downtown residential area and threatened what remained. These events created both the stimulus and the opportunity for grassroots electoral challenges to the machine's hegemony. Mayor Corning, who had won reelection in 1969 by a landslide (37,896 to 15,212), won by his smallest margin ever in 1973 (25,390 to 21,838). Also in 1973 an insurgent Democrat lost by only 14 votes in her bid to represent a downtown residential neighborhood in the city council.

Thus the surge in neighborhood association establishment in Albany after 1975 seems to have responded to the conjuncture of changes in the local political environment, but what accounts for the even later formation of neighborhood associations in the suburban ring? One obvious explanation is that it is tied to the timing of suburban population growth. We have found, however, that Albany's suburbs grew most rapidly between 1950 and 1960, with a fairly steady growth rate since that time. No special surge in population or housing construction occurred around 1980. Nor did suburban job development accelerate at that time. A critical moment in that respect occurred in 1966, when the first major suburban shopping mall, anchored by Macy's and Sears Roebuck, opened in the suburban town of Colonie. Subsequently, as in most of the United States, retail, wholesale, and manufacturing employment have grown more rapidly in the Albany suburbs than in the central cities of Albany, Schenectady, or Troy. Again, no particular acceleration of this trend occurred around 1980 that would account for the rush to form neighborhood associations. If their founding was a response to suburban growth, it was a delayed one and was perhaps a response to accumulated issues.

Such a delayed response is a curious phenomenon. We tend to think of growth politics in terms of local oppositions to plans for change in the immediate future—a proposed new highway or landfill or apartment complex. One analyst, Rudel (1989), painted quite a different picture for semirural areas in western Connecticut that only recently have experienced high-density growth. In these places the notion of planning itself is new and, to some degree, has been imported by newcomers. Only after the subdivisions are in, Rudel suggested, do residents notice inadequate drainage or overcrowded roadways or other problems of development. Then, after the fact, residents become conscious of growth as a political question. We cannot test Rudel's interpretation here, but it represents an important hypothesis for future research.

Founding Issues

We can gain a better sense of the origins of these neighborhood associations by looking into the issues that initially motivated their organization. By a large majority the neighborhood associations were formed as a result of a specific issue (in 59 of 72 cases). We coded the founding issues into four categories, with a residual "other" category. The frequency of these types is reported in Table 19.2. The most common was a specific industrial or commercial development (23 cases). Specific residential developments (9 cases) also were common. Both of these typically involved disputes over the rezoning of land to a higher density or from residential to nonresidential use. Transportation issues were the concern in 10 cases, as diverse as the establishment of "trackless trolley routes" (the McKownville Improvement Association, dating back to 1924!), the truck traffic in a residential area, and the construction of a major highway. All transportation issues mentioned by respondents referred to "intrusions" into the neighborhood environment, rather than to availability of transportation

Table 19.2 Founding Issue by City/Suburban Location

Founding Issue	City	Suburb	Total
Industrial/commercial development	17.9%	48.5%	31.9%
Residential development	17.9%	6.1%	12.5%
Transportation problem	10.3%	18.2%	13.9%
General concerns	5.1%	3.0%	4.2%
Other reasons	28.2%	9.1%	19.4%
No specific reasons	20.5%	15.2%	18.1%
Total	39	33	72

services. In 3 cases the respondents cited simply a "general concern" with the neighborhood as their reason for forming, and the remaining 14 were formed for miscellaneous reasons.

These results reveal the central importance of land-development questions for neighborhood associations, as distinct from, for example, public services or social activities. By and large, neighborhood associations sprang up in response to specific land-use changes that concerned residents.

This generalization applies particularly to the suburban neighborhood associations. Of 28 suburban neighborhood associations that noted a specific issue, 16 listed industrial and commercial projects of various sorts. These included, for example, expansion of an industrial zone adjacent to a residential area and the placement of a neighborhood shopping center.

City neighborhood associations show a different pattern: Nearly half (19 of 39) of the city cases were formed on the basis of "other" issues or on no specific issue. This trend was most apparent during the 1975-1979 period of rapid growth of city neighborhood associations, when 10 of 18 cases had "other" or no issues as the basis for their formation. We interpret this finding as an illustration of two sorts of processes. The first is the delayed response to federal redevelopment programs. The second is what can best be summarized as an epidemic of neighborhood organizing: neighborhood associations formed in response to the creation of other neighborhood associations.

Numerous examples reveal the importance that federal funding has had in stimulating neighborhood association formation since 1975. The neighborhood association may seek specifically to provide services by using federal funds or to serve as a pressure group to influence how those funds are used in the city or simply to be the formal mechanism by which the community participation requirements for funding are fulfilled. We note that the first funds from the Community Development Block Grant program were distributed in Albany in 1975 (at about $2 million per year, increasing to about $4 million per year in 1979). The South End Neighborhood Association was formed in 1978 to promote the construction of moderate-income housing in the neighborhood with federal community development funding. In Schenectady the Mount Pleasant Neighborhood Association was formed in 1975 to deal with community development funds; Troy's South Troy United Neighborhood Association was formed in 1978 for the same purpose.

The epidemic, or imitation effect, results from the activities of a central core of pioneer organizations. In Albany several early organizations established a Neighborhood Resource Center in 1975 with the specific mission of encouraging new neighborhood associations. In the following year, they created a formal coalition: the Council of Albany Neighborhood Associations (CANA). Among respondents in our survey, the representatives of three neighborhood associations cited the importance of such support (Delaware Avenue, Krank Park, and Inner Brick Square). None of these reported a specific founding issue, and the Inner Brick Square Neighborhood Association (established in 1981) specifically identified CANA as its principal stimulus. This concrete evidence of the process through which imitation may have occurred convinces us that this explanation should be taken seriously.

Current Issues

However a neighborhood association was formed, a more important question is what issues it deals with on a continuing basis. For most neighborhood associations, the issue that stimulated its organization is long past, but after the initial issue is resolved, the neighborhood association may find new matters to confront. What are these issues?

To gain information on this question, our interviewers offered respondents a series of 18 different types of issues and asked how often these had been discussed at board meetings in the past year. Response categories ranged from 1 (not at all) to 4 (almost every meeting). Table 19.3 lists the issue areas and indicates the proportion of neighborhood associations that reported their being discussed "several times" or "almost every meeting." Because such striking differences appear between city and suburban issues, each geographic region are discussed separately.

For city neighborhood associations, the top issues (in order) are "streets and sidewalks," "cleanliness of the area," and "traffic congestion and parking." Other very common issues are "parks and playgrounds," "condition of housing," and "police protection." Of these, only traffic congestion and parking was discussed frequently by a majority of suburban neighborhood associations (at the extreme, only 18.8% of suburban associations frequently discuss police protection). Clearly, most of these are specifically urban issues, responsive to the higher urban crime rate and shared responsibility for the immediate environment. City neighborhood associations are more likely than suburban associations to discuss frequently all of these issues.

By contrast, land development—in the form of "residential land-use changes," "industrial-commercial encroachment on residential land use," and perhaps "traffic congestion and parking"—clearly is the central issue in suburban neighborhood associations, as posited in the growth machine model. This does not mean that development questions are unimportant to the city neighborhood

Table 19.3 Proportion of Neighborhood Associations Reporting Frequent Discussion of Various Types of Local Issues

Type of Issue	City n = 39	Suburb n = 33	Total n = 72
Traffic congestion and parking	82.9	64.5	75.0
Residential land use changes, housing developments	70.0	71.0	70.4
Encroachment on residential land use	66.7	71.9	69.0
Streets and sidewalks	85.4	46.9	68.5
Personal safety	75.6	58.1	68.1
Parks and playgrounds	82.5	48.4	67.6
Cleanliness of the area	85.4	32.3	62.5
Condition of housing	80.5	32.3	59.7
Reputation of the area	70.0	45.2	59.2
Police protection	78.0	18.8	52.1
Visual aspects or architectural standards	52.5	46.9	50.0
Noise level	52.5	35.5	45.1
Fire protection	37.5	20.0	30.0
Garbage collection	43.9	3.7	27.9
Shopping facilities	30.8	20.7	26.5
Street lighting	37.5	32.3	23.9
Type of people who live in the area	35.0	9.4	23.6
Health care facilities	15.4	0.0	8.8

associations: Two thirds or more cited such issues. In some ways, indeed, every issue—whether noise or public safety or some other—may be associated with development. What is distinctive is that neighborhood association leaders in the suburbs are more likely to *describe* their issues as development issues. Moreover, development issues are the nearly exclusive concern in the suburbs, whereas they are among many issues of concern in the cities.

One also might have expected a strong emphasis on school issues, especially in the suburbs. School-related questions (e.g., class size, tax rates, school closings), however, were mentioned by only two respondents. This does not represent a disinterest in problems of public education but, rather, a very clear demarcation of responsibilities between neighborhood associations and other organizations that are concerned with schools (especially the parent-teacher organizations).

Working with 18 distinct issues is difficult. We inspected the Pearson correlations among them and also used exploratory factor analyses to determine which kinds of issues seem to come up together in neighborhood association meetings. Although these are not interval-scale variables, Pearson correlations commonly are applied to ordinal scales with as many as four response categories. We present the correlation matrix in Table 19.4.

The 18 indicators have been combined into four indices, defined in the following list. The conceptual distinctions among these four are not always obvious. For example, we have already intimated that traffic congestion may

Table 19.4 Correlation Among Types of Issues Discussed at Neighborhood Association Meetings

	1	2	3	4	5	6	7	8	9	10	11	12	13	14	15	16	17	18
1 Condition of housing	—																	
2 Reputation of area	.46	—																
3 Cleanliness of area	.69	.42	—															
4 Noise level	.20	.40	.17	—														
5 Visual aspects	.31	.36	.27	.11	—													
6 Types of people	.44	.39	.43	.29	.03	—												
7 Police protection	.41	.36	.53	.19	.20	.29	—											
8 Fire protection	.25	.17	.23	.25	.24	.11	.44	—										
9 Personal safety	.30	.30	.29	.38	.34	.15	.45	.17	—									
10 Street lighting	.36	.27	.46	.17	.26	.21	.49	.25	.33	—								
11 Parks and Playgrounds	.31	.25	.38	.15	.24	.20	.33	.14	.10	.36	—							
12 Streets and sidewalks	.42	.37	.39	.06	.26	.16	.53	.23	.29	.46	.25	—						
13 Garbage collection	.51	.26	.53	.33	.25	.36	.53	.35	.36	.63	.17	.37	—					
14 Traffic congestion	.23	.28	.23	.53	.33	.11	.28	.28	.39	.17	.15	.23	.35	—				
15 Shopping facilities	.14	.29	.14	.36	.39	-.02	.03	.35	.04	.25	.30	.28	.14	.30	—			
16 Health care facilities	.28	.25	.28	.20	.32	.19	.36	.52	.20	.32	.18	.33	.27	.24	.45	—		
17 Encroachment	-.25	.00	-.25	.04	.26	-.23	-.13	-.15	.04	-.15	-.16	-.08	-.13	.25	.27	-.02	—	
18 Residential change	-.16	-.02	-.17	-.04	.22	-.26	-.18	-.04	-.05	-.24	-.29	-.14	-.19	.16	.20	-.13	.50	—

be a development issue, but it is listed here under the category of "collective consumption." The basis for our choice is the correlations of discussion of traffic congestion with other issues.

1. *Safety* includes four issue areas: police protection, fire protection, personal safety, and street lighting. These are positively intercorrelated, especially police and fire protection (.44), police and personal safety (.45), and police and street lighting (.49).

2. *Collective consumption* includes six kinds of services or facilities that affect the convenience of living in the neighborhood (although some are privately, rather than publicly, provided): parks and playgrounds, streets and sidewalks, garbage collection, traffic congestion, shopping facilities, and health care facilities. Correlations among these issues are all positive and range from .14 (between shopping facilities and garbage collection) to .45 (between shopping facilities and health care facilities).

3. *Lifestyle* includes six issues in which the amenities or quality of life in the neighborhood are emphasized: condition of the housing, reputation of the area, cleanliness of the area, noise level, visual aspects and architectural standards, and types of people who live in the area. Several of these issues are highly correlated with one another (e.g., condition of housing and cleanliness at .69; reputation of the area and types of people at .39). Visual aspect is not much associated with noise level or types of people (correlations less than .10), but it has correlations above .30 with both condition of housing and reputation.

4. *Development* includes only two issues, though these are among the most consistently discussed by neighborhood associations: impact of commercial, industrial, institutional activities on residential land use; and residential land-use changes, such as construction of new housing developments and condominium conversion. These two issues are *negatively* correlated with the majority of other issues, making them quite a distinct pair.

We have used these categories to code responses to the question, Which three issues have board members been most concerned about in the past year? We proceed now to this item because for each of these three issues, we have additional information concerning the political process. In analyzing the distribution of these top three issues, note that there are potentially 216 cases (72 neighborhood associations times 3 issues). Table 19.5 shows the percentage of neighborhood associations in which a particular type of issue was listed among the top three, for the city and suburban associations separately. Issues that could not be classified clearly are combined into an "other" category. In large part, Table 19.5 confirms the previous findings regarding the "most discussed" issues. Thus development issues are especially important to the suburban associations (31.3% of suburban neighborhood associations listed development as one of their top three issues), and collective consumption, lifestyle, and safety issues are relatively more important in the cities. These differences are statistically significant (based on standard *t* tests) for both the development and lifestyle issues.

What accounts for this emphasis on development issues in the suburbs? One factor might be the higher income levels of suburban residents, which is

Table 19.5 Type of Issue Ranked Among the Three Most Important in the Past Year, by City/Suburban Location

Type of Issue	City	Suburb	Total
Lifestyle	26.5%	10.1%	19.0%
Safety	14.5%	9.1%	12.0%
Collective consumption	28.2%	20.2%	24.5%
Development	18.8%	31.3%	24.5%
Other	12.0%	29.3%	19.9%
Total	117	99	216

related to the privatization of services and de-emphasis of collective services. Another factor might be the more recent development of suburban areas, making new construction a more potent issue. In principle, these two interpretations could be evaluated by multivariate analysis, in which the effects of suburban location are examined with controls for the median family income and age of the housing stock or other variables such as occupational composition or home ownership. We have chosen not to report more complete models because of the small number of independent observations (74 interviews) and the high correlations between suburban location, income, and age of housing (approximately .70).

Opponents and Allies

The final step in our analysis is to examine neighborhood associations' perception of their political situation on these issues. Who are the important actors on their side, and who are their main opponents? The interview question noted that "neighborhood organizations often find themselves working together with other groups on some issues, while engaged in conflict at other times. Taking each of the three issues you have mentioned, please indicate the major groups with which you have cooperated and with which you have been in conflict."

We provided no set response categories. Responses subsequently were coded into the following categories: state government (or specific state agencies or officials), local government (or specific local agencies or officials), businesses or land developers, other local voluntary organizations, and other individual residents in the neighborhood.

The distribution of responses is shown in Table 19.6, cross-tabulated by the type of issue and city/suburban location. This table has four panels, one for each major issue category. Within each panel, we report the number and proportion of neighborhood associations that noted a particular actor as cooperating or conflicting. For example, reading from Table 19.6, on only one lifestyle issue cited by a city neighborhood association (2.9% of lifestyle issues cited by city

Table 19.6 Conflicting and Cooperating Groups by Type of Issue and City/Suburban Location

Type of Issue	Cooperating Groups			Conflicting Groups		
	City	Suburban	Total	City	Suburban	Total
Lifestyle						
State government/officials	1 (2.9)	1 (11.1)	2 (4.7)	1 (2.9)	—	1 (2.3)
Local government/officials	23 (67.6)	5 (55.6)	28 (65.1)	11 (32.4)	5 (55.5)	16 (36.4)
Business and developers	2 (5.9)	2 (22.2)	4 (9.30)	11 (32.4)	3 (33.3)	14 (32.6)
Local voluntary associations	10 (29.4)	—	10 (23.3)	1 (2.9)	—	1 (2.3)
Other individual residents	2 (5.9)	—	2 (4.7)	4 (11.8)	1 (11.1)	5 (11.6)
Safety						
State government/officials	2 (10.5)	1 (11.1)	3 (10.7)	—	1 (11.1)	1 (3.6)
Local government/officials	14 (73.5)	5 (55.6)	19 (67.9)	6 (31.6)	3 (33.3)	9 (32.1)
Business and developers	1 (5.3)	1 (11.1)	2 (7.1)	—	—	—
Local voluntary associations	3 (15.8)	—	3 (10.7)	1 (5.3)	—	1 (3.6)
Other individual residents	1 (5.3)	2 (22.2)	3 (10.7)	1 (5.3)	1 (11.1)	2 (7.1)
Collective consumption						
State government/officials	4 (12.1)	1 (5.0)	5 (9.4)	3 (9.1)	.	3 (5.7)
Local government/officials	22 (66.7)	9 (45.0)	31 (58.5)	15 (45.5)	10 (50.0)	25 (47.2)
Business and developers	7 (21.2)	5 (25.0)	12 (22.6)	5 (15.2)	4 (20.0)	9 (17.0)
Local voluntary associations	3 (9.1)	7 (35.0)	10 (18.9)	2 (6.1)	1 (5.0)	3 (5.7)
Other individual residents	1 (3.0)	1 (5.0)	2 (3.8)	4 (12.1)	1 (5.0)	5 (9.4)
Development						
State government/officials	—	4 (12.9)	4 (7.5)	—	2 (6.5)	2 (3.8)
Local government/officials	11 (47.8)	17 (54.8)	28 (51.9)	11 (47.8)	19 (61.3)	30 (55.6)
Business and developers	7 (31.8)	2 (6.5)	9 (17.0)	12 (54.5)	13 (41.9)	25 (47.2)
Local voluntary associations	6 (27.3)	11 (35.5)	17 (32.1)	1 (4.5)	2 (6.5)	3 (5.7)
Other individual residents	1 (4.5)	1 (3.2)	2 (3.8)	1 (4.5)	—	1 (1.9)

NOTE:Numbers in parentheses are the percentages of neighborhood associations (of those reporting a particular type of issue) that identified the group as cooperating (or conflicting).

neighborhood associations) were state officials noted as a cooperating group. These data are provided separately for city, suburban, and all neighborhood associations.

By far the most important protagonist of neighborhood associations turns out to be local government. Local government was cited as a cooperating organization in more than half of the issues (53%) and as a conflicting organization in nearly half (43%). In 15 suburban cases and 26 central-city cases, local government was cited on both sides. For example, the neighborhood association may perceive a municipal agency as an opponent but perceive the city council or a particular council member as an ally. Local government was most likely to be considered cooperative on safety and lifestyle issues but was most likely to be considered an opponent on land-development issues. This pattern holds among both city and suburban neighborhood associations. Partly because land-development issues are less predominant in the central city, however, local government is slightly more likely to be considered an ally for city neighborhood associations and less likely to be seen as an opponent.

The importance of local government is reinforced by responses to another question on the frequency of interaction with representatives of various other organizations. The highest frequency of interaction was reported with city government: 38% reported "much" interaction (the top category on a 5-point scale).

The next most common player, especially as a conflicting group, is businesses and land developers. This result certainly supports the growth machine paradigm of local politics as business development versus the neighborhoods. Business is most commonly perceived as an adversary on land-development issues (47% of such issues) but also is cited on 33% of lifestyle issues as an adversary.

Perhaps the surprise here, given the literature's emphasis on conflicts, is the number of cases in which business and land developers are seen to be allies of neighborhood associations—in 14% of the cases, even including nine land-development issues. Seven of these land issues are in central-city areas, and they deserve more detailed attention. In two cases the specific issue was historic preservation of buildings in the neighborhood, which was being pursued by private developers with the support of neighborhood association leaders. In two other cases the issue was housing rehabilitation through use of public subsidies.

Again we have complementary evidence from another question. Respondents were asked whether their organization tended more often to agree, to agree and disagree about equally, or more often to disagree with "realtors and property developers" in their areas. Although 48% reported that they more often disagree, an important minority of *15% said that they more often agree* with realtors and property developers.

One interesting example is a neighborhood association in downtown Schenectady in which the majority of the members were local businesspersons. This association sought to have vacant land developed for business uses, as well as to promote construction of low-income and senior citizen housing. In this process they worked closely with a large local construction company.

Sometimes such relationships are initiated by real estate developers. For example, in one interview that we conducted, a representative of a firm that has large financial commitments to housing renovation in downtown Albany described a number of improvement activities in cooperation with neighborhood associations. These included support of renovation of a historic landmark, sponsorship of summer camps and youth employment programs, neighborhood clean-ups, and the like. Clearly these activities were directed at generating a positive public image, as well as at influencing the reputation of gentrifying areas in which the firm was investing.

Such examples are a reminder that the business sector is a needed partner in neighborhoods in which the residents seek constructive changes. These cases may be more prevalent in central cities. When asked whether their orientation was more toward preserving the neighborhood pretty much as it is or toward making some major changes in the neighborhood, 20% of city neighborhood associations responded that they seek major changes, compared with only 3% of suburban associations.

Other groups are named much less frequently. Other local voluntary organizations are sometimes supporters (22% of issues), especially on land-development issues (32%); yet they rarely are viewed as opponents. State government officials and local residents are mentioned in fewer than 10% of the cases as either allies or opponents.

Multivariate analysis is a method to sort out the results in Table 19.6. Overall, what is the effect of the type of issue on whether local government or developers are considered to be allies or opponents? Are there any independent city/suburban differences? Table 19.7 presents results of four logistic regressions in which the type of issue is represented by dummy variables (coded 0 and 1, with safety issues treated as the omitted category) and location is dichotomized as city or suburb (where suburb = 1 and city = 2). Logistic regression is similar to ordinary least squares in that the direction and statistical significance of the coefficients allow one to weigh the relative effects of predictor variables. The difference is that the dependent variable is the log of the odds of being in one category versus the other category of a dichotomous dependent variable. In Table 19.7 we report t values for testing statistical significance despite the fact that our sample is close to being the entire population of neighborhood associations in the Albany area. The purpose is to provide a standard metric by which to evaluate whether a variable has an independent effect.

To illustrate the interpretation of logit equations, we can transform the logit coefficients. For example, the logit coefficient for the effect of an issue that concerns development on the likelihood that developers will be named as a conflicting group is 1.034. Taking the antilogarithm of this value yields 2.812 as an effect on the odds ratio. Because safety issues are the omitted category, the odds of developers being named as a conflict group are nearly three times higher for development issues than for safety issues. Similar calculations could be performed for other coefficients.

First, local government is somewhat more likely to be considered an *ally* on lifestyle (two-tailed $p < .10$) and collective consumption (two-tailed $p < .20$)

Table 19.7 Logistic Regressions Predicting Local Government and Developers as Cooperating or Conflicting Groups, by Type of Issue and Location (regression coefficients with *t*-values in parentheses)

	Local Government		Developers	
	Cooperating	*Conflicting*	*Cooperating*	*Conflicting*
Lifestyle issue	.329	.080	−.077	.713
	(1.63)	(0.39)	(−0.23)	(2.72)
Collective consumption	.261	.278	.475	.306
	(1.40)	(1.48)	(1.83)	(1.12)
Development issue	.178	.412	.306	1.034
	(0.97)	(2.22)	(1.13)	(4.28)
Location	−.269	.177	−.109	−.041
	(−1.90)	(1.23)	(−0.54)	(−0.23)
Constant	.528	.144	−1.505	−.156
	(1.87)	(0.51)	(−3.59)	(−0.45)

issues than on other issues, though these effects are not quite significant at the .05 level. Location also has a significant effect: City neighborhood associations are less likely to see local government as an ally than are suburban associations. Local government is significantly more likely to be considered an *opponent* on development issues than on other issues. This effect is relevant to the assumption inherent in the growth machine model that local officials are allied to prodevelopment forces.

Second, business and developers are more likely to be perceived as a cooperating group on collective consumption issues, which include some facilities (shopping and health care) that are privately provided. They are significantly more likely to be seen as opponents on development issues than on other issues, and this is the strongest effect shown in Table 19.7. No city/suburban difference appears in perceptions of developers as allies or opponents.

Challenges to the Growth Machine Paradigm

This study leads to some conclusions about neighborhood organizing in our specific case and to some implications for theories of growth politics in general.

In the Capital District, we find that neighborhood associations are a remarkably recent phenomenon: The great majority formed after 1975 in the central cities and after 1980 in the suburbs. A careful review of the city of Albany suggests a range of external and internal influences on the appearance of neighborhood associations. The major influence here probably was the strength of the local Democratic party machine, which was jealous of any independent political force. Eventually federal antipoverty and community-development programs, offering new resources for neighborhoods to compete for and introducing requirements for grassroots representation, became one foundation for neighborhood organizing. A second factor was gentrification and the commercial revitalization of downtown Albany following construction of Rockefeller's new government building complex in the heart of town. These events injected new issues and, in gentrifying neighborhoods, a new type of local activist, whose influence was magnified through the Neighborhood Resource Center.

We would argue that the *political machine* has been supplanted by a *growth machine*. Although we have not traced the history of neighborhood organizing in surrounding towns, we suspect that further research would reveal similar kinds of factors. The formation and character of neighborhood associations depend on the local political context, the intervention of external agencies, the composition of the population, and the objective growth patterns that people experience.

We can conclude more definitely that these associations are formed to be single-issue political actors, most commonly in response to specific changes or proposals for change in land use. In their foundation, neighborhood associations are profoundly identified with the politics of growth.

We must distinguish, however, between city and suburban neighborhoods. In the heyday of 1975-1980, when most city neighborhood associations were formed, a large proportion had no particular founding issue. We cannot discount the role of fashion and learning in this process. Even in the suburbs, where respondents had no difficulty identifying a founding issue, the timing of new organizations (concentrated in a 5-year period after most city associations had become active) hints at the role of imitation.

After the initial issue is resolved, neighborhood associations deal with quite a broad range of issues. In suburban communities land-development questions remain uppermost, a result that we can attribute to the recency of development. In the cities, however, land use is only one of many issues; much attention is given to safety, public amenities, and services of various sorts. Suburban location is negatively related to lifestyle issues even when controlling for income and age of housing effects.

We believe that in both suburban and city neighborhoods the intention of neighborhood organizing is to protect the residential environment. In the suburbs much of this environment is privately owned and controlled, and nonresidential activities are more spatially segregated. City residents deal with surroundings over which they have less personal control, so they experience a wider range of problems as collective issues. These observations indicate that questions that are related only indirectly to land development need to be addressed in the growth machine model. Residents defend many dimensions of neighborhood, such as the land and buildings, the kinds of people who live there, their access to daily necessities, and personal safety.

In a more recent formulation (Logan & Molotch, 1987) these are described as *use values,* in contrast to the *exchange values* that the rentier coalition seeks through more intensive land development. The issue is not simply the growth of the city but whether growth will favor one or the other set of values. This formulation has the advantage that it can encompass those situations in which residents advocate growth (particularly in declining neighborhoods, where people understand the issue not as growth but as abandonment). More research is needed on the range of values pursued by residents in different sorts of neighborhoods.

Complexity in the range of issues is matched by variation in the lineup of allies and opponents of neighborhood associations. Business interests and real estate developers typically are identified as opponents of neighborhood associations on issues of land use, and this statistically significant relationship provides good evidence for the basic model shared by Elkin (1985), Molotch (1976), Logan and Molotch (1987), Mollenkopf (1983), and Fainstein and Fainstein (1983). Yet some exceptional cases also are theoretically important. In some inner-city neighborhoods, developers are perceived as *allies* in promoting positive changes. This would be a surprising alliance: Most of the authors cited here agree with Fainstein and Fainstein (1983, p. 257) that capitalists respond to working-class or minority concerns only when "confronted with a threat to the social relations of production." Our view is that depressed neighborhoods may embrace developers by default: The political system is closed to them, and their only apparent hope for change is through private investment.

Another provocative finding is that city government is perceived to be a more important political actor than real estate interests. Furthermore, local government is seen as often as an ally (but especially on lifestyle and collective-consumption issues) as an opponent (especially on development questions). If these perceptions were accurate, they would undercut the growth machine paradigm and lend more credence to a pluralist model of local politics. Perhaps the most important possibility in this respect is that city governments are split, or at least are perceived to be split, on some of the issues important to neighborhood associations. This could take the form of a minority representation among elected officials—a voice, but not a decisive voice, in favor of neighborhood interests. Or structural divisions may occur. For example, Elkin (1985) argued that city bureaucracies often are more supportive of residents' interests than are elected officials and that in the current "federalist" period, these bureaucracies increasingly are autonomous. If so, there may be room for successful maneuvering in defense of use values—even in cities in which one has no difficulty in identifying a dominant progrowth coalition.

Our respondents were almost uniformly optimistic about their activity: Most of them (80%) asserted that their organizations had been "very" or "moderately" successful in achieving their goals during the past year, and even more (89%) believed that they had been very or moderately successful in dealing with the issue that stimulated their initial organization. One tends to discount such reports as image management by committed leaders; yet our data show a rapid expansion of this form of grassroots organization, focused especially on land-development issues, in which the leaders perceive that they have important allies in local government. Is this a true gauge of political realities, or is it a reflection of regime strategies of negotiation, posturing, co-optation, and compromise?

Rather than think of residents as a single class of actors, it may be more useful to expect that the city government will protect *some* neighborhoods from *some* kinds of changes. More research should be directed toward identifying when and why this occurs. By stressing the middle-class electoral constituency that most local regimes rely on, both Elkin (1985) and Fainstein and Fainstein (1983) implied that middle-class neighborhoods will tend to be protected. We concur that there is likely to be a class bias in land-use politics. We would be cautious in interpreting this bias, however. Our fieldwork provided clear evidence that neighborhood association leaders do not take cooperation for granted. Middle-class residential leaders, especially those from gentrifying neighborhoods, approach city government most often as outsiders demanding a response, wary of the intentions of city officials and unwilling to rely on them to define the issues. In the era of the political machine, neighbors approached ward leaders as supplicants, expecting their protection. Today's neighborhood associations attempt to take a more active role, wielding power in their own right. In the Albany case this is the most important change brought about by the movement.

References

Arnold, J. L. (1979, November). The neighborhood and city hall: The origin of neighborhood associations in Baltimore, 1880-1910. *Journal of Urban History, 6,* 3-30.

Barber, B. (1965). Participation and mass apathy in associations. In A. Gouldner (Ed.), *Studies in leadership: Leadership and democratic action* (pp. 477-504). New York: Russell & Russell.

Boyte, H. (1980). *The backyard revolution: Understanding the new citizen movement.* Philadelphia: Temple University Press.

Brown, C. (1986). Machine politics. In A. F. Roberts & J. A. Van Dyk (Eds.), *Experiencing Albany* (pp. 67-73). Albany, NY: Nelson A. Rockefeller Institute of Government.

Burnett, A. (1983). Neighborhood participation, political demand making, and local outputs in British and North American cities. In A. Kirby, P. Knox, & S. Pinch (Eds.), *Public service provision and urban development* (pp. 316-362). New York: St. Martin's.

Cox, K. R. (1973). *Conflicts, power, and politics in the city.* New York: McGraw-Hill.

Cox, K. R. (1981). Capitalism and conflict around the communal living space. In M. Dear & A. J. Scott (Eds.), *Urbanization and urban planning in capitalist society* (pp. 431-456). New York: Methuen.

Crain, R. L., & Rosenthal, D. B. (1967). Community status as a dimension of local decision making. *American Sociological Review, 32*(6), 970-984.

Crenson, M. (1983). *Neighborhood politics.* Cambridge, MA: Harvard University Press.

Davidson, J. L. (1979). *Political partnerships: Neighborhood residents and their council members.* Beverly Hills, CA: Sage.

Dye, T. (1970). City-suburban social distance and public policy. In J. Goodman (Ed.), *Perspectives in urban politics* (pp. 363-373). Boston: Allyn & Bacon.

Elkin, S. L. (1985, Spring). Twentieth century urban regimes. *Journal of Urban Affairs, 7,* 11-28.

Fainstein, N. I., & Fainstein, S. S. (1983). Regime strategies, communal resistance, and economic forces. In S. Fainstein, N. Fainstein, R. Hill, D. Judd, & M. Smith (Eds.), *Restructuring the city: The political economy of urban development* (pp. 245-282). New York: Longman.

Fisher, R. (1984). *Let the people decide: Neighborhood organizing in America.* Boston: Twayne.

Form, W. (1954, May). The place of social structure in the determination of land use. *Social Forces, 32,* 317-323.

Goering, J. M. (1979, October). The national neighborhood movement: A preliminary analysis and critique. *American Planners Association Journal, 45,* 506-514.

Greer, S. (1962). *The emerging city: Myth and reality.* New York: Free Press.

Guest, A. M. (1985, August). *The mediate community: The nature of local and extra-local ties within the metropolis.* Paper presented to the American Sociological Association, Washington, DC.

Guest, A. M., & Oropesa, S. (1984, December). Problem-solving strategies of local areas in the metropolis. *American Sociological Review, 49,* 828-840.

Henig, J. R. (1982). *Neighborhood mobilization: Redevelopment and response.* New Brunswick, NJ: Rutgers University Press.

Janelle, D. G. (1977). Structural dimension in the geography of locational conflicts. *Canadian Geographer, 21*(4), 311-328.

Kennedy, W. O. (1983). *O Albany*. Albany, NY: Viking.

Ley, D., & Mercer, J. (1980, April). Locational conflict and the politics of consumption. *Economic Geography, 56*, 89-109.

Logan, J. R., & Molotch, H. (1987). *Urban fortunes: The political economy of place*. Berkeley: University of California Press.

Long, N. (1958, November). The local community as an ecology of games. *American Journal of Sociology, 64*, 251-261.

Lowe, P. D. (1977). Amenity and equity: A review of local environmental pressure groups in Britain. *Environment and Planning, 9*(1), 35-58.

Mansbridge, J. J. (1980). *Beyond adversary democracy*. New York: Basic Books.

McEneny, J. J. (1981). *Albany: Capital city on the Hudson*. Woodland Hills, NY: Windsor.

Mollenkopf, J. H. (1983). *The contested city*. Princeton, NJ: Princeton University Press.

Molotch, H. (1976, February). The city as a growth machine. *American Journal of Sociology, 82*, 309-330.

Rossi, P., & Dentler, R. (1961). *The politics of urban renewal*. Glencoe, IL: Free Press.

Rudel, T. (1989). *Situations and strategies in American land-use planning*. New York: Cambridge University Press.

Schoenberg, S. P., & Rosenbaum, P. L. (1980). *Neighborhoods that work: Sources for viability in the inner city*. New Brunswick, NJ: Rutgers University Press.

Smith, M. P., & Keller, M. (1983). Managed growth and the politics of uneven development in New Orleans. In S. Fainstein, N. Fainstein, R. Hill, D. Judd, & M. Smith (Eds.), *Restructuring the city: The political economy of urban development* (pp. 126-166). New York: Longman.

Swanstrom, T., & Ward, S. (1987, September). *Albany's O'Connell organization: The survival of an entrenched machine*. Paper presented at the Annual Meeting of the American Political Science Association, Chicago.

Taub, R. P., & Surgeon, G. P. (1977, March). Urban voluntary associations, locality based and externally induced. *American Journal of Sociology, 83*, 425-442.

Thomas, C. J. (1986). *Between citizen and city: Neighborhood organizations and urban politics in Cincinnati*. Lawrence: University of Kansas Press.

Wood, J. (1981). *Leadership in voluntary organizations*. New Brunswick, NJ: Rutgers University Press.

Wood, R. (1958). *Suburbia: Its people and their politics*. Boston: Houghton Mifflin.

PART SEVEN

Urban Housing

Introduction to Urban Housing

Patricia Baron Pollak

Housing, like all issues of modern urbanization, is complex. The housing stock of a community, including its quality, quantity, array of conditions, spatial organization, and tenure patterns, is based on the interrelationship of social, economic, political, and legal forces. The resulting housing pattern of a community is no more the result of one force or factor and can no more be wholly understood through examination from one perspective, area, or discipline alone than can any other issue or problem of modern society. Like all other areas of modern urban society in the United States, housing is inherently a very complicated arena.

To individuals and households, housing is residential location. As such, it is more than a major economic investment. The setting of people in a place is more than shelter, more than physical space. It surely is more than a physical environment alone. Yet the physical environment has a profound effect on our lives. Our housing is the assignment of people to place. Where we live affects our access to employment, to educational facilities, to social and recreational activities, to goods and services (e.g., shopping, medical care, child care, church), and to those with whom we (and our children) come into contact on both a casual and a profound basis. Our housing puts us in place. And, to a great extent, our "place" influences our opportunities. Our housing is an economic, social, psychological, and political resource for us. It affects the goods and services available to us and delimits our choices for those who represent us in important public policy decisions.

To a community, housing and, hence, the spatial ecology of a community's population are also multifaceted. Housing patterns influence school enrollment, the need for and consumption of public and private services, the need for and use of transportation, and access to employment, shopping, and services. A community's housing stock—its quality, quantity, and condition—is a measure of individual, family, household, and community well-being. A community's housing stock is also a large part of a community's fiscal base. The economic value of real property (including the housing stock) is the basis on which a municipality can raise revenue for the provision of both necessary and discretionary services and amenities. A community's housing stock is, therefore, an important resource, from a variety of perspectives, to a community.

The housing stock of a community in the United States today is not a random array. It is to a very great extent the result of policy decisions. Housing can be planned from the start, with available sites designated for large or small lots or for single or multifamily dwellings; and areas already developed can be enhanced with public services and amenities or can be relatively ignored. Newly paved roads and sidewalks, new street lighting, special plantings, well-kept parks, well-run schools with special services, and amenities such as municipal pools and skating rinks can greatly enhance the "quality of life" in a residential area. Similarly the lack of maintenance and municipal attention and the infusion of public resources can speed the deterioration of an area and support the decision of those households with adequate resources to seek out and move to "better" urban or suburban neighborhoods. The patterns and conditions of residential neighborhoods are not solely the result of unencumbered "free-market" behavior, but rather the result of decisions made by our representatives on behalf of all of us.

Housing patterns are also the result of myriad other decisions, including about the development of municipal infrastructure (roads, sewer, and water) and about locations appropriate for residential, industrial, and business activity. Classic location theory relates the decisions of households and firms to locate in proximity to the scarcest of urban resources, the city center. Yet in recent history that location has not always been viewed as optimal for residential location (Alonso, 1964). Many households have valued space and amenities away from the city center instead. Because significant developments in transportation and communications have made it possible and often desirable for businesses to locate away from downtown areas, households valuing a shortened trip to work also need no longer prize a close-to-city-center location. In addition, an increasing number of people work either for themselves, or for others, from their homes. As a result, residential location patterns are not as predictable as simple models of locational decisions might explain. And decisions about most of what goes on in an urban area, or in any other area, for that matter, affect housing.

The underlying assumption of planning (or the making of decisions for) the allocation of resources in a community is that if left without such guidance, the collectivity of decisions made by individuals, households, business, and industry would most likely not be in the best interest of all of us. That is, without some degree of manipulation, the greater public welfare would be diminished. Zoning, for example, one of the principal tools of planning, emanates from the police power of the state. The legitimacy of zoning has been reaffirmed repeatedly to protect the public health, safety, and morals and the general welfare of the community (Haar & Kayden, 1989). A community's zoning ordinance allocates areas or districts within a community to particular land uses. The zoning ordinance sets aside specific districts within a community for residential, commercial, and industrial uses. Within each of these broad categories (and there can be others, such as agricultural in rural areas), refinements often are made to designate particular subtypes of the specified land use. For example, within the broad

category of "residential land use" are often separate single-family and multifamily housing districts. Within "single-family" may be small lot, medium lot, and
large lot districts. In each of these the zoning ordinance will specify a minimum
lot size for an approvable building site. Some districts may specify a larger setback
(or front yard) than the others. And other requirements within the general category of "single-family" may vary from one district to another. The operational
concept is that with these regulations the greater public welfare will be at least
enhanced, if not protected.

The myriad issues of urban housing may be examined in many other ways.
If we sidestep the issue of the federal role in housing, and because this volume
is devoted to the urban arena that seems appropriate, we can look to those areas,
particularly the purview of municipal capacity. Numerous municipalities and
private nonprofit organizations in this country are developing means to ensure
the development of at least some affordable housing within their communities
(Stegman & Holden, 1987). Municipalities are developing housing trust funds
(Connerly, 1993), municipal loan funds (Institute for Community Economics,
1987), and home ownership assistance programs (Stegman & Luger, 1993).
Private nonprofit organizations are also taking an active role in providing low-
and moderate-income housing through such vehicles as mutual housing associations, limited equity cooperatives, and community land trusts. When we
consider the distribution of housing and the limits of locational choice, municipal zoning regulations have a major impact on the patterns of housing throughout a community. There is increasing awareness of the importance of how a
municipality defines a family for local zoning purposes and the impact of that
definition on the availability of housing (Ritzdorf, 1988). These definitions can
limit severely the nature of a household occupying or wishing to occupy a single-
family dwelling. Issues of code enforcement; rent control; the provision of roads
and transportation services (Rosenbloom, 1982); trash collection and disposition; neighborhood amenities such as parks, recreation, appropriate police, fire,
and emergency services; the location and adequacy of schools (Feld, 1989); and
medical facilities are also areas within municipal control that have significant
impacts on the residential environment. All of these affect the supply, availability, and affordability of housing and, hence, the quality of life that accrues to
individuals, families, and households of the many variations within American
society today.

One can also consider housing from the householder's perspective. The
availability and distribution of mortgage financing is a particularly important
issue as we examine who is able to obtain financing for the purchase of a home
(FannieMae, 1992). One also can consider housing issues as they pertain to particular populations. How communities accommodate group homes for people
with disabilities (Salsich, 1986), even given the sanctions for not doing so provided
by the Federal Fair Housing Amendments Act of 1988, affects the lives of so
many. So does the approach toward the availability of temporary and permanent
housing for homeless individuals and families (Rossi, 1989), transitional housing for individuals and families (Cook et al., 1988), and special needs emergency
shelters for battered women and their children (Sprague, 1991). A growing

body of literature examines the particular problems of women in the housing market (Smith & Thomson, 1987) and the need for available, affordable, and appropriate housing for the elderly (Hancock, 1987). And there is also a growing body of literature on "new" or alternative uses for the single-family housing stock. This literature examines ways communities and nonprofit groups and organizations can reuse housing already available in communities to better meet the needs of households today (Gellen, 1985). Obviously there are many issues and many perspectives from which to examine housing in American cities today. Housing is a very complex arena indeed.

A community's decisions, including those on the zoning of land, the development of water and sewer lines, transportation routes and services, the allocation of resources for schools, and public amenities, all affecting a community's housing stock, its availability and affordability to households of various economic means, are policy decisions. Because, as Dye (1972) has so succinctly pointed out, "public policy is whatever government chooses to do or not to do," housing policy decisions and all of the decisions that affect housing are no different from other decisions that affect the allocation of resources. Decisions made will have a cost, whether direct or indirect. Decisions about the direct expenditure of funds and the indirect costs that stem from investment or the lack thereof affect a wide arena in a community. These policy decisions, like all policy decisions, are political.

Housing policy decisions are, therefore, extremely important to many individuals, groups, and organizations, both public and private, in a community. Because housing affects so many areas of our lives and our community, many people are concerned with the decisions that are made. Reference group theory states that a decisionmaker makes a decision in relation to his or her relationship to others. Some decisionmakers seek to aggregate the interests of those concerned with the decision. Others seek to be consistent with a particular perspective or worldview. Still others make decisions based on a history of relationship with other decisionmakers and trade support for individual decisions over time. Most decisionmakers at one time or another make decisions in all of these ways. The important point for housing and for all local public policy decisions is that regardless of decision-making method or style, decisionmakers do not operate in isolation. Decisionmakers rely on the input, opinions, and positions of others. And for decisions that will directly or indirectly affect a community's housing stock, it is clear that there can be many interested parties indeed.

The two selections in Part 7 focus on the making of housing policy. They share a common theme in the illustration of the highly complex and politicized nature of public policy decision making in today's urban environment. In Chapter 20, "The Limits of Localism," Peter Dreier and W. Dennis Keating address the success that populist Mayor Raymond Flynn had in developing policies that improved housing conditions for Boston's poor and working-class residents. The chapter, however, as much describes how a particular mayor was able to confront what the authors term "a very serious housing crisis" as it does examine the politics of public policy decision making with the housing arena as the central focus. Dreier and Keating clearly cast housing policy decisions as local public policy decisions. Interest groups (or stakeholders) directly and indirectly concerned with the situation are identified, and their interests are made explicit for us. The authors aptly describe the decision-making environment and the authority within which Mayor Flynn could operate. The relationship between Mayor Flynn and other decision-making bodies, particularly the Boston Redevelopment Authority and the bureaucracy of city hall, are identified. Through coalition building, the development of constituencies, the accommodation of interest groups, and, at times, interim compromise to demonstrate outcomes, Boston's Mayor Flynn was able to build a constituency and develop the outcomes for policies supportive of low- and moderate-income housing. The reading is an excellent case study illustrating the multitude of parties and interests and the complexity of tackling multifaceted housing issues.

In Chapter 21, "Local Government Support for Nonprofit Housing," Edward G. Goetz examines the role of the nonprofit housing sector. Through survey research, the depth and breadth of community development corporations (CDCs) in U.S. cities are revealed, as are their significant activity in the production and management of affordable housing. Yet this chapter also is as important for the story it tells about the political nature of housing in American cities today as it is for what it reveals about the nature of nonprofit housing development. The chapter goes beyond a description of the role of CDCs in housing production to examine how and why some have greater impact than others. Goetz's survey explores the organization of the nonprofit housing sector and links the degree of that organization to the ability of CDCs to garner a range of support for their efforts. The results of the research reveal that (a) "the more highly organized community-based housing interests were perceived by respondents as exerting more influence on local housing policy," (b) "it is the political organization of low-income housing advocates that creates the policy response," and (c) "support for community-based housing is, at least in part, a political arrangement between local housing advocates and public officials." This selection, too, is as important for what we learn from it about the way urban policy outcomes are derived as it is illuminating about one particular element of the greater housing picture.

Taken together, the two selections in Part 7 teach important lessons about the complexity of the housing supply in our communities. They demonstrate the highly politicized nature of public policy decision making and illustrate their points with examples from the housing arena. Although they tell only two stories, they incorporate a great deal of information about the nature of housing policy making in our cities, and the lessons are broadly applicable.

References

Alonso, W. (1964). The historic and the structural theories of urban form: Their implications for urban renewal. *Land Economics, XL*(2), 227-231.

Connerly, C. E. (1993). A survey and assessment of housing trust funds in the United States. *Journal of the American Planning Association, 59*(3), 306-319.

Cook, C. C., et al. (1988). *Expanding opportunities for single parents through housing.* Minneapolis: Minneapolis/St. Paul Family Housing Fund.

Dye, T. R. (1972). *Understanding public policy.* Englewood Cliffs, NJ: Prentice Hall.

FannieMae. (1992). Discrimination in the housing and mortgage markets. *Housing Policy Debate, 3,* 2.

Feld, M. (1989). Equity empowerment in planning: Lessons from the Yonkers case. *Journal of Planning Education and Research, 8*(3), 167-175.

Gellen, M. (1985). *Accessory apartments in single family housing.* New Brunswick, NJ: Rutgers University, Center for Urban Policy Research.

Haar, C. M., & Kayden, J. S. (1989). *Zoning and the American dream: Promises still to keep.* Chicago: American Planning Association Planners Press.

Hancock, J. A. (1987). *Housing the elderly.* New Brunswick, NJ: Rutgers University, Center for Urban Policy Research.

Institute for Community Economics. (1987). *The community loan fund manual.* Greenfield, MA: Institute for Community Economics.

Ritzdorf, M. (1988). Not in my neighborhood: Alternative lifestyles and municipal family definitions. *Lifestyles: Family and Economic Issues, 9*(3), 264-276.

Rosenbloom, S. (1982, Summer). Federal policies to increase the mobility of the elderly and the handicapped. *Journal of the American Planning Association, 48,* 335-350.

Rossi, P. (1989). *Without shelter: Homelessness in the 1980s.* New York: Priority.

Salsich, P. W. (1986, Fall). Group homes, shelters, and congregate housing: Deinstitutionalization policies and the NIMBY syndrome. *Real Property, Probate and Trust Journal, 21,* 413-434.

Smith, R. L., & Thomson, C. L. (1987). Restricted housing markets for female-headed households in U.S. metropolitan areas. In W. van Vliet et al. (Eds.), *Housing and neighborhoods: Theoretical and empirical contributions* (Chap. 16). Westport, CT: Greenwood.

Sprague, J. F. (1991). *More than housing: Lifeboats for women and children.* Stoneham, MA: Butterworth.

Stegman, M. A., & Holden, J. D. (1987). *Nonfederal housing programs: How states and localities are responding to federal cutbacks in low-income housing.* Washington, DC: Urban Land Institute.

Stegman, M. A., & Luger, M. I. (1993). Issues in the design of locally sponsored home ownership programs. *Journal of the American Planning Association, 59*(4), 417-432.

Suggestions for Further Reading

Apgar, W. C., et al. (1990). *The state of the nation's housing, 1990.* Cambridge, MA: Joint Center for Housing Studies.

Birch, E. L. (1993). Stop the world . . . and look what planners can do! *Journal of the American Planning Association, 59*(4), 413-416.

Bratt, R. (1989). *Rebuilding a low-income housing policy.* Philadelphia: Temple University Press.

Golant, S. M. (1992). *Housing America's elderly: Many possibilities/few choices.* Newbury Park, CA: Sage.

Guilderbloom, J. I., & Applebaum, R. P. (1988). *Rethinking rental housing.* Philadelphia: Temple University Press.

Henig, J. R. (1981). Gentrification and displacement of the elderly. *The Gerontologist, 21*(1), 67-75.

Hoch, C., & Slayton, R. (1989). *New homeless and old: Community and the skid row hotel.* Philadelphia: Temple University Press.

Jaffe, M., & Smith, T. P. (1986). *Siting group homes for developmentally disabled persons* (Planning Advisory Service Report #397). Chicago: American Planning Association.

Mallach, A. (1984). *Inclusionary housing programs: Policies and practices.* New Brunswick, NJ: Rutgers University, Center for Urban Policy Research.

Mandelker, D. R. (1970). A rationale for the zoning process. *Land-Use Controls Quarterly, 4,* 1-7.

Mitchell, J. P. (Ed.). (1985). *Federal housing policy and programs: Past and present.* New Brunswick, NJ: Rutgers University, Center for Urban Policy Research.

Mulroy, E. A. (1990). *Women as single parents: Confronting institutional barriers in the courts, the workplace, and the housing market.* Dover, MA: Auburn House.

Pollak, P. B. (in press). Rethinking zoning to accommodate the elderly in single-family housing. *Journal of the American Planning Association.*

Schwartz, D. C., et al. (1992). A new housing policy for the 1990s. *Journal of Urban Affairs, 14*(3/4), 239-262.

Stone, M. E. (1990). *One-third of a nation: A new look at housing affordability in America.* Washington, DC: Economic Policy Institute.

Swanstrom, T. (1989). No room at the inn: Housing policy and the homeless. *Journal of Urban and Contemporary Law, 35,* 81-105.

Varady, D. P. (1980, July). Housing problems and mobility plans among the elderly. *Journal of the American Planning Association,* pp. 301-314.

Wolch, J., & Dear, M. (1993). *Malign neglect: Homelessness in an American city.* San Francisco: Jossey-Bass.

20

The Limits of Localism

PROGRESSIVE HOUSING

POLICIES IN BOSTON, 1984-1989

Peter Dreier

W. Dennis Keating

During the 1980s, American cities were confronted with worsening housing problems when the federal government greatly reduced its responsibility for addressing low-income housing issues (National League of Cities, 1989). President Reagan's conservative policy was to reduce federal involvement and expenditures for low-income housing and to delegate responsibility for this problem to state and local government (Hays, 1985). The results included a decline in home ownership, drastic cutbacks in federally subsidized low-income housing production and rehabilitation programs, a shortage of housing for the poor, and increased homelessness. In the past, state and local governments have been heavily dependent on federal housing subsidies to support low-income housing programs. In the recent era of federal fiscal austerity and major cutbacks in federal domestic social programs, few city governments have attempted to develop their own comprehensive housing policies aimed at resolving the housing crisis at the municipal level.

In contrast to most other major American cities, Boston stood out as a progressive city in housing policy under populist Mayor Raymond L. Flynn, who was elected in 1983 and reelected overwhelmingly in 1987 despite serious political opposition and limited funding. (He was reelected again in 1991 but did not finish his term: President Clinton appointed him U.S. Ambassador to the Vatican.) In this chapter we explore the question of the extent to which a progressive municipality can address housing affordability equitably in the face of conservative federal housing policies. The record of the Flynn administration is contrasted with that of his liberal predecessor, Kevin White, who served as Boston's mayor from 1967 to 1983.

What do we mean by a *progressive* housing policy? In his study of five cities, Clavel (1986, pp. 10-12) identified four elements of progressive municipal housing policies: (a) expanded public regulation of private property, (b) promotion of alternatives to the private market, (c) advocacy for the interests of poorer city residents, and (d) increased citizen participation, especially at the neighborhood level. Our criteria for evaluating the Flynn administration's progressive housing policies, following Clavel, are as follows: (a) Did its policies improve housing conditions for Boston's poor and working-class residents? (b) Were neighborhood and housing activist groups empowered to participate actively in decision making in housing? (c) Did the administration develop structural changes to promote nonprofit low-income housing? and (d) Did municipal leaders actively advocate progressive housing policies at the local, state, and federal levels?

Flynn's politics were an example of *urban populism,* in which a local political regime sought to promote the interests of working-class and poor citizens and their neighborhoods, while continuing to promote urban growth. Elkin (1987) characterized this as the effort to organize the urban citizenry to better regulate the "commercial republic" to promote the public interest. The extent to which a populist municipal regime can implement such a policy while maintaining a sustained economic growth through mostly private investment has generated great debate. Peterson (1981) argued that city governments have limited influence on economic development and generally must support the private market if they are to remain competitive. We argue that Flynn's policies demonstrated that, at least in a strong local development climate, urban populism can promote more democratic housing and development policies successfully.

The White Regime

Kevin White was first elected mayor of Boston in 1967 in a divisive campaign in which he defeated Ed Logue, Boston's urban-renewal czar, and Louise Day Hicks, head of the Boston School Committee and foe of desegregation. White began as a protenant, proneighborhood urban populist with reforms that included "little city halls" and rent control (Ferman, 1985; Mollenkopf, 1983). By the mid-1970s, White, who had been a liberal Democrat, had become much more conservative. He dropped his neighborhood-oriented policies. He introduced vacancy decontrol, which effectively ended rent control, and supported unbridled new downtown development, winning him the backing of the real estate industry (Weinberg, 1981).

Boston had experienced a steady economic decline from 1950 through the 1970s. The city's population dropped from a high of 801,000 in 1950 to 563,000 in 1980. As manufacturing and blue-collar employment fell, the city's tax base declined, and public services declined along with it. Those concerned with reversing Boston's economic decline advocated a transformation of the central city and its downtown area into a commercial center (Meyerson & Banfield, 1966). The urban-renewal strategy in Boston during the 1960s was designed to achieve this goal and to attract the middle class to live in Boston. By the late 1970s, the successful effects of this strategy had begun to unfold as a boom in development began. Although downtown Boston prospered, however, many of the city's neighborhoods deteriorated, and poverty and unemployment were widespread.

The 1980 passage of Proposition 2½ in Massachusetts and the election of President Ronald Reagan meant hard times for Boston. The constitutional ceiling on real property taxes meant a one-third reduction in the city's revenue from this source. Among U.S. cities, Boston is the most reliant on property taxes. About half of the land in Boston is exempt from taxation, largely because of institutional and governmental ownership, and tax exemption was provided as an incentive for about half of the new development between 1975 and 1981.

In 1981, 19% of the city's employees were laid off. Federal aid to Boston ($167.4 million in 1981) declined by 36% in 1982 (Slavet, 1985). In the midst of this fiscal crisis, White turned to the *Vault,* Boston's corporate leadership (Boston Urban Study Group, 1983; Dreier, 1983), for assistance in obtaining state authorization for new local taxes and in reducing city services.

Particularly in his fourth (and last) term, White became identified with downtown development. In contrast to his initial image as a reformer, White relied heavily on the private market and growth to promote the development of the city. He did not attempt to redistribute the benefits of this growth to Boston's disadvantaged people and neighborhoods but instead relied upon the trickle-down approach to economic growth. In fact, he eliminated such regulatory policies as rent control that were designed to prevent hardship caused by the rise of housing costs in the private market. White ended his initial efforts at broadening citizen participation in government. White's changed politics and the city's revitalized economy set the stage for a revival of populist politics in Boston in 1983. Beset by corruption charges against his administration, the fiscal crisis, and a neighborhood backlash against his reduction of city services and employees, White decided not to run again in 1983.

The Election of Flynn in 1983

In the nonpartisan election, seven candidates vied to replace White as mayor. The two winners were both progressives—City Councilor Raymond Flynn and State Representative Mel King. Flynn, a former state representative, came from the Irish working-class neighborhood of South Boston. He ran on a proneighborhood platform and made housing a key issue. Flynn supported tenants' rights legislation (including rent control), a proposal to link downtown and neighborhood development by charging commercial developers fees to support below-market housing, and increased neighborhood participation in developmental decision making. King, the first black candidate to proceed to a mayoral runoff election in Boston, had more radical views about housing and neighborhood issues. Flynn defeated King decisively by a 2 to 1 margin in the November 1983 election, and he assumed office in January 1984.

Flynn took office having promised to balance more equitably the benefits of downtown growth and development and the interests of neighborhoods. The theme of his neighborhood-based campaign was economic justice. In making this his platform, Flynn became one of the few urban populist, progressive candidates who have challenged traditional growth machine politics. Flynn focused on issues of economic inequality. According to Swanstrom (1988, p. 123):

> Urban populist mayors seek to redefine the agenda of city politics. First, they seek to displace ethnic and racial divisions with economic division: the haves versus the have nots. Second, once the terrain is shifted to economics, populist politicians seek to shift the issue from one of growth to one of distribution. In the absence of federal initiatives, populist politicians attempt to devise local solutions to the problems of uneven development.

Flynn's policies contrasted sharply with those of Kevin White, who embraced the growth machine model of urban politics.

The development of growth machine politics at the local level has a long history (Logan & Molotch, 1987). Inherent in the growth machine model of politics is the assumption that the benefits of growth in the private sector (e.g., downtown development aimed at white-collar employment, tourism, upper income housing) will trickle down to the lower income population. In this view, redistributive policies are considered irrational and objectionable because they involve increased regulation and taxation of business, which will provide a disincentive to business and will result in the failure to attract the middle class to the central city. It is assumed that progressive municipal regimes will be at a competitive disadvantage (Peterson, 1981).

On assuming office, Flynn faced many problems. He inherited a $40 million operating deficit from White. As a populist candidate who was not supported by Boston's growth coalition, which was symbolized by the Vault (Dreier, 1983), Flynn had either to reassure the business, development, and real estate interests that he was not antibusiness or to develop a strategy to neutralize their opposition. Flynn had to be careful to avoid the fate of other populist municipal officials whose administrations had been weakened or defeated by local growth coalitions. For example, Cleveland's Mayor Dennis Kucinich was driven from office in 1979 after completing only one term; the city had been forced into default by local banks, and Kucinich barely had survived a recall election (Logan & Molotch, 1987; Swanstrom, 1985, 1988).

Boston's Housing Crisis

Flynn inherited a very serious housing crisis. Boston's economic and population growth fueled a strong housing market that threatened to displace many of the city's poor and working-class residents. By 1983 Boston already had experienced several years of sustained real estate appreciation, which had led to gentrification. This trend, symbolized by a wave of condominium conversions and rising rents, began in the late 1970s in the neighborhoods closest to the downtown area, but by the early 1980s, it had spread to outlying white and minority working-class neighborhoods.

Boston increasingly was attractive to yuppies and empty nesters who competed with poor and working-class residents for scarce housing (Ganz, 1985). Because 70% of the residents were renters (about evenly divided among private multifamily buildings, public/subsidized housing, and owner-occupied two- and three-deckers), they were particularly vulnerable to displacement from rising housing values. Even many white, middle-class homeowners, who typically benefit from rising housing values, began to worry that despite the appreciation of their own homes, Boston's hot housing market would prevent their children from being able to stay in their neighborhoods. Indicators of Boston's housing crisis included the following:

- Home prices more than doubled between 1975 and 1983, from an average of $25,000 to $56,800 for a single-family home. According to the National Association of Realtors (1985), the Boston metropolitan area ranked seventh nationwide in average single-family home prices in 1983; by 1985 it ranked first.
- Rents in the private rental market rose dramatically. Between 1977 and 1981 rents rose 48%, whereas renters' incomes rose only 35%. By the first quarter of 1984 the median monthly rent was $528—a 16% increase in 1 year. Renters' average household income ($12,000 in 1980) was only 60% of owner-occupiers' household income. By 1980 almost half (48%) of Boston renters paid over one quarter of their incomes for rent (Achtenberg, 1984). The percentage of Boston renters paying over half of their income for rent rose from 11% to 21% between 1980 and 1985. The rental vacancy rate was only 2.5% in the fall of 1983.
- Condominium conversion in Boston began in the late 1970s. In 1975 Boston had only 1,568 condominiums among its approximately 240,000 housing units. By 1983 the number had reached 14,377—most of which had been apartments. Only one fifth of Boston renters could afford the typical condominium, for which the average price in 1983 was $62,375 (Boston Redevelopment Authority [BRA], 1987).
- In 1980, after years of mismanagement, the Boston Housing Authority (BHA) was placed in court-ordered receivership. More than 4,500 of its 16,500 public-housing units were vacant—many of them vandalized. Beset by rampant crime and daily chaos, the BHA's bureaucracy was so troubled that it failed to spend the federal and state funds it had received to modernize deteriorating projects. The city's low-income housing shortage was so severe that despite this dismal situation, 7,000 families were on the BHA's waiting list in 1982.
- In addition to the BHA's units, Boston had approximately 23,000 units of federal- and state-subsidized rental housing. A substantial portion (at least 4,000 units) of the developments subsidized by the Department of Housing and Urban Development (HUD), however, was troubled—either foreclosed and owned by HUD or at risk of foreclosure. The policy of the Reagan administration was to auction off troubled projects to the highest bidder. An even larger problem was the potential expiration of rent subsidies and use restrictions in most HUD-subsidized projects.

- By 1983 homelessness had become a visible problem in Boston. In a census by the city's Emergency Shelter Commission (1983), the homeless population was estimated at 2,700. This figure included only the visible homeless.
- Boston's low-income neighborhoods were littered with abandoned buildings that were both havens for crime and fire hazards. Most of these were in formerly white (now minority) neighborhoods. This problem was the result of several factors: the exodus of whites from these neighborhoods during and since the 1960s; bank redlining and realtor blockbusting; and city assessing practices (prior to Proposition 2½) that placed unfair property tax burdens on property owners in these areas. Neither the city government nor any other group had an accurate count of the number (or location) of abandoned buildings or units. The city government did not even know how many buildings the city owned through foreclosure or how many owners owed back taxes. In 1985 the Flynn administration undertook a survey and counted approximately 3,000 units in 800 buildings (Boston Neighborhood Development and Employment Agency, 1988).
- Although Boston had only one fifth of the metropolitan area's population (3 million), it had almost half of the area's subsidized housing and low-income population. Many suburban communities resisted low-income housing through large-lot "snob zoning," which put additional pressure on Boston's housing market.

Housing Politics

Constituencies

Like most local governments, the local government in Boston had very limited powers and resources at its disposal to address this crisis—primarily regulatory powers, some discretionary funds, and control of public property. But the ability of the Flynn administration to use even these tools was shaped by a number of political and administrative factors.

In Boston, as in most cities, the constituencies for housing policies include a number of varied elements with very different stakes. The relative political influence of these groups helps shape the "room for maneuver" within local government.

Within the broader business community, the real estate industry, in particular, has an important stake in city policy. It has benefited from the "new Boston" downtown building boom and the resulting skyrocketing land and housing values. Landlords, developers, management firms, and brokers exercise considerable political influence. Through the Greater Boston Real Estate Board (GBREB), they oppose measures that threaten to reduce real estate development. The GBREB opposed the general thrust of Flynn's housing platform, and its members donated heavily to his opponents. Although Flynn won the mayoral race, the real estate industry's influence was apparent through the newly elected city council. Only 4 of the 13 members of the city council were endorsed by tenant groups.

Housing activists represent another political constituency. They favor regulatory and developmental policies that support the preservation and production of low- and moderate-income housing. Compared with their counterparts in other cities, Boston's housing advocates are numerous and sophisticated. Flynn, as a candidate and as mayor, was able to win the support of their constituency largely on the basis of his housing policy platform and programs.

Housing activists include tenant groups, advocates and providers for the homeless, community development corporations (CDCs), social service agencies, church-based groups, and senior citizen organizations. Unlike the real estate industry, the housing activists are not united within one organization and tend to join forces on an ad hoc basis. The Massachusetts Tenant Organization (MTO) represents tenants in private and state-subsidized apartments, and public-housing tenants have their own group. In general, tenant groups refrained from direct involvement in electoral politics. It was not until 1981, when MTO formed a political action committee to endorse and campaign for city council candidates, that private-housing tenants engaged directly in citywide electoral politics.

Neighborhood associations, as varied as Boston's neighborhoods, are a third constituency. Their geographic turf ranges from blocks with a few hundred residents to large neighborhoods with 20,000 residents. These voluntary organizations tend to be dominated by homeowners. They emerged to pressure city government for improved municipal services.

Housing and development issues were not the major concerns of neighborhood associations throughout the early 1980s. In general, neighborhood associations opposed rent control and public and subsidized housing, which they viewed as contributing to neighborhood decline and housing abandonment. The real estate and development boom put housing issues on their agenda. In general, they sought a greater voice in reviewing housing developments proposed for their neighborhoods. Some simply opposed any new developments, particularly those involving low-income or special-needs housing (e.g., housing for the mentally ill). In low-income neighborhoods, the blight of abandoned buildings and vacant land became issues.

In the past, neighborhood groups tended to voice their concerns on an ad hoc, project-by-project basis, typically by appearing at hearings of the Zoning Board of Appeals to support or oppose variances for new developments. Their influence was based primarily on their informal ties to local politicians, many of whom emerged from these voluntary neighborhood groups. Neighborhood associations had not formed any citywide umbrella group to represent their concerns.

The City Hall Bureaucracy

As newly elected mayor, Flynn did not have a firm grip on the city hall bureaucracy, particularly in the area of housing and development. Flynn had an ambitious agenda, but the bureaucratic structure he inherited was not equipped to carry it out. Flynn inherited a bureaucracy in which at least 10 departments were involved in the development, regulation, or management of housing and no previous overall coordination had been attempted. The Citizens Housing and Planning Association (CHPA), a watchdog and advocacy group, warned that the new mayor would "inherit a chaotic jumble of institutions, incoherent of structure, and confused of purpose" (CHPA, 1983).

The Boston Redevelopment Authority (BRA) is the key municipal agency responsible for housing and redevelopment. The BRA was controlled by a five-member board that had been appointed by Mayor White, and the members' terms were not coterminous with Flynn's. Although the BRA board agreed to hire Flynn's choice for director, its legal and political independence led to compromises.

Under White, the BRA had abandoned neighborhood planning and had focused almost exclusively on downtown development. The BHA was in court-ordered receivership. The rent board was understaffed. The inspectional services department was under federal investigation for corruption. The agencies lacked elementary information. The city's zoning code was 25 years old and did not reflect new development pressures; thus almost every development required a variance. The city agencies that owned land and buildings did not control the Community Development Block Grant (CDBG) funds that could be used for rehabilitation or write-downs for new construction. No clear guidelines had been established to indicate how agencies should relate to neighborhood groups.

Flynn was reluctant to endorse CHPA's proposal for a far-reaching reorganization, fearing that it would paralyze the development process for several years. Instead of dramatically reorganizing during his first term, Flynn moved cautiously to gain greater control over the city hall housing and development bureaucracy, to impose policy and administrative coordination, to establish clear policies for relating to neighborhood organizations, to consolidate agencies with redundant functions, and to reform departments plagued with corruption and mismanagement.

The degree of Flynn's success or failure in overcoming the obstacles posed by the city's housing politics—the various factions of the housing lobby (real estate industry, housing activists, and neighborhood groups) and the city hall bureaucracy, as well as the city council, the courts, and the media—is best measured by examining the Flynn administration's housing policies and their implementation.

Housing Policies

Linkage and Inclusionary Housing

Boston's strong real estate market and the severe decline of federal housing funds led housing advocates to seek new revenues and techniques for creating affordable housing by extracting additional public benefits from private developers. *Linkage* (requiring large-scale commercial developers to subsidize affordable housing) and inclusionary housing became two hotly contested mechanisms for achieving this goal (Dreier, 1989).

In Boston the debate over linkage became a symbol of the widening gap between downtown development and neighborhood neglect. Linkage was first proposed in Boston in early 1983 by Massachusetts Fair Share (MFS), an Alinsky-style community group based in working-class neighborhoods, and a *Boston Globe* columnist. Following precedent established in San Francisco, linkage was promoted as a way to mitigate some of the housing market pressure caused by the escalation of jobs in the downtown area and the population boom (Keating, 1986). MFS and the *Globe*'s advocacy led lame-duck Mayor White to appoint a 30-member advisory committee in June 1983 to study the issue and make a recommendation. The committee was weighted with developers and city hall staff and included only two advocates of a strong linkage policy. The GBREB opposed the entire concept of linkage, warning that it would destroy the city's booming development climate.

During the 1983 mayoral contest, all but one of the seven major contenders endorsed some version of linkage; Flynn and King advocated the strongest versions (Muzzio & Bailey, 1986). When in October 1983 it looked as if the eventual winner would seek to enact some form of linkage, White's committee issued its report (Advisory Group, 1983). It recommended a linkage policy requiring downtown office developers to pay $5 per square foot over a 12-year period, with the first 100,000 square feet exempted. This formula, in terms of its present value, actually amounted to only $2.40 per square foot. The money was to be placed in the Neighborhood Housing Trust, which would be established to allocate the funds for affordable housing projects. Payments would begin 2 years after the building permit for a project was issued. Flynn, MFS, and other advocates criticized the advisory committee's recommendation, calling for a full $5-per-square-foot formula instead. Just before Flynn was to take office, White pushed his advisory committee's proposal through the BRA and the zoning commission.

As mayor, Flynn accepted this compromise linkage formula for a year and a half, waiting to see whether the policy would have the negative consequences that the real estate industry had warned it might have. It soon became clear that developers viewed the linkage fee as simply another cost of doing business in Boston's hot real estate market. With one of the lowest office vacancy rates in the country (8% in mid-1987) and Class A office rents of more than $30 per square foot, Boston developers simply passed on the fee to their commercial tenants.

The Boston Linkage Action Coalition wanted a fee increase to $10 per square foot, immediate payment (rather than phased), and abolition of exemptions (Smith, 1988). In early 1986 Flynn successfully proposed increasing the linkage fee to $6 per square foot to be paid over a 7-year period beginning on issuance of a building permit. This, in effect, doubled the existing linkage formula.

By October 1989 more than $76 million in linkage funds had been committed by 41 downtown developers. Flynn created the Neighborhood Housing Trust to allocate the linkage funds on the basis of proposals from developers for low- and moderate-income housing. Priority was given to nonprofit developers. As of October 1989, $28 million had been approved for allocation, primarily for "gap financing" of housing developments, which encompassed about 2,900 units (BRA, 1989; Brooks, 1989). Nonprofit housing developers received 84% of the allocated linkage funds.

Developers have learned to live with linkage. Trade unions support it as long as the projects provide jobs. Housing activists rely on it for funding. But linkage has not been without its skeptics. Boston's preservationist community expressed concern that the Flynn administration could be tempted to approve megaprojects simply for the linkage fees and jobs they provide and to overlook the environmental, design, and other aspects of development with long-term consequences. Although this issue was raised with the development of several major's office projects that were approved in Flynn's first 2 years, subsequent projects met with few objections from preservationists. Residents of several neighborhoods adjacent to proposed large-scale projects objected to the potential impact that these megaprojects would have on their neighborhoods. Midway through his first term, Flynn initiated a major downtown-development plan to reduce the zoning density of the central business district and adjacent areas (King, 1990). Flynn also appointed advisory committees of neighborhood residents to review megaprojects. As a result, concern that Flynn might "sell the skyline" for linkage and jobs benefits was reduced.

In 1986 housing activists and the Flynn administration began to push for another linked-development policy—inclusionary housing—which would require housing developers to set aside affordable housing units in otherwise market-rate projects. This push was triggered by the realization that publicly subsidized housing development was inadequate to meet lower income housing needs in Boston's expensive market. Inclusionary housing had been pioneered in suburban California and New Jersey (Mallach, 1984).

Opposition to inclusionary housing was likely to be even broader than was opposition to linkage. Linkage affected a small number of major developers who were building large downtown office towers. Inclusionary housing, however, would affect a much larger number of more diverse housing developers. Also it was not at all clear that many neighborhood associations, which in the past had opposed subsidized housing, would support a policy that would bring more low-income housing into their neighborhoods. In July 1986 Flynn submitted an inclusionary housing policy to the BRA that was strongly opposed by the GBREB. It called for private developers to set aside 10% of all housing units (in projects of 10 units or more) for low- and moderate-income residents.

The Flynn administration used its political ties to mobilize support for inclusionary housing not only from housing activists but also from neighborhood associations, the building trade unions, Governor Dukakis's administration, and some developers. Thus Flynn added new players to the growth-with-justice coalition.

City officials began to encourage housing developers to comply voluntarily with the inclusionary zoning guidelines while the policy was being publicly debated. The developers of the largest private project in the city's history reluctantly agreed to comply with the inclusionary housing guidelines, as well as to target additional funds for affordable housing in the adjacent South Boston neighborhood. By early 1987 these voluntary agreements had undercut the GBREB's claim that inclusionary housing would harm the housing-development boom.

Flynn won the policy debate but lost the legal war. In June 1987 the U.S. Supreme Court decided two cases involving land-use regulation that addressed the issue of "regulatory takings" (*First Evangelical Lutheran Church v. County of Los Angeles,* 1987; *Nollan v. California Coastal Commission,* 1987). In effect, these rulings could limit the ability of government to zone land use without adequate compensation for the loss (or "taking") of property rights (Fulton, 1987). Rather than risk a legal challenge, the city continued to seek only developers' "voluntary" compliance with the inclusionary housing guidelines. By early 1989, 231 low- and moderate-income housing units and more than $2 million in developer contributions had been produced as a result of this policy.

Disposition of City Land, Buildings, and Funds

Important resources available to local governments for housing development include the inventory of publicly owned buildings and land, as well as discretionary funds (including CDBG funds and, in Boston, linkage fees). Grantsmanship in obtaining competitive state and federal housing subsidies also is a measure of the effectiveness of the city's housing policy.

The Flynn administration inherited a sizable inventory of city-owned properties. The White administration's policies toward these city-owned properties reflected its development priorities. For example, the city government did not aggressively seek to foreclose on tax-delinquent owners of abandoned buildings or vacant land in order to assemble parcels for development, and properties that were acquired by the city were auctioned off to the highest bidders. Most of these properties were in low-income neighborhoods. When Flynn took office, almost none of the auctioned properties had been transformed into housing.

In the early 1980s the White administration developed a plan for the rehabilitation of closed schools, primarily into housing, without any neighborhood participation. The White administration's track record consisted primarily of turning the surplus schools in upscale neighborhoods into market-rate housing and leaving the abandoned schools in poorer neighborhoods to remain undeveloped.

Mayor Flynn promised to reverse this legacy of neglect and gentrification of city-owned property. Flynn took three important steps to resolve bureaucratic obstacles that he inherited. First, he put an end to the auctioning of city-owned property to the highest bidder. City-owned properties, he declared, would be sold for affordable housing. Development would take place through public competition with neighborhood involvement. Second, Flynn formed a coordinating committee of relevant city departments to develop an inventory of city-owned properties and to establish an information clearinghouse for the public so that they could identify the owners of vacant buildings and lots in their neighborhoods. The labor-intensive project of identifying, mapping, and computerizing this information took more than 2 years. Finally Flynn consolidated two city agencies—one responsible for managing city-owned property and the other responsible for allocating CDBG funds for neighborhood development—into one community-development department, the Public Facilities Department (PFD). Although he rejected the recommendation for a single superagency, he did streamline the administration of development policies.

After the two development agencies (PFD and the BRA) produced an inventory of all city-owned properties (identifying more than 3,000 buildable parcels, 747 of which were city owned), Flynn announced a goal of breaking ground on every parcel of city-owned property by the end of 1991. By the end of 1988, more than 4,000 units of affordable housing had been constructed on city-owned land.

The Flynn administration also used city-owned land to expand the number of homeless shelters, transitional housing for women, lodging houses, and special-needs housing for the mentally ill and alcoholics. In Flynn's first 4-year term the number of beds in such shelters doubled to more than 2,000. The siting of these projects usually met with neighborhood opposition. This experience led Flynn to object to demands from some neighborhood groups that they be given a final veto over the disposition of city property.

How did the Flynn administration achieve these affordable housing objectives on city-owned property? Five factors can be identified.

First, the city government, forgoing revenue, sold the city's property for a nominal amount to reduce development costs. Second, developers were required to build mixed-income projects with the profits from market-rate units, helping to subsidize the below-market units. For example, during 1986-1988, 12 surplus schools, designated as such after Flynn took office, were rehabilitated into 472 housing units, of which more than 60% would house low- and moderate-income persons.

Third, the city aggressively pursued competitive state and federal funds. The Flynn administration developed a close relationship with Governor Dukakis's administration. Massachusetts has the most comprehensive housing programs of any state government. Boston received a significant share of the state's housing subsidies.

Despite the drastic decline in federal housing funds and despite the fact that Flynn, a Democrat, was a constant critic of the Reagan administration's housing programs, the Flynn administration did well. Flynn was supported by Boston's powerful representation in the U.S. Congress. Whereas the White administration used Urban Development Action Grant (UDAG) funds primarily for commercial projects, the Flynn administration sought UDAGs for housing projects targeted for low-income residents. It also obtained HUD discretionary funding for rental rehabilitation and housing for the homeless.

Fourth, the allocation of the city's discretionary CDBG and linkage funds was prioritized for affordable housing developments on city-owned property. Fifth, priority was given to nonprofit developers. When Flynn took office, Boston already had a fledgling network of CDCs. The Flynn administration made a decision to nurture and expand this network. By late 1987, Boston had more than 30 nonprofit housing groups. During Flynn's first term, more than two thirds of the housing units approved for development on city-owned property were sold to nonprofit organizations to undertake affordable housing projects.

To expand the capacity of CDCs to undertake large-scale projects, the Boston Housing Partnership (BHP) was formed. The BHP is an outgrowth of a joint effort among business leaders, government officials, and the CDCs. The BHP's role is to help the CDCs improve their development capacity by taking advantage of economies of scale. Its first project was the rehabilitation of 700 vacant or substandard units in 60 buildings, many of them city owned, into low-income housing. Its second project was the rehabilitation of 938 units foreclosed by HUD. Its third project was the development of limited equity (controlled resale and profit) cooperatives on city-owned land in several neighborhoods.

The BHP is perhaps the most successful public-private community housing partnership in the country (Bratt, 1989). Through the BHP, the CDCs rehabilitate, own, and manage the developments. The BHP involves at least 20 sources of financing. The city contributed CDBG and linkage funds to hire staff and acquire properties—including several city-owned buildings—and tax abatements; the state provided tax-exempt mortgage financing and rental subsidies; the federal government provided tax credits for corporate investors and Section 8 rent subsidies for tenants; several local foundations and the national Local Initiatives Support Corporation (LISC) provided funds to hire the core staff; and the United Way contributed funds for CDC staff. The BHP's success led the Flynn administration to draft federal legislation modeled on the BHP (Flynn, 1987).

Tenants' Rights: Rent Control and Condominium Conversion

Since the mid-1960s, Boston's major housing battleground has been the regulation of rents, evictions, and condominium conversions. It has become the key litmus test for identifying political candidates as "conservative" or "liberal."

Boston enacted a strong rent-control law in 1969 that covered all private rental housing except owner-occupied two- and three-unit buildings. Subsidized and public housing also were exempted. By 1975 political support for strong rent control had eroded; rent control had become a convenient scapegoat for housing abandonment and high property taxes on homeowners. Mayor White and the city council adopted "vacancy decontrol," which permanently removed an apartment from regulation after a tenant left. As a result, by 1983 a vast majority of the once-regulated apartments had been gradually exempted from rent control—from more than 100,000 units to fewer than 25,000 units. Only those tenants who had lived in their apartments after 1976 were protected by rent control. White further demonstrated his opposition to rent control by appointing people who were opposed to rent control as members of the five-person rent control board and by understaffing the agency.

In the late 1970s a wave of condominium conversions fueled another round of tenant protest. In 1979 Flynn (then a city councilor) proposed a ban on condominium conversions, a policy that had little support among his colleagues. A compromise was reached that provided tenants with advance notice before they could be evicted for condominium conversion and relocation expenses.

For the 1981 city council race, the city's tenant groups formed a political action committee in support of protenant candidates. They endorsed a six-person *tenant ticket* for the nine council seats, all elected at large citywide. Only two of their candidates won, but one of them—Flynn—received more votes than any other candidate.

The tenant mobilization pressured the new council to strengthen the tenant-protection laws slightly. It changed the vacancy-decontrol law to a "rent-grievance" system: Tenants in decontrolled apartments (units previously covered by rent control) could initiate a grievance if annual rent increases exceeded 15%. Tenants could be evicted only for "just cause."

Flynn was endorsed by MTO when he ran for mayor. A cornerstone of his platform was an overhaul of the tenant-protection laws, a return to full rent control, and either a ban on evictions for condominium conversion or a ban on conversion itself. Shortly after assuming office, the Flynn administration introduced comprehensive tenant-protection legislation.

In October 1984 the city council rejected Flynn's plan. In its place the council substituted a stronger rent-grievance system, banned condominium evictions for low-income and elderly tenants, extended (up to 3 years) the notice period for other tenants facing condominium conversion, and increased moving expenses (from $750 to $1,000) for tenants displaced by conversions.

The compromise measure accurately reflected the balance of political forces at that time. The 1983 city council elections inaugurated a new system wherein nine members represented geographic districts and four were elected at large. Only three of the nine council districts had strong tenant organizations. Only one of the at-large councilors supported Flynn's plan, reflecting the power of the real estate lobby in the city council.

In mid-1985 Flynn convinced the city council to give the rent board the authority to regulate condominium conversions by requiring landlords to obtain a permit before a conversion could take place. The GBREB, however, successfully challenged Flynn's policy in court. Then, in 1986, Flynn and tenant activists successfully pushed the city council to enact a condominium conversion eviction ban. Flynn filed legislation in the state legislature to give Boston the authority to enact the condominium permit law overturned by the supreme judicial court. It was enacted in early 1987. In summer 1988, with a slightly more progressive council in place, the city council enacted a permit system, giving the rent board the authority to regulate condominium conversions. Ironically, by 1989 there was an oversupply of condominium units owned by investors because of past speculative conversions.

The Flynn administration also supported state legislation to place HUD-subsidized developments under rent control if the owners exercised their option to prepay the federally subsidized mortgage. Without such protections, up to

10,000 subsidized units in Boston could eventually be at risk if owners were to prepay in order to take advantage of Boston's strong housing market and convert to condominiums. When the state legislature failed to act, the local government in Boston passed its own legislation. Boston was the first city in which such a regulatory policy was enacted. It may have encouraged owners to sell to tenants at a discount.

In his first 5 years in office, Flynn substantially improved the city's tenants'-rights laws. Rather than the dramatic sweeping change he sought in his first year in office, however, the improvements came incrementally.

The tenants'-rights laws involve government regulation of private property. Opposition from landlords and GBREB and support from tenant groups was expected. Although Flynn was unable to enlist the Vault and major employers to support his tenants'-rights agenda, he helped shift the balance of political forces by enlisting the support of labor unions, religious leaders, and some neighborhood association leaders who previously had been neutral on the issue.

The Impact of Flynn's Policies

The resources available to local government to solve housing problems are very limited. During a period of drastic federal housing cutbacks, local governments lack the revenue base to address adequately the housing needs of their low- and moderate-income residents. For a variety of political, fiscal, and administrative reasons, when federal largesse is withdrawn, most city governments, liberal and conservative, have not developed comprehensive and sustained alternatives.

The Flynn administration was an exception to this rule. Perhaps more than the local administration in any other major American city, Flynn's administration in Boston actively addressed the city's housing crisis. Using existing tools and resources—and seeking to invent and create new ones—Flynn's administration made an aggressive effort to develop a housing policy that sought both to protect and to produce housing for poor and working-class residents. These efforts met political resistance, legal challenge, and bureaucratic obstacles, but they nevertheless reflected a strong commitment to serve the needs of Boston's poor and working class.

As a skillful politician, Flynn was able to promote this agenda and remain extremely popular. Through his populist appeal and policies, he was able to broaden and redefine the growth coalition. He was able to accommodate the development community (if not landlords), the business community, and the construction trade unions by promoting *managed growth* and *balanced development*. Flynn walked a tightrope between confrontation and compromise with the powerful business and development community while promoting a progressive housing agenda that helped unite minority, ethnic, and neighborhood groups around common interests.

It is possible to find other cities with some of the same policies as those carried out by the Flynn administration (Nenno, 1986). In no other city, however, has an attempt been made to pull together all of these and other policies into a comprehensive program. Thus the efforts in Boston to develop a progressive housing approach despite Reagan's austerity policies reveals a great deal about the potential and the limits of local housing policy.

What criteria should be used to evaluate a progressive local housing policy? We suggest the following four standards.

Better Housing Conditions

Are Boston's poor and working-class residents better housed than they would have been had the free, unregulated market been allowed to operate without local government's intervention? Clearly the policies that we have outlined improved the housing conditions for Boston's residents (Clay, 1988). Renters are more secure and pay less than they would have if the tenants'-rights law had not been strengthened. The supply of affordable housing has been expanded, providing greater choice. Public-housing residents increasingly live in better conditions.

At the same time, however, the city government lacks the legal tools and financial resources to stem the tide of gentrification. Private market forces have pushed housing prices in the unregulated sector beyond what most Boston residents can afford. The inventory of HUD-subsidized housing is at risk, and local government lacks the resources to fill the federal subsidy gap. The waiting list for public housing has doubled, which reflects a growing desperation among the very poor. City resources (even with the state's support) are not adequate to build enough new low-income housing to accommodate this demand (Slavet, 1988).

In these terms the Flynn administration's efforts served as a holding action to slow the tide of gentrification brought on by market forces. The administration also used market forces—the city's strong economic climate—to extract concessions from the private sector (through voluntary partnerships and involuntary regulations). Otherwise, the benefits of economic growth would not have trickled down to the poor and working class. To add to the benefits of these housing policies, the Flynn administration's other programs—such as the Boston jobs policy and the Boston compact between employers and the public schools—helped residents gain access to jobs that improved their opportunities in the housing market (Dreier, 1989).

On balance, however, the city's legal and economic resources simply were too limited to stem the forces of the private labor and housing markets that create a wide gap between available incomes and housing prices. The housing conditions for Boston's poor and working-class residents are better than they would have been without the Flynn administration's policies, but, overall, housing prices have increased faster than the incomes of these groups.

Empowerment

Although Kevin White rhetorically supported decentralized government, he never gave power to Boston's neighborhoods. In contrast, a cornerstone of

Flynn's populism was the empowerment of neighborhoods. He created neighborhood councils (NCs) to work with the city on the development of city-owned
property and to review all proposed development projects. He also created
planning and zoning advisory committees (PZACs) to assist in the modernization of the city's zoning code.

The best example of Flynn's policy is the Dudley Street Neighborhood
Initiative (DSNI), a foundation-funded group in the poorest section of Roxbury,
Boston's largest minority area. The city supported DSNI's plan to redevelop 30
acres of vacant land—half of it city owned—through a community land trust.
The city delegated the power of eminent domain to the DSNI to enhance its
development power, the first time that such an action was taken in the United
States. The neighborhood will develop the plan and participate in its implemen-
tation over a 5-year period through a CDC.

The Flynn administration's effort to empower neighborhood groups met
with some initial skepticism and resistance. Existing neighborhood leaders
worried that the NCs and the PZACs would supplant them as power brokers
in city hall. But, in a number of cases, the Flynn-appointed NCs and PZACs
opposed the administration's development efforts. The most contentious issue
was Flynn's unwillingness to grant them veto power over development decisions. Flynn argued that such vetoes potentially could conflict with the administration's responsibility to site low-income housing, homeless shelters, and
group homes for the mentally ill. In most situations, however, the administration allowed the NCs, PZACs, and other community groups sufficient input in
the development review that the veto issue never came to a head. In 1989 Flynn
opposed a campaign for virtually absolute community control of development
while offering to increase neighborhood influence on the zoning board of appeals.

The Flynn administration also provided funding for housing activist
groups. Not only CDCs but also tenant groups in public housing and in private
housing (the MTO) received funds to carry out their organizational agendas.
Occasionally they differed with the Flynn administration, usually over tactics
and strategy, not policy. MTO was a major ally for tenants'-rights legislation.
MFS was the catalyst for the Linkage Action Coalition's support for stronger
linkage and for inclusionary housing policies. The CDCs received funding to
carry out low-income housing developments, but they also supported the
administration's other housing policies. Flynn's support of DSNI reflected an
unprecedented delegation of power to a neighborhood organization.

The most controversial issue between Flynn's administration and Boston
neighborhood groups was the attempt of the Greater Roxbury Incorporation
Project (GRIP) to create a separate municipality called "Mandela." Flynn strongly
opposed what he called a racially divisive proposal, and the proposal was
defeated overwhelmingly in 1986 and 1987 referenda.

Advocacy

Does the local government accept the need to "live within its limits" during
a period of austerity, or does it challenge those limits by advocating for greater

resources at the state and federal levels? In Boston the Flynn administration changed the composition and rules of the growth coalition. Flynn became a prominent force in state and national housing politics, lobbying for new policies and greater resources for the poor—especially for low-income housing.

In 1989, in an unusual example of a city government acting affirmatively to enforce the Community Reinvestment Act (CRA), which was designed to spur lender investment in poorer neighborhoods, the Flynn administration initiated a redlining study (Finn, 1989) that ultimately confirmed racial discrimination in mortgage lending in Boston. It then created a municipal linked-deposit program, which leveraged city funds to reward those lenders committed to invest in Boston's poorer neighborhoods, and established a Community Banking Commission to oversee CRA compliance. Using these tools and working with the community organization, the Flynn administration pressured the banks to develop a $400 million community-reinvestment plan, including new branches, a below-market mortgage pool, and other programs targeted to low-income and minority areas.

At the state level, Flynn was the only major elected official to endorse and rally support for the "right-to-housing" campaign—an initiative opposed by both the Dukakis administration and the real estate industry. A coalition of religious, housing, and human service organizations, led by the Massachusetts Coalition for the Homeless, initiated a campaign to put a right-to-housing amendment to the state constitution before the voters through a referendum. Flynn also endorsed and testified on behalf of state legislation, supported by housing activists but opposed by the real estate industry, to impose rent control and condominium conversion restrictions on housing for which HUD subsidies were expiring.

At the national level Flynn became the recognized spokesman for the nation's local officials on housing and poverty issues. Soon after taking office, Flynn became active in the U.S. Conference of Mayors and the National League of Cities, two groups whose political effectiveness had waned significantly during the Reagan administration. Along with Mitch Snyder and the National Coalition for the Homeless, Flynn became one of the leading advocates for what became the 1987 McKinney Act, the first federal legislation during the Reagan era to fund services for the homeless. Flynn's staff drafted the Community Housing Partnership Act, modeled on the BHP, to provide federal funds to nonprofit housing developers (Dreier, 1987). The bill was incorporated in the omnibus housing bill that was pending in the U.S. House of Representatives as of August 1990.

Flynn's credibility as an advocate for affordable housing was based on the perception that, under his leadership, Boston was on the forefront of housing policy. As the mayor of the nation's 20th largest city, he carved out a niche for himself as a populist spokesperson for the urban poor, for federal aid to cities, and for progressive housing policies.

Models

Local government lacks the resources to solve all of the housing needs of its constituents, but it can develop policies and programs that can become models for federal housing programs. It can show that new, innovative concepts are feasible and can be replicated elsewhere if given adequate support.

The Flynn administration's most lasting legacy in this regard was its support for the nonprofit (or "social") housing sector. Boston is considered the most successful city in nurturing a network of sophisticated CDCs. Ironically the support for BHP by many of Boston's top business leaders has provided enormous credibility for a nonprofit approach to housing development that challenges the underpinnings of a market-driven housing system. Not only do Boston's nonprofit groups build, rehabilitate, and manage housing, but they also insulate their developments from Boston's speculative housing market by imposing rent limits and resale controls on their housing.

There are, of course, limits to what Boston's nonprofit groups can achieve, even with substantial city government, state government, private foundation, and business support. There are not enough resources to address the housing needs of all of Boston's low- and moderate-income residents adequately. Despite the government subsidies, these projects still require highly leveraged private financing. The capacity of Boston's nonprofit groups is still uneven, particularly in the management of the housing they develop. These, however, are problems that can be overcome with time and adequate funding. By promoting the nonprofit sector and its approach to housing, the Flynn administration nurtured a housing-delivery system that outlasted Flynn's mayoralty, as well as will serve as a model for other cities and for federal housing policy.

Progressive Municipal
Housing Policy: The Lesson of Boston

The case of Boston under populist Mayor Ray Flynn illustrates that a progressive municipal government can institute a redistributive housing policy aimed at offsetting and regulating market forces to provide better below-market-rate housing for its poor and working-class citizens. This is the reverse of what occurs in most American cities, including Boston under the administration of liberal Mayor Kevin White when, in his administration's later stages, deregulation and the promotion of market forces were promoted. White's policies exacerbated the housing crisis of Boston's less well-off citizens. Populist Flynn's policies helped alleviate that crisis.

This conclusion counters Peterson's (1981) claim that city governments cannot successfully engage in and sustain redistributive social policies that would discourage private investment. In the case of Boston under Flynn, Boston's economic transformation into one of the leading markets in the United States for private development and investment meant that its locational advantage allowed a reform mayor to institute progressive policies without the fear that they would provoke a serious threat to his political future. In addition, Flynn did not take an antigrowth stance but, rather, sought balanced growth and economic justice. He built a broadened growth coalition that supported development that included neighborhood-based development and neighborhood participation in planning and development. This inclusion allowed the city government to intervene actively through administrative, regulatory, tax, and subsidy policies to promote equitable, rather than uneven, development and affordable housing in Boston's working-class and poor neighborhoods.

No single city government alone, however, is capable of solving a city's housing crisis, at least if it is of the magnitude of Boston's in the late 20th century. Despite its own innovations, contributions from the private sector, and strong support from a liberal state administration, Flynn's regime was able to make continued, but only modest, progress in meeting Boston's affordable housing needs.

Because of fiscal constraints, local governments—whether progressive, liberal, or conservative—cannot realistically be expected on their own to solve the nation's housing problems. Nor can they expect to receive sufficient aid from state governments to offset the loss of housing assistance from the federal government. During the Reagan administration, HUD's budget was reduced from $33 billion in 1981 to $8 billion in 1988. The production of new subsidized housing fell dramatically as Reagan's HUD tried to eliminate federal housing-production programs in favor of short-term housing vouchers. HUD also championed the privatization of federally subsidized housing. Although Reagan's administration did not succeed in its stated goal of removing the federal government from housing, it did cripple federal housing programs and did as little as possible to address the burgeoning problem of homelessness that is partially a result of its anti-low-income-housing policies (Hartman, 1986).

What is required is a combination of progressive local housing policies and the reestablishment of a serious federal commitment to affordable housing (Angotti, 1986). That federal commitment must go beyond the market-based policies of past administrations, liberal and conservative alike. There must be a recognition of affordable housing as an entitlement (Davidoff, 1983; Institute for Policy Studies Working Group on Housing, 1989). Although the landmark 1949 and 1968 national housing legislation rhetorically supported this concept, there has never been a commitment by the executive or legislative branches to provide sufficient funding to fulfill the stated national goal of decent, affordable housing and a nondiscriminatory market (Dolbeare, 1983).

Unless the federal government is fully committed to a major housing program, cities will continue to suffer from housing shortages, displacement, racial discrimination, uneven development, and homelessness. In March 1988 the congressionally convened National Housing Task Force recommended that the federal government at least once again provide modest support for new housing for low- and moderate-income Americans, especially first-time home buyers (National Housing Task Force, 1988; Suchman, 1988). This recommendation falls far short of what is needed. Only a major reordering of federal spending priorities and resolution of federal budgetary deficits could produce truly adequate resources for the federal aid necessary for local governments in progressive cities like Flynn's Boston to address their affordable housing problems successfully in an equitable way.

References

Achtenberg, E. P. (1984). *Preserving affordable rental housing in Boston.* Boston: Rent Equity Board.

Advisory Group. (1983). *Report on linkage between downtown development and neighborhood housing.* Boston: Author.

Angotti, T. (1986). Housing strategies: The limits of local actions. *Journal of Housing, 43*(5), 197-205.

Boston Neighborhood Development and Employment Agency. (1988). *Abandoned property in Boston.* Boston: Author.

Boston Redevelopment Authority (BRA). (1987). *Boston housing: Facts and figures.* Boston: Author.

Boston Redevelopment Authority (BRA). (1989). *Building affordable homes: Linkage.* Boston: Author.

Boston Urban Study Group. (1983). *Who rules Boston: A citizens' guide.* Somerville, MA: Institute for Democratic Socialism.

Bratt, R. G. (1989). *Rebuilding a low-income housing policy.* Philadelphia: Temple University Press.

Brooks, M. E. (1989). *A citizen's guide to creating a housing trust fund.* Washington, DC: Center for Community Change.

Citizens Housing and Planning Association (CHPA). (1983). *Boston's development and housing: A reorganization proposal.* Boston: Author.

Clavel, P. (1986). *The progressive city: Planning and participation, 1969-1984.* New Brunswick, NJ: Rutgers University Press.

Clay, P. L. (1988). *Housing in Boston: A five year retrospective.* Boston: University of Massachusetts, McCormack Institute of Public Affairs.

Davidoff, P. (1983). Decent housing for all: An agenda. In C. Hartman (Ed.), *America's housing crisis: What is to be done?* (pp. 186-201). Boston: Routledge & Kegan Paul.

Dolbeare, C. (1983). The low income housing crisis. In C. Hartman (Ed.), *America's housing crisis: What is to be done?* (pp. 29-75). Boston: Routledge & Kegan Paul.

Dreier, P. (1983, October 10). The Vault comes out of darkness. *Boston Business Journal, 3,* p. 1.

Dreier, P. (1987, Fall). Community-based housing: A progressive approach to a new federal housing policy. *Social Policy, 18,* 18-22.

Dreier, P. (1989). Economic growth and economic justice in Boston: Populist housing and jobs policies. In G. Squires (Ed.), *Unequal partnerships: The political economy of urban redevelopment in postwar America* (pp. 35-58). New Brunswick, NJ: Rutgers University Press.

Elkin, S. L. (1987). *City and regime in the American republic.* Chicago: University of Chicago Press.

Emergency Shelter Committee. (1983). *Seeing the obvious problem.* Boston: Author.

Ferman, B. (1985). *Governing the ungovernable city.* Philadelphia: Temple University Press.

Finn, C. (1989). *Mortgage lending in Boston's neighborhoods 1981-1987.* Boston: Boston Redevelopment Authority.

First Evangelical Lutheran Church v. County of Los Angeles, 107 S.Ct. 2378 (1987).

Flynn, R. L. (1987, November/December). Federal/city/partnerships: A new housing agenda. *Journal of Housing, 44,* 233-240.

Fulton, W. (1987, December). Exactions put to the test. *Planning, 53,* 6-10.

Ganz, A. (1985, June). Where has the urban crisis gone? How Boston and other large cities have stemmed economic decline. *Urban Affairs Quarterly, 20,* 449-468.

Hartman, C. (1986). Housing policies under the Reagan administration. In R. G. Bratt, C. Hartman, & A. Meyerson (Eds.), *Critical perspectives on housing* (pp. 362-377). Philadelphia: Temple University Press.

Hays, A. (1985). *The federal government and urban housing: Ideology and change in public policy.* Albany: State University of New York Press.

Institute for Policy Studies Working Group on Housing. (1989). *The right to housing: A blueprint for housing the nation.* Washington, DC: Author.

Keating, W. D. (1986). Linking downtown development with broader community goals: An analysis of linkage policies in three cities. *Journal of the American Planning Association, 52*(2), 133-141.

King, J. (1990, May). How the BRA got some respect. *Planning, 56,* 4-9.

Logan, J. R., & Molotch, H. L. (1987). *Urban fortunes: The political economy of place*. Berkeley: University of California Press.

Mallach, A. (1984). *Inclusionary housing programs*. New Brunswick, NJ: Rutgers University, Center for Urban Policy Research.

Meyerson, M., & Banfield, E. C. (1966). *Boston: The job ahead*. Cambridge, MA: Harvard University Press.

Mollenkopf, J. H. (1983). *The contested city*. Princeton, NJ: Princeton University Press.

Muzzio, D., & Bailey, R. W. (1986, Winter). Economic development, housing, and zoning: A tale of two cities. *Journal of Urban Affairs, 8*, 1-18.

National Association of Realtors. (1985). *Quarterly home sales report*. Washington, DC: Author.

National Housing Task Force. (1988). *A decent place to live*. Washington, DC: Author.

National League of Cities. (1989). *A time to build up: A survey of cities about housing policy*. Washington, DC: Author.

Nenno, M. K. (1986). *New money and new methods: A catalog of state and local initiatives in housing and community development*. Washington, DC: National Association of Housing and Redevelopment Officials.

Nollan v. California Coastal Commission, 107 S.Ct. 3143 (1987).

Peterson, P. (1981). *City limits*. Chicago: University of Chicago Press.

Slavet, J. S. (1985). *Boston's recurring fiscal crisis: Three decades of fiscal policy*. Boston: University of Massachusetts, McCormack Institute of Public Affairs.

Slavet, J. S. (1988). Housing issues in Boston: An analysis of old and new problems. In *Program results and proposals for a future agenda*. Boston: University of Massachusetts, McCormack Institute of Public Affairs.

Smith, M. P. (1988). The uses of linked development policies in U.S. cities. In M. Parkinson, B. Foley, & D. R. Judd (Eds.), *Regenerating the cities: The UK crisis and the U.S. experience* (pp. 93-109). Glenview, IL: Scott, Foresman.

Suchman, D. R. (1988, June). Renewing the nation's commitment to housing: Recommendations of the National Housing Task Force. *Urban Land, 47*, 2-5.

Swanstrom, T. (1985). *The crisis of growth politics: Cleveland, Kucinich, and the challenge of urban populism*. Philadelphia: Temple University Press.

Swanstrom, T. (1988). Urban populism, uneven development, and the space for reform. In S. Cummings (Ed.), *Business elites and urban development: Case studies and critical perspectives* (pp. 121-152). Albany: State University of New York Press.

Weinberg, M. W. (1981, Spring). Boston's Kevin White: A mayor who survives. *Political Science Quarterly, 96*, 87-106.

21

Local Government Support
for Nonprofit Housing

A SURVEY OF U.S. CITIES

Edward G. Goetz

Nonprofit and voluntary associations are important elements of the delivery system for human and social services in the United States (Salamon & Abramson, 1982). The nonprofit sector has grown during the past 50 years in step with the increasing responsibilities of the welfare state (Salamon, 1986). Despite recent cutbacks in government activity in a range of services, nonprofits continue to play large roles in delivering basic human and social services to Americans.

The growth of the nonprofit sector has been very pronounced in the community development arena. Community development corporations (CDCs) have mushroomed in number and in scope since their introduction in the 1960s (Vidal, 1989). Although CDCs undertake a number of different activities, the largest subset is concerned with the production and management of affordable housing (National Congress for Community Economic Development, 1989). They are also taking on a growing importance in local efforts to produce low-income housing (Stegman & Holden, 1987). Already, in many markets, nonprofit CDCs are the only actors involved in developing low-income housing (Pickman, Roberts, Leiterman, & Mittle, 1986).

The growing interest in nonprofits is partially an evolutionary change and partially a response to specific contemporary problems. The expiration of affordability requirements contained in subsidy loans that the federal government has made to private developers since 1960 has resulted in a potentially sizable loss of low-income housing units.

Between 750,000 and 2 million low-income units are in jeopardy of being lost to the market through the repayment of federal loans, expiring use restric-

AUTHOR'S NOTE: This research was funded through a University of Minnesota Graduate School Grant-in-Aid. I would like to acknowledge the assistance of Patty Beech and Norah Davies and thank the anonymous reviewers for their comments.

tions, and defaults (Clay, 1987; National Low Income Housing Preservation Commission, 1988; U.S. General Accounting Office, 1986).

Nonprofit housing corporations increasingly are being looked to as a solution to this problem because nonprofits are less likely to convert units to market rate after the expiration of subsidies (Dreier, 1987). During the 1960s and into the 1980s, CDCs gained a prominence in affordable housing production because they offer advantages that for-profit producers do not. CDCs provide for greater tenant control of housing, they target lower income households better than for-profits do (Bratt, 1987a), and they have better access to and relationships with low-income communities (National Congress for Community Economic Development, 1989). On the basis of 1989 rates of production, Vidal (1989) estimated that CDCs, nationally, produce close to 45,000 units per year.[1]

CDCs and Local Housing Service Delivery Systems

Although nonprofits are increasingly active locally, there is little systematic knowledge about how cities are incorporating CDCs into housing policy implementation or whether they are providing the assistance necessary for CDCs to operate effectively. Bratt (1989) provided a description and analysis of the state of Massachusetts and its programs in support of community-based housing. Massachusetts has created a well-developed system of support that "includes all of the necessary funding and subsidy mechanisms" (Bratt, 1989, p. 281) to maintain a community-based housing strategy. In Massachusetts and in most other locations, the ties between nonprofit CDCs and the government are very strong. Nonprofit organizations are reliant on government for a large portion of revenues (Lipsky & Smith, 1989). A study of New York City CDCs found that 76% receive funding from the state government, 67% from the city, and 17% from federal sources (New Ventures, 1989). A national study found that 92% of the surveyed CDCs received federal funds (the largest portion of which was from the Community Development Block Grant [CDBG], which is allocated locally), 53% received state funds, and 42% received local government funds (Vidal, 1989).

As Lipsky and Smith (1989) suggested, the relationship between government and nonprofit groups is one of mutual dependence: The nonprofit organizations depend on the government to provide financial support, and the government relies on nonprofit groups to deliver an array of social and human services. Despite this mutual dependence, there is no uniform method by which local governments provide support to nonprofit CDCs. In this chapter, I examine the relationship between government and nonprofit CDCs in U.S. cities with populations over 100,000 and describe the ways local governments link themselves with CDCs to provide affordable housing. Whereas most studies of the nonprofit sector focus on CDCs, the unit of analysis here is the city, and the research question relates to how cities incorporate the nonprofit sector into their local housing delivery system.

Needs of Nonprofit Housing Corporations

Nonprofit producers of affordable housing require support in four specific ways: ongoing administrative support, predevelopment capital, project capital, and technical assistance.

Nonprofit CDCs cannot rely on project revenues for ongoing administrative funds. Given the mission of CDCs to produce affordable housing for low-income households, the cash flow from most CDC housing projects is minimal, if it exists at all. Finding resources to pay for staff and administrative expenses is therefore a continuing problem for CDCs. Administrative funding was the most mentioned funding priority in surveys of CDCs in New York City (New Ventures, 1989), the Minneapolis-St. Paul area (Common Space, 1988), Los Angeles (Kahane, Neff, & Barad, 1988), and the state of Minnesota (Minnesota Housing Partnership, 1989).

The typical sources of administrative support for CDCs are the government and private foundations. Yet these sources must be tapped anew each year or two as grants expire. A small percentage of CDCs have regular and reliable sources of support such as a sponsor organization—often a religious one—or an endowment fund. The majority of CDCs, however, are in a continual search for administrative funds to support the office activities of the organization.

The second regular need of CDCs is capital. CDCs need access to sufficient amounts of low-interest or no-interest acquisition and/or construction financing. A special form of project capital needed by most CDCs is *predevelopment financing*—funding to cover expenses incurred on a project before it is begun. In fact, Los Angeles-area CDCs mentioned the lack of predevelopment and bridge financing as the largest impediment to greater productivity (Kahane et al., 1988). Unlike for-profit developers, CDCs do not have a ready reserve or cash balance to cover these predevelopment expenses.

Beyond predevelopment financing, CDCs require low-interest financing for hard costs related to property acquisition and construction. Again, the objective of providing low-cost housing to lower-income households precludes heavy reliance on conventional financing. CDCs therefore must seek out low- (or no-) cost financing from other sources. The most prominent source of project capital financing is the public sector (Kahane et al., 1988; New Ventures, 1989; Pamuk & Christensen, 1989; Vidal, 1989). Because the federal government has reduced the amount of low-income housing assistance it provides, cities and states have attempted to make up some of the difference with innovative local programs. Another source of project capital is foundation money. Sometimes special nonprofit intermediaries, such as the Local Initiatives Support Corporation (LISC) and the Enterprise Foundation, act to channel low-interest financing to nonprofits (Stegman & Holden, 1987). When this occurs, CDCs have a regular source of project capital.

Finally CDCs typically are in need of some degree of technical assistance and expertise (Minnesota Housing Partnership, 1989). Many CDCs have small staffs (partly because of the difficulty in obtaining administrative support) and thus have to contract out for design, architectural, engineering, and/or construction supervision expertise (Common Space, 1988; Kahane et al., 1988). Legal expertise and assistance in project planning and syndications and property management are also needed by nonprofits. By paying for this expertise on a project-by-project basis, the CDC amortizes the cost into the project itself, rather than funds one or more staff positions on a regular basis.

Local governments and nonprofit CDCs are in a position of mutual dependence (see Salamon, 1986) in the pursuit of low-income housing objectives. For nonprofit CDCs to provide affordable housing to low-income households, they must spend time tracking down sources of administrative support, project financing, and technical assistance. This can be a time-consuming and resource-intensive process. For the local government, CDCs represent an increasingly useful means of implementing local affordable housing policy. In this study I examine the extent to which local governments and CDCs have wedded their common needs and objectives into ongoing and systematic working relationships.

Data

Survey questionnaires were sent to housing officials in 173 cities with populations over 100,000.[2] Responses were received from 133 jurisdictions, for a 77% response rate. Responding cities closely matched the distribution of all larger cities in regional breakdown, population size, and central city/suburban status.

Of the cities in the sample, 95% reported having nonprofit CDCs working on affordable housing (see Table 21.1). This finding indicates that CDCs have become a nearly universal phenomenon in cities with populations over 100,000. In 45% of the cities, there are fewer than 5 CDCs, and 35% of the cities report having from 5 to 10 nonprofit housing developers. The remaining 16% of the cities report more than 10 organizations.[3] The number of nonprofit housing developers is, as would be expected, significantly correlated with the size of the city ($r = .74, p < .001$). Nonprofit developers are more common in the Midwest (an average of 10.46 per city) and the Northeast (7.72) than in the South (4.45) and the West (5.65).

The number of units produced by the nonprofit CDCs in 1989 varies widely from 0 (in 18 cities, including the 7 cities that report having no CDCs) to a high of 2,200 (in Nashville). Nonprofits produced more than 100 units in 1989 in 36 cities (29%), and in 9 of those cities, CDCs produced more than 500 units. The number of units produced is also positively correlated with city size ($r = .42, p < .001$) but does not vary significantly by region. The cities in this sample report more than 17,000 units developed by CDCs in 1989. Extrapolated to all 177 cities with populations greater than 100,000, the production figure for CDCs in 1989 stands at more than 23,000 units.

Table 21.1 The Number of Nonprofit Housing Developers and the Number of Housing Units Produced by Nonprofit Developers in Responding Cities, 1989

Number of CDCs	Number of Cities	%	Units Produced	Number of Cities	%
0	7	5	0-10	38	29
1-4	59	45	11-50	39	30
5-10	45	35	51-100	17	13
11+	20	16	101-500	27	21
			501+	9	7
N = 131				N = 130	
Missing cases = 2				Missing cases = 3	

Table 21.2 Support for Nonprofit Housing Developers Provided by the Responding Cities

Form of Support	Number of Cities	%
Administrative funding	75	59
Project financing	104	82
Predevelopment financing	65	52
Technical assistance	70	56

NOTE: N = 126 cities that reported having nonprofit housing developers.

Local Government Support for Nonprofit Developers

Respondents were asked to indicate whether their jurisdictions provided each of the four means of support described in the previous section. Table 21.2 provides a summary of the responses.

Funding for nonprofit housing developers' administrative costs is provided by 75 of the responding cities (59%). The most common form of support provided by local governments is development capital; of the cities in the sample, 82% report making project financing available to nonprofit housing developers. Even predevelopment financing, the least common form of support, is offered by more than half (52%) of the cities. Finally 55% of the cities report providing technical assistance to nonprofit housing developers operating in their jurisdictions.

The cities that provide funding support for CDCs do so primarily from their CDBG program. Table 21.3 reveals the source of funds for those cities that provide predevelopment, development, or administrative funding. As Table 21.3 shows, close to 90% of the cities offering financial support use CDBG funds for this purpose. The second most common source of funding is the local redevelopment agency, typically through tax increment financing. This source is used for administrative funding by 10% of the cities and for predevelopment financing by 17%. Municipal general operating funds are used by fewer than

Table 21.3 Sources of Funding for Support to Nonprofit Housing Developers

Funding Source	Administrative Support		Predevelopment Financing		Project Capital	
	n	%	n	%	n	%
CDBG	67	94	57	89	87	88
Redevelopment agency	7	10	11	17	13	13
City general funds	6	8	4	6	5	5
State	2	3	4	6	5	5
Other city	1	1	3	5	14	14
Other federal	0	—	2	3	19	19
N	71		64		99	
Missing cases	4		1		5	

NOTE: Percentages add up to more than 100 because of multiple responses.

10% of the cities, as are state funds. Other federal funds (usually the rental rehabilitation program) and other local funds (housing trust funds, bond issues) are most commonly used as sources of project capital for nonprofit developers. Even then, they are used for those purposes by fewer than one city in five.

The technical assistance provided by cities is as diverse as general organizational planning and construction supervision and architectural design. The most common form of assistance is project-specific financial analysis, which is provided by 47% of the cities.[4] Construction monitoring and project supervision are offered by 34% of the cities, design and architectural assistance by 31%, construction specification and document-writing assistance by 25%, and general organizational planning and fund-raising assistance are provided by 11% of the cities.

The Politics of Local Nonprofit Housing

The survey findings have shown that CDC housing developers are active in 95% of U.S. cities with populations of over 100,000. In this section I describe the politics of nonprofit housing in these cities. It is reasonable to expect that in those cities with fewer nonprofit developers, CDCs more likely will be citywide in scope. Conversely, in cities that have many CDCs, the CDCs are likely to be neighborhood based in order to divide the city into service areas. It is also expected that the more CDCs in existence, the more likely they are to join a formal coalition to further political and organizational objectives. Coalitions can serve both the technical and political needs of member CDCs. They often provide technical assistance and informational services to their members, political support, and a means of political participation for nonprofit organizations incorporated under statute 26 U.S.C. §501 (c) (3) 1988, which legally prohibits them from directly engaging in political action.

Both of these expectations are supported by the survey findings. On the one hand, in cities with fewer than 5 CDCs, only 35% of the respondents claimed that CDCs in their cities were primarily or exclusively neighborhood based. On the other hand, in cities with more than 10 CDCs, 70% of the respondents indicated the CDCs were neighborhood based ($\chi^2 = 21.7, p < .001$). In addition, officials in only 10% of the cities with fewer than 5 CDCs report the existence of a CDC coalition, compared with 84% of the cities with more than 10 CDCs ($\chi^2 = 39.4, p < .001$).

Respondents were asked not only if CDCs were organized into a coalition but also if there was an active citywide low-income housing advocacy group. The existence of such a group was reported in 83 cities (64%). The great majority of these groups (73%) have formed since 1980 (59% since 1985). Responses to the questions about a CDC coalition and the low-income advocacy group were combined to create a variable measuring the level of organization of community-based housing interests that ranged in value from 0 to 2. That variable, ORGANIZE, is strongly and positively correlated with the number of CDCs present in a city ($r = .47, p < .001$).

In turn, the more highly organized community-based housing interests were perceived by respondents as exerting more influence on local housing policy. Respondents were asked how influential low-income housing advocates were in making the city's affordable-housing policy. Answers ranged from *influential on a variety of issues* (coded 4) to *not involved in policy making* (coded 1). The level of influence, as judged by city officials, is positively correlated with the degree of organization of community-based housing interests ($r = .51, p < .001$).

The level of influence by low-income housing advocates is, in turn, strongly related to the amount of support provided to nonprofit developers by the local government. The four methods of support examined earlier—administrative funding, predevelopment financing, project financing, and technical assistance—were combined into one variable, SUPPORT, ranging in value from 0 to 4 to reflect the number of such policies pursued by each city. That variable is positively correlated with the reported influence of low-income housing advocates at $r = .43$ ($p < .001$).

A path analysis reveals a straightforward linear model in which the number of CDCs in a city predicts the level of organization within the community-based housing sector (beta coefficient significant at $p < .001$). The level of organization significantly predicts the degree of influence in local housing policy making (at $p < .001$), but the number of CDCs has no direct effect on influence. Finally the degree of influence on policy making significantly predicts the amount of government-sponsored support for CDCs (at $p < .001$), with no direct effects from either the number of CDCs or the level of organization. This model is unchanged when controls for city size and region are introduced.

These findings support a model of political influence in which the organization of community-based housing interests, partly determined by the size of the nonprofit housing sector, is able to influence local housing policy in order to provide the most needed forms of support for nonprofit housing developers. It is not the size of the nonprofit housing sector per se that determines the policy response (there is no direct influence from the number of CDCs to policy outcomes); rather, it is the political organization of low-income housing advocates that creates the policy response. The integration of CDCs into local housing delivery systems is widespread and at least partially a result of the political organization of the community-based housing sector.

These findings suggest that support for community-based housing is, at least in part, a political arrangement between local housing advocates and public officials. How these political arrangements are made at the local level, however, is subject to wide variation. The next section presents a tentative typology of the arrangements that exist in different jurisdictions between the community-based housing sector and the local government.

Methods of Incorporating CDCs Into Low-Income Housing Delivery Systems

Four models of local government/CDC cooperation can be isolated. These models differ primarily along dimensions that reflect the basic needs of CDCs—that is, arrangements for administrative funding, availability of project capital and the existence of financial intermediaries, and the provision of technical assistance to CDCs. The models are as diverse as local state sponsorship of CDCs, in which the local government provides for each of the primary needs, and a state of disorganization in which there exists no systematic source of support for CDCs in any of the need areas. In addition, the models differ on the extent to which the nonprofit sector is organized and the nature of local-state/CDC relations. The four models are local-state sponsorship, partnership, community-based network, and preorganization.

These models exist in varying degrees in localities with active nonprofit housing activity, and they may, in fact, exist simultaneously in the same city. The typology is meant to characterize the dominant form of relationships between public and nonprofit sectors in the housing arena. The distinctions made in these models are an attempt to capture variations in institutional and programmatic relationships created by local public and nonprofit sectors in low-income housing production.

Model 1: Local-State Sponsorship

In this model the local state provides for each of the primary needs of the nonprofit housing sector. It is the regular and ongoing provision of organizational support to nonprofits and its source in the local state that set this model apart from the others.

San Francisco provides the best example of this model. Through the Mayor's Office of Housing, the city provides CDBG funds for the ongoing operating expenses of 11 CDCs. These CDCs act as regular subgrantees in the city's CDBG program, providing housing development (new construction and rehabilitation), management, architectural, and housing counseling services.

The city has funded eight of these nonprofits to produce low- and moderate-income housing in six target neighborhoods. During the past 12 years, these

CDCs have played the major role in housing rehabilitation and affordable new construction in the city. The operating expense grants that the city provides to these nonprofits represent, on the average, 54% of the administrative funds used by these eight development corporations.[5]

In addition, the city sets aside $2 million to $4 million annually in project capital pools for the exclusive use of the nonprofit developers. The 1989 San Francisco CDBG program funded a site-acquisition pool in the amount of $2.5 million and a Community Housing Rehabilitation Program in the amount of $1.9 million.

The city also funds the operating expenses of two technical service corporations that, in turn, provide free assistance to the nonprofit development corporations funded by the city. These technical corporations, their operating expenses underwritten by the city, are able to provide no-cost architectural, engineering, design, and construction expertise to the development corporations, thus reducing the cost of housing production.

The city's CDBG program, which is the source of administrative and program funding for the nonprofits, provides the regular linkage between the local state and the CDCs. Nonprofits do operate in San Francisco outside this state-sponsored network. BRIDGE, a regional nonprofit, is the largest of these; the Catholic Archdiocese, through its nonprofit Catholic Charities, is also active in local low-income housing. These nonprofits, however, represent the minority in both number of CDCs and amount of housing developed.

Model 2: Partnership

The partnership model refers to those cities in which the local public, private, and nonprofit sectors have come together to form institutional and programmatic partnerships to provide low-income housing. Boston and Chicago are two examples of cities operating under this model. The institutional partnerships play the role that the state plays in Model 1; that is, the partnerships channel project capital and technical assistance to nonprofit development corporations. Unlike the local-state sponsorship model, the partnership model does not provide the same level of ongoing administrative support for nonprofits.

Stegman and Holden (1987, p. 103) called the Chicago and Boston Housing Partnerships "prototypes of a new kind of local institution." These partnerships bring together public, private, and nonprofit actors engaged in housing production. The city participates by contributing capital for project development. In Boston, both the city and the state governments provide capital to the partnership. The private sector is represented by the mortgage financing of local lenders and the investments of insurance companies and other private investors. The nonprofit sector is represented by neighborhood-based housing development corporations. In Chicago, LISC, a national nonprofit housing investor, is a partnership member. The partnership, with representatives from each of these sectors, allocates the financing resources to the CDCs that are the actual project developers.

The partnerships thus serve as locally constituted financial intermediaries, attracting both public and private financing and channeling that capital to nonprofit developers. In addition, the Chicago partnership provides the nonprofit developers with various forms of technical assistance from project-feasibility analysis to the assembly of a development team. The Boston Housing Partnership provides its nonprofit developers with predevelopment funds in addition to regular project capital.

In both the partnership model and the local-state sponsorship model, locally constituted intermediaries are channeling resources to the nonprofits. In fact, the existence of these intermediaries and the reliability of their support provide the primary advantages of these models. In Model 1 the intermediary is the local state; in Model 2, it is the partnership. In the final two models such locally constituted intermediaries do not exist, nor do other regular and reliable sources of development support.

Model 3: Community-Based Network

In cities in which the community-based network model exists, there is no formal, ongoing relationship between the local government and the CDCs. Despite the lack of such a linkage, local CDCs are organized into a coalitionlike body. The greatest utility to member CDCs of the coalition is the organization's political advocacy (Union Institute, 1989). Without a locally constituted intermediary to provide and channel project capital, let alone administrative support, CDCs rely on their own organization for political, and sometimes technical, resources. There can be, but not often is, regular or extensive public and private sector involvement in the CDC network.

St. Paul and Minneapolis provide examples of the community-based network model. In St. Paul, 21 organizations, including 12 nonprofit housing developers, have organized the Coalition for Community Development. Organized in 1984, the coalition brought together neighborhood-based developers to discuss the basic needs of the organizations and how to better communicate those needs to the local public and private sectors. The coalition's activities thus have been more political or advocacy-based from the beginning. For example, the coalition has negotiated Community Reinvestment Act compliance with local lenders and provided assistance in drafting the city's vacant-housing policy initiative. In addition to this advocacy, the coalition provides limited technical assistance to members.

In Minneapolis the situation is similar. CDCs organized into the Minneapolis Nonprofit Consortium in 1980. The impetus for organizing was a threat by the city to withdraw state sponsorship of a number of nonprofits. The consortium traded that administrative support for a regular source of project capital for development projects. As a result of the consortium's lobbying effort, the city sets aside a pool of funds for low-income rental housing rehabilitation. Nonprofits must compete with for-profits for this fund, however. Currently the consortium serves primarily as a political and organizational resource for its members.

Model 4: Preorganization

In these cities—represented in this sample by, among others, San Antonio, New Orleans, and Omaha—CDCs exist and are active but are not organized. Public support of nonprofit activity is sporadic and is dependent mainly on the individual political resources and initiative of CDCs. Public support that is provided is likely to be project specific and not programwide. Similarly private-sector participation is not organized. As in the community-based network model, no locally constituted financial intermediary serves as a clearinghouse for project capital or a source of technical assistance to nonprofit developers.

Discussion

The data presented here reveal that nonprofit housing development occurs in almost all U.S. cities with populations over 100,000. Further, the practice of local government support for nonprofit housing developers is widespread. Development capital for nonprofit CDCs is offered by 82% of the responding cities, 59% provide administrative funding to CDCs, and just over 50% provide technical assistance and predevelopment financing. Over 76% of the responding cities offer two or more of these benefits to local CDCs, and only 11% offer none.

There can be little doubt that CDCs are heavily integrated into housing policy approaches of local governments. What these data do not reveal, however, is the depth of the support provided by the state. As Bratt (1989, p. 281) argued, government support of community-based housing "however comprehensive and exemplary, may face serious problems" without deep subsidies. Without such data, an evaluation of government support for CDCs remains incomplete.

The findings do suggest that the political organization of the community-based housing sector is an important reason why local governments provide support to community-based housing. Path analysis showed that the greater the level of organization within the community-based housing sector, the greater the influence on housing policy making and the more forms of support for CDCs provided by local government.

Political arrangements for the linkage of CDCs to local government are organized in various ways. Four models were offered to describe the way CDCs are (or are not) incorporated into the local housing service delivery system. The models are differentiated on a number of dimensions.

First, these models represent varying levels of need satisfaction for nonprofit developers. Nonprofits, as argued earlier, need access to supports as diverse as administrative funding and project-based technical assistance. Under the local-state sponsorship model, each of these needs is addressed at least partially by the local state. Under the preorganization model, none of these needs are guaranteed by any actor. In between these two extremes, the partnership model and the community-based network model provide varying degrees of project and technical support. Regular and reliable satisfaction of needs allows CDCs to concentrate organizational resources on project-related activities. Theoretically this focus should improve the organization's productivity.

Second, the models differentiate themselves by the level of organization within the nonprofit sector. In Models 1, 2, and 3, the nonprofit sectors have organized into political and/or administrative entities. Only in the preorganization model is there no sense of collective action among CDCs.

Finally these models describe different programmatic relationships between local government and the CDC sector. In Models 1 and 2 there is a formal and program-based relationship between the state and the nonprofit housing sector. Under such an agreement there is a basis for expectation for both parties, and the form of program cooperation is set within institutional rules. Structured relations with the local state can be a mixed blessing for CDCs, however. Although state sponsorship can provide regular sources of support, such a close relationship can create political strains. Nonprofits may find themselves in the position of tempering political criticism of their local sponsor, should the ties between them become too close and the nonprofit organizations become too dependent on state sponsorship. In Models 3 and 4 the cooperation between the local government and the CDCs is unstructured.

Representing political as well as programmatic arrangements between local public and nonprofit actors, these models are subject to change. In the late 1970s, Minneapolis operated in a manner similar to a local-state sponsorship model. In 1980 the city opted to eliminate ongoing administrative support of CDCs, and the city came to fit Model 3. More recently attempts have been made to change Los Angeles from Model 4, in which CDCs are sporadically tied into local government support, to a Model 2 partnership in which the public, private, and nonprofit sectors provide capital and technical support to affordable housing by nonprofit CDCs. This effort thus far has proven ineffectual because both the city and the nonprofits have refused to endorse the partnership idea.

A definitive assessment of the forces that lead to the adoption of one model or another awaits further research. Nevertheless the accommodations between local public and nonprofit actors in low-income housing are, it seems, primarily political in nature. The strength of the community-based movement in general, and housing CDCs in particular, the salience of the low-income housing issue, and the existence of public officials sympathetic to community-based housing all might increase the likelihood of a more institutionalized (Model 1 or Model 2) accommodation of CDCs.

Nonprofit Organizations
as a Political Resource

Nonprofit CDCs have emerged as important actors in housing policy. This is, in part, because of the reduction of tax incentives and subsidies to for-profit developers (Bratt, 1987b) and the crisis of expiring use restrictions that force a rethinking of housing policy strategy (Dreier, 1987). Yet this analysis suggests that the devolution of housing policy authority to the local level also has played a role in increasing the importance of nonprofit community-based development. The findings suggest that CDCs offer tangible political benefits to local policymakers. Because of their community base, CDCs represent a potential constituency for local officials. Providing support to nonprofit CDCs is not simply a technical decision about how to implement housing policy but also is a political decision that provides officials the opportunity to respond to neighborhood-based constituencies. The devolution of housing policy to cities and states has made this political connection stronger now than in the past. Further, the organization of low-income housing advocates and providers helps solidify their political importance as a constituency and brings greater local government support for nonprofit housing.

Notes

1. Pamuk and Christensen (1989) estimated that CDCs in the San Francisco Bay Area have accounted for 42% of the affordable housing in that region since 1980. A study of New York CDCs estimated that close to 5,000 units were under development in 1988 (New Ventures, 1989).

2. The mailing list of the National Association of Housing and Redevelopment Officials was used as the primary source for locating housing officials. In larger cities, up to three separate agencies were represented on the mailing list: the housing authority, the redevelopment agency, and, typically, a community development (CD) department. When the choice existed, the CD department was chosen to receive the questionnaire. CD departments are more centrally located, bureaucratically, than are the more specialized housing authorities and redevelopment agencies; thus it was assumed that officials in this agency would be more informed about the range of issues covered by the questionnaire.

3. The most CDCs in one city were in Los Angeles, which reported 45 nonprofit developers. According to New Ventures (1989), however, New York City (not among the respondents to this survey) has 70 active housing CDCs.

4. The following percentages in the text are based on the cities that report providing technical assistance ($n = 70$). With nine missing cases, the denominator for this analysis is 61.

5. These figures are from San Francisco's 1989 CDBG program year budgets. In fact, if one outlying corporation is taken out of the analysis (this corporation relies on the city for only 2% of its administrative costs), the average figure rises to over 60%.

References

Bratt, R. G. (1987a). Dilemmas of community-based housing. *Policy Studies Journal, 16,* 324-334.
Bratt, R. G. (1987b). Private owners of subsidized housing vs. public goals: Conflicting interests in resyndication. *Journal of the American Planning Association, 53,* 328-336.
Bratt, R. G. (1989). *Rebuilding a low-income housing policy.* Philadelphia: Temple University Press.
Clay, P. L. (1987). *At risk of loss: The endangered future of low-income rental housing resources.* Washington, DC: Neighborhood Reinvestment Corporation.
Common Space. (1988). *Survey of nonprofit developers in Minneapolis and St. Paul.* Minneapolis: Author.
Dreier, P. (1987). Community-based housing: A progressive approach to a new federal policy. *Social Policy, 18*(2), 18-22.
Kahane, F., Neff, J., & Barad, B. (1988). *Survey of Southern California nonprofit housing organizations.* Los Angeles: Corporation Fund for Housing.
Lipsky, M., & Smith, S. R. (1989). Nonprofit organizations, government, and the welfare state. *Political Science Quarterly, 104,* 625-648.
Minnesota Housing Partnership. (1989). *Survey of low-income housing providers in Minnesota.* Minneapolis: Author.

National Congress for Community Economic Development. (1989). *Against all odds: The achievements of community-based development organizations.* Washington, DC: Author.

National Low Income Housing Preservation Commission. (1988). *Preventing the disappearance of low-income housing* (A report to the House Subcommittee on Housing and Community Development and the Senate Subcommittee on Housing and Urban Affairs, United States Congress). Washington, DC: Author.

New Ventures. (1989). *The activities and accomplishments of New York City's community development organizations.* New York: Author.

Pamuk, A., & Christensen, K. (1989). Preliminary findings on San Francisco Bay Area nonprofit housing developers. *Berkeley Planning Journal, 4,* 19-36.

Pickman, J., Roberts, B. F., Leiterman, M., & Mittle, R. M. (1986). *Producing lower income housing: Local initiatives.* Washington, DC: Bureau of National Affairs.

Salamon, L. M. (1986). Government and the voluntary sector in an era of retrenchment: The American experience. *Journal of Public Policy, 6*(1), 1-20.

Salamon, L. M., & Abramson, A. J. (1982). The nonprofit sector. In J. L. Palmer & I. V. Sawhill (Eds.), *The Reagan experiment.* Washington, DC: Urban Institute Press.

Stegman, M. A., & Holden, J. D. (1987). *Nonfederal housing programs.* Washington, DC: Urban Institute Press.

Union Institute. (1989). *Survey of nonprofit associations: Review of the data.* Washington, DC: Author.

U.S. General Accounting Office. (1986). *Rental housing: Potential reduction in the privately owned and federally assisted inventory.* Washington, DC: Government Printing Office.

Vidal, A. C. (1989). *Community economic development assessment: A national study of urban community development corporations.* New York: New School for Social Research, Community Development Research Center.

PART EIGHT

Urban Growth

Introduction to Urban Growth

Nico Calavita

There is a new mood in America. Increasingly, citizens are asking what urban growth will add to the quality of their lives. . . . Today, the repeated questioning of what was once generally unquestioned—that growth is good, that growth is inevitable—is so widespread that it seems to us to signal a remarkable change in attitudes in the nation. (Reilly, 1973)

So begins *The Use of Land,* the 1973 Task Force Report sponsored by the Rockefeller Brothers Fund and edited by William K. Reilly, who later was to become the beleaguered Environmental Protection Agency administrator for much of the Bush administration. The sponsors, the members, and the staff of the task force were hardly radical, but their message was: The way America was developing was unacceptable, and new approaches were necessary. Citizens were rebelling against growth, and that, indeed, signaled "a remarkable change in attitudes in the nation."

Criticisms of Growth

The questioning of urban growth in the 1970s was part of a larger criticism of economic growth and consumption as indicators of social and individual success. Economic growth had been considered good for all of this millennium, and when, in the 20th century, socialism mounted its challenge to capitalism, the issue was not the desirability of economic growth; both systems measured their success on how fast their economies grew. The conflict between capitalism and socialism was about which class deserved to enjoy the benefits of that growth. Economic growth and consumption then had been used for so long and so widely as indicators of social and individual success that to question them was to question our collective identity and individual self-esteem. But in the early 1970s those yardsticks of success were subject to question.

Doubts about the desirability of urban growth were raised too. Until then it was generally believed that urban growth was highly desirable and that its benefits far outweighed any associated costs. It was believed that the increases in real estate and sales taxes accompanying new development would amply compensate for the cost of public facilities made necessary by the new growth. It was also generally believed that growth would enhance the economic base of a community and bring more and better jobs. Urban growth, everybody knew, benefits everyone.

Beginning in the late 1960s, this view came to be challenged. Antigrowth sentiment was engendered by growing environmental consciousness both at the local and global levels and, to a lesser extent, by concerns about the fiscal impact of growth. People became concerned about loss of open space, destruction of landforms, and the cutting down of forests. Alerted by Rachel Carson's *Silent Spring* (1962), many became concerned about the effects that new technologies and new products such as pesticides had on wildlife, the food chain, and human health. At the fiscal level the long-term impacts of new growth were being analyzed more closely, and the benefits of growth were being set against the need for new capital facilities that new development creates. Generally this new attitude toward growth was "part of a rising emphasis on humanism, on the preservation of natural and cultural characteristics that make for a humanly satisfying environment" (Reilly, 1973, p. 34). The result was growth management.

Growth Management

Growth management is "generally defined as the regulation of the *amount, timing, location,* and *character* of development" (Levy, 1991, p. 218). Old planning tools, such as general plans, capital improvement programs, and zoning and subdivision regulations, were still the main instruments used to regulate development, but now they were structured not only to accommodate growth but also to tame it in ways that were environmentally and fiscally sound. Two basic approaches were adopted: (a) adequate public facilities ordinances, through which growth was phased in in accordance with the availability and adequacy of public facilities, an approach first tested in 1969 in Ramapo, New York, and upheld by the New York Court of Appeals in 1972 (*Golden v. Planning Board of the Town of Ramapo,* 1972) and (b) annual permits limitations, which limited the number of units that could be built in 1 year, with developers competing— the "beauty contest," as developers derisively referred to the process—for the right to build those units. Developments generally were rated on the basis of two broad criteria, first used in Petaluma, California: (a) "availability of public utilities and service" and (b) "quality of design and contribution to public welfare and amenity." In 1975 the U.S. Court of Appeals upheld the Petaluma Growth Control Ordinance. Growth controls had become as American as motherhood and apple pie.

But there were problems. Critics argued that by limiting residential development and not economic growth, growth controls attack the symptoms, and not the causes, of growth; that by limiting only residential development, the cost of housing would increase; and that by controlling growth in some but not all localities in a region, growth would spill over into areas without growth controls, worsening growth problems in those localities and in the region as a whole. These and other criticisms led Paul Niebanck (1986), a growth management expert, to charge:

> At its dead end worst it has meant the denial of entry into a community to all but the most affluent population groups; the relegation of local planners to the role of border guards; the conversion of the natural environment into an article of private consumption; and the abandonment of the idea of responsible membership in the larger society. (p. 4)

Ballot Box Planning

Despite its problems, growth management continued to be used as a mechanism to limit growth in many localities, especially in California, where the passage of Proposition 13 in 1973 severely limited the ability of local governments to raise property taxes to pay for the costs of development. Many of these growth limitation measures were not passed by the local decision-making bodies, but through initiatives placed on the ballot by citizen groups angry about the effects that rapid growth was having on their quality of life. At least in California, the much-touted "growth revolt" (California Office of Planning and Research, 1980) was made possible by the initiative and the referendum. "Ballot-box planning," as the use of the initiative, referendum, and recall for land-use and growth issues has been called by Orman (1984), generally is seen as a citizen response to the perceived failure of elected officials, unduly influenced by developers, to manage growth effectively (Caves, 1992; Glickfield, Graymer, & Morrison, 1987; Orman, 1984). As Orman (1984, p. 4) has indicated: "Where local legislatures can act to meet development problems head on, there will be no need for turning to ballot measures. Where they can't, citizens will have available their constitutionally protected right to reclaim the legislative mantle and have a shot at those problems themselves." It is not surprising that under such circumstances (when "legislators are dominated by privileged special interests"; Cronin, 1989, p. 11), citizens turned to the initiative process. But why do citizens believe that their elected officials are dominated by privileged special interests? Why are these interests so powerful? What is their power base? How do they operate? Growth machine theory helps us answer those questions.

Growth Machine Theory

Harvey Molotch, a sociologist at the University of California, Santa Barbara, published an article in 1976 and a book (with John Logan) in 1987 that made a strong case for a model of local politics based on the power of the "growth machine" (a coalition of land-based interests generally able to legitimate the ideology of growth and to manipulate the planning process to foster growth and increase land-use intensities). Thus "the political and economic essence of virtually any given locality, in the present American context, is growth" (Molotch, 1976, p. 310). This type of growth comes inevitably at the expense of the quality of life of the general population.

This is possible because in the United States, "weak regulation combined with local autonomy makes each property owner a potential speculator whose investment return depends on decisions made by the local government. . . . Given the opportunity to get rich through changes in spatial relations, those with property interests remain the most consistently motivated and powerful urban actors" (Molotch & Vicari, 1988). Economic and population growth bind together individuals who benefit from such growth and set them apart from others who use the city "principally as a place to live and work" (Logan & Molotch, 1987, p. 50). Those who benefit from growth include not only land-owners but also bankers, developers, contractors, and realtors, as well as related businesses such as title search companies, architectural, legal, and engineering consultants, utility companies, and local newspapers. For them, the city is a growth machine, to be used to increase the wealth of "those in the right position to benefit" (Logan & Molotch, 1987, p. 50).

Government constitutes a key partner of the growth machine. It is "the arena in which land-based interest groups compete for public money and attempt to mold those decisions which will determine land-use outcomes" (Molotch, 1976, p. 312). At the citywide level the primary role of government is to promote growth by creating a good "business climate," by providing the necessary infra-structure, facilities, and services, by establishing redevelopment projects, by providing land on favorable terms, by training workers, and by giving tax breaks to incoming business. Perhaps even more importantly, government can "connect civic pride to the growth goal, tying the presumed economic social benefits of growth in general to growth in the local area" (Logan & Molotch, 1987, p. 60). Thus growth-inducing schemes are justified on the basis that they increase the tax base and improve fiscal health, or that they provide jobs especially for minorities. Such declarations might be justified in some cases, but, in general, "growth is at best a mixed blessing and the growth machine's claims are merely legitimating ideology, not accurate descriptions of reality" (Logan & Molotch, 1987, p. 85).

Intensification of land uses generally leads to a deterioration in the quality of life for those who use their localities not as a mechanism for enrichment, but as a place to live, work, and play. They will not become rich through continuous growth but, more likely, they will see their quality of life deteriorate. Growth usually is accompanied by increases in air and water pollution, traffic congestion, crime, social unrest, cost of housing, and a decrease in environmental amenities (Feagin, 1988). The promise of jobs is often elusive. Molotch (1976) and Logan and Molotch (1987) have argued that new jobs usually are captured by new-comers who, because of their skills, education, or the "right" ethnic background, are able to bypass the local residents, especially the disadvantaged. They also find little evidence that growth brings higher wages or is a panacea for urban poverty.

It is not surprising, then, that citizen and environmental groups periodically oppose growth and development in their community, leading to growth conflicts, especially sharp in areas that experienced rapid growth during the 1970s and 1980s, such as California and Florida. But what chance of success do citizen groups have in fighting the growth machine?

Growth Conflicts and Their Outcomes

In Chapter 22, "The Growth Machine Versus the Antigrowth Coalition," Ronald K. Vogel and Bert E. Swanson present a case study of growth conflicts in Gainesville, Florida, and document the almost insurmountable odds faced by opponents of growth. The authors question Logan and Molotch's suggestion that growth machines can be challenged from the bottom up. They found that, at least in Gainesville, the growth machine seems to have a systemic advantage over the antigrowth coalition. Even though the antigrowth coalition was able to win local offices, that effort was not sustained, and the attempt to change growth policies in Gainesville failed.

Vogel and Swanson also analyze why growth management, with its promise of stewardship, did not appease both pro- and antigrowth forces in Gainesville. On the one hand, antigrowth forces believed that growth management should include the notion that growth may be limited or stopped. Progrowth forces, on the other hand, viewed growth management as a growth *accommodation* mechanism that would help ensure that a minimal level of public facilities would be provided with development, that "concurrency," the term that characterizes growth management in Florida, could be ensured. Given these contrasting interpretations, it can be assumed that growth management is shaped by the power that the two camps have at particular times. There is no doubt, however, that growth forces generally have greater monetary and staying power than their opponents. They are strongly motivated, dedicating time and resources to push for growth and greater development intensities; their livelihood depends on it. Antigrowth forces, however, have only limited time and resources. On occasion, members of citizen groups opposing growth might neglect their families and jobs in defense of their quality of life, but eventually they have to abandon or limit the time dedicated to their cause.

Vogel and Swanson are not so sure, however, that even if citizen groups were to gain power and sustain it, the community would be better served. More likely, there would be an increase in segregation and unequal distribution of services. This result is possible because citizen, neighborhood, and environmentalist groups that challenge the growth machine from time to time usually are made up of upper class individuals. Low-income and minority groups, who tend to participate less and possess more limited educational or monetary resources, are likely to lose, both from progrowth and antigrowth policies: Thus more powerful groups—either pro- or antigrowth—will identify their interests with the community interest and will influence the political process to their advantage.

The pessimism of Vogel and Swanson is to be contrasted with an article in which De Leon and Powell (1989) analyze the 1986 victory of Proposition M (limiting the amount of office space in downtown San Francisco) and conclude that "successful growth control movements are possible in large U.S. cities." These opposing findings might be the result of the limited time frame of analysis of the two studies. Calavita (1992) maintains that it is important to study growth conflicts over a longer period of time to make sense of the alternating fortunes of pro- and antigrowth forces. His analysis of San Diego for 1970-1990 found that growth control forces can, through initiatives, significantly affect the *quality* and *location* of development, whereas the growth machine "will control the playing field when long-term growth is challenged." This difference became especially clear in 1988, when the growth machine defeated four ballot measures in San Diego even though the problems of several years of rapid growth had led to a citizen revolt in the mid-1980s, forcing the City of San Diego Council to pass an Interim Development Ordinance in 1987.

Do Growth Controls Control Growth?
(Probably not, but they can make it better)

Even when growth-control policies are passed, they do not seem to slow population growth. Logan and Zhou (1989, p. 468) analyzed growth-control policies for 1970-1980 and surveyed planning officials in 387 suburbs. They found that growth-control policies have little effect on population growth, median family income, median rent, and percentage of black residents. This result might occur because (a) "the formal adoption of growth controls and environmental limitations do not guarantee their implementation" (p. 468) and (b) how plans are implemented will depend on the balance of forces in the community.

One reason for the political "hegemony" of the growth machines is their "staying power." If developers lose at the policy-making stage, they will not give up. Although proponents of growth-control measures probably will go home after a victorious campaign, developers will persist in their efforts to have their project approved or density increased.

Research by Warner, Molotch, and Lategola (1992) also indicates that growth measures do not appear to have much of an effect on housing construction and population growth. This finding might be due not only to implementation failures but also to growth controls that are not really meant to limit growth but that are established for "symbolic" reasons, to appease an angry populace. What these studies found is that growth controls, or the threat of growth control initiatives, can improve the quality of development, as Calavita (1992) found in the San Diego case. If the quantity of development is being threatened and if the regulatory environment is tough, then developers are more willing to make concessions to provide additional amenities in their projects.

The amount of money that developers pay for public facilities (alternatively called exactions, development impact fees, or linkage fees) varies widely among localities. In some localities developers are made to pay not only for roads, parks, libraries, and other public facilities but also for other needs that development generates, such as day-care centers and low-income housing. In cities with a long tradition of strong controls, such requirements have become routine.

It should be remembered, however, that even in cities with stiff requirements, "questions persist about the adequacy of the fees to cover the related costs of development" (Stone & Martinek, 1991) because fees "never pay the total cost of development" (Nelson, 1991); rather, they "recoup some portion" of government's costs to provide capital facilities (Nicholas, 1992, p. 557). Localities where growth is challenged "demand (and receive) more" (Molotch, 1993), but not enough to cover all costs of growth. In so doing they are able to protect their quality of life better than localities where development proceeds with few restrictions.

Developers will play along and pay high fees as long as economic growth will continue at high levels, generating strong demand for housing or commercial space. If economic growth slows to a trickle, however, as in Southern California during the early 1990s, the growth machine will stage a counterattack, blaming excessive fees and regulations for the economic downturn and demanding the cutting of fees (Colgan, 1993).

Downtown Progrowth
Coalitions and Redevelopment

The growth machine model of city politics grew out of the analysis of booming areas such as Southern California and Florida. But the challenge to progrowth forces by citizen groups did not begin in the 1970s in areas exploding with growth. On the contrary, conflicts over growth and development started in the 1960s and early 1970s in the older cities of the Northeast and the Midwest as a result of urban renewal programs.

The story of the urban renewal program is well known. Ostensibly the program was to clear slums and replace them with sparkling, sanitary, low-cost housing. But in reality it was a program whose main purpose was to refurbish the central business district of many of the older, larger cities in the U.S. where property values had been sinking as economic activities moved to the suburbs. Downtown property owners started as early as the 1920s to campaign to use the power and dollars of government to revitalize their downtowns (Weiss, 1980). The 1949 and 1954 Housing Acts provided them with more than they hoped for—billions of federal dollars given to sympathetic local urban renewal agencies that used their power of eminent domain to tear down existing low-income housing and replace it with office towers, public buildings, shopping centers, and upper class housing. Progrowth coalitions were formed between local politicians, businesses, financial and real estate interests, and building trade unions to push for the economic redevelopment of the central business district. In the process, they destroyed hundreds of thousands of low-income units and small businesses.

In the early 1960s, criticisms of the urban renewal program started to mount, from both liberals and conservatives. Anderson (1964) called it "the federal bulldozer," and Glazer (1965) "the asphalt bungle." These criticisms were accompanied by "an explosion of protest and community activism" (Mollenkopf, 1983, p. 17), not only on the part of the poor, mostly black residents, but also on the part of educated middle-class professionals who recently had chosen to live in central cities.

Through neighborhood activism, progrowth coalitions were undermined and large-scale clearance projects were ended. In 1973 the Urban Renewal Program was terminated, and in 1974 it was incorporated into the Community Development Block Grant Program (CDBG). Community development has differed from urban renewal because of the former's emphasis on rehabilitation and conservation, citizen participation, community improvement, and drastically reduced funding.

The Politics of Redevelopment

The defeat of the progrowth coalitions contributed to the displacement of private investment from Northeastern to Southwestern cities where "conservative progrowth coalitions reign unchallenged" (Mollenkopf, 1983). These cities, with the urban renewal program gone, turned to "redevelopment," which under state laws usually grants localities the power of eminent domain and the power of "tax-increment financing." This taxation power permits redevelopment agencies to issue bonds against the expected property increases within the redevelopment areas.

Given the importance of progrowth coalitions in these cities and their eagerness to push for growth in general and downtown redevelopment in particular, it might be expected that little leeway would exist to balance those demands with the needs of the community. In Chapter 23, "Growth Politics and Downtown Development," Robyne S. Turner suggests instead that cities dominated by the imperative of economic growth in some cases might follow policies that seek to pursue economic development while at the same time trying to ameliorate the costs of growth through linkage policies that require private developers to contribute to public benefits through programs for child care, affordable housing, public open space, and job training. In examining four Florida cities, Turner finds that they employed different approaches to facilitate downtown development. Orlando was the most progressive in its efforts to integrate equity issues into its planning by promoting affordable housing, preserving neighborhoods, and providing amenities; Fort Lauderdale was the least likely to use governmental power to distribute some of the benefits of redevelopment to the low-income downtown residents. Why these variations? Politics, Turner tells us. Growth politics is not monolithic, as it varies on the basis of important political and administrative dynamics

Conclusion

In trying to understand how and why cities and regions grow the way they do, it becomes necessary to understand urban politics in its relation to the economy. The issue is whether the economic growth imperative is so strong that very little autonomy is left to government to pursue other policies aimed at protecting neighborhoods or improving upward mobility of excluded groups. What we have learned is that urban politics generally is dominated by powerful progrowth coalitions but that under certain cultural, political, and organizational conditions, such domination can be challenged.

References

Anderson, M. (1964). *The federal bulldozer: A critical analysis of urban renewal, 1949-1962.* Cambridge: MIT Press.

Calavita, N. (1992). Growth machines and ballot box planning: The San Diego case. *Journal of Urban Affairs, 24,* 1-24.

California Office of Planning and Research. (1980). *The growth revolt: Aftermath of Proposition 13?* Sacramento: Author.

Carson, R. (1962.) *Silent spring.* Boston: Houghton Mifflin.

Caves, R. (1992). *Land-use planning: The ballot box planning revolution.* Newbury Park, CA: Sage.

Colgan, J. (1993, September/October). Impact fees and the new reality. *California Planner,* pp. 8-9.

Cronin, T. E. (1989). *Direct democracy: The politics of initiative, referendum, and recall.* Cambridge, MA: Harvard University Press.

De Leon, R., & Powell, S. (1989). Growth control and electoral politics: The triumph of urban populism in San Francisco. *Western Political Quarterly, 42,* 307-331.

Feagin, J. R. (1988). Tallying the social costs of urban growth under capitalism: The case of Houston. In S. Cummings (Ed.), *Business elites and urban development.* Albany: State University of New York Press.

Glazer, N. (1965). The asphalt bungle. *New York Herald Tribune Book Week.* January 3, p.1.

Glickfield, M., Graymer, L., & Morrison, K. (1987). Trends in local growth control ballot measures in California. *Journal of Environmental Law, 6,* 111-158.

Golden v. Planning Board of the Town of Ramapo, 30 N.Y. 2d 359, 285 N.E. 2d 359 (1972).

Levy, J. (1991). *Contemporary urban planning.* Englewood Cliffs, NJ: Prentice Hall.

Logan, J. R., & Molotch, H. (1987). *Urban fortunes.* Berkeley: University of California Press.

Logan J. R., and Zhou, M. (1989). Do suburban growth controls control growth? *American Sociological Review, 54,* 3, 461-471.

Mollenkopf, J. (1983). *The contested city.* Princeton, NJ: Princeton University Press.

Molotch, H. (1976). The city as a growth machine. *American Journal of Sociology, 75,* 309-330.

Molotch, H. (1993). The political economy of growth machines. *Journal of Urban Affairs, 15,* 29-53.

Molotch, H., & Vicari, S. (1988). Three ways to build: The development process in the United States, Japan, and Italy. *Urban Affairs Quarterly, 24,* 188-214.

Nelson, A. (1991). *Development impact fees as a win-win solution.* Unpublished manuscript, Georgia Institute of Technology, City Planning Program, Atlanta.

Nicholas, J. (1992). On the progression of impact fees. *Journal of the American Planning Association, 58*(4), 517-524.

Niebanck, P. (1986). Introduction. In D. Porter (Ed.), *Growth management: Keeping on target?* Washington, DC: Urban Land Institute.

Orman, L. (1984). Ballot-box planning: The boom in electoral land-use controls. *Bulletin of the Institute of Environmental Studies, 25*(6), 1-15.

Reilly, W. K. (Ed.). (1973). *The use of land: A citizens' policy guide to urban growth.* New York: Thomas Y. Crowell.

Stone, K., & Martinek, D. (1991, November). The economic consequences of unmanaged growth: Western city. *League of California Cities,* 6-9.

Warner, K., Molotch, H., & Lategola, A. (1992). *Growth control: Inner workings and external effects.* Presented at a California Policy Seminar, University of California, Berkeley.

Weiss, M. (1980). The origins and legacy of urban renewal. In P. Clavel, J. Forester, & W. Goldsmith (Eds.), *Urban and regional planning in an age of austerity.* New York: Pergamon.

Suggestions for Further Reading

Bernard, M., & Rice, B. R. (Eds.). (1983). *Sunbelt cities: Politics and growth since WWII.* Austin: University of Texas Press.

Cummings, S. (1988). *Business elites and urban development: Case studies and critical perspectives.* Albany: State University of New York Press.

Finkler, E. (1972). *Nongrowth as a planning alternative: A preliminary examination of an emerging issue* (ASPO Report No. 54). Chicago: American Society of Planning Officials.

Fulton, W. (1991). *Guide to California planning.* Point Arena, CA: Solano.

Lord, G., & Price, A. (1992). Growth ideology in a period of decline: Deindustrialization and restructuring, Flint style. *Social Problems, 39*(2), 155-169.

Mollenkopf, J. (1989). Who (or what) runs cities, and how? *Sociological Forum, 4*(1), 119-137.

Nicholas, J. (1992). On the progression of development impact fees. *Journal of the American Planning Association, 54*(4), 517-524.

Ryan, J. (1991). Impact fees: A new funding source for local growth. *Journal of Planning Literature, 5*(4), 401-407.

22

The Growth Machine Versus
the Antigrowth Coalition

THE BATTLE FOR OUR COMMUNITIES

Ronald K. Vogel

Bert E. Swanson

The Growth Machine Thesis

A predicate for most American communities is to grow—whether through economic growth, city size (population, land area, employment, budgets), or capital infrastructure and services. Form and Miller (1960) have suggested that industry shapes the community. Some cities grow and others decline (by any of these measures) when businesses move to reduce their costs, improve access to markets, and enhance their profits. Since World War II, Sunbelt cities have benefited from the rise of a service economy, but Snowbelt cities have experienced the decline of heavy manufacturing. Of course, the different kinds of growth or decline determine the distribution of benefits and costs.

Growth is not the goal in all communities. Many urban problems are associated with growth, including traffic congestion, air and water pollution, increased crime, social unrest, higher cost of living, and a decline in the quality of life (Feagin, 1988). Those who are dissatisfied with these conditions frequently oppose growth. Elected officials have tried population caps (St. Petersburg and Boca Raton, Florida), moratoriums on growth (Broward and Monroe Counties, Florida), and impact fees ("Symposium: Development," 1988) to limit, slow, or regulate growth (see Freilich & Einsweiler, 1975). Many city governments have adopted "linkage" policies that specify that developers must make contributions to the public good—such as building low-income housing—in exchange for the right to develop (Clarke, 1987; "Symposium: Linkage," 1988).

AUTHORS' NOTE: A summer research grant by the University of Louisville Commission on Academic Excellence contributed to the completion of this project.

Molotch (1976) posed the debate over growth as a political struggle pitting "growth machines" against the scattered resistance of ragtag "antigrowth coalitions." His study stimulated a spate of research, including studies of the relationship between business power and policy outputs (Lyon, Felice, Perryman, & Parker, 1981), case studies of the growth machine under challenge by antigrowth coalitions in the Sunbelt (Trounstine & Christensen, 1982) and the Snowbelt (Swanstrom, 1985), surveys of the attitudes of citizens and elected officials toward growth (Baldassare, 1981; Maurer & Christenson, 1982), and studies of alternative growth strategies pursued by communities (Fainstein, Fainstein, Hill, Judd, & Smith, 1986).

Recently Logan and Molotch (1987) refined and elaborated the growth machine thesis, distinguishing the motives of growth advocates who pursue "exchange value" of land from antigrowth activists who desire "use value." They identify the process by which local "rentiers" facilitate corporate movement by providing a positive business climate and the physical infrastructure and services needed by business. Questioning the wisdom of these "subsidies," they found that population growth and economic development are mixed blessings, often a redistribution from the "have nots" to the "haves."

Taken together, Logan and Molotch (1987), Stone and Sanders (1987), and Cummings (1988) have provided a cogent, well-developed response to Peterson's (1981) thesis that cities have a unitary interest in growth and economic development. Local political regimes are not subject simply to market forces. They can and do affect the course of events in their communities, though systemic power (Stone, 1980) remains an important constraint. The varying outcomes associated with different growth machine configurations and growth strategies are documented in this research. The authors are critical of neoconservative policies and question whether the private sector can or should be relied on to solve urban problems.

Having repudiated neoconservative answers at a theoretical level and demonstrating the ineffectiveness of their policies in practice, scholars turned their attention toward efforts to defeat the growth machines that benefit from prevailing policies. Logan and Molotch end their book with a call to arms to those who might take on the growth machine: "U.S. cities must stop competing among themselves for capital and use their relatively high levels of legal autonomy to compete as a collective force against the growth machine system that has captured them all" (p. 292). Some communities are better situated to "change the conditions they impose on capital for the right to a site" (p. 293). If costs were increased in the "privileged zones most attractive to capital," sources of capital would have to pay or go elsewhere (p. 294). Thus developers also would be expected to pay their own way in other communities that are not so strategically located. Logan and Molotch suggest that growth machines can be challenged from the bottom up, arguing: "People *can* capture control over the places in which they live and critically judge the value of what they make and the community conditions under which they produce it" (p. 296).

Growth Management

Logan and Molotch (1987) are not against all growth in the community. "Declining cities experience problems that might be eased by replacement investments. Even in growing cities, the costs of growth can conceivably be limited by appropriate planning and control techniques" (p. 85). But they are skeptical.

> For many places and times, growth is at best a mixed blessing and the growth machine's claims are merely legitimating ideology, not accurate descriptions of reality. Residents of declining cities, as well as people living in more dynamic areas, are often deceived by the extravagant claims that growth solves problems. These claims demand a realistic evaluation. (p. 85)

When these evaluations are undertaken, they usually demonstrate that "the promised benefits of growth . . . have been greatly exaggerated by the local growth activists" (p. 88).

In many parts of the country (including Florida, California, Oregon, Virginia, and Vermont), the "hard look" that Logan and Molotch call for is supposed to be taking place in the form of growth management. *Growth management* is touted as a rational planning process to arrive at community decisions regarding growth rates, the mix of residential, industrial, and commercial development, the trade-offs between "use" and "exchange" values, the provision of public services, and the protection of the environment.

Growth management is "a system for guiding, directing, limiting, and encouraging growth so that we can meet the inevitable demands for housing, infrastructure, and other growth support systems" (DeGrove, 1983, p. 1). DeGrove, while head of Florida's Department of Community Affairs, stated: "In the past developers went about their business in ignorance. This was by no means a deliberate conspiracy to ruin the environment. But environmentally, they didn't do well. . . . Today's developer has no excuse" (*Florida Trend,* 1984, p. 99). But DeGrove also made clear that growth management is not antigrowth. "I'm not advocating no growth. I think growth management should encourage growth" (*Florida Trend,* 1984, p. 99).

For DeGrove, the "striking feature" of land and growth management programs is the development of a planning system that is "linked" to new or existing regulatory systems. He believed that past efforts at environmental regulation were marked by duplication, overlap, and needless delays for the private sector. He said the decade of the 1980s could be "a time when environmentalists shed their passion for ad hoc solution in exchange for a new growth management system that draws the regulatory process within the framework of a new kind of comprehensive plan" (DeGrove, 1984, p. 6). Burt (1983, p. 1) captured the essence of DeGrove's position:

> He does not talk of limiting growth, calling that impossible, but of managing it. He advocates an idea fast becoming popular among planners: the designation of urban areas into which all new development would be squeezed. . . . It would force urban

redevelopment by denying space for development to sprawl across the countryside. . . . In return for that . . . he would pledge the retention of open spaces and natural vistas. With this, he believes developers and environmentalists might realize a mutuality of interests as happened in Oregon, and create a coalition of public support.

DeGrove sought to ensure that the infrastructure to service growth was in place and that developers and new residents shouldered the costs of growth through impact fees. He was aware that citizens were skeptical, fearing that "they'll get the higher density but in the long run they won't get the tradeoff of open space and protection of environmental areas. They're afraid the system finally will break down and they'll get the high density everywhere" (Burt, 1983, p. 1).

Although Logan and Molotch (1987) found that growth controls do not pose any real danger to "value-free development," they are not without hope that growth controls may one day be a potent weapon to limit the power of growth machines:

> Beyond fostering loud but largely symbolic debates, growth control programs have had little effect; at most they have influenced the distribution of some residential development, with regulation deflecting the affluent towards places improved through the restrictions. None of this means that *eventually* the seesawing between developers and growth controller won't be more definitely resolved, and with important use and exchange value consequences. (p. 162)

Can growth management, as it is being instituted in Florida, be the instrument Logan and Molotch favor to force growth advocates to "pay" for their growth and prevent negative externalities from being forced on the community? On the surface, growth management would appear to provide a compromise between market-driven growth and no growth that would satisfy the concerns of environmentalists and the desire of developers to profit from growth. Government "manages" growth through comprehensive planning, designating areas suitable for growth and ensuring the placement of an adequate infrastructure to service that growth, at the same time ensuring that the quality of services is maintained and the environment is protected.

The Study

The purpose of this study is threefold. First, we examine the perceptions of local actors on growth. The way growth is defined has important effects on growth policies. Second, we appraise the prospects of the antigrowth coalition in the struggle to defeat the growth machine. The new urban politics called for by Logan and Molotch (1987) requires not only that antigrowth coalitions be able to win local offices but also that they be able to sustain themselves and enact policies that change the way growth occurs in the community. Finally, we consider the potential to end the battles over growth by using growth management. If growth management is a compromise, is it recognized as such by both the growth machine and antigrowth coalition actors?

Although the issue of growth versus no growth should be a dominant cleavage in any community (if we are to give credence to the growth machine thesis at all), Logan and Molotch believe that the development of an effective antigrowth coalition is more likely to occur in certain communities. For this research we chose a place that would be favorable to the Logan and Molotch thesis.

The basis of the antigrowth coalition, according to Molotch (1976), is "a mixture of young activists (some veterans of the peace and civil rights movements), middle-class professionals, and workers, all of who see their own rates as well as life-styles in conflict with growth" (p. 327). Leadership is provided by "government employees and those who work for organizations not dependent on local expansion for profit" (p. 327). Growth machines have been defeated in Palo Alto, Santa Barbara, Boulder, and Ann Arbor, which are all university towns (p. 328).

We study Gainesville, Florida, a university town in which growth has become a divisive issue. Gainesville has been growing at about 1% annually. About 80,000 people reside in the central city, with a comparable population in the surrounding urban area. The latter, managed and planned by the county government, is growing more rapidly. Half of the area's labor force works for government; the University of Florida is the major employer.

After prohibitions on faculty participation in public affairs were lifted in the 1960s (McQuown, Hamilton, & Schneider, 1964), a shift in community power occurred; reformers defeated the "main street crowd" for election to the city commission. The old guard believed that the market should dictate growth with little or no government intervention. Although the new members of the city commission were not antigrowth, they were critical of unbridled growth policies and thus worked to put sound planning principles into practice. Strict zoning, an adequate infrastructure, and services were the hallmark in the city—but not in the county, where most of the growth was occurring. A new balance was established in the city when the growth forces adjusted to the new rules of the game. Although individual electoral contests remained heated, growth ceased to be a major issue. Increased enrollments at the university ensured continued growth for some time.

In the early 1980s, an antigrowth coalition, based in neighborhoods and environmental groups, rose up to challenge growth and the related negative impacts. Members of the coalition gained a working majority on the city commission. However, the defeat of the mayor in the 1986 spring elections cost the "antigrowthers" their majority on the city council. (The office of mayor is mainly a symbolic one. The mayor is selected by the council on a rotating basis.) "Progrowthers" remained in control of the county commission, although two members of the antigrowth coalition were elected, one in 1984 and the other in 1986. In the fall of 1988, one of these county commissioners was defeated.

In many ways the antigrowth mayor epitomized the type of antigrowth leader Molotch wrote about. He was vocal in opposing American involvement in El Salvador, advocating disinvestment in South Africa, and protesting the nuclear arms race. He was a former musician and had managed a record store. (He has since returned to music.) His politics were progressive. He fought for passage of a conflict-of-interest policy for the city's Hazardous Materials Committee, worked against an effort to place a six-lane highway through a residential neighborhood, and attempted to reduce densities in new developments. He closely scrutinized the budget and opposed expenditures that induced growth or subsidized development. He played the role of consumer advocate and watchdog in overseeing the municipal utility, water, and wastewater extensions into the unincorporated area.

The city and county commissions have disagreed on a variety of policies, including consolidation, annexation, land-use planning, and the provision of services for the urbanized area. County officials blame the strict antigrowth policies of the city for forcing development beyond city limits. City personnel counter that in the county, which is dominated by progrowth forces, development occurs unchecked, with little regard to its impact on the environment or the city.

The county government now provides a number of urban services in the unincorporated area, and an urban service tax has been imposed. Efforts by city government to annex unincorporated areas have been resisted by the county commission, which fears the loss of tax revenues. Some city residents question whether annexation would benefit the city. They fear that the cost to upgrade the infrastructure may not be offset by additional revenues. Some rural residents object to county provision of municipal services because of higher tax rates, whereas city residents complain about double taxation. A task force was established to study the desirability and feasibility of a "Greenbelt" as one means of creating an urban boundary for the city.

Political Protagonists' Perceptions

Political protagonists at the leadership level have stated often and clearly their policy positions on the growth issue. They also have gone to some length to characterize negatively their opponents' positions. These views have been gleaned from political campaign rhetoric, interviews with the principals, and statements made during various public forums on land use.

The Growth Machine View

The growth machine is forthright about its own position. To support growth today does not mean the same as it meant several years ago. There is greater sophistication and sensitivity to the environment today. To favor growth no longer means advocating uncontrolled, unlimited growth. Rather, members of the Chamber of Commerce argue that growth and a better quality of life can be pursued simultaneously. Whereas in the past, members of the business community resisted new taxes to support an expanded infrastructure and services required by growth (they believed growth paid for itself), today they recognize that a growth strategy requires tax increases to pay for the urban infrastructure—roads, sewers, water, and schools. Members of the growth machine do not distinguish population growth from economic development.

Members of the growth machine believe they have not been given sufficient credit for developing a well-planned community with good residential neighborhoods and economic stability. They brought the university and a major teaching hospital to town. They built shopping malls, adding to the community's tax base and providing jobs. The growth machine forces take credit for road improvements and for providing better air service than would have been achieved strictly by market forces.

Antigrowth activists are characterized as elitists, unconcerned about the need for more jobs, opportunities for minorities, or the placement of an adequate infrastructure to support the population. They are viewed as emotional, impractical, and uninformed. As a prominent growth machine spokesperson explained in an interview, "Now that they are in the community, they want to slam the gate closed." Antigrowthers are blamed for urban sprawl because they restricted development opportunities in the city. They are seen as far too negative when addressing environmental problems, seeking to affix blame on companies for past practices that were thought safe at the time. Perhaps most serious, they are viewed as "doomsayers" who spread dissension and divide the community. The growth machine grudgingly credits the antigrowth coalition with contributing to better planning and forcing the community to recognize the importance of natural resources.

Members of the business community will support candidates who are concerned with neighborhoods and the environment. For example, the challenger to the antigrowth mayor was considered by some to be an antigrowther who had broadened his perspective. He was an advocate of growth management. In fact, some business interests supported a more vocal progrowther in the primary and shifted their support to the "growth management" candidate only after the progrowther was defeated. Even though members of the growth machine opposed some of the growth management candidate's positions, he was considered to be reasonable and to have an open mind.

A pattern, however, can be seen in the business opposition to candidates or incumbents who question growth or whose support is not ensured. During the course of this study, a scandal dubbed "Chambergate" hit the headlines. Disputes over growth led the president of the Chamber of Commerce to attempt to silence the critics of growth. The Chamber targeted the two antigrowth county commissioners and two faculty members—one the president of the Sierra Club, the other an urban geographer—who were raising significant concerns about development in the county. The scenario included monitoring faculty involvement in public affairs and having the respective faculty members' department chairs and other higher college officials refute their antigrowth statements. The Chamber's plan included manipulation of key figures in the black community who would advance the cause of growth.

A transcript of a tape-recorded meeting on the plan was leaked to the press. Although he had not attended the meeting, the university president was implicated in the plan. He denied the allegations and indicated his support for academic freedom. A storm of protest and outrage forced the president of the Chamber of Commerce to resign. He apologized for his actions and said he had exaggerated the university president's support and role in his plans. The Florida Board of Regents conducted a brief investigation and concluded that the university president was innocent of any improprieties, and the matter was laid to rest. The faculty union was somewhat disgruntled with the investigation and concerned that sufficient support was not being given to academic freedom. Additional questions were raised about the extent of university involvement in local economic development. Although a new Chamber of Commerce president and executive committee were installed, there is little evidence that the basic growth objectives were modified, though the incident did result in the elimination of the political action committee. The geographer later addressed the Board of Regents, claiming he was being penalized in merit reviews in his department as a result of his activities.

The Antigrowth Coalition View

The antigrowth coalition, made up of neighborhood organizations, environmentalists, and watchdogs of elected officials, is equally strident toward its protagonists. Although they acknowledge that the growth machine has developed the community and enjoyed a number of successes, they question whether continued growth is beneficial. They admonish the growth machine for delaying adoption of the county's comprehensive plan, especially the conservation element. They believe the "carrying capacity" of the land has been reached and fear additional population growth will only reduce the quality of life.

Antigrowthers seek to debunk the myth promulgated by the growth machine that the marketplace will provide economic benefits to everyone while protecting the environment and contributing to stable neighborhoods. They oppose taxing present homeowners to subsidize the costs of growth and the profits of the real estate industry. Leaders of the antigrowth coalition objected to the antigrowth label in interviews. One argued:

> Antigrowth is not a fair word nor an accurate description for a complex discussion of what kind of growth should occur and the rate of that growth. Growth gets simplified for political purposes. It is not as simple as growth versus antigrowth.

He preferred terms such as *environmentalist* and *fiscal conservative* to define his position. He believed that the antigrowth label was really a "straw man" intended to discredit antigrowth forces.

> We are labeled "antigrowth" but we are acting and behaving in the public interest to be sure questions about development are being answered satisfactorily. . . . Greedy

people think we are antigrowth. They want growth at any cost. We raise questions and [so are accused of] stand[ing] in the way of progress.

The problem as he saw it was that business leaders believe that "if you don't wave the flag of progress every damn minute, city hall will collapse in economic ruin."

Another member of the antigrowth coalition stated in an interview, "I am progrowth, but growth of what? I am not against growth in the quality of life, better jobs, the environment—all positive measures of improvement." He argued that some decisions to further growth interests were made at the expense of the community. He said that the university administration's decision to allow students to live off campus and to have cars contributed to urban sprawl and traffic congestion. He also questioned the wisdom of committing resources to a regional airport and a research and development park, in the light of the low probability of success of these projects.

Members of the antigrowth coalition believe they are acting in the community's overall long-range interest. They wish to preserve and enhance the quality of life—the natural environment, jobs, and leisure-time activities. They distinguish population growth, which they generally oppose, from economic growth, which they may support if current residents are not required to bear the costs of additional development and do not suffer any reduction in the quality of life. Some members of the coalition focus on preservation of the natural environment, whereas others are more concerned about the impact of growth on neighborhoods.

The antigrowth coalition takes credit for preventing some developments that would have been harmful to particular neighborhoods. Even Chamber of Commerce members asked for the antigrowth mayor's assistance in blocking development in their neighborhoods. Antigrowthers believe their efforts have improved the quality of development, stimulated the city to purchase land for open space, forced consideration of energy conservation, and focused attention on the value of the environment.

The Chambergate scandal confirmed the convictions of antigrowthers that business has only learned to use the rhetoric of "growth management" to mask "unrestrained growth" driven by the marketplace. Business leaders support growth management to the extent that government facilitates market forces and puts into place the physical infrastructure necessary to ensure continued growth or subsidize growth that the market could not otherwise support. Business leaders are not ready to accept real community direction and management of growth through public budget, utility, and land-use policies unless their own developments are facilitated.

Can the Antigrowth Coalition Win and Sustain Its Victory?

The pro- and antigrowth factions continue to vie in elections, in formulating the comprehensive plan, and in a variety of other decisions. The antigrowth coalition prevailed in city government for two years (1984-1986), eventually losing to "moderate growth management" candidates. Partly in reaction to electoral politics in the city, the antigrowth coalition gained a foothold on the county commission, capturing two of five seats (one in 1984 and one in 1986), and then losing one of these (1988).

The antigrowth coalition has well-developed communication networks, although strong formal organizations are lacking and the leadership is fluid. Networks form around self-selected single-issue experts on water pollution, environmental contamination, and the like. As communication increases on an ever-widening number of concerns among the many single-issue groups, broad coalitions form. The antigrowthers maintain that one of their greatest resources is their belief that "their cause is right." The Chamber of Commerce acknowledges that they are a force to be reckoned with and that their dedication causes the business community to take notice. Leaders of the antigrowth coalition have been described as charismatic, articulate, sincere, effective, and evangelical in their efforts to slow or end growth.

The growth machine includes many civic and trade associations with professional staff who serve the business sector. The Chamber of Commerce probably is the most important, although the real estate and homebuilder organizations also are quite active. Business leaders have regular contact with one another through the many service organizations in the community. When they disapprove of a government policy, they spread the word quickly. Members of the business community are not monolithic—some focus on downtown, others on development outside the city. They tend, however, to move in the same direction because their organizations provide a forum and build support for progrowth candidates. Local campaign finance records reveal consistently high levels of financial support for progrowth candidates from these groups (Turner, 1986).

In addition, many progrowth advocates hold leadership positions on government advisory boards and influence the civic and educational institutions (e.g., hospital boards, community college, charities). One member of the antigrowth coalition described in an interview how the new university president addressed a joint service club meeting (an occasion usually used to boost the football team) and declared war on the antigrowth coalition. He attacked city government (dominated at the time by antigrowth commissioners), saying that the attitude of city officials must change.

To the antigrowth activists it appeared that the growth machine had decided that if it could not run the city, it would dismantle it. The state legislative delegation passed local bills that stripped selected functions from the city. A separate library district was created. Efforts continue to set up an airport authority, a joint city-county planning agency, consolidated police protection under the county sheriff, and an independent utility authority (the utility is owned by the city). One antigrowth leader reported in an interview that "we used to joke that there would be no city government left but two functions, swimming pools and buses," both losers financially.

The initiation of a local "visions" process (a retreat of leaders who meet to plan the community's future) by the state legislature is indicative of the growth machine's resources and determination to eliminate the antigrowth contingent from city government. The initial visions proposal called for the participants in the goals conference to be graduates of the Chamber of Commerce "Leadership Gainesville" program. This requirement was eliminated, and the Alachua County state legislative delegation selected the participants. A progrowth steering committee was appointed and arranged a weekend retreat to set community goals under an American Assembly process (see Vogel & Swanson, 1988). The growth goals were:

I. It is likely that the rate of population growth in Alachua County will be low (less than 1 percent per year) unless efforts are made to attract industry into the area. We must attract clean, high quality industry. The county is in the "sun belt," and it has an attractive environment and friendly residents. But a number of factors are working against growth, including declining enrollment at the University of Florida, a perception of an antigrowth attitude on the part of the City of Gainesville, intergovernmental conflicts, and changes in the national economy.

II. Efforts should be undertaken to promote an economically sound and environmentally sensitive growth. The preferred future for Alachua County is a moderate rate of growth (2-4 percent per year) that is well planned and fully supported with public facilities. Protection of the environment must be an extremely high priority in all development activity throughout Alachua County to ensure a high quality of life. (Visions 2000, 1986)

The leaders of Visions 2000 made no secret of their dissatisfaction with the antigrowth coalition, and they claimed the defeat of the antigrowth mayor in 1986 as one of their greatest accomplishments.

The growth machine seems to have a systemic advantage over the antigrowth coalition. Even when statistics demonstrated that growth is not automatic (less than 1% per year), a leader of the growth machine forces asserted in an interview that there is a "fundamental and statistical probability that growth will occur, it is happening." Antigrowth leaders explain this as a "conceptual bias" in favor of development, pointing out that even our language is prejudiced toward growth; we refer to land as undeveloped or developed. The expectation is that a higher usage is desirable.

It is this growth ideology that most upsets the antigrowth element. Antigrowth actors constantly face the charge that they are kooks and obstructionists who prevent the emergence of consensus in the community. This view was eloquently stated by the antigrowth mayor in his last public address after he was defeated:

Gainesville is a finely balanced community with passionately held opposing viewpoints. Elections in Gainesville prove that point because they are won and lost on the slimmest of margins. But . . . that's ok, because differences of opinion about the future of our city are very important. . . .

> By all indications we are a community of great diversity and differences of opinion, and yet there has been *tremendous* effort exercised by certain public officials and certain private interest groups to compel everyone to accept a single future for Gainesville based on slogans or catchy names, like "Good for Gainesville," or "Growth With Conscience" or "Visions 2000." (Gordon, 1986)

The defeat of the antigrowth county commissioner in 1988 further illustrates the odds against the antigrowth coalition. First elected in 1984, he was joined in 1986 by another environmentalist who ran on the issue of polluted domestic water well-fields. They generally were unable to implement their agenda. A sole opponent challenged the first commissioner in the Democratic primary. A newcomer to the community, the challenger had joined the university's Center for Growth Management (in the law school) and had served on the Zwick Committee, a state commission that had formulated the Growth Management Act of 1985. He believed he would be more effective in promoting growth management than the incumbent, who, by the time of the primary, had alienated the civic elites, developers, and the local newspaper.

The incumbent environmentalist focused on the costs of growth, attacking his opponent, the growth management candidate, for his willingness to raise taxes to support growth instead of relying on impact fees. The challenger lined up vigorous business support and captured the endorsement of the local newspaper. He was able to raise a record $42,000 campaign chest (compared with the nearly $17,000 raised by the incumbent commissioner). The incumbent won the Democratic primary by 1,000 votes out of 20,000 cast; each vote cost him $1.62, compared with $4.58 for the challenger.

That should have been the end of the contest because most elections are decided in the Democratic primary; the Republicans rarely mount a challenge. However, the Democratic party leadership indicated displeasure with the results. A committee was formed (mostly bankers, developers, realtors) to back a Republican progrowth challenger. The challenger, raising three times as much money as the incumbent and receiving the newspaper endorsement, won by a 2,000-vote margin out of 60,000 votes cast in the general election. He is the first Republican elected to the county commission. It is doubtful that this election has settled the growth issue. Clearly the progrowth forces won this battle; they retained a county commissioner who was a realtor and a loyal supporter of growth, and they also ran and won a campaign for a moderate "environmentalist," though she claims she is "nobody's third vote." Those defeated have already begun to organize for the next election.

The experience of the antigrowth coalition in Gainesville is not unlike that of antigrowthers in other communities. A "slow-growth revolt" in California has resulted in several direct referendums on growth being placed on the ballot (Trombley, 1988). Peirce (1988) reports "the grassroots-led slow growthers found themselves totally outspent, outclassed, and outfoxed by a development community that saw a genuine threat to its livelihood and started pouring unprecedented millions of dollars into county ballot fights" (p. 3C).

Growth Management Assessed

Growth management offered some promise of eclipsing the growth-no growth battles. It is a distinctive approach to community development in which new development is tied to infrastructure placement and the provision of services. To the extent that growth management is instituted, policymakers must identify environmentally sensitive areas and set aside land for residential, industrial, commercial, and recreational use. Impact fees or other taxes are used to recover the economic costs of growth to the community.

In this sense, growth management represents a compromise between growth at any cost and no growth based on environmental concerns. Environmentalists tolerate some "growth" in return for "management." Developers accept some "management" in return for knowing they may proceed with development without costly delays or interference. Growth management also eliminates the more egregious practices of developers who would rape the land or despoil neighborhoods. Developers must comply with hundreds of regulations as diverse as providing for water drainage and placing "buffers" around development.

But growth management has not ended the battles between growth machines and antigrowth coalitions. The state-forged alliance between large developers and environmentalists forms the basis of the growth-management compromise. Neighborhood activists, however, are reluctant parties to this alliance. They continue to be concerned with protecting the integrity of their neighborhoods and are unwilling to accept higher density levels as a trade-off for preserving green space somewhere else.

An antigrowth leader explained in an interview that he opposed compromise short of stopping most development, arguing that one cannot attain a "balanced" growth policy (growth management) that "has been out of balance for 60 years." He complained that after only 1 year in power, "suddenly we need balance!" Growth management was not a compromise from his vantage point, or if it was, it was made at the expense of city neighborhoods that bore the costs of higher densities to protect outlying areas from development. Although Molotch and Logan (1984) pointed out tensions in the growth machine, they overlooked this potential tension within the antigrowth coalition that can be exploited. Growth management may yet drive a wedge between environmentalists and neighborhood activists.

Growth management is unacceptable to neighborhood activists because it is based on the premise that growth is inevitable. In fact, growth management potentially allows population to increase so long as services can be provided to meet the growth. The object of growth management is to restrict sprawl, not high densities. Thus, if technology permits higher population levels, there is no rationale for limiting growth. The fact that growth is a political question and may be limited for reasons other than the ability to provide services is not acknowledged. Growth management advocates reject the notion that growth may be limited or stopped. This is why Molotch in the past has dismissed "planning" (which would include growth management) as nothing more than a sophisticated growth strategy not to be confused with no growth (Molotch,

1976, p. 316). From this perspective, at most, growth management represents a shift from conservative market-oriented progrowth to the "liberal" growth ideology described by Swanstrom (1985), with government intervention to promote, as well as regulate, growth.

The cases of the defeated antigrowth mayor (neighborhood activist) and county commissioner (environmentalist) best illustrate the difficulty of defining and evaluating growth management. Both actors called themselves supporters of growth management. Each was defeated by a so-called "growth management" candidate who had the overwhelming support of the growth machine. The defining characteristic of growth management is that growth is tied to the placement of the physical infrastructure and services to meet the population's needs. It only requires "concurrency"—the provision of some minimal level of services and amenities at a time approximating the occurrence of development. Thus growth management candidates can favor or oppose more growth. The label indicates only that they believe that plans for the community should include provision of services to those who will come in the future. Growth management can serve the interests of the growth machine or the antigrowth coalition.

Growth management is fraught with misunderstandings that prevent consensus. Pro- and antigrowth activists mean very different things by the term *growth management*. Antigrowthers view growth management as a buzzword that everyone can support. Questions by city or county commissioners about the costs and benefits of a proposed development are viewed by progrowth forces as evidence of an antigrowth attitude and as an attack on the free market system. The business community puts the emphasis on growth, whereas the antigrowth coalition emphasizes management (see also Rubin, 1988).

The protagonists continue to eye each other with suspicion as they pursue their interests and attempt to ensure that the other side does not gain a power advantage over the shaping of land-use policy. The reality of growth management seems to depend on the balance of power—not just when the plan is adopted, but at the implementation stage as well. In the past, local commissioners have had discretion to amend and/or interpret planning requirements at the time of site platting. This judgment means that proponents of both sides have an incentive to compete with each other in local elections to ensure that their side's interest or interpretation will prevail.

Conflict persists because little evidence shows that concrete application of growth management can make growth pay its own way. The growth machine appears to compromise by adopting liberal growth management that costs them little. Developers do not lose much by accepting growth management because their growth strategy (at least in Florida) calls for clean industry and protection of the amenities that make Florida attractive. The Florida Advisory Council on Intergovernmental Relations (1986) estimates that it costs more than $20,000 to provide services to every new household. A growing number of Florida communities have adopted impact fees, but they fall short of the actual costs of growth. Tough enforcement of growth management provisions will be resisted as unreasonable or harmful to economic development. Politically it seems unlikely that growth can be made to pay for itself, because communities and states compete against each other to offer the most favorable business climate.

In South Florida, where development occurred unchecked in the 1950s through the 1970s, growth management is little more than an effort to rationalize development after the fact by providing the infrastructure and services for the population already there (see Logan & Molotch, 1987, pp. 159-160). Although neighborhood activists may challenge decisions that harm their neighborhood, the antigrowth coalition has lost before it ever gets organized. The damage is done; the community is developed except for some "in-fill." Here growth management forces can only attempt to make any additional development pay its way and prevent further encroachment on environmentally sensitive lands and agriculture. This is not as difficult as it sounds because the only lands left to protect are the Everglades and a few isolated pockets. The fundamental character of the community is established. (Many growth management advocates believe a major hurricane, though costly in terms of lives and property lost, may offer Florida a second chance at development.) Although conservative progrowth and antigrowth sentiments are still present, little is left to argue over. Liberal growth management is the dominant ideology, and the issue of growth ceases to be as conflictual.

In areas where undeveloped land is still plentiful (north and west Florida), the growth issue is much more volatile, and who wins public office is still a matter of concern. Antigrowthers have an incentive to back candidates, as do conservative progrowthers who are seeking development opportunities. Herein, growth management occupies a middle position and has few adherents among profit seekers or those seeking to protect a way of life except for its rhetorical value.

But growth management should not be discounted. Recent reforms at the state level suggest that the Growth Management Act of 1985 may become a "hammer" to ensure that localities do not grow beyond their capacity to provide services. Once the state approves local growth-management plans, development will be barred if the infrastructure to support it is not in place. In January 1988, Republican Governor Martinez called the new state growth-management legislation "truth in development" because local governments will be prevented from dismissing the costs associated with developments they have approved. According to the governor, communities unwilling to foot the bill should curtail growth. One should not underestimate, however, the growth machine's ability to shift the costs of development to the public sector and those less able to pay. Growth management has the potential to limit subsidies for growth. It is unlikely to end the debate over the growth issue. Growth management, if implemented, is more than rhetoric but less than the neighborhood activists can accept.

Growth Management as a Means of "Keeping and Using the Books"

Our concern is not so much that growth strategies in cities are pursued and that private actors profit from these strategies, but rather that local policymakers do not have an acceptable mechanism and procedures to take account of the community interest. According to Long (1958), an "ecology of games" is pursued by different occupational groups in the community, not one of which is synonymous with the broader community interest (see also Warren, 1988). This precludes a comprehensive approach to the community and its problems.

Long's diagnosis of the problem (the lack of any communitywide games) is that leaders and citizens have lost their sense of community and responsibility and are preoccupied with the pursuit of their own self-interest. For Long, the solution is to restore the sense of community and responsibility. This requires not only overcoming metropolitan fragmentation but also recognizing that leaders have a responsibility for the polity and the economy (Long, 1980).

Long (1986) believes that a useful approach toward this end would be for city governments to keep and use social and economic accounting books. But, according to Logan and Molotch (1987), local government is unable to pursue the community interest because it is the captured instrument (or at least a willing partner) of the growth machine, which is composed of factions in collective pursuit of self-interest. Unless local politics becomes more competitive (i.e., antigrowth coalitions come to power from time to time), even if cities were to keep books, they would just be more efficient in pursuing private growth interests.

Although Logan and Molotch (1987) believe that the community would be better served if antigrowth coalitions came to power, we are less sure. Neighborhood politics is not necessarily closer to the public interest than business privilege. Neighborhood politics usually reinforces segregation and unequal distribution of services across neighborhoods. Whereas business may lose if neighborhoods gain clout, upper status groups will still be favored (Stone, 1980). In the battle between the growth machine and the antigrowth coalition, the question may be, Will minorities and low-income citizens receive benefits under either set of policies?

A key problem is the lack of consensus on what is in the best interest of the community. In the absence of a communitywide approach to problems, various sectors of the community are free to define the community interest as they see fit. Local government is likely to become an instrument of the politically attentive and their narrow views.

Growth often is seen as a panacea to community problems, whether fiscal, economic, or social. Logan and Molotch (1987) raise serious questions about the ability or desirability of the "growth game" to solve community problems. Although political campaigns certainly indicate that growth is a key issue in local elections, the debate often takes place without reference to empirical evidence. In fact, with no one monitoring the quality of life in the community, it is difficult to believe the database is even available to assess accurately the impacts of growth on the community. The most useful approach would be for each community to assess the benefits and costs of continued growth. Such an approach may shed light on whether growth rates are affected by who is in power—the growth machine or the antigrowth coalition. It is questionable whether antigrowthers can actually stop growth beyond engaging in symbolic politics. But then the growth machine may be playing a similar game (see Wolman, 1988) and also may be unable to affect growth rates.

Growth management is a technical, rational approach wherein the public interest is defined in procedural terms. We believe the public interest must be defined in a more substantive fashion. But here, perhaps, growth management offers a starting point. By requiring local policymakers not only to "keep the books" but actually to use them (see Swanson & Vogel, 1986), growth management may lead to a more substantive view of the public interest. Local policymakers are forced to address the costs and benefits of their development and the larger vision they are guiding the community toward, and then to set up minimal standards for the quality of life in their community. If policymakers seek population growth, they must consider the impacts of the growth on the community and identify sources of funds to pay for additional services. Unfortunately, however, growth management does not require local policymakers to separate the issue of job creation from population growth. Nor are they required to coordinate their economic development plans with their growth management plans (Swanson & Turner, 1989).

In this chapter we explored local actors' perceptions of growth. We found that the way growth was defined had serious implications for policy making. We also examined the ability of the antigrowth coalition to sustain its victory. We believe that antigrowth coalitions may only temporarily win control of government. While in office, however, they can raise serious questions concerning the desirability of growth. Even out of power, they may continue to make enough noise that elected officials must proceed cautiously in supporting growth. Finally we considered the potential of growth management to resolve the battles over growth and to lead to the definition and pursuit of the community interest. We found that disagreements about growth and its benefits limit the ability of growth management to resolve the conflict between the growth machine and the antigrowth coalition. But growth management may have the effect of educating members of the growth machine to the effects of their policies. This is especially true when policymakers are confronted with the budgetary consequences of their growth policies.

Growth management changes the nature of the question from Is growth good or bad? to What kind of growth? How much growth? Where will the growth go? When will the growth occur? Who will benefit and pay for the growth? and What impact will the growth have on the community? Future debates about growth probably will be couched in the language of growth management. Whether growth management will lead to rational policy making or is only a way to rationalize what the prevailing interests want remains to be seen.

References

Baldassare, M. (1981). *The growth dilemma.* Berkeley: University of California Press.
Burt, A. (1983, October). The next Florida. *Florida Environmental and Urban Issues, 11,* 1.

Clarke, S. E. (1987). More autonomous policy orientations: An analytic framework. In C. N. Stone & H. T. Sanders (Eds.), *The politics of urban development* (pp. 105-124). Lawrence: University of Kansas Press.

Cummings, S. (Ed.). (1988). *Business elites and urban development*. Albany: State University of New York Press.

DeGrove, J. (1983, January 5). Presentation to House Select Committee on Growth Management, Tallahassee.

DeGrove, J. (1984). *Land growth and politics*. New York: American Institute of Planning.

Fainstein, S., Fainstein, N. J., Hill, R. C., Judd, D. R., & Smith, M. P. (Eds.). (1986). *Restructuring the city: The political economy of urban development*. New York: Longman.

Feagin, J. R. (1988). Tallying the social costs of urban growth under capitalism: The case of Houston. In S. Cummings (Ed.), *Business elites and urban development* (pp. 205-234). Albany: State University of New York Press.

Florida Advisory Council on Intergovernmental Relations. (1986). *Impact fees in Florida*. Tallahassee: State of Florida.

Florida Trend. (1984, May 22). John DeGrove loaded for bear and aiming at developers. pp. 98-99.

Form, W. H., & Miller, D. C. (1960). *Industry, labor, and community*. New York: Harper & Row.

Freilich, R., & Einsweiler, R. (1975). *Urban growth management systems*. New York: American Society of Planning Officials.

Gordon, G. (1986, May 22). *Remarks of Mayor-Commissioner Gary Gordon*. Gainesville, FL: Gainesville City Commission.

Logan, J. R., & Molotch, H. L. (1987). *Urban fortunes: The political economy of place*. Berkeley: University of California Press.

Long, N. (1958). The local community as an ecology of games. *American Journal of Sociology, 64*, 251-261.

Long, N. (1980, May). The city as a local political economy. *Administration and Society, 12*, 5-35.

Long, N. (1986, Spring). Getting cities to keep books. *Journal of Urban Affairs, 8*, 1-7.

Lyon, L., Felice, L. G., Perryman, M. R., & Parker, E. S. (1981, May). Community power and population increase: An empirical test of the growth machine model. *American Journal of Sociology, 86*, 1387-1400.

Maurer, R. C., & Christenson, J. A. (1982, June). Growth and nongrowth orientations of urban, suburban, and rural mayors: Reflections on the city as a growth machine. *Social Science Quarterly, 63*, 350-358.

McQuown, R., Hamilton, W., & Schneider, M. (1964). *Political restructuring of a community* (No. 27). Gainesville, FL: University of Florida, Public Administration Clearing Service.

Molotch, H. (1976, September). The city as growth machine: Toward a political economy of place. *American Journal of Sociology, 82*, 309-332.

Molotch, H., & Logan, J. (1984, March). Tensions in the growth machine: Overcoming resistance to value-free development. *Social Problems*, pp. 483-499.

Peirce, N. R. (1988, November 27). If government fails to control growth, voters will force the issue. *Tampa Tribune-Times*, p. 3C.

Peterson, P. E. (1981). *City limits*. Chicago: University of Chicago Press.

Rubin, H. J. (1988, Winter). The Danada farm: Land acquisition, planning, and politics in suburbia. *Journal of the American Planning Association, 54*, 79-90.

Stone, C. N. (1980). Systemic power in community decision making: A restatement of stratification theory. *American Political Science Review, 74*, 978-990.

Stone, C. N., & Sanders, H. T. (Eds.). (1987). *The politics of urban development*. Lawrence: University of Kansas Press.

Swanson, B. E., & Turner, R. (1989, February). *Economic development and comprehensive planning: Implementing a state mandate*. Paper presented at the Georgia Political Science Association Convention, Savanna.

Swanson, B. E., & Vogel, R. K. (1986). Rating American cities: Credit risk, urban distress, and the quality of life. *Journal of Urban Affairs, 8*, 67-84.

Swanstrom, T. (1985). *The crisis of growth politics*. Philadelphia: Temple University Press.

Symposium: Development impact fees. (1988, Winter). *Journal of the American Planning Association, 54*, 3-78.

Symposium: Linkage fee programs. (1988, Spring). *Journal of the American Planning Association, 54*, 197-224.

Trombley, W. (1988, July 31-August 3). A slow-growth revolt. *Los Angeles Times*.

Trounstine, P. J., & Christensen, T. (1982). *Movers and shakers: The study of community power*. New York: St. Martin's.

Turner, R. (1986, April). *Measuring changes in political support for growth through campaign financing*. Paper presented at the Florida Political Science Association Convention, Sarasota.

Visions 2000. (1986). *The future of Alachua County, Florida*. Gainesville, FL: Author.

Vogel, R. K., & Swanson, B. E. (1988). Setting agendas for community change: The community goal setting strategy. *Journal of Urban Affairs, 10,* 41-61.

Warren, R. (1988). The community in America. In R. L. Warren & L. Lyon (Eds.), *New perspectives on the American community* (5th ed., pp. 152-157). Chicago: Dorsey.

Wolman, H. (1988). Local economic development policy: What explains the diversion between policy analysis and political behavior? *Journal of Urban Affairs, 10,* 19-28.

23

Growth Politics and Downtown Development

THE ECONOMIC IMPERATIVE

IN SUNBELT CITIES

Robyne S. Turner

Downtown development in fast-growing Sunbelt cities exemplifies the factors contributing to growth politics: rapid growth, dominance of business interests, and development facilitated by local government (Bernard & Rice, 1983; Mollenkopf, 1983). As in cities nationwide, downtown development agendas in Sunbelt cities often are heavily influenced by a private-sector political coalition or growth machine (see Logan & Molotch, 1987; Mollenkopf, 1983; Molotch, 1976; Swanstrom, 1985). Because growth is the mainstay of the local economy in many Sunbelt cities (Fishkind, Milliman, & Ellson, 1978), the likelihood is that the demands for continued economic growth in downtown, where land prices increase rapidly, will act as a major constraint on policy selection (see Wong, 1990).

Growth management is a familiar option in fast-growing locations as a means to mitigate the dysfunctions of growth, as well as to enhance growth (DeGrove, 1984). Growth management, along with other long-term mechanisms such as strategic planning, linkage policies, and comprehensive planning, is a policy tool that is used to encourage private-sector economic growth but that often neglects the need for communitywide benefits. The local political climate is the source of policy support, but city governments find it politically difficult to adopt strategies that balance the needs of the community with the demands of the private sector for economic growth (Fainstein & Fainstein, 1987; Friedland, 1983). Sunbelt cities are more likely to resist any public/private benefit linkage because the development coalition facilitates private development interests with little concern for the impact on neighborhoods or residents or the diminishment of the quality of life (Eisinger, 1988; Levine, 1989).

AUTHOR'S NOTE: This research was funded, in part, with an Internal Research Grant and faculty fellowship from Florida Atlantic University. I would like to thank Ron Vogel and Susan Clarke for helpful comments on an earlier version.

Yet the Sunbelt case cities presented here reveal that cities with similar conditions and planning parameters opt for different policy approaches to downtown housing and economic development. The identification of linkage policies, development standards, and growth management as a means to balance the impacts of development in those downtowns suggests a variety of relationships between the public sector and the private-development sector within growth politics. Because fast-growing urban areas are economically tied to development, distributions of the costs and benefits of downtown development to both commercial areas and residential neighborhoods generally are skewed to favor the private sector (Peterson, 1981). The political reality that economic prosperity is favored will not disappear. However, there is room to maneuver the impacts of development within those policy constraints. Growth politics, as a political reaction to competing interests, can vary (see Stone, 1984).

A city's policy and administrative approach to commercial development, affordable housing, and public amenities are examined as the basis for understanding the political implications of the distribution of downtown development impacts. The makeup of political coalitions, administrative and planning efforts, political leadership, and public involvement all contribute to policy adoption and success. Even in rapidly growing urban areas, downtown growth can be managed without sacrificing continued economic development.

The cases presented here provide evidence of alternative scenarios of growth politics, challenging the narrow expectation that the competition for downtown development resources is limited to zero-sum political choices between commercial and neighborhood interests. The politics of downtown development is presented in a comparative assessment of four case cities in Florida: Jacksonville, Orlando, Tampa, and Fort Lauderdale.

Downtown
Development as Growth Politics

The likelihood of adoption and the eventual success of downtown development policies are affected by their use in the political process as systemic rules (Stone, 1980). These policies represent the willingness and commitment of a city to link communitywide or neighborhood benefits with private-sector development opportunities.

The degree to which equity considerations are included in strategy selection, on the basis of who pays and who benefits, suggests different styles of politics within the broad parameters of growth politics. A stronger linkage between private and public development costs and benefits may reflect a progressive or managed-style growth politics (Clavel, 1986). The absence or less use of these linkages, skewing the short-term benefits to the private sector, suggests a facilitative style of growth politics.

Costs and Benefits of Downtown Development

Economic development may be an imperative in Sunbelt cities, but policies can be distinguished by the strategy to distribute the costs and benefits to the residential and commercial sectors (Eisinger, 1988). The supply-side strategy relies on using public revenues to offer economic development incentives that reduce business costs as a means to attract incrementally the existing supply of business firms to relocate to the community. An alternative strategy to facilitating development by cutting private costs is to use demand-side strategies that strategically use public resources to invest in local economic development capacity building. This is a means to stimulate business demand through an attractive location and to support existing local firms and labor supplies. These policies are targeted to provide long-term community benefits as a measured return on long-term public costs as investments.

Evidence suggests that Sunbelt cities generously support economic development, particularly with traditional subsidy policies (Bowman, 1988; Eisinger, 1988). The ability of a progrowth coalition to dominate the politics of development, especially in the downtown, is heightened in Sunbelt cities, where there is a lack of politicization and a history of reform-style politics through business interests (Kantor, 1987; Mollenkopf, 1983).

The impact of development on downtown residents and workers may be to displace neighborhoods for commercial development. A traditional political resistance movement, such as an antigrowth coalition (Molotch, 1976), is likely to be ineffective, or even absent, in downtown (Vogel & Swanson, 1989). As will be shown, downtown residents in the case cities who are likely to be affected by commercial development are poor black renters. This is not the typical configuration of a growth-resistance coalition, which is often characterized as affluent, white, educated homeowners (deHaven-Smith, 1987; Molotch, 1976).

Local government is more likely to accept this impact than to adopt policies that provide long-term benefits such as real economic improvement for professionals and inner-city residents (Barnekov & Rich, 1989; Levine, 1989). In a progressive political climate, however, strategic or growth-management processes can be mechanisms to accommodate private development while ensuring long-term community benefits through targeted programs or public amenities financed, in part, by private-sector contributions on the basis of the impact of commercial development (Turner, 1990b; Vogel & Swanson, 1989).[1] Linking development with community needs is a method to mitigate such negative impacts of downtown commercial development as higher housing prices, limited employment opportunities for downtown residents, and commercial encroachment into downtown neighborhoods (Squires, 1989). Linkage policies require private developers to contribute to public benefits through programs for child care, affordable housing, public open space, and job training (Merriam, Brower, & Tegeler, 1985; Porter, 1985; Wrenn, 1987).[2]

But policies that balance the impacts of development also affect the bottom line of private-sector projects. This becomes difficult to implement in a climate of growth politics in which political resistance to such efforts is high. Boston and San Francisco are contrary examples of development politics. In those cities, demand for commercial space exceeds the supply of land. Linkage policies are adopted even in the face of resistance by the private sector (Porter & Lassar, 1988). Yet a high level of political mobilization also results in strong political reaction by public officials in the face of neighborhood demands, which is not typical of Sunbelt cities (Mollenkopf, 1983).[3]

The experiences of the case cities is presented to bear out the expectation that political climate is a stronger factor than growth rates in predicting the use of linkage-type policies (Goetz, 1991). Politics not only matters, but its variance also can be categorized on the basis of factors of the political and administrative climates.

Public/Private Relationships in Downtown

Businesses and land-based interests are the source of economic prosperity for local economies. This reality may constrain public officials in their selection of policy, but the choice is guided by politics (Wong, 1990). Elkin (1985, p. 13) stated that "land interests and city politicians regularly try to build up institutional arrangements designed to reduce the probability that major land use changes become a matter of public dispute. Such efforts are crucial to creating a workable alliance." The degree to which administrative and policy mechanisms are used to facilitate private-sector interests through the public sector is derived through political relationships. Stone (1987a, p. 16) suggested that "to talk about coalition formation and conflict management is not to reduce policy making to those activities but to suggest that the understanding of development policy that guides a set of decision makers will be influenced by the political arrangements that surround the process of deliberation." The relationships between public and private sectors are applicable to the broad theory of urban regime politics (Elkin, 1985; Stone, 1987b, 1989).[4] Although regime theory is not entirely appropriate in the political subsector of downtown development, its premise has guided the development of the classifications of the political relationships that affect downtown policy selection. These classifications reflect the role that government takes to distribute the costs and benefits of downtown development and the political relationships that generate that distribution. The classifications are as follows:

1. *Caretaker.* Local government has a laissez-faire approach to the private development agenda, offering little policy and/or revenue assistance but also no opposition.
2. *Facilitator.* Local government supports private-sector development in a booster capacity, offering subsidies and tax policies that create an attractive business climate.
3. *Satisficer.* The public sector addresses demands from both the business sector and community residents, but the policy strategy is reactive, ad hoc, and short term.
4. *Director.* Government seeks to manage development strategically, guided by the impacts of private development on communitywide needs over the long term.

Table 23.1 Development Policy Relationships

Public/Private Relationship	Development Policy	Policy Criteria
Caretaker	No linkages	Supply-side
Facilitator	Subsidies	Supply-side
Satisficer	Incentives	Demand-side
Director	Community linkages, growth management	Demand-side

These categories expand the expectation that development politics is merely a zero-sum competition for resources between the private sector and other downtown community interests. Instead they provide a continuum of public-sector responsibility to reflect social concerns in cost-benefit considerations (see Beatley, 1988). The role of local government is to promote the city by using supply-side or demand-side policies (see Table 23.1).[5]

In the next section we explore the justification for these classifications and include an assessment of development-policy results as factors of the political relationships between public and private sectors and a classification of each case city.

Downtown Development in the Case Cities

Why have these cities adopted and implemented different downtown development processes and policies? Local government can influence (positively or negatively) the marketability of downtown through the planning and administrative regulation process. The political choice to implement such demand-side policies as growth management and linkage policies can enhance the viability of downtown as a development location while balancing the costs and benefits of the effects of growth politics.

Three categories of factors must be considered if one is to understand downtown development in each case city. First, the demographic and market conditions are compared as a possible explanation for different policies between the cities. Second, the cities are compared in terms of the effect that the administrative process has on mitigating the impacts of development. Third, each city is examined in terms of its political leadership, tensions, and participation as likely explanations for policy divergence.

Downtown Conditions

The case cities are typical Sunbelt cities where growth is rapid and such dysfunctions as sprawl and traffic congestion are common (Florida State Comprehensive Plan Committee, 1987). These cities are subject to Florida's aggressive comprehensive planning and growth-management laws.[6] But the implementation of these regulations is guided by political forces within each city. There are strong political demands to use project-by-project criteria to evaluate development projects instead of relying on growth management parameters. As in other Sunbelt cities, the case cities have experienced an exodus of commercial and retail activity from downtown and into suburban areas (see Frieden & Sagalyn, 1989).

The city of Orlando is by far the strongest-growing urban area of the four cases, even though its urban-area growth is similar to the Tampa area's (see Table 23.2). Although Fort Lauderdale had a drop in population throughout the 1980s, it continues to be the central city in a rapidly growing urban area. The rate of development in commercial and retail space, as well as in the daytime workforce, is very similar between the case cities. All four downtowns began the 1980s with similar rates of poverty and minority populations. With the exception of Fort Lauderdale, the downtown areas had similar median incomes and number of housing units. All four cities have well-developed residential neighborhoods in the downtown or on the periphery, with Fort Lauderdale and Orlando having high concentrations of low-income minorities and affluent whites (see Table 23.2). Not surprisingly, the cities have downtown occupancy patterns that are highly segregated by race and income (Turner, 1990a).

As these four cities entered the 1980s, existing residents were threatened by displacement from the encroachment of downtown commercial activity. Orlando and Jacksonville had the greatest downtown commercial growth throughout the 1980s (see Table 23.2). Downtown housing is affordable because it is likely to be in poor condition. Commercial encroachment is expected because demand for residential use of the land, as is, is low, deflating prices. As downtown growth increases, residential properties will no longer be affordable, from a rental standpoint, or desirable, from a tax standpoint (Feagin, 1986). As this economic conflict intensifies, the politics of the growth machine is expected to allow displacement of land uses with low economic production and current use value, such as affordable housing, and replace them with high use values, such as commercial development (Logan & Molotch, 1987). This exchange, however, has not occurred in every case city. One explanation for the divergence is that the public-sector administrative process intercepts private political pressure for commercial development. The administrative process is a conduit for the variations in the politics of growth.

Administrative Processes

Decision making for downtown development in the four cities has been or is currently in a separate board or authority (see Table 23.3).[7] When housed in a quasi-governmental administrative unit, the staff has the latitude to deal one-on-one with private developers, removing some development decisions from the city council process. Ideally this process can expedite public decisions about development, relying on the technical expertise and entrepreneurship of the public-sector staff (Barnekov & Rich, 1989; Kantor, 1987). But this process can expedite either purely private-sector interests or a variety of downtown interests, depending on the political climate.

Table 23.2 Downtown Conditions and Policies

Descriptor	Orlando	Jacksonville	Tampa	Fort Lauderdale
Population				
Urban area, 1988	623,425	677,007	825,411	1,213,655
Growth rate, 1980-1988	32.4%	18.5%	27.6%	19.2%
Citywide, 1988	158,921	639,146	285,225	150,553
Growth rate, 1980-1988	24.0%	18.2%	5.0%	-1.8%
Downtown, 1980				
Minority population	8,885	11,743	7,449	14,885
Poverty rate	32.5%	45.6%	42.1%	26.1%
Median income	$6,610	$5,307	$5,293	$10,102
Renter/owner median income	$6,250/$10,000	$5,000/$8,350	$4,750/$10,650	$9,250/$16,500
Housing units	9,215	9,166	4,891	17,661
Renter	72.6%	60.0%	44.5%	63.7%
Downtown, 1989				
Commercial-office (million sq. ft.)	5.9	6.1	6.4	2.6
Office vacancy rate	20.0%	16.0%	17.0%	16.3%
Daytime employment	38,000	75,000	31,000	21,000
Retail space (million sq. ft.)	0.78	1.00	0.80	NA[a]
Commercial under construction (million sq. ft.)	1.200	1.600	0.432	0.900
Added commercial space, 1980-1988 (million sq. ft.)	4.14	3.50	1.78	1.72

SOURCES: Population data for 1980-1988 and 1988 were obtained from the Bureau of Economic and Business Research (1989); data for downtown 1980 were obtained from the U.S. Bureau of the Census (1980); data for downtown 1989 were obtained from various public and private sources.
NOTE: a. NA = data not available.

Table 23.3 Downtown Policies

	Orlando	Jacksonville	Tampa	Fort Lauderdale
Downtown Development Authority	Yes	Yes	No	No
Boundaries include residential areas	Yes	Yes	No	No
Tax increment financing (TIF)	Yes	Yes	Yes	Yes
TIF for housing	Yes	Yes	No	No
Design standards	Yes	No	Yes	No
Downtown affordable housing	Yes	Yes	No	No

Separate revenue sources are provided for distribution within these downtowns. Special ad valorem taxes and the more popular tax increment financing (TIF) are publicly supervised revenues that can be used to facilitate purely private commercial development or other multibenefit projects such as mixed-use facilities, joint public-private equity projects, and market rate or affordable housing (see Table 23.3).

The downtown district boundaries that define revenue-raising and expenditure target areas vary in each city. They may include only the central business district (CBD), as in Fort Lauderdale and Tampa, or may include residential areas adjacent to the CBD, as in Orlando and Jacksonville. Orlando uses this expanded boundary as a means to facilitate the use of excess TIF revenues for housing purposes in low-income and minority downtown areas. Jacksonville has taken responsibility for promoting the redevelopment of existing residential properties within its downtown development authority (DDA) boundaries by designating six residential districts eligible for assistance using downtown TIF or other city funds (Jacksonville DDA, 1988). In Fort Lauderdale and Tampa, low-income and minority neighborhoods adjacent to downtown fall under the jurisdiction of separate planning processes and are not eligible to be recipients of downtown TIF revenues.

Even with fragmented authority and political interests, it is possible for local government to use the downtown development planning-and-approval process to include the needs of less advantaged downtown residents. The effectiveness of that political activity, however, may be related to the degree to which downtown development is administratively integrated with or separated from the city government processes and staff responsible for existing human services programs.

For example, in Tampa many plans have been created for the downtown, but there is no mechanism to promote consistency between disparate processes. Even the comprehensive plan process, administered through a joint city/county body, specifically exempts the downtown from its responsibility (Hillsborough County, 1989a). Tampa followed a path of unfettered private development without regarding how to mitigate the problems of rapid growth. The transfer of the DDA decision making to the mayor's office in 1986 began to integrate the planning needs for downtown but continues to neglect the impact of downtown development on nearby low-income neighborhoods (Hillsborough County, 1989b).

Orlando, on the other hand, integrated its staffing and planning processes, creating consistency between the products of their efforts. The downtown development board (DDB) is staffed by persons from the planning, community redevelopment, and housing departments, in addition to regular DDB staff. The result has been attention to a wide variety of downtown beneficiaries and a long-term strategy to manage the strong pressures for growth without diminishing that demand.

Jacksonville's downtown development decision making was consolidated in the DDA, but only after Mayor Jake Godbold had threatened to dismantle the authority unless it followed his lead to attract new and, perhaps, out-of-town downtown developers aggressively. The DDA, with Mayor Godbold, became the primary source of development authority, but it also has developed informal relationships with other city departments. This was especially evident when the next mayor, Tommy Hazouri, formed the 1989 Housing Task Force, which linked the DDA to the housing department and economic development office to address downtown housing development, among other issues, strategically.

Finally Fort Lauderdale has the least integration of downtown development implementation across city government program offices and the most isolated structure of downtown revenues of the case cities. The DDA is a totally autonomous body that has been dominated by private-sector persons who perceive little merit in development regulations.[8]

Distribution of Development Impacts

None of the case cities have adopted mandatory linkage programs that require payments from commercial developments. Orlando and Tampa, however, have adopted design standards that link such public amenities as art, open space, facade treatment, and ground-floor retail to commercial projects through their downtown development standards. In addition, Orlando has a separate design-approval process for downtown projects that supersedes the traditional building and zoning department route. This parallel process was created to hold downtown developments to tougher criteria in order to produce managed growth and quality developments. The private sector has come to appreciate this result as a predictable environment in which its property values increase.

Orlando used two other integrative mechanisms as means to provide development benefits to downtown neighborhood interests. City staff in many departments, including the DDB, have worked together to build affordable housing adjacent to downtown through a separate nonprofit entity—the Orlando Neighborhood Improvement Corporation, Inc. To date, more than 100 units of subsidized low-income housing have been added to existing neighborhoods (Mitchell, 1990). This project is, in part, possible through the city's resolve to protect the integrity of downtown residential neighborhoods. The city downzoned existing residential zoning designations in downtown that acted as a means to prevent speculative buying and commercial encroachment into viable neighborhoods of varying races and income levels.

Jacksonville has not employed design standards but has created three district action plans in the downtown to direct the use of TIF revenues and guide growth. Fort Lauderdale, however, has not used TIF and has no separate downtown plan, no design standards, and no public/private development ventures. Resistance has been strong to any local government involvement in private development ventures. Thus few policy or planning parameters exist to guide the physical development of downtown or to direct any impacts of that growth.

Classifications of Downtown Politics

Why have the four cities followed these different paths to facilitate downtown development? In a word, politics.

Specifically it is the political climate or style of growth politics in each city as classified by the relationship between the public and private sectors in downtown development. Following is a discussion of the political motivations, traditions of decision making, and leadership from the public sector as these constitute the style of growth politics in each city.

Orlando: Progressive Director

Orlando's style of growth politics is classified as a progressive director because local government has taken significant policy steps to mitigate the impacts of downtown development through growth management using public- and private-sector efforts. Policies are progressive in that they reflect commitment and action to preserve neighborhoods, to provide public amenities, and to promote affordable housing in and around the downtown core area, even in the face of extreme growth pressures. Orlando recognized that growth was an inevitability, but the public sector used the political process to link public and private benefits. This policy direction is due primarily to the political climate in which Mayor William Frederick has been able to move beyond the traditional supply-side approach to economic development.

Mayor Frederick is credited by many with the vision and political will to promote growth management in downtown. The intent has been to create consensually a downtown environment that is attractive to developers and the public alike without wholesale displacement of neighborhoods and low-income problem areas. In his 1988 state-of-the-city address, Mayor Frederick said that community consensus is "built on a foundation of uncommon effort and agreement" and that this consensus is important because it indicates "old fashioned leadership and courage" to guide the city (Frederick, 1988, pp. 3, 10). The integrative nature of framing downtown issues and implementing a balanced policy package was a political choice accomplished with the proactive efforts of city leaders.

The mayor took the lead to form a citywide housing task force, recommending the policy to use downtown TIF revenues for low- and moderate-income housing in the downtown district. The strategy for long-term downtown stability is guided by this political-consensus process as the local style of growth politics. Popular civic vision guides the political climate because of a void in political power in Orlando. This open access to influence is due, in part, to a changing local economy and a cooperative decision-making atmosphere between civic networks (Aschoff, 1990).

Tampa: Private-Sector Facilitator

Unlike Orlando, Tampa's politics has a history of domination by long-standing private-sector interests that expect government to facilitate developments downtown and to be a booster (Kerstein, 1991). The previous mayor, Bob Martinez (later governor of the state), catered to the private developers and the interests of old-guard political cronies to create a downtown development process without any government regulations. The mayor supported limited downtown planning efforts to develop site concepts for promoting and marketing the downtown, but he refused to restrict the development process by requiring any regulations. Public facilities and a retail area were funded with revenues that were off the city budget.[9] This satisfied a fiscally conservative constituency, facilitated private development, and complimented the political aspirations of the mayor (see Kerstein, 1991).

Old-guard interests were happy with the facilitator approach to downtown development by the city, but they criticized the congested and unbalanced pattern. In response the next mayor, Sandra Freedman, eliminated the DDA and consolidated all development decision-making power in her office and staff departments.

This new mayor, however, had an additional, albeit parallel, agenda to support communitywide and neighborhood interests. Her broad electoral support was based on themes of fairness and fiscal conservatism.[10] Thus she guided the adoption of Tampa's first downtown design standards, which was a politically controversial agenda.

Tampa was going through a political transition, and the downtown private actors were split, clouding their support of the downtown initiatives of the mayor (Lubove, 1990). Freedman appointed long-term corporate downtown interests to her newly formed Advisory Committee on Downtown. They added political clout and were willing to support some restrictions that would enhance their existing downtown property holdings (Kerstein, 1989). Other, younger, downtown interests, however, resisted the new regulations. The Downtown Partnership, a marketing-oriented spin-off of the defunct DDA, was willing to work with the mayor and the advisory committee to reach compromises and eventually to support the downtown development ordinance. Consensus was not achieved as easily in Tampa as it was in Orlando. The various elements of the private sector have differing expectations and no cooperative network for decision making except the limited-access "old-boy network."

Jacksonville: Facilitator/Satisficer

Jacksonville may be a model for Tampa's political future because it has successfully made the transition from a city dominated by an old-guard private-sector network to a broader based private/public agenda intent on creating a growth environment. Mayor Godbold guided Jacksonville's development philosophy through the 1980s in a fashion similar to that of Tampa's Martinez; he aggressively sought to expand downtown opportunities for the private sector and to facilitate new development. This mayoral leadership is credited with attracting significant new developments, including the Rouse festival market at Jacksonville Landing, the centerpiece of the downtown renaissance. Leadership is assisted through the strong-mayor style of consolidated government in Jacksonville; however, that system also creates the need for strong political coalitions.

The private sector is the dominant actor in the politics of Jacksonville, particularly through the chamber of commerce. But the city's mayors were elected on populist platforms, creating winning coalitions through blue-collar and minority participation. This prompted Mayor Hazouri to take on housing problems as a primary issue. He established a permanent citywide housing commission with a special committee on downtown housing that addressed needs as diverse as market rate and low-income subsidized units (Jacksonville Mayor's Commission on Housing, 1989).

The private sector, however, has been successful in keeping the city from too much regulatory interference or benefit redistribution as the focus of downtown redevelopment. The Rouse festival shopping project leveraged community development block grant funds as a loan guarantee, but only minimal benefits were required to be set aside for low-income residents. Both Mayors Godbold and Hazouri fully supported business interests downtown while offering parallel policy processes to appease downtown neighborhood interests.

Fort Lauderdale: Caretaker/Facilitator

The development of Fort Lauderdale's downtown has mirrored the prevailing political philosophy in the city: The best government is the least government, and the more fiscally conservative, the better. Fort Lauderdale's laissez-faire approach to development stemmed from the prevailing opinion among influentials that the city should take on as few urban characteristics as possible, even in the face of staggering growth rates in the urban area. The city down-zoned all of its residential property in the 1970s and refused to facilitate its downtown as a regional center, acting with a negative supply-side development policy strategy. New public projects such as county administration facilities, the Riverwalk esplanade, and a performing arts center exemplify the extent of local government involvement downtown. These projects have facilitated new private growth but distribute few benefits to low-income downtown residents.

The city's caretaker style of growth politics relies on separate spheres of development decision making. Two agendas for downtown represent the private-sector commercial interests and the city's attempt to provide public facilities and maintain minimal services for middle-class downtown residents. Although numerous downtown plans have been prepared, the decisionmakers at the DDA and city hall have not adopted or implemented any (South Florida Regional Planning Council, 1989). Even the growth-management plan gives little attention to downtown.

The city changed to a district representation system in 1988. The one minority council person who represents downtown neighborhoods has had minimal success in drawing council attention to the likely encroachment and displacement that nearby low-income rental neighborhoods will face as commercial development continues. There have been few opportunities to build local capacity. Thus the public sector maintains a minimal role in distributing development impacts.

SUMMARY

Florida cities are under extreme growth pressures, yet only Orlando has followed a course of growth management that attempts to balance the needs of commercial development with neighborhood and community interests. Because all four cities are required to prepare a comprehensive plan and growth-management strategies, they all have the opportunity to manage growth. But that management does not necessarily produce the same outcomes (Turner, 1990b). The political dynamics in each city contribute to the degree to which development costs and benefits are linked. The political leadership of Orlando took that city in a different direction from that of the strong leadership in Tampa, suggesting that leadership can invoke various results (Jones, 1989; Judd & Parkinson, 1990). Fort Lauderdale's lack of leadership has created a void in which the private sector directs itself, leaving politically vulnerable low-income neighborhoods unprotected from development encroachment. In Jacksonville, however, that private-sector power is curbed by the parallel power of a populist political constituency—the result of the fragmented politics of that city.

Local governments in Orlando and Jacksonville support private development downtown. They also are politically sensitive to the needs of neighborhoods. The political climate creates the need to extract communitywide benefits from commercial growth. The inclusion of low-income neighborhoods in policy development and a commitment to affordable housing within the downtown district are a result of public-sector pressure and community expectations. Tampa has had a strong progrowth agenda that has not been sensitive to the impacts of downtown development. As is starting in Fort Lauderdale, changes in the private-sector power structure and in the mayor's seat are prompting some small efforts to attend to public amenities and affordable housing.

Can Development Impacts Be Managed?

These cases suggest that public-sector responses can vary in a progrowth environment. Florida has been and continues to be oriented toward strong growth and private development. The introduction of statewide comprehensive planning regulations for cities and counties has provided an opportunity for progressive change. The divergent growth-management strategies in downtown development that are classified here are evidence that growth politics is not monolithic but hinges on important political and administrative dynamics.

The economic imperative is real in urban cities, especially in rapidly growing Sunbelt cities. Economic development will consume resources if it is allowed to compete politically with neighborhoods, housing, and general public benefits in a zero-sum game. Downtown areas are especially vulnerable to policy constraints imposed by the private sector, because they are struggling to compete with booming suburbs. Yet political choices remain as to how to distribute the costs and benefits of development. The public sector must confront whether downtown success should be achieved with or without balancing the impacts of that growth on housing in downtown neighborhoods, public access to policy decisions, and aesthetic satisfaction.

This chapter has contributed an alternative view of that downtown competition. Political choice can be released from the constraint of the economic imperative if the political perspective of policies such as growth management is widened to balance subsidies as supply-side strategies with capacity building as demand-side strategies, emphasizing the potential positive impact on commercial development. By integrating equity issues into the long-term policy agenda and acting on that agenda, cities can improve the distribution of downtown development costs and benefits. The challenge for city government is to implement these ideas to further that balanced agenda within the pressures of traditional growth politics.

Notes

1. This concept also is developed by Robinson (1989) as the dichotomy of the corporate center and alternative populist, balanced by the corporate-distributive approach. In public administration this concept is described as managing institutional diversity (White, 1989). See Reed, Reed, and Luke (1987) for a discussion of strategic planning. Levine (1989) defined three emerging development policy trends as strategic planning, linkage policies, and community economic development.

2. Public amenities also may include benefits that can be enjoyed by the general public, such as public art, open space, galleries, and plazas, which can be included in building designs, as well as benefits that improve societal conditions, such as housing, day care, and transportation (see Lassar, 1989).

3. The results of a recent survey by Goetz (1991) suggest that community political participation is positively associated with the presence of development strategies that include equity considerations, such as linkage policies (*type II* economic development strategies). High growth rates, which would signal strong demand, were not found to be associated with the use of these policies.

4. Stone (1987b, 1989) and Elkin (1985) expanded this concept as a broad perspective of city development and created several classifications based on public- and private-sector roles that can be broadly expressed as a regime. Specifically see Elkin (1985, pp. 11-13), Stone (1989, pp. 1-6), and DiGaetano (1989, pp. 263-264).

5. Information and data for this study were collected from April through July 1989 through personal, on-site interviews with public- and private-sector officials involved with and responsible for downtown development policy in each city. Additional data were gathered from many public- and private-sector documents (Fort Lauderdale Planning Department, 1989; Jacksonville Planning Department, 1989; Orlando Planning Department, 1982, 1989; Tampa Mayor's Downtown Advisory Committee, 1988; Tampa Community Redevelopment Agency, 1983) and through telephone conversations with city and county government agencies and private companies.

6. All Florida cities must prepare a growth-management plan that addresses land use, infrastructure and capital facilities, transportation, and housing (Local Government Comprehensive Planning and Land Development Regulation Act, *Florida Statutes*, chap. 163 amended, 1985). There is no requirement for a separate downtown component of the growth-management plan. The process, however, is an opportunity for cities to give attention to specialized downtown needs and to include public benefits as a component in their plan documents through designation as a redevelopment area or urban activity area.

7. Florida cities have statutory authority to create quasi-independent special districts for the purpose of developing strategies for redevelopment in areas found by the city to have slum or blighted conditions or a shortage of affordable housing for low- and moderate-income persons. Although the statutes call for a broad community application, these activities, as designated under the Community Redevelopment Act (*Florida Statutes*, § 163.340, 1969), are often placed within a DDA.

8. The DDA of Fort Lauderdale was created through special state statute. Although the city council makes the appointments to the authority, it cannot remove them. The revenue structure of the DDA comes from a special assessment of downtown commercial property owners, with the rate set by their vote.

9. Two TIF districts were created to support a retail area and a convention center. An urban development action grant was used to support a mixed-use project.

10. She won 77 of 78 precincts in the 1987 election for mayor. Coincidentally this was the first city election under the new district council representation system (Kerstein, 1991).

References

Aschoff, S. (1990). Orlando. *Florida Trend, 32*(10), 48.

Barnekov, T., & Rich, D. (1989). Privatism and the limits of local economic development policy. *Urban Affairs Quarterly, 25,* 212-238.

Beatley, T. (1988). Development exactions and social justice. In R. Alterman (Ed.), *Private supply of public services* (pp. 83-95). New York: New York University Press.

Bernard, R., & Rice, B. (Eds.). (1983). *Sunbelt cities.* Austin: University of Texas Press.

Bowman, A. (1988). Competition for economic development among Southeastern cities. *Urban Affairs Quarterly, 23,* 511-527.

Bureau of Economic and Business Research. (1989). *Florida estimates of population: April 1, 1988.* Gainesville: University of Florida.

Clavel, P. (1986). *The progressive city.* New Brunswick, NJ: Rutgers University Press.

DeGrove, J. (1984). *Land, growth, and politics.* Chicago: American Planning Association Planners Press.

deHaven-Smith, L. (1987). *Environmental publics* (Monograph 87-2). Cambridge, MA: Lincoln Institute of Land Policy.

DiGaetano, A. (1989). Urban political regime formation. *Journal of Urban Affairs, 2,* 261-281.

Eisinger, P. (1988). *The rise of the entrepreneurial state.* Madison: University of Wisconsin Press.

Elkin, S. (1985). Twentieth century urban regimes. *Journal of Urban Affairs, 7,* 11-28.

Fainstein, N., & Fainstein, S. (1987). Economic restructuring and the politics of land use planning in New York City. *Journal of the American Planning Association, 53,* 237-248.

Feagin, J. (1986). Urban real estate speculation in the United States. In R. Bratt, C. Hartman, & A. Meyerson (Eds.), *Critical perspectives on housing* (pp. 99-118). Philadelphia: Temple University Press.

Fishkind, H., Milliman, J. W., & Ellson, R. W. (1978). A pragmatic econometric approach to assessing economic impacts of growth or decline in urban areas. *Land Economics, 54,* 442-460.

Florida State Comprehensive Plan Committee. (1987). *Keys to Florida's future.* Tallahassee: Author.

Fort Lauderdale Planning Department. (1989). *Future land-use element: Comprehensive plan.* Fort Lauderdale, FL: Author.

Frederick, W. (1988). Orlando state of the city '88. *Quality Cities, 62*(1), 2-5, 9-11.

Frieden, B., & Sagalyn, L. (1989). *Downtown, Inc.* Cambridge: MIT Press.

Friedland, R. (1983). The politics of profit and the geography of growth. *Urban Affairs Quarterly, 19,* 41-54.

Goetz, E. (1991). *Type II policy* and mandated benefits in economic development. *Urban Affairs Quarterly,* *26,* 170-190.

Hillsborough County, City-County Planning Commission. (1989a). *Future of Hillsborough comprehensive plan for Tampa: Future land-use element.* Tampa, FL: Author.

Hillsborough County, City-County Planning Commission. (1989b). *Future of Hillsborough comprehensive plan for Tampa: Housing element.* Tampa, FL: Author.

Jacksonville Downtown Development Authority (DDA). (1988). *Initial action plan for Jacksonville's core business district.* Jacksonville, FL: Author.

Jacksonville Mayor's Commission on Housing. (1989). *A housing strategy.* Jacksonville, FL: Author.

Jacksonville Planning Department. (1989). *Comprehensive plan 2010: Housing element.* Jacksonville, FL: Author.

Jones, B. (Ed.). (1989). *Leadership and politics.* Lawrence: University of Kansas Press.

Judd, D., & Parkinson, M. (Eds.). (1990). *Urban affairs annual review: Vol. 37. Leadership and urban regeneration.* Newbury Park, CA: Sage.

Kantor, P. (1987). The dependent city: The changing political economy of urban economic development in the United States. *Urban Affairs Quarterly, 22,* 493-520.

Kerstein, R. (1989, March). *The political-economy of urban development in Tampa.* Paper presented at the Annual Meeting of the Urban Affairs Association, Baltimore.

Kerstein, R. (1991). Growth politics in Tampa and Hillsborough county. *Journal of Urban Affairs, 13,* 55-76.

Lassar, T. (1989). *Carrots and sticks: New zoning downtown.* Washington, DC: Urban Land Institute.

Levine, M. (1989). The politics of partnership. In G. Squires (Ed.), *Unequal partnerships* (pp. 12-34). New Brunswick, NJ: Rutgers University Press.

Logan, J., & Molotch, H. (1987). *Urban fortunes.* Berkeley: University of California Press.

Lubove, S. (1990). Tampa Bay. *Florida Trend, 32*(10), 47.

Merriam, D., Brower, D., & Tegeler, P. (Eds.). (1985). *Inclusionary zoning moves downtown.* Chicago: American Planning Association Planners Press.

Mitchell, S. R. (1990). The housing connection. *Planning, 56*(9), 18-19.

Mollenkopf, J. (1983). *The contested city.* Princeton, NJ: Princeton University Press.

Molotch, H. (1976). The city as growth machine. *American Journal of Sociology, 82,* 309-331.

Orlando Planning Department. (1982). *Downtown Orlando redevelopment area plan.* Orlando, FL: Author.

Orlando Planning Department. (1989). *Growth management area plan: Downtown, updates.* Orlando, FL: Author.

Peterson, P. (1981). *City limits.* Chicago: University of Chicago Press.

Porter, D. (1985). The office/housing linkage issue. *Urban Land, 44*(9), 16-21.

Porter, D., & Lassar, T. (1988). The latest on linkage. *Urban Land, 48*(12), 7-11.

Reed, C., Reed, B. J., & Luke, J. S. (1987). Assessing readiness for economic development strategic planning. *Journal of the American Planning Association, 53,* 521-530.

Robinson, C. (1989). Municipal approaches to economic development. *Journal of the American Planning Association, 55,* 283-295.

South Florida Regional Planning Council. (1989). *Downtown Fort Lauderdale trends and conditions.* Hollywood, FL: Author.

Squires, G. (Ed.). (1989). *Unequal partnerships.* New Brunswick, NJ: Rutgers University Press.

Stone, C. (1980). Systemic power in community decision making. *American Political Science Review, 74,* 978-990.

Stone, C. (1984). City politics and economic development: Political economy perspectives. *Journal of Politics, 46,* 286-299.

Stone, C. (1987a). The study of the politics of urban development. In C. Stone & H. Sanders (Eds.), *The politics of urban development* (pp. 3-24). Lawrence: University of Kansas Press.

Stone, C. (1987b). Summing up: Urban regimes, development policy, and political arrangements. In C. Stone & H. Sanders (Eds.), *The politics of urban development* (pp. 269-290). Lawrence: University of Kansas Press.

Stone, C. (1989). *Regime politics: Governing Atlanta, 1946-88.* Lawrence: University of Kansas Press.

Swanstrom, T. (1985). *The crisis of growth politics.* Philadelphia: Temple University Press.

Tampa Community Redevelopment Agency. (1983). *CRA plan for the downtown community redevelopment area.* Tampa, FL: Author.

Tampa Mayor's Downtown Advisory Committee. (1988). *Downtown land-use policy plan support documents.* Tampa, FL: Author.

Turner, R. S. (1990a). *Downtown development in four Florida cities: Final report.* Boca Raton: Florida Atlantic University.

Turner, R. S. (1990b). New rules for the growth game. *Journal of Urban Affairs, 12*(1), 35-48.

U.S. Bureau of the Census. (1980). *Detailed population characteristics*. Washington, DC: Government Printing
    Office.
Vogel, R. K., & Swanson, B. E. (1989). The growth machine versus the antigrowth coalition. *Urban Affairs
    Quarterly, 25,* 63-85.
White, L. (1989). Public management in a pluralistic arena. *Public Administration Review, 49,* 522-532.
Wong, K. (1990). *City choices*. Albany: State University of New York Press.
Wrenn, D. (1987). Making downtown housing happen. *Urban Land, 46*(4), 16-19.

PART NINE

Urban Education

Introduction to Urban Education

Jeffrey A. Raffel

In almost every big city, dropout rates are high, morale is low, facilities are often old and unattractive, and school leadership is crippled by a web of regulations. . . . The failure to educate adequately urban children is a shortcoming of such magnitude that many people have simply written off city schools as little more than human storehouses to keep young people off the streets. (Carnegie Foundation for the Advancement of Teaching, 1988, p. xi)

There is general agreement that America's urban schools are failing (Galster & Hill, 1992). Urban schools are the battleground for many of the issues (e.g., poverty, racial conflict, illiteracy) that vex our nation. As Kenneth Sirotnik states in Chapter 24, "Improving Urban Schools in the Age of 'Restructuring' ": "Put simply, these schools are at the confluence of racism and poverty in our society." But although there is agreement that our urban schools must change, there is little agreement on the nature of the change needed and no common vision of how these changes should or could occur.

The Challenges

American public schools are under considerable scrutiny and criticism. They are accused of perpetuating a stifling bureaucracy, of not educating students to deal with the realities of the new global economy, of failing half of the nation's students who do not go on to postsecondary education, and of being inferior to the schools in other developed nations. Although these accusations have been challenged (Bracey, 1993; Huelskamp, 1993), these optimistic responses have centered generally on the ability of America's *best* schools and

AUTHOR'S NOTE: I thank Debra Brucker and Carol Sirkowski for their helpful comments on a draft of this chapter.

students to measure up. Few would argue, however, that very many of either are found in today's cities.

In Chapter 25, "Visions of America in the 1990s and Beyond," Jean Schensul and Thomas Carroll describe a major challenge facing urban schools today, which is likely to intensify in the years ahead—an increasingly diverse student population with great cultural and linguistic variation. This challenge magnifies a recent trend. A majority of the clients of our nation's urban schools today are minorities. The demographic change in cities during the last few decades is reflected in the changing clientele served by urban schools. By 1985 only a few of the nation's large cities had a majority of white students. Most of students were black or Hispanic in school districts such as Boston, Buffalo, Cincinnati, Milwaukee, Norfolk, Pittsburgh, Rochester, and San Francisco. In our largest cities—Baltimore, Chicago, Cleveland, Dallas, Detroit, New Orleans, Philadelphia, St. Louis, and Washington, DC—hardly any students were white. In New York City only 200,000 of the more than 900,000 students were white. In Los Angeles 100,000 of the more than 500,000 students were white.

Many of the students who attend urban schools today bring with them special needs. Many of the students are poor. Many are from single-parent households. Many do not speak English as their primary language. (Schensul and Carroll note: "In Los Angeles, more than 80 languages are spoken in the school system.") Many have physical or mental disabilities and are classified as requiring special educational services.

Sirotnik lays out "the litany of our nation's demography" in his chapter, starting with the high number of births to unmarried women and ending with gross differences in TV watching among the races. Reacting to such a litany, the authors of a recent status report on urban education in the nation conclude:

> The changing demographic face of the nation's cities has placed tremendous burdens on the nation's schools. Many children come to urban schools not prepared to learn; many are negatively affected by poverty, drugs, alcohol, poor health; many have physical and mental disabilities derived from these causes. To the extent that schools reflect society, our urban schools now face problems that are quantitatively and qualitatively more difficult than those faced in previous decades. (Raffel, Boyd, Briggs, Eubanks, & Fernandez, 1992, p. 264)

The problems of urban schools are manifested in many ways—from bricks and mortar to sticks and stones. The Carnegie Foundation for Teaching reported in 1988 that half of New York City's 1,050 school buildings are at least 50 years old. The facilities are so dilapidated that the board of education said it would take $4.2 billion during the next 10 years to refurbish the physical plant. The school facilities repair bill for the nation was estimated at $30 billion. (As one manifestation of this problem, in autumn of 1993 the New York City schools opened weeks late because of problems related to asbestos inspection in and removal from its schools.) Along with the crumbling physical structure has come a crumbing social structure. Violence in and around urban schools is being perceived increasingly as a major problem. In the 1992-93 school year the Philadelphia Board of Education reported more than 700 weapons found in the vicinity of the city's public schools.

The challenges to urban schools are tremendous. Our society, through its laws, attitudes, and inaction, has concentrated much of its social problems in the cities. The solutions to these problems, including how to improve urban schools, are not clear. The following approaches are now being considered.

Changing Urban Schools

Increasing Resources for Urban Schools

Unfortunately, despite the apparent needs, our nation has not focused its resources in urban areas (Kozol, 1991). America's schools are financed primarily through the local property tax system, a system that perpetuates the separation of resources from needs (Cook, 1989). Approximately 45% of the revenues for public education comes from local sources (generally property taxes), 49% from state governments, and 6% from the federal government. Many states have school finance systems that do not offset the inequalities built into the local property tax system or that make these inequities even worse. Court suits based on inequities in state finance systems have been brought successfully in many states, but the impact on urban school districts and students has been muted. Political conflicts over solutions may rage on for years, as they have in Texas and New Jersey, because providing poor school districts with more funds requires a combination of transferring revenues from well-off districts and increasing taxes.

Although increasing the funds available to urban school districts may be helpful, money is a necessary but not sufficient solution to urban school woes. Summaries of educational research indicate that educational expenditures are not regularly related to student achievement (Hanushek, 1986) and even that lowering class size, America's favorite educational remedy, is not likely to have much impact on school success (Odden, 1990).

Compensatory Programs

The traditional remedy to educational problems in urban schools has been compensatory education. Under the notion that disadvantaged children need more advantages to compensate for what they have missed, the federal government funds a major compensatory program (Known as Chapter One) for elementary school students and the well-known Head Start program for preschoolers. Both programs have been oft evaluated and criticized, and both appear to have some modest impact on educational success. These approaches assume that what we do in our schools is fundamentally valid but insufficient to help students lacking family and environmental support. This incremental approach has been criticized, however, as noted below.

Improving Teaching

Many analysts of urban schools view the problem as a technical one; teachers do not teach urban children through appropriate methods (e.g., see Walberg, 1984). Thus new methodologies, teacher retraining (called "professional development" and "in-service training" in educational jargon), and approaches to facilitate these become the major change points. Advocates of changes in teaching methodologies concentrate their efforts on changing teacher classroom activity, often without a concern for the role of other forces in the educational environment. A related approach is to change schools by establishing explicit standards in various subject areas and to develop related performance measures for students to accomplish. A number of states (e.g., Vermont, Delaware) are basing their school reform hopes on this approach, in coordination with federal efforts in this direction (e.g., the National Council of Teachers of Mathematics).

In Chapter 26, "The Evolution of a University/Inner-City School Partnership," Andrea Zetlin, Kathleen Harris, Elaine MacLeod, and Alice Watkins describe a case analysis of an attempt to establish professional development schools built on a partnership of a university and two inner-city schools in Southern California. Their analysis helps us understand the difficulty in changing schools and "the very real forces working against successful partnerships." (Such forces are the ones that lead many to argue for more radical or system-changing solutions to force innovation, responsiveness, and a lessening of bureaucratization in urban schools.)

Building on Community and Culture

In the "Importance of Community" section of Sirotnik's Chapter 24, we find the arguments for changing schools on the basis of a vision of a better linkage between the community and the urban school. Two social scientists have led movements for a school and community-based change in urban schools. James Comer of Yale University has developed a program to better link the home and the school and build "social capital." Henry Levin of Stanford University has spearheaded an effort to create accelerated schools that build on the strengths, not the deficits, of disadvantaged children. The early reports from experiments based on these approaches are positive. As Sirotnik notes from the work of Ogbu, one of the forces that such schools must overcome is the "development of anti-achievement norms ('oppositional social identity') by minorities when faced with social stigmas of 'acting white.' "

Related to the culturally based approaches is the concern for the environment that leads to dysfunctional behavior and school problems. Simply put, until America invests in its urban children and their health and well-being, their schoolwork is unlikely to improve sufficiently. Thus Raffel et al. (1992) argue, "The success of urban schools is dependent upon our nation's health, criminal justice, economic, and other systems. Too many leave the nation's problems on the doorstep of the urban school" (p. 264).

Changing Schools Through School Desegregation

Believing that equality can be built only upon schools that do not isolate minorities, advocates of school desegregation have called for metropolitan school plans. Gary Orfield (1978), a leading proponent of this approach, calls for developing plans that link cities and suburbs educationally. But many blacks resent the notion that their children can learn onlyif they are sitting next to a white child, and they believe that such plans put the burden of busing and adjustment on their children; and many whites vehemently oppose the idea of "forced busing" from suburbs to cities (Cook, 1989). The U.S. Supreme Court has made metropolitan remedies difficult to achieve with its *Milliken v. Bradley* decision.

In Chapter 27, "Race, Urban Politics, and Educational Policy Making in Washington, DC," Floyd W. Hayes, III examines the history of school desegregation and tracking students by ability in the Washington, DC, school system. Hayes describes the reactions of many black citizens to both school desegregation and tracking, and in doing so he highlights the problems that urban school districts have in balancing the goals of quality education, integrated education, and responsiveness to parents. Hayes's view, that "judicial activism may have been the single most significant factor in the ultimate decline of public education," represents a growing skepticism in the black community about school desegregation. Many would disagree with his analysis. Few will disagree with his conclusions, that for "urban school systems throughout America, the great challenge of the future will be to reinstate quality education and thereby rebuild confidence in public education."

Changing Governance

Chubb and Moe (1990) believe that this goal is impossible. They have launched the strongest challenge yet to American public school governance by arguing that public schools by their very nature lead to overbureaucratization and school ineffectiveness. According to these researchers, as school districts try to cope with the various pressures placed on them from teachers, parents, businesses, and community groups, the response is to add bureaucratic rules that stifle creativity and responsiveness. And certainly newspaper headlines seem to bare the major problems of governance facing urban school districts. Urban superintendents are becoming an endangered species. The New York City school district has had a succession of short-term leaders since Richard Green's untimely death from a heart attack in office. (In Cleveland a few years ago, the superintendent put a gun to his head, thus ending his personal and organizational misery.)

Many panaceas for urban school renewal based on governance changes have been offered. The most controversial has been school choice. Advocates argue that power must shift from bureaucratic dinosaurs of urban school districts to parents through such a system. Some call for a voucher system in which parents would be given a grant to send their children to the public or private school of their choice (Chubb & Moe, 1990). Others claim that choice within the public school system is sufficient to force the bureaucracy to respond to diverse needs. They point to the success of choice and alternative schools in Community District No. 4 in East Harlem in New York City. Related to these concepts is controlled choice, in which students are not assigned to an attendance area school, but rather select among the schools of the district. Cambridge and Springfield, Massachusetts, have had such systems for years. Many school districts do not have choice systems but do have individual schools, primarily magnet schools, that serve to attract students outside attendance areas for special programs. Christine Rossell (1990) presents substantial quantitative evidence that systems with magnet schools may accomplish more desegregation over the long run than mandatory plans, as well as offering superior education and parental choice.

A less radical measure to change governance—school restructuring—has been suggested as a means to improve urban schools. Exact definitions and notions may differ here, but in general, supporters seek to move decision-making authority over personnel, curriculum, and budget to the school level and have teachers and parents play a major role in these decisions (see Clune & Witte, 1990, for a number of excellent papers on restructuring and choice alternatives). The Chicago school system has taken a giant step in this direction by giving broad authority to parents in its more than 500 schools, including the authority to select principals, approve budgets, and develop curriculum. But early reports on the effects of this restructuring are mixed, especially in the light of the district's struggle to deal with a $300 million budget deficit in autumn of 1993. The work of Brown University's Dean Ted Sizer has led to school restructuring under the aegis of the Coalition of Essential Schools. In these schools a conscious effort is made to involve teachers in transforming the school on several dimensions, but most notedly in their own roles, with the principle of "teacher as coach, student as worker."

Conclusions

As I argue, in this introduction to the field of urban education, and as the authors of the readings to follow indicate, there are no easy answers to the question of what will improve urban education. Nor can the problems of urban schools be solved apart from dealing with the demographic, economic, and social problems of urban areas. Today's students will have many opportunities to make contributions to help change urban schools tomorrow, and these readings will help them consider how this task might be accomplished.

References

Bracey, G. W. (1993). The third Bracey report on the condition of education. *Phi Delta Kappan, 75*(2), 104-117.

Carnegie Foundation for the Advancement of Teaching. (1988). *An imperiled generation: Saving urban schools*. Princeton, NJ: Princeton University Press.

Chubb, J. E., & Moe, T. M. (1990). *Politics, markets, and America's schools*. Washington, DC: Brookings Institution.

Clune, W. H., & Witte, J. E. (Eds.). (1990). *Choice and control in American education* (Vols. 1 & 2). London: Falmer.

Cook, K. (1989). Public education: Opportunity and its limits. In D. B. Robertson & D. R. Judd (Eds.), *The development of American public policy: The structure of policy restraint* (pp. 245-278). Glenview, IL: Scott, Foresman.

Galster, G. C., & Hill, E. W. (1992). *The metropolis in black and white: Place, power, and polarization*. New Brunswick, NJ: Rutgers University, Center for Urban Policy Research.

Hanushek, E. A. (1986). The economics of schooling: Production and efficiency in public schools. *Journal of Economic Literature, 24*, 1141-1177.

Huelskamp, R. M. (1993). Perspectives on education in America. *Phi Delta Kappan, 74*(9), 718-721.

Kozol, J. (1991). *Savage inequalities*. New York: Crown.

Odden, A. (1990). Class size and student achievement: Research-based policy alternatives. *Educational Evaluation and Policy Analysis, 12*(2), 213-227.

Orfield, G. (1978). *Must we bus? Segregated schools and national policy*. Washington, DC: Brookings Institution.

Raffel, J. A., Boyd, W. L., Briggs, V. M., Jr., Eubanks, E., & Fernandez, R. (1992). Policy dilemmas in urban education: Addressing the needs of poor, at-risk children. *Journal of Urban Affairs, 14*(3/4), 263-289.

Rossell, C. H. (1990). *The carrot or the stick for school desegregation policy: Magnet schools or forced busing*. Philadelphia: Temple University Press.

Walberg, H. (1984). Improving the productivity of America's schools. *Educational Leadership, 41*(8), 19-27.

Suggestions for Further Reading

Cibulka, J. G., Reed, R. J., & Wong, K. K. (Eds.). (1992). *The politics of urban education in the United States*. London: Falmer.

Council of Great City Schools. (1991). *Strategies for success: A plan for achieving the national urban education goals*. Washington, DC: Author.

National Commission on Children. (1991). *Beyond rhetoric: A new American agenda for children and families*. Washington, DC: Author.

National School Boards Association. (1988). *A survey of public education in the nation's urban school districts*. Alexandria, VA: Author.

24

Improving Urban Schools
in the Age of "Restructuring"

Kenneth A. Sirotnik

A school is a school is a school, is it not? It is a building or cluster of buildings with walls surrounding people who are using up time, space, materials, and other resources. A school is an organization with aims and purposes ostensibly educational in nature; teaching and learning, more or less, go on in these places. It is a complex organization, marked at once by rather tight controls at some levels (e.g., certification and fiscal accountability at state, district, and building levels) and yet rather loose controls at others (e.g., activities behind the classroom door). It is, of course, an open system, affecting and heavily affected by its environment (its school district, its community, and its regional, state, national, and even international contexts). It is, therefore, a political system, pressuring and being pressured by its constituencies and interest groups, negotiating everything from its purposes and functions as an organization to the details of its daily operations. It is also a cultural system, with norms, roles, and expectations—written and unwritten—for what people should do in the system and how it has always been done. The people are older and younger, they are the educators and the to-be-educated: teachers, administrators, counselors, aides, librarians, support staff, and students. And these people have lives, both within and without the school; their existence can be more or less isolated, more or less supported, depending on the presence or absence of significant others inside and outside the system.

In a few words, schools and school systems are messy, complicated, and highly interactive. Yet they are remarkably stable and extraordinarily difficult to change in fundamental ways (Elmore & McLaughlin, 1988; Sarason, 1990). Nonetheless, changing them in fundamental ways is precisely what the rhetoric of *restructuring* appears to be about. Notwithstanding the occasional camouflaging of parental choice in talk of restructuring, the term has been used by many educators and policymakers to signify systemic change, an inside-out, bottom-to-top change that alters significantly the major commonplaces of schools and schooling (e.g., see Cohen, 1988; Cuban, 1989; Walberg, Bakalis, Bast, & Baer, 1989). Who is in charge and who makes what decisions, what and how do children learn, how and for what purposes are schools organized, who does what in and out of schools, and how do we know if it is all worthwhile? These and other questions are answered in talk of restructuring in ways that suggest fundamentally different conditions and circumstances in schools.

For example, restructuring suggests that leadership, decision making, and accountability should be concentrated at the building level instead of at state or central office levels. In other words, with the leadership and support of board members, superintendent, and central office staff, principals, teachers, and parents should be empowered to work together—and to be responsible to one another — as they create better schools for students. These better schools should be characterized by fundamental changes in curriculum and curricular practices, such as (a) the absence of structures and practices (e.g., tracking or static ability and age grouping) that limit educational success for many students and (b) the presence of a variety of teaching and learning strategies and activities designed to leave no child out, to achieve high levels of learning for all children, and to promote a quality common curriculum that goes well beyond basic skills of literacy and numeracy. Moreover, these schools should demonstrate assessment systems high in fidelity relative to expected and valued learning—higher order and critical thinking, creative and expository writing, and verbal communication.

Restructuring, clearly, is not for the timid or the tired. Cynics, skeptics, and pessimists need not apply. Restructuring is only for those who are not dissuaded by the many short-lived educational innovations and fads that have come and gone (sometimes more than once) over the years, making little if any impact on schools and classrooms since the turn of the century (see the analyses of Berman & McLaughlin, 1978; Charters & Jones, 1973; Cuban, 1984; Sarason, 1971/1982; Slavin, 1989). And these cycles of reform have paled in scope next to the present proposals for restructuring. The bold, energetic, and optimistic, then, have their work cut out for them.

It is hoped they will pay some attention to what can be learned from the past. There is a vast body of literature on the theory of organizations and change, generally (see Bolman & Deal [1984] and Scott [1981] for reviews of much of this work). In particular, schools have received their fair share of attention in the literature on innovation, implementation, improvement, and change (e.g., Fullan, 1982; Goodlad, 1975; Huberman & Miles, 1984; Joyce, Hersh, & McKibbin, 1983; Leming & Kane, 1981; McLaughlin, 1976, among many others). In view of the nature of schools as organizations, the application of sociology and cultural anthropology to the study and improvement of schooling has been extensive (see Waller, 1932; and more contemporary accounts such as Hurn, 1985; Jackson, 1968; Lortie, 1975; Meyer & Rowen, 1978; Ogbu, 1982; and Sarason, 1971/1982).

In working on the problems of significant school restructuring, what we know about schools as organizations and about attempts to change them and their curricula can be combined with what we know about knowing—that is, how and why people act on information and knowledge in the context of their work (Sirotnik, 1989). Knowledge is not something envisioned and generated by researchers, to be disseminated and implemented by practitioners. Educators are, in Schon's (1983) words, "reflective practitioners," knowledge producers, as well as knowledge users.

Most important, educators must be seen as "critical inquirers" (Sirotnik & Oakes, 1986). *Inquiring* educators question what they are doing, how they are doing it, and why they are doing it in the first place. They question the existing curriculum. They do not automatically embrace the latest educational elixir, be it "new" math, programmed instruction, homogeneous grouping, cooperative learning, outcome-based education, management by objectives, building-based management—or restructuring. Instead they consider extant knowledge in the context of their current practices. They question what they are doing now and why it came to be that way. They inform themselves, not only with relevant research and recommended practices, but also with information available in their own school setting—indicators of student progress, school and classroom practices, and attitudes and sentiments of key constituencies, such as educators, students, and parents. They experiment, they take risks, they do not require immediate closure, and they tolerate ambiguity as they invent, modify, or revise their educational programs and practices. In short, inquiring educators must be intellectually and actively "playful" (March, 1972); they must learn to suspend belief in the way things are and to explore creatively the way things might become.

Critically inquiring educators go a significant step further: They seek to uncover implicit values and beliefs underlying current educational practices and programs. They ask questions about whose interests are and are not being served by the way curricula are organized and by the ways students are taught. They pledge allegiance to public education in a democratic society; they consider the consequences of compulsory schooling and the moral imperative that the best education be received by each and every student. They are concerned, therefore, with competing conceptions of social justice and how to balance the interests of individuals with the common good. Such questions are, indeed, difficult to ask, let alone answer (see Goodlad, Soder, & Sirotnik, 1990). Nonetheless it is the duty of professional educators to continually test their assumptions and practices against the ultimate criterion of all our children learning.

The renewing school—or restructured school and its restructured curriculum, if you like—is where educators are empowered to be critical inquirers. It is a school where leadership, responsibility, and accountability reside not only with administrators but also with teachers and parents. It is a school where decisions in the best interests of all students are made collaboratively by those—educators, students, and parents—with vested interests in all students. It is a school where leadership from the district results in the resources necessary (e.g., time, training) to allow reflective and critical practice to become commonplace in the day-to-day activities of school.

Urgencies for Urban Schools

Such are the visions of reconceptualized and reconstructed schools of the future. Like so many other discussions of school improvement and change, these visions have not rested on separate arguments for rural, suburban, metropolitan, or urban schools. After all, a school is a school is a school—or is it?

If a school is a school is a school, why do some educators refuse to work in some schools and not others; why are parents so concerned about choice; why is "white flight" dramatically changing the demographics of our urban centers? The bottom line is that some schools, usually our urban schools, end up, in effect, as crucibles attempting to withstand the heat of political and economic inequities. Put simply, these schools are at the confluence of racism and poverty in our society.

The litany of our nation's demography cannot be repeated too often:[1]

1. Births to unmarried women in their later teens number 20.5 per 1,000 for whites, compared with 79.4 for blacks.
2. The proportion of children under 18 living in poverty is 16% for whites, compared with 43% for blacks and 40% for Hispanics; in households with a female as single parent, these figures rise to 45% for whites, compared with 67% for blacks and 72% for Hispanics.
3. Unemployment rates of young adults (20-24 years old) are around 10% for whites, compared with nearly 25% for blacks and about 13% for Hispanics; 54% of high school dropouts are unemployed during the year they leave school.

Educational indicators are similarly revealing:

1. High school dropout rates are 30% among Hispanics, near 15% among blacks, and slightly over 10% among whites.
2. The proportion of 13-year-old students who are 1 or more years below expected grade level is 22% for whites, compared with 33% for blacks and 41% for Hispanics.
3. Nearly 90% of whites can read at an intermediate level of proficiency, whereas the figure is less than 70% for Hispanics and about 65% for blacks.
4. Writing-performance scores for whites are above average, whereas those for blacks and Hispanics are below average; likewise for mathematics achievement.
5. Literacy tests given to young adults demonstrate that over 60% of whites can perform at adequate levels or higher, compared with slightly over 20% of blacks and under 40% of Hispanics.
6. And 39% of white high schoolers (11th grade) watch 3 or more hours of TV per day, compared with 63% of blacks and 45% of Hispanics.

All of these comparisons are magnified in their urgency for urban schooling when we realize that even a decade ago, nearly three quarters of all blacks and about half of all Hispanics populated the inner cities of this nation. Today estimates suggest that most of our nation's largest school districts are over 70% minority (nonwhite) in composition.

Leadership and a Normative Commitment

To be sure, we have some evidence to suggest that urban schools and the people in them can rise above this statistical portrait of conditions for educational failure. The occasional heroic efforts of an administrator or a teacher in these places are the stuff that movies are made of. Unfortunately, even if heroes could be transported, there are not enough Marva Collinses, Jaime Escalantes, and George McKennas to go around. It is not heroes, but hard and continuous work by many, that will be necessary to turn most urban schools around.

But work to what end? Education is a normative endeavor (Goodlad et al., 1990). Talk of restructuring schools and their curricula is premature unless fundamental beliefs, values, and human interests are made explicit and purposive. Such matters do not ordinarily receive sustained inquiry in districts and schools. Moreover, normative issues are easy to avoid in relatively homogeneous settings reflecting dominant cultural patterns. Thus, although the problems already faced by urban schools may be just around the corner, the more homogeneous suburban schools can conceivably sidestep such issues and not experience problems (at least in the short run). Not so in the case of schools in urban settings. Restructuring or significantly altering the educational programs and practices of urban schools runs immediately and squarely up against the incongruities between a national rhetoric of quality and equality and the local realities of inequity and mediocrity (or worse).

Dynamic and normatively consistent leadership, therefore, would appear to be especially crucial in setting and sustaining vision and action toward the significant improvement of an urban school. The contemporary literature on organizations, of course, is replete with advice on the importance of knowledgeable, visionary, and empowering leadership in bringing about organizational renewal and change (e.g., see Bennis, 1989; Block, 1988; Cleveland, 1985; Kanter, 1983). The experiences and findings of those deliberately setting out to infer the characteristics of urban schools whose students have beat the odds—that is, scored higher than expected on standardized achievement tests—confirm the importance of strong and informed leadership (Andrews & Soder, 1987; Brookover, Beady, Ford, Schmitzer, & Wisenbaker, 1979; Edmonds, 1979). And there is a substantial and varied body of literature on the importance of educational leadership up and down the schooling hierarchy, argued on grounds far broader than test-score outcomes alone (e.g., Barth, 1980; Bates, 1984; Foster, 1984; Goodlad, 1984; Kerr, 1987).

Green's (1987) concept of the "conscience of leadership" would seem especially useful in talk of restructuring urban schools. Regardless of the domain of leadership (state, district, school, or classroom), "making ethical sense of professional practice requires a grasp of the point of the profession. If we can clearly see the point of our profession, then the ethical sense of day-to-day practice will become more evident" (Green, 1987, p. 105).

Perhaps the more homogeneous and affluent suburban districts and schools can afford to skirt the moral imperatives of educating and being educated in public schools, as they set about their tasks of restructuring. Educators, at least professional educators as defined by Green (1987), in urban schools can ill afford to whitewash the restructuring conversation with talk only of altered leadership and decision-making roles and cutting-edge developments in curriculum and instruction. The educational interests of all students to whom educators are responsible must be held up explicitly as ends against which means can be aligned. This normative agenda should have been yesterday's; it cannot wait until tomorrow. If the demographers are correct, we are on the verge (if not past it) of losing an entire generation of the urban poor and minority youth of this country (Carnegie Foundation for the Advancement of Teaching, 1988; Hodgkinson, 1989; William T. Grant Foundation Commission on Work, Family, and Citizenship, 1988).

What might this normative agenda look like? A proposal of conceptual and philosophical prerequisites for meaningful change in urban schools is examined in detail elsewhere (Sirotnik, 1990). Briefly we first must assume no systematic differences in human learning potential other than those attributable to individual variation itself. Second, we must assume that schools can make a difference—that schooling environments can be created where most students can achieve high levels of learning in a valued, common curriculum. Taken together, these assumptions suggest that quality educational experiences can be created in public schools such that most students—regardless of race, ethnicity, gender, economic status, or any other irrelevant characteristic—can achieve high levels of mastery. Both equity and excellence, therefore, are addressed in these assumptions, with the implication that there can be no educational excellence without educational equity. Moreover, without deliberate attention to high-quality curriculum and instruction, students may experience equity while they experience something considerably short of excellence.

A Commitment to Evaluation

These assumptions imply standards of educational practice, and educational practice must be held accountable. In a restructured school, however, accountability means responsibility—educators being responsible to each other, to their students, and to their community constituency. These responsibilities include maintaining a critical and reflective stance in relation to their work and the normative grounding for this work. In line with Green's (1987) comments, the *professional* educator is a critical inquirer or, if you would prefer, an evaluator (Sirotnik, 1987). Evaluation is an ongoing part of educational work in classrooms and in schools. It is both formative and summative. And it both guides and is guided by purpose and moral commitment to the concept of public schooling in a democratic society.

These ideas can be illustrated by operationalizing the normative content in the above working assumptions. Excellence, for example, can be indicated by conditions, practices, and outcomes in schools and classrooms that are associated with high levels of learning for most students in all valued goal areas of the common curriculum. Equity, then, can be indicated when there are no systematic differences in the distribution of these conditions, practices, and outcomes based on race, ethnicity, gender, economic status, or any other irrelevant grouping characteristics. An evaluation system, therefore, would be on the lookout for (a) increasingly favorable information on the conditions, practices, and outcomes and (b) decreasing differences based on this information between gender, racial, ethnic, and economic status groups.

Implementing this kind of evaluation system is serious business. The information required is comprehensive (Sirotnik, 1990). For example, information on conditions includes instructional resources, allocation of instructional full-time equivalents, use of experienced versus inexperienced staff, availability and use of significant time for staff planning and critical inquiry, and the like. Tracking placements, ability grouping, higher order thinking strategies, uses of direct instruction, and so forth are the kinds of practices that need to be monitored. Student outcomes must go far beyond those assessed by standardized tests if excellence is what we have in mind (Sirotnik & Goodlad, 1985). Just a beginning set of information might include attendance; dropout, suspension, and expulsion rates; grade point average; course enrollment patterns; standardized and criterion-referenced achievement measures, and assessments of critical thinking skills, communication (verbal and written) skills, citizenship, and effort.

This information would need to be organized in a longitudinal and relational database. Districts would need to "bite the bullet" and disaggregate, reporting the data by gender, race, ethnicity, and economic status *at the building level*. And schools would need the support and resources by which such information could be incorporated into an ongoing evaluation or process of critical inquiry. Perhaps one measure of successful restructuring would be the extent to which this body of information (a) is not used as a weapon against educators, (b) does not produce defensive reactions by educators, (c) is welcomed and incorporated constructively into dialogue, decision making, and action taking at the building level, and (d) stimulates further evaluative inquiry.

The Importance of Community

Districts and their schools are complex organizational systems. This discussion has focused so far on educators and the circumstances and conditions within their places of work, but clearly schools are not closed systems. Districts and schools interact profoundly with their environment. In particular, they are influenced strongly by the social, political, and economic conditions and circumstances of their constituencies—students, their families, and the culture(s) of the local community. To ignore the intimate connections between school and community in the reform and restructuring of urban schooling is to condemn such attempts to almost certain failure.

By and large, schools still are organized for teaching and learning as though there was a relatively intact support structure for students when they go home— two parents, some books, some monitoring of how time is spent, some help on homework, some emphasis on and expectations for academic success, and some continuity from one day to the next. This kind of support structure, as we have seen, is fading rapidly for most students in this country. For poor and minority students in our urban schools, the demographic realities and fragmented home life could not be less conducive to success in ordinary school-based education.

It is this sharp disjuncture between organized schooling and the realities of urban life that has led such educators as James Comer (1980) to worry quite actively about the educational fate of urban children. Coming at it from his background as a child psychiatrist, Comer (1988) has hypothesized:

> The contrast between a child's experiences at home and those in school deeply affects the child's psychosocial development, and . . . this in turn shapes academic achievement. The contrast would be particularly sharp for poor minority children from families outside the mainstream. . . . The failure to bridge the social and cultural gap between home and school may lie at the root of the poor academic performance of many of these children. (p. 43)

Hypotheses such as this, derived from a more cultural-anthropological viewpoint, enjoy a fair amount of empirical support as well (e.g., the case studies reported by Clark [1983], Grant & Sleeter [1988], and Payne [1984]). The conceptual and empirical work of Fordham and Ogbu (1986) suggests additional exacerbating factors—namely the development of antiachievement norms ("oppositional social identity") by minorities when faced with the social stigmas of "acting white."

Whatever the hypothesis or theoretical framework, however, the disproportionate failure among poor and minority children in urban schools is a dismal reality. Ignoring the intimate connections between psychology and culture—and between society, family, and schooling—is not likely to change this reality. Comer's reported success with the New Haven intervention project in two inner-city schools provides some hope and direction for a more enlightened approach. The program was based on a rather comprehensive model of urban school restructuring, including not only site-based decision-making teams of educators, students, and parents but also formal and informal means for mobilizing community resources and involvements—mental health teams, parents' groups, social activities, and special projects (Comer, 1988). Among the important lessons learned were that (a) this kind of restructuring will take a long time—more than 15 years in the New Haven project and (b) considerable and sustained work and effort on staff development, school-community and human relations, dealing with conflict, and the like will be ongoing agenda items.

Underlying this kind of project, of course, is a belief in multiculturalism and the working assumption that two or more cultures can find common ground for educational aims and outcomes, while still respecting and maintaining crucial features of their own and each other's identity (Banks, 1981). In other words, the notions that separate cultures can and must become one (assimilationism) or that they can and must remain separate (cultural pluralism) are rejected in favor of the obvious middle position. Thus, as Banks (1986) tells us:

> The challenge to Western societies and their schools is to try to shape a national culture from selected aspects of traditional cultures co-existing in delicate balance with a modernized post-industrial society. . . . Westernized nation-states will be able to create societies with overarching goals that are shared by all of their diverse groups only when these groups feel that they have a real stake and place in their nation-states and that their states mirror their concerns, values, and ethos. (p. 61)

This is, indeed, a considerable challenge. Once again, our public schools—and particularly our urban schools—must be called on as social agencies for change, as institutions that "dare to build a new social order" (Counts, 1932). The point is this: There is simply no way to tinker with restructuring urban schools and their curricula without encountering the hard work of fundamental social reconstruction. And this hard work necessarily includes encountering and dealing constructively with suspicion and resistance from all interest groups, including minorities. For example, there is little evidence to suggest that white, mainstream society will voluntarily support fundamental changes to serve the interests of minorities. Educators and policymakers can expect, therefore, that minorities may view with considerable distrust (and rightly so) any new proposal that "changes the rules of the 'game' "—be it vouchers, choice, common curricula, alternative assessment practices, site-based management, or restructuring.

This underscores again the importance of critical inquiry and involving all stakeholders in critically questioning *any* proposed educational changes, the human interests served (or not served) by these changes, and, ultimately, how these changes will impact favorably on furthering both educational equity and excellence.

SUMMARY

To be sure, regularities and commonplaces within and among public schools make them familiar places to educational reformers. But a school is *not* a school is *not* a school. In particular, our inner cities and the communities and schools within them present unique challenges for those wishing to significantly alter, improve, and sustain—restructure, if you wish—these places for teaching and learning.

It will take a strategic combination of resources, talent, leadership, and intelligence. It will take committed educators who have and know how to use a varied pedagogical repertoire; who greet individual and group differences as a pedagogical challenge, not as a necessary evil; and who recognize multicultural education, not as a set of topics, but as a process of teaching and learning. It will take educators, parents, and other community constituencies who, in collaboration, are willing to take risks, look at their work critically and constructively, and learn from their mistakes. It will take state and local officials and educational leaders who have the courage, commitment, and foresight to support these kinds of school improvement and renewal activities. And it will take time—much more than usually is allocated to the latest educational bandwagon that rolls through town.

Such will be the brave new world of restructuring urban schools—an agenda not just for the 1990s, but for new generations well into the 21st century.

Note

1. The statistical profiles to follow are based mostly on 1985 analyses published by the Office of Educational Research and Improvement (1988). Additional sources of information include the Carnegie Council on Adolescent Development (1989), Ornstein (1984), and Usdan (1984).

References

Andrews, R. L., & Soder, R. (1987). Principal leadership and student achievement. *Educational Leadership, 44*(6), 9-11.

Banks, J. A. (1981). *Multiethnic education: Theory and practice.* Boston: Allyn & Bacon.

Banks, J. A. (1986). Ethnic diversity, the social responsibility of educators, and school reform. In A. Molnar (Ed.), *Social issues and education: Challenge and responsibility* (pp. 59-77). Alexandria, VA: Association for Supervision and Curriculum Development.

Barth, R. S. (1980). *Run school run.* Cambridge, MA: Harvard University Press.

Bates, R. J. (1984). Toward a critical practice of educational administration. In T. J. Sergiovanni & J. E. Corbally (Eds.), *Leadership and organizational culture* (pp. 260-274). Urbana: University of Illinois Press.

Bennis, W. (1989). *Why leaders can't lead.* San Francisco: Jossey-Bass.

Berman, P., & McLaughlin, M. W. (1978). *Federal programs supporting educational change* (Vol. 8). Santa Monica, CA: RAND.

Block, P. (1988). *The empowered manager.* San Francisco: Jossey-Bass.

Bolman, L. G., & Deal, T. E. (1984). *Modern approaches to understanding and managing organizations.* San Francisco: Jossey-Bass.

Brookover, W., Beady, C., Ford, R., Schmitzer, J., & Wisenbaker, J. (1979). *School systems and student achievement: Schools make a difference.* South Hadley, MA: Bergin.

Carnegie Council on Adolescent Development. (1989). *Turning points: Preparing American youth for the 21st century.* New York: Carnegie Corporation of New York.

Carnegie Foundation for the Advancement of Teaching. (1988). *An imperiled generation: Saving urban schools.* Princeton, NJ: Author.

Charters, W. W., & Jones, J. E. (1973). On the risk of appraising non-events in program evaluation. *Educational Researcher, 2,* 5-7.

Clark, R. M. (1983). *Family life and school achievement: Why poor black children fail.* Chicago: University of Chicago Press.

Cleveland, H. (1985). *The knowledge executive.* New York: E. P. Dutton.

Cohen, M. (1988). *Restructuring the educational system: Agenda for the 1990s.* Washington, DC: National Governors' Association.

Comer, J. (1980). *School power: Implications of an intervention project.* New York: Free Press.

Comer, J. (1988). Educating poor minority children. *Scientific American, 259*(5), 42-48.

Counts, G. S. (1932). *Dare the school build a new social order?* New York: John Day.

Cuban, L. (1984). *How teachers taught: Constancy and change in American classrooms 1890-1980.* New York: Longman.

Cuban, L. (1989). The "at risk" label and the problem of urban school reform. *Phi Delta Kappan, 70*(10), 780-784, 799-801.

Edmonds, R. (1979). Effective schools for the urban poor. *Educational Leadership, 37*(1), 15-24.

Elmore, R. F., & McLaughlin, M. W. (1988). *Steady work: Policy, practice, and the reform of American education.* Santa Monica, CA: RAND.

Fordham, S., & Ogbu, J. E. (1986). Black students' school success: Coping with the "burden of 'acting white.' " *Urban Review, 18*(3), 176-206.

Foster, W. P. (1984). Toward a critical theory of educational administration. In T. J. Sergiovanni & J. E. Corbally (Eds.), *Leadership and organizational culture* (pp. 240-259). Urbana: University of Illinois Press.

Fullan, N. G. (1982). *The meaning of educational change.* New York: Teachers College Press.

Goodlad, J. I. (1975). *The dynamics of educational change.* New York: McGraw-Hill.

Goodlad, J. I. (1984). *A place called school.* New York: McGraw-Hill.

Goodlad, J. I., Soder, R., & Sirotnik, K. A. (1990). *The moral dimensions of teaching.* San Francisco: Jossey-Bass.

Grant, C. A., & Sleeter, C. E. (1988). *After the school bell rings.* Philadelphia: Falmer.

Green, T. F. (1987). The conscience of leadership. In L. T. Sheive & M. B. Schoenheit (Eds.), *Leadership: Examining the elusive* (pp. 105-115). Alexandria, VA: Association for Supervision and Curriculum Development.

Hodgkinson, H. L. (1989). *The same client: The demographics of education and service delivery systems.* Washington, DC: Institute for Educational Leadership, Center for Demographic Study.

Huberman, A. M., & Miles, M. (1984). *Innovation up close: How school improvement works.* New York: Plenum.

Hurn, C. J. (1985). *The limits and possibilities of schooling.* Boston: Allyn & Bacon.

Jackson, P. (1968). *Life in classrooms.* New York: Holt, Rinehart & Winston.

Joyce, B. R., Hersh, R. H., & McKibbin, M. (1983). *The structure of school improvement.* New York: Longman.

Kanter, R. M. (1983). *The change masters.* New York: Simon & Schuster.

Kerr, D. H. (1987). Authority and responsibility in public schooling. In J. I. Goodlad (Ed.), *The ecology of school renewal, 86th Yearbook of the National Society for the Study of Education* (pp. 20-40). Chicago: University of Chicago Press.

Leming, R., & Kane, M. (1981). *Improving schools: Using what we know.* Beverly Hills, CA: Sage.

Lortie, D. (1975). *School teacher: A sociological study.* Chicago: University of Chicago Press.

March, J. G. (1972). Model bias in social action. *Review of Educational Research, 42*(4), 413-429.

McLaughlin, M. W. (1976). Implementation as mutual adaptation. *Teachers College Record, 77*(3), 339-351.

Meyer, J. W., & Rowen, B. (1978). The structure of educational organizations. In M. W. Meyer (Ed.), *Environments and organizations: Theoretical and empirical perspectives* (pp. 78-109). San Francisco: Jossey-Bass.

Office of Educational Research and Improvement. (1988). *Youth indicators 1988: Trends in the well-being of American youth.* Washington, DC: U.S. Department of Education.

Ogbu, J. E. (1982). Cultural discontinuities and schooling. *Anthropology and Education Quarterly, 13,* 290-307.

Ornstein, A. C. (1984). Urban demographics for the 1980s: Educational implications. *Education and Urban Society, 16,* 477-496.

Payne, C. M. (1984). *Getting what we ask for: The ambiguity of success and failure in urban education.* Westport, CT: Greenwood.

Sarason, S. B. (1982). *The culture of the school and the problem of change* (rev. ed.). Boston: Allyn & Bacon. (Original work published 1971)

Sarason, S. B. (1990). *The predictable failure of educational reform: Can we change course before it's too late?* San Francisco: Jossey-Bass.

Schon, D. A. (1983). *The reflective practitioner.* New York: Basic Books.

Scott, W. R. (1981). *Organizations: Rational, natural, and open systems.* Englewood Cliffs, NJ: Prentice Hall.

Sirotnik, K. A. (1987). Evaluation in the ecology of schooling: The process of school renewal. In J. I. Goodlad (Ed.), *The ecology of school renewal, 86th Yearbook of the National Society for the Study of Education* (pp. 41-62). Chicago: University of Chicago Press.

Sirotnik, K. A. (1989). The school as the center of change. In T. J. Sergiovanni & J. H. Moore (Eds.), *Schooling for tomorrow: Directing reforms to issues that count* (pp. 89-113). Boston: Allyn & Bacon.

Sirotnik, K. A. (1990). Equal access to quality in public schooling: Issues in the assessment of equity and excellence. In J. I. Goodlad & P. J. Keating (Eds.), *Access to knowledge: An agenda for our nation's schools* (pp. 159-185). New York: College Board.

Sirotnik, K. A., & Goodlad, J. I. (1985). The quest for reason amidst the rhetoric of reform: Improving instead of testing our schools. In W. J. Johnston (Ed.), *Education on trial: Strategies for the future* (pp. 277-298). San Francisco: Institute for Contemporary Studies Press.

Sirotnik, K. A., & Oakes, J. (1986). Critical inquiry for school renewal: Liberating theory and practice. In K. A. Sirotnik & J. Oakes (Eds.), *Critical perspectives on the organization and improvement of schooling* (pp. 3-93). Boston: Kluwer-Nijhoff.

Slavin, R. (1989). PET and the pendulum: Faddism in education and how to stop it. *Phi Delta Kappan, 90,* 750-758.

Usdan, M. D. (1984). New trends in urban demography. *Education and Urban Society, 16,* 399-414.

Walberg, H. J., Bakalis, M. J., Bast, J. L., & Baer, S. (1989). Reconstructing the nation's worst schools. *Phi Delta Kappan, 70,* 802-805.

Waller, W. (1932). *The sociology of teaching.* New York: John Wiley.

William T. Grant Foundation Commission on Work, Family, and Citizenship. (1988). *The forgotten half: Noncollege youth in America.* Washington, DC: Author.

25

Visions of America in the 1990s and Beyond

NEGOTIATING CULTURAL
DIVERSITY AND EDUCATIONAL CHANGE[1]

Jean J. Schensul

Thomas G. Carroll

The 1990s will see greater ethnic and cultural diversity in the United States than in any other period in American history. The consequences of this diversity are already being felt in every sector of society, including the workplace, the arts, the health care system, and the schools. Educators increasingly are called on to solve problems stemming from different and sometimes conflicting interests of ethnically, culturally, and socially distinct populations. Finding creative solutions to problems of access, cultural and economic inequities, culturally and socially relevant curriculum content, and linguistic differences is the challenge of the 1990s and beyond.

Discussions of cultural diversity are not new to U.S. policymakers. Indeed, much of our sociocultural history revolves around efforts to use education to address cultural diversity in American life. The circumstances of these efforts have varied, and the institutional responses to diversity were notably different, but two of these periods—the 1920s and the 1960s—are particularly noteworthy. During the 1920s the population grew primarily as a result of the immigration of hundreds of thousands of workers and their families from the countries of northern, central, and southern Europe. This period was marked by a national labor shortage and the immigration of predominantly rural agriculturalists and small-town semiskilled wage laborers to the United States. These immigrants differed by country of origin, educational and socioeconomic status, and the resources they brought with them into the country. Entering the class structure at different points, they gained upward mobility at slower or more rapid rates.

National policies favored an assimilationist approach; new arrivals were expected to learn English, obtain work, and integrate into American communities with relatively little help, either from community residents or from public or private institutions. With the creation of child labor laws and enforcement of public education, children were required to attend schools; and those schools functioned to "mainstream" students and to divest them of the language and culture of their countries of origin. Those who did not assimilate were relegated to marginal status, along with most African Americans and American Indians.

The 1960s marked a turning point in American social and cultural life. During the 1960s and 1970s, African Americans, Latinos representing many different ethnic groups, American Indians—and later, women, and special interest groups including gay men and women and the elderly—organized nationally to demand a more equitable distribution of economic and educational resources and better access to positions of power and influence. One important organizing principle was the call for public promotion of group identity and cultural pluralism and the right to claim a cultural legacy lost through earlier periods of assimilation. The primary means for regaining group identity were educational and cultural institutions. This period in American cultural life saw the development of ethnic and women's studies centers; community-based, culturally appropriate service-delivery models; the development of multicultural curriculum modules for use in public school settings; and a vast and rich array of literary and musical productions that continue into the present.

Anthropologists were actively involved in these movements, participating in curriculum development at all levels, supporting new community-based institutions, acting as cultural brokers to link schools more effectively to politically aware and culturally diverse communities, and insisting on a dialogue between unresponsive mainstream institutions and critical African American, Latino, and other politicized groups. The goal was to achieve a more inclusive society that would place high value on cultural diversity while recognizing national commonalities.

A period of retrenchment followed, characterized by neglect of equal opportunity and affirmative action principles at all levels, a concerted attack on selected legislation intended to increase options for women, reduced interest in ethnic studies programs, and an increasing number of incidents of interethnic and interracial prejudice and violence in local communities and on college campuses (Select Committee on Children, Youth, and Families, 1987). This trend has been accompanied by advocacy of a return to a national policy of assimilation to the dominant western European cultural tradition, the result of which may well be increased marginalization of ethnic and minority groups throughout much of the country (Schensul, 1985).

The Effects of Increased Immigration From Third World Countries

In the 1990s, however, this policy of assimilation is being seriously challenged. The increase in demographic diversity during the past decade is the result of immigration from non-European countries, a consequence of the Immigration

Act of 1965 (U.S. Bureau of the Census, 1987). United Nations statistics show more than 14 million foreign-born people living in the United States in the early 1980s, and the number of new immigrants as totaling between 11 million and 13 million for the decade (Population Reference Bureau, 1989, p. 11). "Immigration is now America's major source of population growth," and the racial and ethnic composition of the nation's population will continue to shift (United Way of America, 1989).

New immigrants add skills and productivity to the U.S. labor force. Further, they bring with them the history, cultural wealth, values, and perspectives of their countries and cultural or ethnic reference groups of origin. The children of immigrants bring cultural and linguistic variation to local school systems. In Los Angeles, more than 80 languages are spoken in the school system; in a Connecticut community college, 41 linguistic and ethnic groups are represented.

The largest numbers of immigrants have been "from the Americas and Asia; 34% are from Asia; another 34% are from Central and South America; 16% are from Europe; 10% are from the Caribbean; and the remaining 6% are from other continents and Canada" (Madrid, 1988, p. 9). Hispanic Americans number more than 15 million, with communities in almost every major urban center of the country, and the Asian community has more than doubled in size during the 1980s (Allen & Turner, 1988; Waldrop & Exeter, 1990).

Nearly one in three Americans will fall within the Census Bureau's designation of "Minority" by the year 2000. "By the end of the next century there will be a new majority population in America—a majority of minorities" (United Way of America, 1989, p. 19). This demographic shift has already occurred in some cities where African Americans, Latinos, and Asians constitute an absolute majority of the local population. By the end of this decade, California, Texas, Florida, and New York will all have minority-majority school populations (Hodgkinson, 1989). Overall, the 1990 census shows a 44% increase among Hispanics since 1980; a 65% increase among Asians, and a 16% increase in the black population.

Although in some cases the income differential among these groups and between them and white populations has narrowed, in others it has widened. For example, the gap between education levels of minority and mainstream adults has widened (among Asians, 53.3% hold managerial positions, and 34% hold a college degree; among blacks, only 11% hold college degrees; and among Hispanics, only 8.6%) (United Way of America, 1989). In addition, the number of minority students entering and completing college and graduate school has declined absolutely and proportionately; and the income level of black, non-Cuban Hispanics and Native American peoples relative to other groups (controlling for inflation) has declined (American Council on Education, 1988; Cutler, 1989). Finally disaggregation of population figures indicates within-group class differences. For example, Koreans, Taiwan Chinese, and Vietnamese migrating to the United States prior to 1975 were generally well educated, from middle-class families, and integrated with supportive community organizations. Hmong, Cambodian, and some Filipino immigrants arriving after 1975 were less well educated, with fewer economic and community resources, and less able to integrate into the economic mainstream (United Way of America, 1989).

Projections for the next two decades indicate that the numbers of African American, Asian, and Latino youth will increase substantially. Schools and workplaces will be increasingly multiethnic, and more responsive and inclusive educational systems and support programs must be put in place to strengthen and maintain the social life of our communities and a competitive position in the world economy (Semerad, 1985).

The members of these ethnic and national groups, and those who won the right to claim their identity through the social justice movements of the 1960s are no longer willing to accept the alternatives of assimilation or marginalization. The impact of these groups on communities, the workplace, and our educational institutions is already profound. Our communities are increasingly multiethnic. As they change in composition, schools, workplaces, and other institutional settings must change, as well, to incorporate the cultural patterns, predominant values, and social needs of a multiethnic population. As ethnic and national groups work to maintain and redefine their identity in the 1990s, we are entering a new period in American cultural life and a new set of challenges for anthropologists, educators, and others concerned with the creation of institutions that respond to cultural diversity.

Implications

Demographic trends will move national institutions to respond in new ways. Because we have no clear-cut, systematic, legislated national cultural policy, our responses, as in the past, will be selected from among the approaches identified above: assimilation, marginalization of those perceived as culturally different, and cultural pluralism and cultural transformation in a national setting. Our institutions can respond by favoring some groups over others or by using a more inclusive approach to education, employment, and public recognition of sociocultural diversity. We may support class differences or attempt to reduce disparities across groups and classes. We may choose to train new employees to conform to current work environments, or our work environments may be organized to respond to the diversity of new employees.

Our communities can respond by favoring use of first language in informal settings, teaching of multiple languages and multilingual, multicultural school and public institutional environments, or by supporting the use of a single national language and set of cultural principles and behaviors. Our educational institutions can draw on and enjoy the literary, media, and expressionist traditions of most of the countries of the world or ignore them. They can respond by promoting multiethnic perspectives in literature, the music and the arts, history, philosophy, and the social sciences (Simonson & Walker, 1988) or by deferring to the Anglo European perspectives of such writers as Allan Bloom (1987) and E. D. Hirsch (1987).

Note

1. This chapter is adapted from Introduction: Visions of America in the 1990s and beyond (August 1990). *Education and Urban Society, 22* (4), 339-345. Copyright © 1990, Sage Publications.

References

Allen, J. P., & Turner, E. J. (1988). Where to find the new immigrants. *American Demographics, 10*(9), 22-27.

American Council on Education. (1988). *One third of a nation: A report of the Commission of Minority Participation in Education and American Life.* Washington, DC: American Council on Education and Educational Commission of the States.

Bloom, A. (1987). *The closing of the American mind.* New York: Simon & Schuster.

Cutler, B. (1989). Up the down staircase. *American Demographics, 11*(4), 32-41.

Hirsch, E. D. (1987). *Cultural literacy: What every American needs to know.* Boston: Houghton Mifflin.

Hodgkinson, H. (1989). *The same client.* Washington, DC: Institute for Educational Leadership, Center for Demographic Policy.

Madrid, A. (1988). Quality and diversity. *AAHE Bulletin, 40*(10), 1-12.

Population Reference Bureau. (1989). *America in the 21st century: A global perspective.* Washington, DC: Author.

Schensul, J. (1985). Cultural maintenance and cultural transformation: Educational anthropology in the eighties. *Anthropology and Education Quarterly, 16*(1), 63-68.

Select Committee on Children, Youth, and Families. (1987). *Race relations and adolescents: Coping with new realities* (Hearing before the Select Committee on Children, Youth, and Families, House of Representatives). Washington, DC: Government Printing Office.

Semerad, R. D. (1985). *Work force 2000.* Washington, DC: U.S. Department of Labor.

Simonson, R., & Walker, S. (1988). *Multi-cultural literacy: Opening the American mind.* St. Paul, MN: Graywolf.

United Way of America. (1989). *What lies ahead: Countdown to the 21st century.* Alexandria, VA: United Way Strategic Institute.

U.S. Bureau of the Census. (1987). *Population profile of the United States 1984/85* (Special Studies Series P-23, No. 150). Washington, DC: Author.

Waldrop, J., & Exeter, T. (1990). What the 1990 census will show. *American Demographics, 12*(1), 20-33.

26

The Evolution of a University/Inner-City School Partnership

A CASE STUDY ACCOUNT

Andrea G. Zetlin

Kathleen Harris

Elaine MacLeod

Alice Watkins

Recently there has been much interest in the creation of professional development schools that would serve as improved teacher training sites. The Holmes Group, a consortium of 97 major research universities working to reform teaching and teacher education, has called for universities and public schools to work together to establish a network of "clinical" schools to improve student learning, revitalize veteran teachers, introduce prospective teachers to the field, and provide a laboratory for research in education (Watkins, 1990). Recognition of the need of such educational partnerships with the dual mission of improving how we teach children and how we prepare future teachers was noted as early as 1900 and voiced repeatedly throughout the 20th century. Dewey, in his book *School and Society,* first cautioned against the lack of a "vital connection" between the "lower" and "higher" parts of our education system (cited in Stoloff, 1989). Later Conant (1963) and then Sarason, Davidson, and Blatt (1986) echoed frustration with the general failure to connect teacher education and schools. Most recently Goodlad (1990a, 1990b) has endorsed the creation of school/university partnerships to provide the "supportive infrastructure" for improving the clinical component of teacher training programs, as well as for reforming schools.

Educators have suggested a number of elements needed for school/university partnerships to succeed, including the clear establishment of common goals, the development of mutual trust and respect among participants, open lines of communication among partners, and shared responsibility and accountability

(Stoloff, 1989). Trubowitz (1986) has described 8 stages that such unions must pass through before they achieve viability as partnerships. Stage 1 is consumed with hostility and skepticism as the partners get to know each other and vent their frustrations with past school-university ventures. Stage 2 is characterized by a lack of trust as the partners build mutual confidence through joint efforts and role merging. Stage 3 is a period of truce with equal participation in school-based projects. Stage 4 brings mixed approval as short-term successes are recognized and as individuals who do not flourish within collaborations find other projects. Stage 5 is characterized by acceptance by both the school and university communities as mutual benefits are realized. Stage 6 is a time of regression as the original collaborative vision is blurred through attrition, faculty promotion, and changes in local funding. Stage 7 brings new members with a transfusion of fresh ideas leading to renewal. Stage 8, the final stage, is characterized by continuing progress in the collaborative effort. Although no time frame is suggested, one can imagine that such a process must evolve slowly over an extended period before Stage 8 is achieved.

Drawing on a combination of ethnographic and interview data, this study provides in-depth information on the development of one collaborative partnership between a university and an inner-city public school. Through a case study approach, the stages involved in the evolution of this partnership and the factors that hindered and facilitated the process have been documented. The methods used to gather information were ethnographic observations of teachers and classrooms and interviews with teachers and university and district administrators.

Ethnography is an important research strategy for understanding the complexity of the psychosocial and cultural forces at work in the setting being studied, as well as for preserving chronological flow (Miles & Huberman, 1984). By systematically observing the school environment, we hoped to learn "reality" from the point of view of the participants involved in the project. We sought to understand the meanings of actions, practices, and events from the teachers, administrators, and university faculty working in the setting. Additionally the longitudinal design of this model provided the opportunity to build working relationships with those we studied. As rapport developed, it provided us with access to their beliefs and attitudes—information that would be difficult to obtain in other ways (Edgerton & Langness, 1978).

The project reported in this chapter is part of a parent project that began during the 1985-86 school year. At that time two California State University School of Education faculty members developed two demonstration training sites in inner-city schools in Southern California. The School of Education is one of the largest and most comprehensive in the country, offering 26 credential programs and 6 master's degree programs. Annually more than 1,400 students earn teaching credentials, many of whom are subsequently employed in the surrounding school districts that serve large numbers of low-income and minority students. The public schools involved in the project—both elementary—were in a region serving almost 60,000 students and featuring majority minority populations. One school had approximately 80% African American and 20% Hispanic students; the other school had approximately 80% Hispanic and 20% African American students.

The overall goals of the project were to (a) train prospective teachers in the kinds of schools where they would most likely be employed, (b) develop a cadre of master teachers who would be good role models, and (c) help regular classroom teachers work more effectively with students who display learning difficulties. During the course of the project, student teachers were regularly placed in these schools, university courses were offered at the school sites to strengthen teaching skills, university faculty made themselves available as resources for teachers and student teachers, and observational research on effective teaching in inner-city classrooms was conducted. After 3 years, changes in personnel at both school sites, including principal and teacher transfers, led to a decision to move the project to a more stable school site.

In 1989 another elementary school in the same region was selected as the new training site. Here the teaching staff was relatively stable: There were typically only three or four teacher transfers each year, and the principal, though newly assigned to the school, had been enthusiastically involved in the project as vice principal at one of the original school sites. This new program was modeled after the program that had been established at the other schools, and this time the project was carefully documented to allow for expansion and replicability.

The school served kindergarten through sixth-grade students and in October 1989 had an enrollment of 805 students—86.6% African American, 12.8% Hispanic, 0.5% Filipino, and 0.1% Anglo. The student population was in transition, and increasing numbers of Hispanic students were being enrolled throughout the school years. On the California Test of Basic Skills (CTBS-U), students across the grades ranked between the 20th and 30th percentiles within the school district in reading, math, and language arts, and third and sixth graders ranked between the 2nd and 16th percentiles statewide.

Two university faculty members—one from the Division of Curriculum and Instruction and one from the Division of Special Education—spent up to 3 days a week at the school, supervising student teachers, acting as resources for teachers, students, and administrators, and conducting university courses. Both constructed detailed field notes of their interactions with staff and students. A third university faculty member became involved at the start of the school year and assumed the role of participant observer, "hanging around" the school campus, observing classrooms, talking with teachers, accompanying classes on school trips, and participating in university courses offered at the school site. Detailed field notes were written of her observations as well. The participant observer also conducted formal interviews with 14 of the 38 teachers, 2 at each grade level who varied in teaching experience (kindergarten through sixth grade). Interviews were conducted also with 3 university administrators, 2 of whom were involved in the project since its inception, and the Dean of the School of Education, who assured its continued funding, as well as the region superintendent and the school principal. All were asked to describe their goals/visions for the partnership, as well as their attitudes toward having student teachers, their expectations for the university, what teachers' greatest needs were, and their ideas about effective teaching in an inner-city school.

Results

Examination of the ethnographic and interview data sources revealed two general themes: (a) the project goals from the perspectives of university and school district personnel and (b) the factors that seemed to affect the partnership effort. Each theme is discussed in detail.

Perspectives of Project Goals

Although university and school staff had met formally before the project commenced and had held lengthy discussions concerning the project aims and expectations, interview data revealed that the various parties held different perceptions of the project goals. These varying perspectives are described, as well as the actions taken toward fulfillment of these "player-specific" goals. The university project faculty perceived the following three goals for the project: (a) to help teachers improve their teaching skills, (b) to implement a collaboration model to help teachers (from regular and special education) work more effectively with students who displayed learning difficulties, and (c) to train teachers to be master teachers who could model innovative teaching techniques and collaborative strategies with colleagues. To accomplish these goals, university project faculty jointly conducted numerous activities—that is, offered university courses at the school site for the teachers, made themselves available as resources for teachers and provided classroom assistance, supervised student teachers in the classrooms, requested and accepted assistance from teachers with regard to supervising student teachers, and conducted student teaching seminars in conjunction with the master teachers.

The region superintendent of the district and the school principal had one primary goal for this project: to develop a pool of potential teachers. Thus they viewed the role of the university faculty as placing and training student teachers who then would be potential teachers for the district. Although neither shared the goals expressed by project faculty, they did provide visible support for one of the overall goals of the project—that is, to train prospective teachers.

The teachers at the school, in keeping with their view of university faculty as resources, perceived three goals for the project: (a) for university faculty to act as liaison between the administration and teachers, (b) for university faculty to help teachers develop innovative instructional strategies, and (c) for university faculty and school staff to work collaboratively to solve school problems. They wanted to accomplish these goals by having courses at the school and by receiving resources and demonstration lessons, as well as moral support, from the university faculty. Although the teachers seemed to realize that the structure of receiving university support necessitated the placement of student teachers, most viewed the placement of student teachers not as an asset, but rather as an added responsibility/burden.

Factors Affecting Collaboration

Both positive and negative factors influenced the collaboration efforts. As described above, a major deterrent to project efforts was the discrepancies between university and school expectations for the project. In part because the project goals were not the same among all project participants, the actions taken or proposed by one party were not always supported by the other parties. For example, the activities conducted by university faculty—courses and helping teachers improve and refine their programs—were not considered a priority by the principal. Rather, the principal narrowly viewed the university faculty as supervisors of student teachers. Also, although the goals of the teachers and the university project faculty centered on restructuring and revitalizing the school, the collaborative approach of the university faculty was not initially supported by the teachers or the school administration.

Another major deterrent was the many unscheduled changes at the school site and with university personnel. At the school, changes occurred in administrative and teaching staff, student population, and classroom rosters. The new principal was transferred a few months into the project, and a former "unpopular" principal was reassigned, resulting in strained relations between administration and staff. The return of the former principal, as well as the school district's decision to shift the school to a year-round schedule, resulted in an uncharacteristic number of requests for teacher transfers, thus destabilizing a previously stable teaching staff. The student population also changed, becoming proportionally less African American as more Hispanic students moved into the community and enrolled in the school. Classrooms had to be reorganized at least three times during the school year to accommodate class size, as well as increasing numbers of limited English proficient (LEP) students. Also unplanned changes occurred in university faculty members involved in the project. Of the two original faculty members, one passed away and one was shifted to an administrative position at the university and was no longer directly involved in the project. Two new faculty members were placed at the school site, and each added his or her own agenda to the project goals.

A third deterrent was strained relationships among participants. In addition to strained relations between the principal and teachers already described, a few teachers showed some reluctance to becoming involved with university faculty. Some teachers were concerned about disruptions caused by having student teachers and university supervisors in their classrooms; others were unwilling to take on the extra work of having a student teacher for the minimal financial compensation offered; and some were reluctant to try new instructional techniques or to question the effectiveness of their teaching style.

Despite these negative factors, many positive outcomes resulted from the efforts of all parties to make the partnership a success. First, an increasing number of teachers began working with the university faculty. Excellent rapport was established between teachers and university staff, and the result was shared resources and the implementation of innovative teaching strategies in classrooms. Together they also succeeded in developing strategies for working more effectively with the principal. Second, through a course offered by university faculty, participating teachers took the initiative for working on schoolwide problems through a collaborative effort; that is, prior to university faculty involvement, collaborative activities among teachers tended to occur randomly. Through the course offered by university faculty, collaboration was viewed as a vehicle through which school problems could be addressed. In this course, teachers identified needs, a problem-solving process was taught and modeled, teachers brainstormed possible solutions and decided on which ones to attempt, and teachers implemented solutions and evaluated results. As a consequence, those teachers who took the university course approached other teachers in the school. Through their collaborative efforts (a) procedures were developed to implement a schoolwide peer tutoring program, (b) general educators started using special educators in the school as resources for students with behavior and/or learning problems, (c) a program for student attendance was developed, and (d) teachers worked together to develop solutions to individual problems they were experiencing with students.

In addition to the positive effects for teachers and students at the school, the university teacher training program also benefited from the collaboration. Master teachers were no longer selected by the school principal. A university course to develop skills on daily classroom supervision of student teachers was offered regularly, and university faculty negotiated with the principal to select master teachers from participants in the course. The traditional student teaching model also was modified, and student teachers worked with their master teachers and university supervisors in different ways. Student teachers were not involved in only one classroom with one master teacher: Student teachers spent time in other classrooms to observe successful learning centers or to receive additional guidance from other master teachers; they participated in weekly seminars led by university faculty and master teachers to discuss strategies for dealing with schoolwide challenges. In addition, university faculty collaborated across disciplines (elementary and special education) and jointly supervised teachers in training for an elementary credential. Thus the university faculty combined their expertise in helping these student teachers understand the problems experienced by inner-city children and methods to help deal with those learning and/or behavior problems. And the school district was positively affected by the project. Over half of the student teachers trained at the school site, on receipt of their credentials, were hired to teach in the district's schools and therefore accomplished the school and district goal of increasing the pool of potential teachers.

Discussion

Although the professional literature contains many calls for universities and schools to work as partners to improve teaching and revitalize teacher education programs, a great deal of preparation must go into this collaborative effort, as indicated by the data reported in this study. Public schools are struggling with many problems, the most notable being frequent changes in administrative and teaching staff, unstable classroom rosters, growing numbers of culturally and linguistically diverse students, increasing referrals for special education services, and unrelenting budgetary cutbacks. Within this atmosphere, school personnel are wary about the benefits of a partnership and question the payoff to them and their schools, given the degree of commitment and energy demanded. Further, universities remain reluctant to embrace the challenges of schools and view as separate endeavors their mission to train teachers and the need to reform/restructure today's schools.

It is within this reality that universities must approach schools to enlist their support to work together. Partnership is a developmental process, and although, as Goodlad (1988) suggested, cooperation between schools and universities should be a natural process as both are in the same business—education—very real forces are working against successful partnerships. Each school and university will bring to the partnership its own set of problems requiring a unique set of solutions. This study suggests that partnerships can result in positive outcomes, although the developers of such unions must be flexible. Each proposed partnership will need to be studied closely to determine what strategies will most likely effect change, given the parameters of the situation.

Much work is still needed at this school to achieve all of the goals established at the outset of the project. School and university participants continue to hold different perceptions of the scope and promise of the project. Thus efforts to share responsibility for tasks and to maximize the interests of all participants in the project—achievements of successful collaborations—have yet to be fully attained. Within the context of Trubowitz's (1986) model, although trying challenges are ahead, the partnership is most likely at Stage 5—acceptance by school and university staff as mutual benefits are realized. But after 1 year of working together at this school, and with a great deal of effort and energy having been expended, participants at both the university and school level are in agreement that progress—though slow—continues to be made toward the successful evolution of a partnership.

References

Conant, J. B. (1963). *The education of American teachers.* New York: McGraw-Hill.

Edgerton, R. B., & Langness, L. L. (1978). Observing mentally retarded persons in community settings: An anthropological perspective. In G. Sackett (Ed.), *Observing behavior: Vol. 1. Theory and applications in mental retardation* (pp. 335-348). Baltimore: University Park Press.

Goodlad, J. I. (1988). School-university partnerships: A social experiment. *Phi Delta Kappan, 69,* 77-80.

Goodlad, J. I. (1990a). Better teachers for our nation's schools. *Phi Delta Kappan, 72,* 185-194.

Goodlad, J. I. (1990b). *Teachers for our nation's schools.* San Francisco: Jossey-Bass.

Miles, M. B., & Huberman, A. M. (1984). *Qualitative data analysis: A sourcebook of new methods.* Beverly Hills, CA: Sage.

Sarason, S. B., Davidson, K. S., & Blatt, B. (1986). *The preparation of teachers: An unstudied problem in education* (rev. ed.). Cambridge, MA: Bookline.

Stoloff, D. (1989, March). *The social context of educational partnerships: A semantic review.* Paper presented at the Annual Meeting of the American Educational Research Association, San Francisco. (ERIC Document Reproduction Service No. ED 311 563)

Trubowitz, S. (1986). Stages in the development of school-college collaboration. *Educational Leadership, 43,* 18-21.

Watkins, B. T. (1990, February 7). Education-school reform group set to endorse plan that would alter teacher training, public schools. *Chronicle of Higher Education,* pp. 15, 20.

27

Race, Urban Politics, and Educational Policy Making in Washington, DC

A COMMUNITY'S STRUGGLE FOR QUALITY EDUCATION

Floyd W. Hayes, III

In the 1950s and 1960s, the movement for public school desegregation came to a head in the context of a convergence of a budding challenge to educational progressivism's theory and practice of societal guidance and the emerging struggle for African American civil rights. Both efforts challenged the administrative authority of educational professionals, experts, and theorists. At the turn of the century, a coalition of educational reformers, local business elites, and the educated middle class succeeded in establishing urban school bureaucracies headed by professional educational managers. The Progressive Era of educational reform is best distinguished by its enthusiasm for a "science" of education, which emphasized expertise, efficiency, and a growing organizational complexity of urban school systems. Because administrative progressives believed that people could manage their own destiny through the conscious application of science to social difficulties, applied educational research—expert knowledge, rather than common sense—came to be viewed as a foundation for social engineering (Tyack & Hansot, 1982). These arrangements, however, severely restricted popular participation in the urban educational policy-making process.

By the 1950s, public discontent mounted with educational professionals and theorists who were accused of lowering educational standards and introducing into the public school curriculum courses that many parents and citizen groups interpreted as replacing reading, writing, and arithmetic with fun, fads, and frolicking (Washington, 1969). Many concerned parents were encouraged and sustained by such widely read and discussed books as *Why Johnny Can't Read,* by Rudolf Flesch (1955), and *Education and Freedom,* by Hyman Rickover (1959). The authors of these texts harshly criticized the limitations of educational progressivism and school professionals and experts.

The Supreme Court's epic 1954 *Brown v. Board of Education* decision, which outlawed public school segregation, and the intensifying civil rights movement of the 1950s and 1960s helped set in motion forces in many urban African American communities that, in turn, came to challenge fundamentally the prerogatives and intent of educational professionals and the exclusionary bureaucratic policy-making process of urban public school systems. For example, in Washington, DC, African American community organizations and activists clashed with educational professionals over the implementation of the school system's desegregation policy of tracking. In opposition to this policy strategy, citizen groups and participants demanded traditional quality education: the distribution and mastery of the fundamental tools of knowledge, academic motivation, and good character building.

The broad goal of this chapter is to examine the politics of educational policy making in Washington, DC, during the 1950s and 1960s. More specifically, the aim is to chronicle the community struggle against the central administration's tracking-policy strategy, a policy strategy that many community participants saw as more damaging to African American children than school segregation in the Old South was. Until 1968, when Washington's first elected school board took office, the city's public schools were the only major federally controlled educational system in the nation. Hence the politics of school desegregation in Washington is significant because, as the nation's capital, the city generally was considered a bellwether for the rest of the country (LaNoue & Smith, 1973). Moreover, the specific events recounted in this chapter have been largely overlooked in the scholarly literature on educational politics in Washington, DC.

Primary data sources for the research presented here include official reports and documents of the District of Columbia public school system. A further step in the process includes an in-depth interview with a key participant in the city's early educational politics examined in this study.[1]

Demographic Transformation

Changing demographic trends in the nation's capital help explain the politicization of educational policy making as the desegregation process unfolded. From 1930 to 1950, while the proportion of the African American population in the Washington metropolitan area as a whole decreased slightly (from 24.9% to 23.6%), the proportion of African Americans in the District of Columbia itself increased from 27.1% to 35.7%. Percentages, however, can be deceiving. In actual numbers the African American population for the Washington metropolitan area increased from 167,409 to 345,954 persons, and the actual numbers of African American District residents rose from 132,068 to 289,600 persons during that 20-year period. A more dramatic population change occurred during the next decade. By 1960, although the proportion of African Americans for the Washington metropolitan area only increased 25.0%, this represented an increase in actual numbers to 502,546 persons. Significantly the proportion of the District's African American population rose to 55.2%, or to 427,100

Table 27.1 Enrollments by Race, Washington, DC, Public Schools, 1951-1981

Year	Total Enrollment	African American Enrollment	White Enrollment	Percentage African American	Percentage White
1951	95,932	50,250	45,682	52.4	47.6
1956	107,312	72,954	34,358	68.0	32.0
1961	127,268	103,804	23,464	81.6	18.4
1966	146,644	133,275	13,369	90.9	9.1
1971	142,899	136,256	6,643	95.4	4.6
1976	125,908	119,814	4,406	95.2	3.5
1981	94,425	89,160	3,321	94.4	3.5

SOURCE: Adapted from *The Burden of* Brown: *Thirty Years of School Desegregation* (p. 16) by R. Wolters, 1984, Knoxville: University of Tennessee Press.
NOTE: Percentages do not add up to 100 after 1973 because Indians, Asians, and Hispanics are counted separately.

persons (Passow, 1967). During the next two decades the proportion and number of Washington's African American residents would continue to mount as increasing numbers of whites relocated to surrounding suburban areas.

The transformation of Washington's racial composition, begun 20 years before the schools were desegregated, set in motion changes in the racial makeup of the city's public school enrollment. A more dramatic transformation took place following school desegregation. Between 1949 and 1953, white enrollment had declined by about 4,000 students; between 1954 and 1958, that enrollment fell by nearly 12,000 students. Overwhelmingly members of a rising middle class of professionals and managers employed chiefly in the burgeoning federal bureaucracy, many whites left Washington to enroll their children in predominantly white public schools in the nearby affluent suburbs in Maryland and Virginia.

The growth of African American enrollment in the District's public schools also was striking. As Table 27.1 portrays, from 1951 to 1956, African American enrollment rose from 52.4% to 68.0%, or from 50,250 to 72,954 students. Five years later, African American students constituted 81.6% of the public school enrollment. There were then 103,804 African American students in a total public school enrollment of 127,268.

Desegregation and the Struggle Against Tracking

In the aftermath of the Supreme Court's decisions on May 17, 1954, in *Brown v. Board of Education* and *Bolling v. Sharpe* (which terminated the dual-school system in Washington, DC), and along with some pressure from the White House, the District government moved to desegregate its public schools. Because of the city's unique political history, its schools were the only major federally controlled public educational system in the nation, and the Eisenhower administration expected Washington to serve as a "showcase" to the nation in making an orderly and prompt transition to a desegregated school system (LaNoue & Smith, 1973).

In 1957, following the departure of Superintendent Hobart Corning, the Washington school board appointed Associate Superintendent Carl Hansen as the new superintendent to lead the school desegregation campaign. Hansen was a Nebraska native who had graduated from the state's university in Lincoln; he had obtained an education doctorate at the University of Southern California and had been a high school teacher and later a high school principal in Omaha before coming to Washington. Thus Hansen was an experienced educator who blended managerial expertise with a strong commitment to desegregated schooling. Indeed, he became one of the nation's foremost spokespersons for school desegregation (Wolters, 1984).

At the commencement of the 1956 school year, the District's educational administration instituted a new program of tracking high school students. Hansen, then associate superintendent responsible for high schools, played a major role in the new policy's implementation. The stated objective of the tracking policy was to allow students with similar academic ability to work together regardless of racial or economic background. On the basis of motivation, teachers' evaluation of students' past classroom performance, and scores on achievement and aptitude tests given by educational researchers in the school system's pupil appraisal department, students were steered into one of four separate curriculum tracks: honors, college preparatory, general, and basic (remedial). High-performing students with IQ scores above 120 were to be placed in a special honors course of study; students with IQ scores between 120 and 75 were assigned either to the college preparatory or the general track if they were performing near grade level; and students with IQ scores below 75 were put into the remedial or basic track. This latter track included students with mental retardation and those performing 3 years below grade level with IQ scores between 85 and 75 (U.S. Congress, 1966; Wolters, 1984).

While the tracking system did result in a modicum of racial and social class mixture in the honors and college preparatory categories, the general and basic tracks, which at the outset consisted overwhelmingly of African American students from economically poor families, came increasingly to include African American youngsters from professional classes as well. From the beginning, the tracking system was the target of numerous critics who expressed the view that students were being labeled and channeled by educational professionals and researchers into a narrow groove from which they would never exit (LaNoue & Smith, 1973; J. R. Mazique, personal interview, March 5, 1986; Wolters, 1984).

Starting with the 1959 school year, on Superintendent Hansen's recommendation the Washington school board extended the tracking system by including elementary and junior high schools. The new policy established three tracks. Beginning with the fourth grade, youngsters with low IQ scores and a significant level of mental retardation were channeled into the basic track. Students ranking normal on aptitude tests and classroom performance were steered into the

general track. Students with IQ scores above 125 and high classroom performance were directed to the honors track (Wolters, 1984).

From its inception, the administratively driven policy of tracking elementary and junior high school students caused conflict. And, as the negative
consequences of the tracking strategy became more and more apparent, broad
community opposition increased precipitously into the 1960s. Three school
board members expressed the apprehension that categorizing youngsters in the
basic and general tracks at the beginning of their educational careers was premature and might retard normal and later academic growth and development
(Wolters, 1984). Community groups protested that the tracking system was a
new tactic for labeling students and for reinstating a structure of dominance
and African American subordination more harmful than racial segregation in
the Old South. Many charged that when educational managers, experts, and
researchers applied to African American students such terms as *culturally deprived,*
economically disadvantaged, victims of developmental disability, or *perma-*
nently handicapped by degenerative evolution, the implication of inherent racial
inferiority would result. Moreover, community participants held the view that
these characterizations helped legitimate a theory and practice of withholding
basic education from African American students labeled *uneducable.* Critics
charged, then, that the tracking system early excluded African American youngsters from academic preparation and that the educational specialists and managers then tested these students on what they had not been taught. This process
allowed educational professionals to achieve a self-fulfilling prophecy (J. R.
Mazique, personal interview, March 5, 1986).

In the early 1960s, community discontent continued to mount as citizen
groups and activists charged that tracking denied African American children a
quality education or even a basic one. Community spokespersons intensified their
criticism of the tracking system, arguing that steering children in low and middle
tracks during elementary school would thwart subsequent educational growth.
Additionally it was reported that in some cases African American children were
summarily defined as mentally retarded without being examined and then
automatically channeled into the low track (U.S. Congress, 1966; J. R. Mazique,
personal interview, March 5, 1986). Finally, angry African American District
residents demanded that all elementary school students, regardless of racial and
class backgrounds, be given a basic education.

In the fall of 1960 the central administration responded to community
concerns by establishing the Amidon School, an experiment in integrated quality
education (Hansen, 1960). The Amidon Plan proved to be a controversial undertaking. Superintendent Hansen announced that the new school, located in the
extensively redeveloped quadrant of southwest Washington, was to be an integrated school that provided a program of educational fundamentals. Students
from all over the District were admitted on a first-come, first-served basis. Superintendent Hansen stated bluntly the guiding assumptions of the Amidon Plan:

> If you teach children directly and in a highly organized way, they will learn better
> and faster and will, if teaching is consistent with what is known about the nature

of learning, grow wholesomely, developing confidence as they acquire competence and gaining in self-respect as they accomplish difficult objectives. (Hansen, 1960, p. 6)

Starting with the first grade, Amidon's curriculum stressed basic subjects—spelling, penmanship, arithmetic, and reading (taught by the traditional phonic method). Grammar, normally taught in the sixth grade, was introduced in the third grade, together with French and Spanish. History and geography were taught individually and not included in "social studies." Although students were grouped together on the basis of similar academic achievement, tracking was not implemented in the Amidon Plan. There were no school parties, no orchestra, and no student government; field excursions also were minimized. In this academically demanding program, homework was the rule (Hansen, 1960).

Controversy, however, could not be avoided. Many liberal educational reformers and theorists condemned the Amidon School as a capitulation to conservatism. Some charged that the new school might succeed only as a result of selecting students from middle-class families who prized quality education and were willing to transport their children personally to and from the school. Other liberal educational reformers attacked the Amidon School's subject-centered curriculum and the institution's emphasis on structured student activity. However, when Hansen later announced that a staggering 94% of Amidon's students—a substantial number of whom came from low-income and working-class families—scored above the national average in both reading and mathematics, many educational experts were put on the defensive (U.S. Congress, 1966). A growing number of parents and community groups argued that the Amidon model of quality education should be implemented throughout the Washington public school system.

By the mid-1960s the issue of public school integration polarized the District and the nation. Many angry African American parents and community organizations charged that as public school integration was pushed forward at all costs as the primary goal of education by educational professionals and even civil rights proponents, good classroom teaching, academic motivation, and positive character development in the public schools increasingly were set aside. Because the federal government controlled the Washington public school system, many African American and white community groups and civic activists lobbied various congressional committees, appealing for improvement in the quality of education for all students in the District's schools.

The League for Universal Justice and Goodwill was one of the community organizations most critical of the tracking system. A grassroots organization headed by the Reverend Walter A. Gray, the league consisted mainly of Baptist and Methodist ministers, African American parents, and an assortment of African American civic groups, including the National Capital Voters Association and the University Neighborhoods Council. The league's primary goal was not particularly racial integration, but rather enhancement of the quality of education in Washington's schools. From its founding in 1960 to 1965, the league continually lobbied local and national policymakers to institute effective measures to stem the tide of educational decline and decay in the District's public schools (J. R. Mazique, personal interview, March 5, 1986).

In 1965 the league petitioned the U.S. House of Representatives Committee on Education and Labor, then chaired by Congressman Adam Clayton Powell, to investigate racial discrimination and the tracking system in the Washington public schools. Eventually 125 leaders of civic, religious, professional, and fraternal organizations signed the September 13, 1965, petition. As a direct result of the league's petition, the House Committee on Education and Labor's Task Force on Antipoverty in the District of Columbia held a series of public hearings in October 1965 and January 1966 to look into the condition of the Washington public schools. Significantly the league's important petition (a formal request by a local community organization) cannot be found in the early pages of the congressional report of the hearings; rather, it is deeply embedded in the more than 850-page document (U.S. Congress, 1966, pp. 260-261).

The petitioners leveled a general accusation of racial discrimination against the District's public school system. Additionally, the petitioners made four specific charges: (a) The tracking system denied equal educational opportunity, (b) school funds were allocated inversely proportional to needs, (c) there was a growing trend toward establishing an all-white administration in the areas of curriculum development and educational policy making, and (d) the District school system flagrantly attacked the personal dignity of African American students, which resulted in rising community hostility and increasing dropouts and delinquency. In addition to urging the congressional committee to undertake a fuller study of the petitioners' complaints, the petition contained a detailed description and elaboration of the petitioners' four major charges against the school system (U.S. Congress, 1966).

The Reverend Gray appeared before the congressional task force and submitted a lengthy written statement that contained a powerful indictment of the tracking system and Superintendent Hansen. The Reverend Gray labeled the tracking system "programmed retardation" and called Hansen "a promoter of educational colonialism and cultural supremacy, the manager of the unfair distribution of funds, violator of public trust and inaugurator of the mass destruction of [the] self-image" of Washington's African American students (U.S. Congress, 1966, p. 266). Gray denounced the tracking system and predicted the progressive deterioration of the District's schools if authoritative measures were not taken to reverse trends he considered disastrous.

The Reverend Gray concluded his statement with a strong recommendation for the immediate abolition of the tracking system and the dismissal of Superintendent Hansen. In addition, he called for a prime emphasis on quality education; a uniform curriculum; annual performance ratings for teachers; the encouragement and promotion of qualified African American teachers; a system of public financial accountability to ensure against discriminatory school-funding policies favoring more affluent schools; the termination of the policy of assigning staff personnel and other professionals to nonteaching services (e.g., testers, counselors), resulting in a spiraling school budget; and the end of the school system's theories and practice of racial and cultural supremacy and attacks on the personal dignity of African American students (U.S. Congress, 1966).

Other league members also spoke during the hearings and directly challenged the administrative hegemony of the District's public schools. Significantly Mrs. Jewell Mazique, a parent and a consultant to the league, expressed the views of many ordinary citizens when she exposed the limits of professional expertise and stated the lay public's role in educational policy making. She asserted:

> Mr. Chairman and members of this committee, we believe that conditions in our local schools, where moral development has declined, and where the quality of education obtained has deteriorated (in spite of rising budgets and Federal aid), command a public airing and an honest investigation of what is wrong. *If the administrators, educational theorists and other professionals cannot supply the answer, then it becomes the public's duty to assist.* We cannot continue indefinitely the downward trend toward imminent collapse of public education and expect our country to maintain its place in the sun. (U.S. Congress, 1966, p. 266)

In her lengthy statement, Mrs. Mazique provided a critical analysis of the tracking system. She, too, referred to tracking as a system of "programmed retardation," exposing the fact that through a kind of unwritten law, children were placed in tracks at the end of kindergarten after the administration of "readiness" tests and evaluation by the school principal, teachers, and counselors. Mrs. Mazique attacked the practice of using the terms *ability grouping* and *tracking* interchangeably, pointing out that the former term allowed for parental choice, while the latter concept denied parental decision making in children's educational development. Charging that the tracking system was grounded in theories of "cultural deprivation," which assumed that African American children were uneducable or lacked sufficient motivation to learn, Mrs. Mazique pointed out that the African American traditional yearning for knowledge dated back to the emancipation of the slaves. If something had occurred in recent years to interrupt that motivation, Mrs. Mazique observed, the problem should be placed at the door of the tracking-policy makers who systematically retarded the educational development of African American and poor white children in Washington, DC. Mrs. Mazique concluded with a demand for the end of the tracking system and the institutionalization of an educational program of excellence throughout the District's schools (U.S. Congress, 1966).

Some educational experts also condemned Washington's public school tracking system, observing that the policy discriminated against students on the basis of race and class. For example, Howard University education professor Elias Blake presented a statistical analysis showing that low track assignment correlated with low family income. Marvin Cline, a Howard University psychology professor, attacked the tracking system as racially discriminatory. He advocated a policy strategy of racial balance in the schools to improve educational quality. For Cline, desegregation was merely a step in the direction of complete racial integration (U.S. Congress, 1966).

Significantly, neither Blake nor Cline had plugged into the network of community groups and concerns set in motion earlier by the Reverend Gray and the League for Universal Justice and Goodwill. When they eventually learned of Blake and Cline and their presentations before the Education and Labor Committee task force, many community participants felt sabotaged. As many community leaders later asserted, Blake's charge that the tracking system targeted low-income families and Cline's emphasis on racial integration and busing subtly shifted the issues away from community hopes and aspirations for enhancing quality education (J. R. Mazique, personal interview, March 5, 1986).

Following an extensive set of hearings, the House Committee on Education and Labor's task force recommended the termination of the District's tracking system. Nothing resulted from the recommendation, however, as most congresspersons considered it advisable to leave educational matters in the hands of local school officials. Because the majority of the District's judicially appointed school board members were loyal to Superintendent Hansen, the tracking system remained unchanged, for a time (Wolters, 1984).

What led ultimately to the abolition of tracking in the District's public schools was a lawsuit filed by Julius Hobson, Sr. Hobson, a civil rights activist and government employee, was also one of the District's most vocal critics of the tracking system and Superintendent Hansen. Hobson, too, had testified before the Education and Labor Committee's task force, where he assailed the tracking system for racial and class bias. In the aftermath of the hearings, Hobson took legal action against the District's school officials, challenging the constitutionality of their educational policies, including tracking. The resulting legal policy decisions, however, contrasted sharply with the hopes and aspirations of the District's African American community for the institutionalization of quality education throughout the schools. In effect, traditional community leaders and organizations lost control of the issues as new persons like Hobson emerged to play major roles and put forward alternative issues. A discussion of these issues follows.

Hobson v. Hansen and the Abolition of Tracking: Contradictions and Dilemmas

In July 1966, Hobson filed a class action in federal district court against Superintendent Hansen, the school board, and the Washington judges who appointed the board. He complained that Washington's public schools unconstitutionally denied African American and impoverished students equal educational opportunities provided to more well-off and white students. In addition, Hobson charged that these educational disparities were the consequences of the tracking system, the principle of neighborhood schools, unfair teacher appointments to schools, and optional school zones for some students. Hobson questioned the constitutionality of the congressional act giving federal district court judges the power to appoint school board members. The last complaint was separated from the others, and a three-judge district court heard it. The nature of Hobson's challenge required that the three-judge panel be composed of circuit judges unrelated to the appointment process. In a decision from which Judge J. Skelly Wright dissented, the panel upheld the constitutionality of the statute, ruling that Congress could delegate appointive authority to federal judges in Washington, DC. A request of the Judicial Conference of the District of Columbia Circuit, however, resulted in a subsequent congressional amendment of the statute to allow for an elected school board (Horowitz, 1977; LaNoue & Smith, 1973).

Judge J. Skelly Wright of the U.S. Court of Appeals heard Hobson's remaining charges. Judge Wright sustained Hobson in *Hobson v. Hansen* (1967). The judge held a lengthy hearing and wrote a far-reaching opinion and decree, ordering the school board to abolish the tracking system, develop a student assignment policy in harmony with the court's order and not based completely on the neighborhood-school principle, provide transportation to African American students wishing to transfer to white schools, terminate optional school zones, and design a program of "color-conscious" faculty assignments to effect teacher integration (Horowitz, 1977; LaNoue & Smith, 1973).

The *Hobson v. Hansen* ruling did not put an end to controversy surrounding the assortment of major educational issues in Washington. Indeed, some dilemmas emerged as a result of the decision. The case is important to Washington's public school history, however, because it set in motion future challenges to educational policy making by accentuating the major issues. The leading actors in the District's educational politics from 1967 to 1971 also emerged as a result of the case and its late effect on the school system (LaNoue & Smith, 1973). These new issues and leaders effectively supplanted goals and struggles of traditional community leaders and organizations, such as the Reverend Gray and the League for Universal Justice and Goodwill, which had fought for quality education since the late 1950s and early 1960s.

Stated in brief, it might be suggested that the *Hobson v. Hansen* decision sought to achieve three objectives: integration, quality education, and equality of educational opportunity. Three prominent actors in Washington's educational politics became publicly associated with each of these objectives. Judge Wright came to represent an ongoing allegiance to integration; Hobson came to represent equal educational opportunity; and Mrs. Anita Allen, recently appointed to the school board, came to represent the quest for quality education. Allen, as a member of Washington's first predominantly African American school board, helped carry the vote to effectuate Judge Wright's decision and to direct Superintendent Hansen not to challenge the ruling. This occurred on July 1, 1967—the day Allen took office. Hansen promptly resigned and joined a dissenting board member in an unsuccessful suit to appeal the *Hobson* decision. Several ineffective superintendents followed Hansen, resulting in the school system's lack of strong executive authority. Moreover, the school board could not provide vigorous leadership because it was embroiled in internal squabbles and faced with community discontent (LaNoue & Smith, 1973).

Although Hobson became Washington's first elected official in more than a century when he was a member of the school board for one year in 1968, he generally played the role of an agitator and outside critic of the school system. Similar to many middle-class reformers, however, Hobson failed to establish strong linkages with the community he was supposed to represent. In his re-election bid in 1969, Hobson lost to Exie Mae Washington, a former domestic worker who was promoted to community leadership status through the poverty program. Hobson also had been damaged in his attempt to encourage educational change because of a weakened school bureaucracy and the lack of strong administrative leadership to respond to his demands (LaNoue & Smith, 1973).

In the final analysis, the three major objectives of the Wright decision in the *Hobson* case could not be accomplished all at once. Judge Wright had strongly emphasized racial integration and set in motion policies and practices that contradicted the Washington African American community's original goal of quality education. The judge's decree, coupled with the research findings of educational policy entrepreneurs (particularly, the Coleman et al. report of 1966), resulted in the goal shifting from quality education to racial balancing and busing. This was not the policy strategy desired by Washington's African American community, although racial balancing and busing fit well within the social management policy agenda of educational reform's liberal professional-managerial elites and many civil rights advocates. The African American community had demanded quality education: the distribution and mastery of the fundamental tools of knowledge (reading, writing, and computational skills), academic motivation, and good character building. For many African American community participants, emphasis on racial balancing and busing set aside the concern for achieving quality education. One community observer flatly characterized the result as a "betrayal in the schools" (Washington, 1969).

Judge Wright's decision in the *Hobson* case also contributed to the continuation of white flight from Washington and its public schools. By the time the decree was handed down, the city's public schools were 92% African American, and with the continued growth in the proportion of African American students, integration became impossible. Even the school board seemed to realize the declining possibility and importance of integration when it released the findings of the Passow report, an exhaustive study and analysis of Washington's schools. One section of the report stated bluntly that it was pointless to speak of racial integration or racial balance in a school system with a student enrollment of more than 90% African American (Passow, 1967).

Conclusion

Prior to the 1950s, American educational politics and policy were dominated largely by the legacy of the turn-of-the-century Progressive Era of school reform. That earlier reform movement sought to remove politics from urban education by centralizing executive power in the person of professional educational managers; establishing bureaucratically organized school systems; and using the rhetoric of "science," expertise, and efficiency as the guiding principles of the educational enterprise. Indeed, both lay officials on urban school boards and the lay public largely were deferential to educational professionals, managers, and their expert knowledge.

However, a managerial style, organizational form, and educational ideology operative in one period can frequently become impediments in the next. An inherent problem with the managerial decision-making process, organizational complexity, and scientistic rhetoric of the educational reform period was that they severely restricted public participation in the urban educational policy-making process. Thus, in the 1950s and 1960s, these arrangements gave rise to a popular challenge to the prerogatives of educational professionals and their exclusive bureaucratic policy-making process. Moreover, ordinary citizens began to confront the hegemonic role of expert knowledge in educational decision making.

In the case of Washington, DC, the introduction of the tracking system into the public schools is an example of managerial-bureaucratic policy making that effectively excluded popular participation. Moreover, the strategy of channeling children into different curricular tracks—based on "scientific" IQ test scores as interpreted by educational experts and classroom performance evaluations by principals and teachers—exemplified educational reformers' rhetoric and practice of social management. In addition, the policy had racial overtones, as it directed a large proportion of African American children to the low track. As a result of being constrained from influencing decisions affecting their interest in protecting their children's future educational development, parental and community groups became increasingly angry and directly challenged the District's educational professionals and specialists. The efforts of the League for Universal Justice and Goodwill and other grassroots organizations to attack politically the formulators and implementors of the tracking system represented a shift from the past era of public school politics when educational professionals largely commanded popular deference to an emerging politics of education distinguished by increased community challenge to public school professionals. This new brand of open and aggressive community politics would come to characterize the politics of urban educational policy making in the 1970s and 1980s (Hayes, 1985).

The present study of the historical trends and developments in the politics of public schooling in the District of Columbia exposes a transition from segregation with parental and community concern for quality education to integration with official emphasis on tracking and later racial busing and balancing, ultimately newer forms of racial discrimination that denied quality education to African American youngsters. The District's public school officials advocated the tracking system as a means of complying with the *Brown* and *Bolling* Supreme Court decisions. In contrast, the African American community attacked this policy, seeking the institutionalization of quality education throughout the public school system. The league and its members defined quality education as the distribution and master of the basic tools of knowledge, academic motivation, and positive character development. They characterized the tracking system as "programmed retardation," charging that this policy strategy was worse than segregation in the Old South. Community spokespersons argued forcefully that tracking would result in the eventual deterioration of public education in Washington.

Asserting that tracking early on (a) retarded a child's later academic growth and development, (b) legitimated a practice of withholding educational fundamentals from African American students, and (c) denied African American children a quality education or even a basic one, the District's African American residents demanded a basic education in all of the city's schools. Indeed, the success of the District's Amidon Plan and its subject-centered curriculum served as proof then and now that a return to the basics in public schools would lead to quality education for all students regardless of race or class background. Recognizing the value of Amidon's curriculum, Washington's community leaders called in vain for similar plans to be implemented throughout the District.

Because of the District's unique political history as the nation's capital, the city public schools were controlled by Congress. Therefore, as community frustration and discontent with public school officials grew, the league petitioned the House Committee on Education and Labor to investigate charges of racial and class discrimination in the District's schools. The hope was that Congress would abolish the tracking system.

Congress did not abolish the tracking policy, maintaining the collective opinion that educational matters should remain in the hands of local officials. The community was defeated in its struggle for quality education even though community leaders had initiated an effective challenge to Washington's public school professionals and the tracking system. Judge J. Skelly Wright's ruling in the *Hobson* case did terminate tracking in the District's schools; however, he ignored the community's cry for quality education as a priority for Washington's schools. Thus, with Wright's decision, the issues shifted from the community's concern for enhancing quality education to the educational professionals' and experts' policy agenda and debate (at the national and local levels) regarding racial balancing and busing. This was the "great betrayal" and represented a regrouping and continuation of educational reform's goal of social management into the 1960s and 1970s. Although racial balancing and busing could not be achieved in Washington's schools, with more than 90% African American students, racial balancing and busing did command the attention of educational professionals and community participants in the surrounding suburban jurisdictions of the greater Washington metropolitan area (see Hayes, 1985). Moreover, in the 1970s an increasingly active judiciary came to sanction and enforce various forms of racial balancing and busing in metropolitan areas across the nation (Kirp, 1982; Willie & Greenblatt, 1981). Such judicial activism may have been the single most significant factor in the ultimate decline of public education.

Although the District's African American community lost its struggle for quality education, its leaders' earlier predictions of the dire consequences of the tracking system's "programmed retardation" cannot be ignored today (see Oakes, 1985). Moreover, Mrs. Mazique's warning before the 1966 congressional task force regarding "the downward trend toward imminent collapse of public education" is confirmed by the contemporary national anxiety and debate about America's current educational predicament (e.g., see Altbach, Kelly, & Weis,

1985; National Commission on Excellence in Education, 1983). Similar to many other urban school systems, the number of National Merit scholars graduating from Washington's high schools is miniscule, and a substantially large proportion of them require remedial assistance on entering college. Washington's public schools are faced with an assortment of educational and social difficulties: dropouts and continued enrollment decline since 1970, drug-related activities, teenage pregnancies, school closings, and growing parental discontent (Sanchez, 1990; Vassell, 1990). Recent indications are that many African American parents are withdrawing their children from the public schools and enrolling them in private and parochial schools. Commenting on the progressive deterioration of public trust in the District's schools, one researcher observed that "a permanent lack of confidence has now replaced the 'crisis of confidence' " (Diner, 1990). For the Washington, DC, public school system, and for urban public school systems throughout America, the great challenge of the future will be to reinstate quality education and thereby rebuild confidence in public education.

Note

1. This study benefited from in-depth interviews with Mrs. Jewell R. Mazique conducted in 1986. She is one of the few, if not the only, remaining key community participant in the events chronicled in this chapter. I am grateful to her for sharing with me her wealth of historical knowledge and personal experience regarding educational politics in Washington, DC, since the 1950s.

References

Altbach, P. G., Kelly, G. P., & Weis, L. (Eds.). (1985). *Excellence in education: Perspectives on policy and practice.* Buffalo, NY: Prometheus.

Bolling v. Sharpe, 347 U.S. 497 (1954).

Brown v. Board of Education, 347 U.S. 483 (1954).

Coleman, J., Campbell, E., Hobson, C. J., McPartland, J., Mood, A. M., Weinfeld, F. D., & York, R. L. (1966). *Equality of educational opportunity* (Report No. OE-38001, Office of Education). Washington, DC: Government Printing Office.

Diner, S. J. (1990). Crisis of confidence: Public confidence in the schools of the nation's capital in the twentieth century. *Urban Education, 25,* 112-137.

Flesch, R. (1955). *Why Johnny can't read.* New York: Harper & Row.

Hansen, C. F. (1960). *The Amidon Plan: For education in the sixties in the D.C. public schools.* Washington, DC: D.C. Government.

Hayes, F. W., III. (1985). *Division and conflict in postindustrial politics: Social change and educational policymaking in Montgomery County, Maryland.* Unpublished doctoral dissertation, University of Maryland.

Hobson v. Hansen, 265 F. Supp. 902 (1967).

Horowitz, D. F. (1977). *The courts and social policy.* Washington, DC: Brookings Institution.

Kirp, D. L. (1982). *Just schools: The idea of racial equality in American education.* Berkeley: University of California Press.

LaNoue, G. R., & Smith, B.L.R. (1973). *The politics of school decentralization.* Lexington, MA: Lexington Books.

National Commission on Excellence in Education. (1983). *A nation at risk: The imperative for educational reform.* Washington, DC: Government Printing Office.

Oakes, J. (1985). *Keeping track: How schools structure inequality.* New Haven, CT: Yale University Press.

Passow, A. H. (1967). *Toward creating a model urban school system: A study of the Washington, D.C. public schools.* New York: Teachers College Press.

Rickover, H. G. (1959). *Education and freedom.* New York: E. P. Dutton.

Sanchez, R. (1990, March 7). New D.C. school figures show large declines. *Washington Post,* pp. A1, A12.

Tyack, D. B., & Hansot, E. (1982). *Managers of virtue: Public school leadership in America, 1820-1980.* New York: Basic Books.

U.S. Congress. House Committee on Education and Labor. (1966). *Investigation of the schools and poverty in the District of Columbia: Hearings before the Task Force on Antipoverty in the District of Columbia* (89th Cong., 1st and 2d sess., 1965-1966, Y4).

Vassell, O. (1990, April 14). Parents group wants board booted. *Washington Post,* pp. A1, A2.

Washington, M. W. (1969, January). Betrayal in the schools: As seen by an advocate of African-American power. *Triumph Magazine,* pp. 16-19.

Willie, C. V., & Greenblatt, S. L. (Eds.). (1981). *Community politics and educational change: Ten school systems under court order.* New York: Longman.

Wolters, R. (1984). *The burden of* Brown: *Thirty years of school desegregation.* Knoxville: University of Tennessee Press.

Concluding Comments

Roger W. Caves

The purpose of this reader has been to introduce students to the study of the urban arena. Each reading has provided some new insights into its various subfields and has attempted to raise questions for future study. As the readings suggest, the urban arena is a broad, complex, and multidisciplinary area of inquiry. It involves a host of disciplines and topics. As former Secretary of the U.S. Department of Housing and Urban Development, Robert C. Weaver, once remarked:

> Urban America is complex. It involves land use, the proliferation of local governments, citizen participation and neighborhood control, the need for areawide approaches to areawide problems, the style and nature of suburbs and tract developments, the revitalization of central cities with a conscious concern for their human attributes as well as their economic viability, crime in the streets and personal safety, education, job opportunities, housing, tax policy, transportation, and aesthetics. (Weaver, 1973, p. 14)

A major point alluded to throughout the reader is that urban problems are interconnected. Many of readings illustrate that it is difficult, if not impossible, to view one urban problem without any discussion of another urban problem. Jane Jacobs acknowledges this complexity in the following passage:

> Cities happen to be problems in organized complexity, like the life sciences. They present "situations in which a half-dozen or even several dozen quantities are all varying simultaneously *and in subtly interconnected ways.*" Cities, again like the life sciences, do not exhibit *one* problem in organized complexity, which if understood explains all. They can be analyzed into many such problems or segments which, as in the case of the life sciences, are also related with one another. The variables are many, but they are not helter-skelter; they are interrelated into an organic whole. (Jacobs, 1961, p. 433)

It has become difficult to separate the destinies of our cities from the destinies of our metropolitan areas. Each depends on the other. Cities are part of a larger metropolitan area, and their individual actions can have serious extraterritorial repercussions. This added complexity definitely complicates formulating policies that affect cities.

Facing Our Urban Challenges

There are no easy answers to dealing with the ills of urban America. There is no panacea. To face these challenges requires educating students in the urban arena. This challenge has been accepted by a number of universities throughout the country. Among the universities offering urban studies/urban affairs curriculums and degrees are the following:

Cleveland State University
Georgia State University
Mankato State University
Michigan State University
Old Dominion University
Portland State University
St. Cloud State University
San Diego State University
University of Akron
University of Delaware
University of Louisville
University of New Orleans
University of Texas at Arlington
Virginia Commonwealth University
Virginia Polytechnical Institute and State University
Wright State University

Avenues for Future Research

The changing nature of our cities calls for continued research in a number of areas. This research could take the shape of theoretical pieces, individual case studies, or empirical analyses. In addition, the importance of comparative research cannot be underestimated. We can learn from each other, and we should not be hesitant about conducting comparative research.

Studying neighborhoods represents a prime area for future research. We need to investigate neighborhoods undergoing change, as well as stable neighborhoods. What factors contribute to neighborhood stability and what factors lead to unstable neighborhoods should be focal areas of research. We also need to examine population mobility, neighborhood revitalization, quality of life, crime and defensible space concepts, and racial unrest in our neighborhoods and cities.

Additional research questions deserving of attention from urban scholars are the following:

How are our urban areas changing?

Who are the winners and losers from neighborhood change?

How can we improve our urban schools?

How can we protect our urban environment?

How are cities meeting their infrastructure needs?

Who will pay for the needed infrastructure improvements?

What is the role of the citizenry in urban governance?

What are the roles of nonprofit organizations in the urban arena?

How can we meet the human services needs of the urban population?

How can we best meet the housing needs of urban America?

How will urban growth be accommodated in the future?

Are current urban programs in the aforementioned areas working?

It is important to remember that we can learn from the past. We should apply knowledge from the past to prepare for the future. I hope readers of this book will better understand the urban arena and its various components and, by doing so, will help build better cities and a brighter future for America's urban citizens. Perhaps the readers today will be better able to understand and resolve the urban issues of tomorrow.

References

Jacobs, J. (1961). *The death and life of great American cities*. New York: Vintage.

Weaver, R. C. (1973). Urban challenges: 1960-1973. In *The American city: A symposium* (pp. 5-43). Austin: University of Texas, Lyndon Baines Johnson School of Public Affairs.

Suggestions for Further Reading

Abrahamson, M. (1974, June). The social dimensions of urbanism. *Social Forces, 28*, 330-351.

Altshuler, A. (1970). *Community control*. New York: Pegasus.

Anderson, M. (1964). *The federal bulldozer*. Cambridge: MIT Press.

Anglin, A., & Holcomb, B. (1992). Poverty in urban America: Policy choices. *Journal of Urban Affairs, 14*(3/4), 447-469.

Barnekov, T., & Rich, D. (1977, June). Privatism and urban development: An analysis of the organized influence of local business elites. *Urban Affairs Quarterly, 12*, 431-460.

Bennett, L. (1993, March). Harold Washington and the black urban regime. *Urban Affairs Quarterly, 28*, 423-441.

Berg, I. (1971). *Education and jobs: The great training robbery*. Boston: Beacon.

Blakely, E. J. (1994). *Planning local economic development* (2nd. ed.). Thousand Oaks, CA: Sage.

Bolotin, F. N., & Cingranelli, D. L. (1983, February). Equity and urban policy: The underclass hypothesis revisited. *Journal of Politics, 45*, 209-219.

Brintall, M. (1989). Future directions for federal urban policy. *Journal of Urban Affairs, 11*(1), 1-21.

Burt, M. R., & Cohen, B. E. (1989). *America's homeless: Numbers, characteristics, and the programs that serve them*. Washington, DC: Urban Institute.

Canty, D. (1969). *A single society.* New York: Praeger.

Caputo, D. A. (1976). *Urban America: The policy alternative.* San Francisco: Freeman.

Castells, M. (1977). *The urban question.* Cambridge: MIT Press.

Caves, R. W. (1992). *Land use planning: The ballot box revolution.* Newbury Park, CA: Sage.

Clark, K. B. (1965). *Dark ghetto.* New York: Harper & Row.

Clay, P., & Hollister, R. (Eds.). (1983). *Neighborhood policy and planning.* Lexington, MA: Lexington Books.

Cummings, S. (Ed.). (1988). *Business elites and urban development.* Albany: State University of New York Press.

Dahl, R. A. (1961). *Who governs?* New Haven, CT: Yale University Press.

Darden, J. T. (1986). The significance of race and class in residential segregation. *Journal of Urban Affairs, 8*(1), 49-57.

Darden, J. T., Duleep, H. O., & Galster, G. C. (1992). Civil rights in metropolitan America. *Journal of Urban Affairs, 14*(3/4), 469-499.

DeHoog, R. H. (1984). *Contracting out for public services.* Albany: State University of New York Press.

DeLeon, R. E. (1992, June). The urban antiregime. *Urban Affairs Quarterly, 27,* 555-580.

Dentler, R. A. (1977). *Urban problems: Perspectives and solutions.* Chicago: Rand McNally.

Downs, A. (1968). Alternative futures for the American ghetto. *Daedalus, 97*(4), 1331-1379.

Downs, A. (1981). *Neighborhoods and urban development.* Washington, DC: Brookings Institution.

Elkins, S. L. (1985). Twentieth century urban regimes. *Journal of Urban Affairs, 7*(2), 11-29.

Elredge, H. W. (Ed.). (1967). *Taming megalopolis* (Vols. 1 & 2). Garden City, NY: Doubleday.

Fischer, C. S. (1976). *The urban experience.* Orlando, FL: Harcourt Brace Jovanovich.

Freedman, L. (1969). *Public housing: The politics of poverty.* New York: Holt, Rinehart & Winston.

Fried, J. P. (1971). *Housing crisis U.S.A.* New York: Praeger.

Funigello, P. J. (1978). *The challenge to urban liberalism: Federal-city relations during World War II.* Knoxville: University of Tennessee Press.

Fusfield, D. R. (1973). *The basic economics of the urban racial crisis.* New York: Holt, Rinehart & Winston.

Gans, H. J. (1962). *The urban villagers.* New York: Free Press.

Gans, H. J. (1967). *The Levittowners.* New York: Vintage.

Gillett, J. W., et al. (1992). The need for an integrated urban environmental policy. *Journal of Urban Affairs, 14*(3/4), 377-399.

Glaab, C. N. (1963). *The American city.* Homewood, IL: Dorsey.

Glasgow, D. G. (1981). *The black underclass.* New York: Vintage.

Glassberg, A. D. (1981). *Representation and urban community.* New York: Macmillan.

Glazer, N., & Moynihan, D. P. (1963). *Beyond the melting pot: Negroes, Puerto Ricans, Jews, Italians, and Irish in New York City.* Cambridge: MIT Press.

Goering, J. (1978). Neighborhood tipping and racial transition: A review of social science evidence. *Journal of the American Institute of Planners, 44*(1), 68-78.

Goldsmith, W. W., & Blakely, E. J. (1992). *Separate societies: Poverty and inequalities in U.S. cities.* Philadelphia: Temple University Press.

Gordon, D. R. (1973). *City limits.* New York: Charterhouse.

Gordon, D. R., et al. (1992). Urban crime policy. *Journal of Urban Affairs, 14*(3/4), 359-377.

Gorham, W., & Glazer, N. (Eds.). (1976). *The urban predicament.* Washington, DC: Urban Institute.

Green, C. M. (1965). *The rise of urban America.* New York: Harper & Row.

Greer, S. (1962). *The emerging city.* New York: Free Press.

Greer, S. (1965). *Urban renewal and American cities.* Indianapolis: Bobbs-Merrill.

Greer, S. (1972). *The urbane view.* New York: Oxford University Press.

Hawley, A. (1971). *Urban society.* New York: Ronald Press.

Hays, R. A. (1985). *The federal government and urban housing: Ideology and change in public policy.* Albany: State University of New York Press.

Henig, J. R. (1982). *Neighborhood mobilization.* New Brunswick, NJ: Rutgers University Press.

Holleb, D. B. (1972). *Colleges and the urban poor: The role of higher education in community service.* Lexington, MA: D. C. Heath.

Hummel, R. C., & Nagle, J. M. (1973). *Urban education in America.* New York: Oxford University Press.

Hunter, F. (1953). *Community power structure: A study of decisionmakers.* Chapel Hill: University of North Carolina Press.

Jacobs, J. (1969). *The economy of cities.* New York: Random House.

Jencks, C., et al. (1972). *Inequality: A reassessment of family and schooling in America.* New York: Basic Books.

Jencks, C., & Peterson, P. (1991). *The urban underclass.* Washington, DC: Brookings Institution.

Kantrowitz, N. (1973). *Ethnic and racial segregation in the New York metropolis.* New York: Praeger.

Kaplan, M., & James, F. J. (Eds.). (1990). *The future of national urban policy.* Durham, NC: Duke University Press.

Katznelson, I. (1981). *City trenches.* Chicago: University of Chicago Press.

Kerner Commission. (1968). *Report of the National Advisory Commission on Civil Disorders.* Washington, DC: Government Printing Office.

Kotler, M. (1967). *Neighborhood government.* Indianapolis: Bobbs-Merrill.

Larson, C. J., & Nikkel, S. R. (1979). *Urban problems: Perspectives on corporations, governments, and cities.* Boston: Allyn & Bacon.

Lazar, I., & Moore, E. (1992). Health policy: Beyond chicken soup and Band-Aids. *Journal of Urban Affairs, 14*(3/4), 291-311.

Levin, H. M. (Ed.). (1970). *Community control of schools.* Washington, DC: Brookings Institution.

Levitan, S. A. (1980). *Programs in aid for the poor for the 1980s.* Baltimore: Johns Hopkins University Press.

Levy, F., Meltsner, A. J., & Wildavsky, A. (1974). *Urban outcomes.* Berkeley: University of California Press.

Liebow, E. (1967). *Tally's corner.* Boston: Little, Brown.

Lineberry, R. L. (1977). *Equality and urban policy: The distribution of municipal public services.* Beverly Hills, CA: Sage.

Loewenstein, L. K. (Ed.). (1971). *Urban studies: An introductory reader.* New York: Free Press.

Long, N. (1958, November). The local community as an ecology of games. *American Journal of Sociology, 64,* 251-261.

Long, N. (1972). *The unwalled city.* New York: Basic Books.

Lynch, K. (1960). *The image of the city.* Cambridge: MIT Press.

McKelvey, B. (1973). *American urbanization: A comparative history.* Glenview, IL: Scott, Foresman.

McLennan, B. N. (Ed.). (1970). *Crime in urban society.* Cambridge, MA: University Press.

Miller, Z. (1973). *The urbanization of America.* Orlando, FL: Harcourt Brace Jovanovich.

Mitchell, R. B., & Rapkin, C. (1954). *Urban traffic: A function of land use.* New York: Columbia University Press.

Modarres, A. (1992). Ethnic community development: A spatial examination. *Journal of Urban Affairs, 14*(2), 97-109.

Mollenkopf, J. (1975). The post-war politics of urban development. *Politics and Society, 5,* 247-296.

Moskowitz, E. S., & Simpson, D. (1983, Summer). Neighborhood empowerment and urban management in the 1980s. *Journal of Urban Affairs, 5,* 183-192.

Moynihan, D. P. (1969). *Maximum feasible misunderstanding.* New York: Free Press.

Mudd, J. (1968). *The urban prospect.* Orlando, FL: Harcourt Brace Jovanovich.

Mudd, J. (1984). *Neighborhood services: Making big cities work.* New Haven, CT: Yale University Press.

National Commission on Urban Problems. (1968). *Building the American city.* Washington, DC: Government Printing Office.

Netzer, D. (1970). *Economics and urban problems.* New York: Basic Books.

Newman, O. (1972). *Defensible space.* New York: Macmillan.

O'Brien, D. J., & Lange, J. K. (1986). Racial composition and neighborhood evaluation. *Journal of Urban Affairs, 8*(3), 43-63.

Palley, M. L., & Palley, H. A. (1977). *Urban America and public policies.* Lexington, MA: D. C. Heath.

Peterson, P. (1981). *City limits.* Chicago: University of Chicago Press.

Peterson, P. (Ed.). (1985). *The new urban reality.* Washington, DC: Brookings Institution.

Reitzes, D. C. (1985). Downtown orientation: An urban identification approach. *Journal of Urban Affairs, 7*(2), 29-47.

Robinson, C. J. (1990). Minority political representation and local economic development policy. *Journal of Urban Affairs, 12*(1), 49-59.

Rosentraub, M. S. (Ed.). (1986). *Urban policy problems.* New York: Praeger.

Rossi, P. H., & Dentler, R. A. (1961). *The politics of urban renewal.* Chicago: Free Press of Glencoe.

Schill, M. H., & Nathan, R. P. (1983). *Revitalizing America's cities: Neighborhood reinvestment and displacement.* Albany: State University of New York Press.

Schnore, L. F. (Ed.). (1975). *The new urban history.* Princeton, NJ: Princeton University Press.

Sennett, R. (Ed.). (1969). *Classic essays on the culture of cities.* New York: Appleton-Century-Crofts.

Shank, A. (Ed.). (1973). *Political power and the urban crisis.* Boston: Allyn & Bacon.

Solomon, A. P., & Vandell, K. D. (1982, Winter). Alternative perspectives on neighborhood decline. *Journal of the American Planning Association, 48,* 81-92.

Stanley, D. R. (1976). *Cities in trouble.* Columbus, OH: Academy for Contemporary Problems.

Steffens, L. (1957). *The shame of the cities.* New York: Hill & Wang.

Stone, C., & Sanders, H. (Eds.). (1987). *The politics of urban development.* Lawrence: University of Kansas Press.

Suttles, G. D. (1972). *The social construction of communities.* Chicago: University of Chicago Press.

Taeuber, K. E., & Taeuber, A. F. (1972). *Negroes in cities.* New York: Atheneum.

Taylor, R. B. (1986). *Urban neighborhoods.* New York: Praeger.

Teaford, J. C. (1993). *The twentieth-century American city.* Baltimore: Johns Hopkins University Press.

Tilly, C. (1974). *An urban world.* Boston: Little, Brown.

Tournier, R. E. (1984). Cities as people, cities as places: Urban revitalization and urban conflict. *Journal of Urban Affairs, 6*(2), 141-151.

Trachtenberg, A., Neill, P., & Bunnell, P. C. (1971). *The city: American experience.* New York: Oxford University Press.

Varady, D. P. (1986). *Neighborhood upgrading.* Albany: State University of New York Press.

Vidal, A. (1992). *Rebuilding communities: A national study of urban community development corporations.* New York: New School of Social Research, Graduate School of Management and Urban Policy.

Warner, S. B., Jr. (1972). *The urban wilderness.* New York: Harper & Row.

Whyte, W. F. (1955). *Street corner society* (2nd ed.). Chicago: University of Chicago Press.

Wilson, J. Q. (1975). *Thinking about crime.* New York: Basic Books.

Wilson, J. Q. (Ed.). (1976). *Urban renewal: The record and controversy.* Cambridge: MIT Press.

Wilson, W. J. (1987). *The truly disadvantaged: The inner city, the underclass, and public policy.* Chicago: University of Chicago Press.

Winter, W. O. (1969). *The urban polity.* New York: Dodd, Mead.

Wirth, L. (1928). *The ghetto.* Chicago: University of Chicago Press.

Wirth, L. (1938, July). Urbanism as a way of life. *American Journal of Sociology, 44,* 3-24.

Zimmerman, J. F. (1972). *The federated city.* New York: St. Martin's.

Name Index

Subject Index

About the Contributors

Carl Abbott is Professor of Urban Studies and Planning at Portland State University. He is the author of several books on the history of cities and city planning. His recent publications include *The Metropolitan Frontier: Cities in the Modern American West* (1993), and *Planning the Oregon Way: A Twenty Year Evaluation* (1994), coedited with Deborah Howe and Sy Adler.

Timothy K. Barnekov is Director of the Center for Community Development and a Policy Fellow in the College of Urban Affairs and Public Policy at the University of Delaware. His publications include *Privatism and Urban Policy in Britain and the United States* (1989), as well as numerous articles on urban economic development policy.

William R. Barnes is Director of the Center for Research and Program Development at the National League of Cities in Washington, DC. He earned a Ph.D. in history from the Maxwell School at Syracuse University.

Charles Bartsch is a Senior Policy Analyst at the Northeast-Midwest Institute in Washington, DC, specializing in economic development and tax issues. He coauthored the book *New Life for Old Buildings.* Currently he is directing a project examining the impacts of environmental contamination on community development initiatives and industrial site reuse and is working with members of Congress and federal agencies to develop policies to facilitate site reuse. He serves on the Sustainable Communities Task Force of the President's Council on Sustainable Development and is completing *Federal Economic Development Programs: Recent Changes,* his most recent volume in an ongoing series of guidebooks.

Nico Calavita is Associate Professor of City Planning in the Graduate City Planning Program at San Diego State University. His research on the politics of growth has been published in the *Journal of Urban Affairs,* and on equity planning and housing policy in the *Journal of Planning Education and Research*. He is also Chairperson, City of San Diego Housing Trust Fund, Board of Trustees.

Thomas G. Carroll serves as Special Assistant for National and Community Service, in the Office of the U.S. Secretary of Education. He is reponsible for coordinating Department of Education service programs with those created by the Corporation for National and Community Service. From 1986 to 1993 he served as Deputy Director, Fund for the Improvement of Postsecondary Education, and from 1979 to 1986 he was a Senior Program Officer at the National Institute of Education. From 1975 to 1979 he was Assistant Professor of Education and Anthropology at Clark University, Worcester, Massachusetts.

Roger W. Caves is Professor and Coordinator of the Graduate City Planning Program and the interdisciplinary undergraduate Urban Studies Program at San Diego State University. His principal interests are urban policy, growth politics, housing and land-use policy, and international planning and policy. He is the author of *Land Use Planning: The Ballot Box Revolution* (1992) and has published in such journals as *Urban Studies, Cities, Land Use Policy, Journal of Urban Planning and Development, Real Estate Issues,* and the *International Journal of Public Administration*. He is currently Vice-Chair of the American Planning Association's Housing and Human Services Division and coeditor of its *Housing and Human Services Quarterly Newsletter*. He is coeditor of Sage Publications' forthcoming book series *Cities and Planning* with Robert Waste and Margaret Wilder.

Thomas A. Clark serves on the faculty of the Graduate Program in Urban and Regional Planning, College of Architecture and Planning, at the University of Colorado at Denver. His principal research interests include regional economic development, urban labor markets, strategic planning, and growth management policy. Recent publications address capital constraints on nonmetropolitan accumulation, the targeting of educational resources for regional development, and metropolitan spatial options.

Susan E. Clarke is Associate Professor of Political Science and Director of the Center for Public Policy Research at the University of Colorado at Boulder. She recently edited and contributed to *The New Localism* (1993) and *Urban Innovation* and has published several articles on urban economic development issues.

Peter Dreier is the E. P. Clapp Distinguished Professor of Politics at Occidental College in Los Angeles. He directs the Public Policy program and is a Senior Fellow at the International and Public Affairs Center (IPAC). He served for 9 years as the Director of Housing at the Boston Redevelopment Authority and was senior policy advisor to Boston Mayor Ray Flynn. He has written widely on urban politics, housing policy, and community development. He serves on the Advisory Board of the Resolution Trust Corporation (RTC), the savings-and-loan clean-up agency.

David R. Elkins is an Instructor in the Department of Public Administration and a graduate student in the Department of Political Science at the University of Kansas. He has published in the *Public Administration Review* and currently is working on his dissertation on urban economic development.

Cees D. Eysberg was educated and now teaches geography at the University of Utrecht. His professional interests include economic geography and the regional geography of the United States. He was a Fulbright scholar at California Polytechnic University, San Luis Obisbpo, in 1984-1985 and was a research asociate at the University of California, Berkeley, in Spring 1987.

Gary L. Gaile is Associate Professor of Geography at the University of Colorado at Boulder. His books include *Spatial Statistics and Models, Spatial Diffusion,* and *Geography in America.* His research interests include urban and regional planning, Third World development, and spatial statistics. He is continuing research in market-based planning approaches.

Edward G. Goetz is Associate Professor in the Housing Program at the University of Minnesota. He is the author of *Shelter Burden: Local Politics and Progressive Housing Policy* (1993), and coeditor of *The New Localism: Comparative Urban Politics in a Global Era* (1993). He has published articles on homelessness and on housing and economic development policy. His current work focuses on the neighborhood politics of subsidized housing and the work of community development corporations.

Scott Greer is Distinguished Professor Emeritus at the University of Wisconsin at Milwaukee. His books include *The Emerging City: Myth and Reality* (1962), *Governing the Metropolis* (1962), *The Urbane View* (1972), and *The Logic of Social Inquiry* (1989). These and others of his seminal works on the modern city have been translated and sold worldwide.

Kathleen Harris is Professor of Education at Arizona State University West. Her work over the last decade has focused on educational collaboration to support teachers and students in culturally and linguistically diverse urban settings.

Floyd W. Hayes, III, is Assistant Professor of Afro-American Studies at San Diego State University. His interests are educational and urban politics and policy.

Franklin J. James is an economist and Professor of Public Policy in the Graduate School of Public Affairs at the University of Colorado at Denver. Formerly, he directed HUD's legislative and urban policy staff and has been a senior researcher at the Urban Institute in the Rutgers University Center for Urban Policy Research. His research focuses on urban policy issues, including economic development and housing policy, and civil rights issues.

Louise Jezierski is Assistant Professor of Sociology and the Program in Urban Studies at Brown University. She is writing a book, *Consent to the City,* based on her research on the politics of postindustrial transformation in Cleveland and Pittsburgh, and has published articles on the implications of public-private partnerships for governance and on postmodern urban theory. Her current research is on identity formation and community empowerment in a comparative study of Latino communities in Boston, Hartford, and Providence.

Marshall Kaplan is Dean and Professor of Public Policy, Graduate School of Public Affairs, University of Colorado at Denver. He has served as Deputy Assistant Secretary in the Department of Housing and Urban Development (1977-1981). He has published numerous books and articles on federal urban policies and urban development.

W. Dennis Keating is Professor of Urban Planning and Law and Associate Dean of the Levin College of Urban Affairs at Cleveland State University. He has written about urban planning and development and housing. His latest book is *The Suburban Racial Dilemma: Housing and Neighborhoods* (1994).

Norman Krumholz is Professor of Urban Planning at the Maxine Goodman Levin College of Urban Affairs at Cleveland State University. Before turning to academia, he had a 20-year career as a planning practitioner in Ithaca and Pittsburgh, and as a planning director for the city of Cleveland from 1969 to 1979. He was President of the American Planning Association in 1987 and received the American Planning Association's Distinguished Leadership Award in 1990. He has published extensively in leading planning and urban affairs journals, and his 1990 book *Making Equity Planning Work* (with John Forester) won the Paul Davidoff Award as best book of the year from the Association of Collegiate Schools of Planning.

John R. Logan is Professor in the Department of Sociology, State University of New York at Albany. He is coauthor (with Harvey Molotch) of *Urban Fortunes: The Political Economy of Place,* winner of the 1990 Award for a Distinguished Scholarly Publication of the American Sociological Association. His current research includes studies of neighborhood change, racial segregation, and ethnic economies in U.S. cities.

Norton E. Long is Professor Emeritus at the University of Missouri at St. Louis. His current interests include education, human capital investment, and economic development. He is author of *The Polity* (1962), and *The Unwalled City* (1972), as well as many classic articles in urban affairs. He chairs a committee of the American Planning Association, with the charge to develop procedures designed to encourage cities to keep meaningful social and economic books.

Elaine MacLeod is a part-time faculty member in the Division of Curriculum and Instruction at California State University, Los Angeles. Her interests include teacher training, implementation of whole language curricula in elementary schools, and educating high-risk students.

Richard P. Nathan is Director of the Nelson A. Rockefeller Institute of Government and Provost of the Nelson A. Rockefeller College of Public Affairs and Policy at the University of Albany. Prior to coming to Albany in 1989, he was Professor of Public and International Affairs at the Woodrow Wilson School at Princeton University. He also has served as Senior Fellow at the Brookings Institution and as a federal government official.

Mary K. Nenno is Associate Director for Policy Development of the National Association of Housing and Redevelopment Officials (NAHRO).

Anthony M. Orum is Head of the Department of Sociology at the University of Illinois at Chicago. His most recent publication on cities is *City-Building in America* (1995), which provides an in-depth history of Milwaukee, Wisconsin, along with comparative analyses of Cleveland, Ohio; Minneapolis-St. Paul, Minnesota; and Austin, Texas.

Patricia Baron Pollak is Associate Professor in the Department of Consumer Economics and Housing at Cornell University. Her recent work has focused on how U.S. communities can be more "family friendly." She is concerned with how local policy decisions affect households, and has written extensively on how communities can use their existing housing stock to create affordable housing units both for the elderly and for others. She is the author of the American Planning Association's technical advisory report on municipal zoning for these options, and AARP's monograph on zoning for shared residences for the elderly. She has directed two U.S. government projects to research and demonstrate how communities can increase housing for their aging populations.

Gordana Rabrenovic is Assistant Professor of Sociology at Northeastern University at Boston. Her research and teaching interests are in the areas of urban sociology, social movements, and voluntary and nonprofit organizations. She currently is working on a book that compares neighborhood mobilization in two different types of cities: declining manufacturing cities and service-sector cities.

Jeffrey A. Raffel is Professor in the College of Urban Affairs and Public Policy at the University of Delaware. He has conducted research on a number of educational policy issues, including metropolitan school desegregation and teacher retention. His articles have appeared in the *Harvard Educational Review, Educational Evaluation and Policy Analysis, Phi Delta Kappan, Urban Affairs Quarterly, Journal of Urban Affairs,* and *Urban Education.* His most recent book is *The Politics of School Desegregation: The Metropolitan Remedy in Delaware,* 1980. He has been a member of Delaware's School Reform Partnership and served as Executive Director of the Committee on School Decision (a committee appointed to work toward a peaceful transition during court-ordered desegregation in the Wilmington metropolitan area).

Daniel Rich is Professor and Dean in the College of Urban Affairs and Public Policy at the University of Delaware. He is also Senior Research Associate in the university's Center for Energy and Environmental Policy and a Visiting Professor in the Centre for Planning at Strathclyde in Scotland. His publications include *Energy and Environment: The Policy Challenge* (1992), *Privatism and Urban Policy in Britain and the United States* (1989), and *Planning for Changing Energy Conditions* (1988).

Mark S. Rosentraub is Professor and Associate Dean, School of Public and Environmental Affairs, Indiana University at Indianapolis. His current research focuses on the effects of the "New Federalism" on the delivery of human services and particularly on changes in the financing of health care and the impact of these shifts on older citizens.

Jean J. Schensul is Executive Director of the Institute for Community Research based in Hartford, CT, and President-Elect of the Society for Applied Anthropology. Her work has focused on health and education research and development in urban areas of the United States, Latin America, Africa, and South Asia. She has published on a variety of topics including AIDS, substance abuse prevention, intercultural communications, qualitative methods, and advocacy research. Her publications include *Collaborative Research and Social Policy* (1987).

Elaine B. Sharp is Professor of Political Science at the University of Kansas. She has published articles on economic development policies and politics, urban fiscal issues, and citizen participation in local politics in a variety of journals. Her most recent books are *The Dilemma of Drug Policy in the United States,* (1994), and *Urban Politics and Administration* (1990).

Kenneth A. Sirotnik is Professor and Chair of Educational Leadership and Policy Studies in the College of Education at the University of Washington. His research, teaching, and other professional activities range widely over many issues, including measurement, statistics, evaluation, technology, educational policy, organizational change, and school improvement. Among his latest books are *The Moral Dimensions of Teaching* (coedited with John Goodlad and Roger Soder) and *Understanding Educational Statistics* (coauthored with James Popham).

Bert E. Swanson is Professor of Political Science and Urban Studies at the University of Florida. He is the author of *Struggle for Equality* (1966) and *Concern for Community* (1970), as well as a co-author of *Rulers and the Ruled* (1964), *Black-Jewish Relations* (1970), *Discovering the Community* (1977), and *Small Town and Small Towners* (1979). He currently is working on futures and comprehensive planning.

Robyne S. Turner is Associate Professor in Political Science at Florida Atlantic University. She has published a number of articles on state and local land-use politics. Her current research includes the effect of growth politics on downtown development and community sustainability.

Ronald K. Vogel is Associate Professor of Political Science at the University of Louisville. His research focuses on metropolitan governance, urban political economy, and national urban policy. He is the author of *Urban Political Economy: Broward County, Florida* (1992), and has published articles in the *Journal of Urban Affairs, Urban Affairs Quarterly,* and *Economic Development Quarterly.* He currently is editing *Handbook of Research on Urban Politics and Policy in the United States* and coediting *Regional Politics: America in a Post-City Era.*

Richard C. Wade is Distinguished Professor of Urban History at the Graduate Center, City University of New York. He is the author of *The Urban Frontier; Slavery in the Cities;* and *Chicago: Growth of a Metropolis.* He was also editor of the *Urban Life in America* series, *The Rise of Urban America,* and *Metropolitan America.* He earned his Ph.D. from Harvard University.

Robert Warren is Professor of Urban Affairs and Public Policy in the College of Urban Affairs and Public Policy, University of Delaware. His primary interests are in urban governance, service delivery, telecommunications, and planning. He has published in such journals as *Coastal Zone Management, Journal of the American Planning Association, Journal of Urban Affairs, Nonprofit and Voluntary Sector Quarterly, Policy Studies Review, Public Administration Review,* and *Urban Affairs Quarterly.*

Robert J. Waste is Albert A. Levin Chair of Urban Studies and Public Service at Cleveland State University. He is the author of *The Ecology of City Policymaking* (1989), *Power and Pluralism in American Cities* (1987), and editor of *Community Power: Future Directions in Urban Research* (1986). His articles have appeared in the *Journal of Urban Affairs, Urban Studies, Urban Resources, Administration and Society, Social Policy, California Journal,* and others. He is also a coeditor for Sage Publications' series *Cities and Planning* with Roger Caves and Margaret Wilder.

Alice Watkins, Ph.D., is Associate Dean of Student Services in the School of Education at California State University, Los Angeles. Her interests include teacher training, educating minority students in the inner city, and developing curricula for children with special needs.

Louis F. Weschler is Professor of Public Administration in the School of Public Affairs at Arizona State University. His recent research projects concern prevention programs for at-risk families in urban settings and neighborhood self-governance. He is a member of the Prevention Research Center and the Morrison Institute for Public Policy at Arizona State University. He consults with cities and counties on improving the quality of services and reducing the consumption costs associated with services.

Margaret Wilder is Professor of Geography and Planning at the State University of New York at Albany. Her research focuses on the effectiveness of community and economic development strategies and their impacts on low-income inner-city communities. Her work has been published in several major urban research journals, and two of her papers on urban enterprise zones won best paper awards from the Urban Affairs Association and the *Journal of the American Planning Association*. She currently is involved in a 3-year project to evaluate the implementation of community development initiatives in Indiana's enterprise zones. She is also a coeditor for Sage Publications' series *Cities and Planning* with Roger Caves and Robert Waste.

Robert C. Wood is the Henry R. Luce Professor of Democratic Institutions and the Social Order at Wesleyan University. He taught at MIT and served as Under-Secretary and Secretary of Housing and Urban Development in the Johnson administration, and was Chair of the Massachusetts Bay Transportation Authority, President of the University of Massachusetts, and Superintendent of the Boston Public Schools.

Andrea G. Zetlin is Professor in the Division of Special Education, School of Education, California State University at Los Angeles. Her research and writing have focused on social competence and family relations of culturally diverse exceptional learners, as well as the development of family, community, and school partnerships.